Human Behavior, Psychology, and Social Interaction in the Digital Era

Anabela Mesquita
CICE - ISCAP/Polytechnic of Porto, Portugal & Algoritmi Centre, Minho University, Portugal

Chia-Wen Tsai
Ming Chuan University, Taiwan

A volume in the Advances in Human and Social Aspects of Technology (AHSAT) Book Series

Information Science REFERENCE
An Imprint of IGI Global

Managing Director:	Lindsay Johnston
Managing Editor:	Austin DeMarco
Director of Intellectual Property & Contracts:	Jan Travers
Acquisitions Editor:	Kayla Wolfe
Production Editor:	Christina Henning
Development Editor:	Brandon Carbaugh
Cover Design:	Jason Mull

Published in the United States of America by
Information Science Reference (an imprint of IGI Global)
701 E. Chocolate Avenue
Hershey PA, USA 17033
Tel: 717-533-8845
Fax: 717-533-8661
E-mail: cust@igi-global.com
Web site: http://www.igi-global.com

Library of Congress Cataloging-in-Publication Data

Human behavior, psychology, and social interaction in the digital era / Anabela Mesquita and Chia-Wen Tsai, editors.
pages cm
Includes bibliographical references and index.
ISBN 978-1-4666-8450-8 (hardcover) -- ISBN 978-1-4666-8451-5 (ebook) 1. Human-computer interaction. 2. Cognitive science. 3. Online social networks. 4. Technology--Social aspects. I. Sarmento, Anabela. II. Tsai, Chia-Wen, 1977-
QA76.9.H85H843 2015
004.01'9--dc23

2015009643

This book is published in the IGI Global book series Advances in Human and Social Aspects of Technology (AHSAT) (ISSN: 2328-1316; eISSN: 2328-1324)

British Cataloguing in Publication Data
A Cataloguing in Publication record for this book is available from the British Library.

For electronic access to this publication, please contact: eresources@igi-global.com.

Advances in Human and Social Aspects of Technology (AHSAT) Book Series

Ashish Dwivedi
The University of Hull, UK

ISSN: 2328-1316
EISSN: 2328-1324

Mission

In recent years, the societal impact of technology has been noted as we become increasingly more connected and are presented with more digital tools and devices. With the popularity of digital devices such as cell phones and tablets, it is crucial to consider the implications of our digital dependence and the presence of technology in our everyday lives.

The **Advances in Human and Social Aspects of Technology (AHSAT) Book Series** seeks to explore the ways in which society and human beings have been affected by technology and how the technological revolution has changed the way we conduct our lives as well as our behavior. The AHSAT book series aims to publish the most cutting-edge research on human behavior and interaction with technology and the ways in which the digital age is changing society.

Coverage

- Information ethics
- Technology and Social Change
- Human Rights and Digitization
- ICTs and human empowerment
- Computer-Mediated Communication
- Cyber Bullying
- Technology Dependence
- Human-Computer Interaction
- Digital Identity
- ICTs and social change

IGI Global is currently accepting manuscripts for publication within this series. To submit a proposal for a volume in this series, please contact our Acquisition Editors at Acquisitions@igi-global.com or visit: http://www.igi-global.com/publish/.

The Advances in Human and Social Aspects of Technology (AHSAT) Book Series (ISSN 2328-1316) is published by IGI Global, 701 E. Chocolate Avenue, Hershey, PA 17033-1240, USA, www.igi-global.com. This series is composed of titles available for purchase individually; each title is edited to be contextually exclusive from any other title within the series. For pricing and ordering information please visit http://www.igi-global.com/book-series/advances-human-social-aspects-technology/37145. Postmaster: Send all address changes to above address.

Titles in this Series

For a list of additional titles in this series, please visit: www.igi-global.com

Rethinking Machine Ethics in the Age of Ubiquitous Technology
Jeffrey White (Korean Advanced Institute of Science and Technology, KAIST, South Korea) and Rick Searle (IEET, USA)
Information Science Reference • copyright 2015 • 342pp • H/C (ISBN: 9781466685925) • US $205.00 (our price)

Contemporary Approaches to Activity Theory Interdisciplinary Perspectives on Human Behavior
Thomas Hansson (Blekinge Institute of Technology, School of Management (MAM), Sweden)
Information Science Reference • copyright 2015 • 437pp • H/C (ISBN: 9781466666030) • US $195.00 (our price)

Evolving Issues Surrounding Technoethics and Society in the Digital Age
Rocci Luppicini (University of Ottawa, Canada)
Information Science Reference • copyright 2014 • 317pp • H/C (ISBN: 9781466661226) • US $215.00 (our price)

Technological Advancements and the Impact of Actor-Network Theory
Arthur Tatnall (Victoria University, Australia)
Information Science Reference • copyright 2014 • 331pp • H/C (ISBN: 9781466661264) • US $195.00 (our price)

Gender Considerations and Influence in the Digital Media and Gaming Industry
Julie Prescott (University of Bolton, UK) and Julie Elizabeth McGurren (Codemasters, UK)
Information Science Reference • copyright 2014 • 357pp • H/C (ISBN: 9781466661424) • US $195.00 (our price)

Human-Computer Interfaces and Interactivity Emergent Research and Applications
Pedro Isaías (Universidade Aberta (Portuguese Open University), Portugal) and Katherine Blashki (Noroff University College, Norway)
Information Science Reference • copyright 2014 • 348pp • H/C (ISBN: 9781466662285) • US $200.00 (our price)

Political Campaigning in the Information Age
Ashu M. G. Solo (Maverick Technologies America Inc., USA)
Information Science Reference • copyright 2014 • 359pp • H/C (ISBN: 9781466660625) • US $210.00 (our price)

Handbook of Research on Political Activism in the Information Age
Ashu M. G. Solo (Maverick Technologies America Inc., USA)
Information Science Reference • copyright 2014 • 498pp • H/C (ISBN: 9781466660663) • US $275.00 (our price)

Interdisciplinary Applications of Agent-Based Social Simulation and Modeling
Diana Francisca Adamatti (Universidade Federal do Rio Grande, Brasil) Graçaliz Pereira Dimuro (Universidade Federal do Rio Grande, Brasil) and Helder Coelho (Universidade de Lisboa, Portugal)

To João, Pedro Ivo, Nuno Miguel and Joãozinho.
To my mother .

Editorial Advisory Board

Table of Contents

Preface xvi

Acknowledgment xxi

Introduction xxii

Section 1
The Impact of the Environment on Human Performance

Chapter 1
The Environmental and Technological Factors of Multitasking 1
Lin Lin, University of North Texas, USA
Bill Lipsmeyer, Sikorsky Aircraft Corporation, USA

Chapter 2
Blogs' Potentialities in Learning: What Are the Key Variables to Promote Cognitive Empowerment 21
Alessandro Antonietti, Catholic University of the Sacred Heart, Italy
Simona C. S. Caravita, Catholic University of the Sacred Heart, Italy
Barbara Colombo, Catholic University of the Sacred Heart, Italy
Luisa Simonelli, Catholic University of the Sacred Heart, Italy

Chapter 3
Assessing the Performance of Decision Support Systems in Military Environment: The 3C Method 45
Jean-Fabrice Lebraty, Lyon 3 University, France
Cécile Godé, Université Lumière Lyon 2, France

Section 2
The Use of Mobile Technology in Human Daily Life

Chapter 4
Giving Up Smoking Using SMS Messages on your Mobile Phone 72
Silvia Cacho-Elizondo, IPADE Business Schoolm, France
Niousha Shahidi, EDC Paris Business School, France
Vesselina Tossan, CNAM, France & EDC Paris Business School, France

Chapter 5
Smartphones and Self-Broadcasting among College Students in an Age of Social Media 95
Franklin Nii Amankwah Yartey, University of Dubuque, USA
Louisa Ha, Bowling Green State University, USA

Section 3
Information Technologies Characteristics, Improvement and Use

Chapter 6
Effects of Display Characteristics on Presence and Emotional Responses of Game Players 130
Zhiming Wu, Sichuan University, China
Tao Lin, Sichuan University, China
Ningjiu Tang, Sichuan University, China
Shaomei Wu, Sichuan University, China

Chapter 7
Towards Intelligent Window Layout Management: The Role of Mental Map 146
Rui Chen, Sichuan University, China
Tao Lin, Sichuan University, China
Tiantian Xie, Sichuan University, China

Chapter 8
Modeling Human Behavior to Reduce Navigation Time of Menu Items: Menu Item Prediction Based on Markov Chain 162
Tian-Tian Xie, Sichuan University, China
Yuxi Zhu, Sichuan University, China
Tao Lin, Sichuan University, China
Rui Chen, Sichuan University, China

Section 4
Relationship between Information Systems and Networks

Chapter 9
Developing a Successful Facebook Fan Page Based on Costumers' Needs 189
Aikaterini Manthiou, NEOMA Business School, France
Lanlung Luke Chiang, Yuan Ze University, Taiwan
Liang Rebecca Tang, Iowa State University, USA

Chapter 10
Using High-Frequency Interaction Events to Automatically Classify Cognitive Load 210
Tao Lin, Sichuan University, China
Zhiming Wu, Sichuan University, China
Yu Chen, Sichuan University for Nationalities, China

Chapter 11
How to Become a Cybercriminal? An Explanation of Cybercrime Diffusion 229
Jean-Loup Richet, University of Nantes, France

Chapter 12
From Relationship to Information: A Study of Twitter and Facebook Usage in Terms of Social Network Size among College Students 241
Chen Yang, Bowling Green State University, USA
Louisa Ha, Bowling Green State University, USA
Gi Woong Yun, Bowling Green State University, USA
Lanming Chen, Bowling Green State University, USA

Chapter 13
Do College Students Benefit from Their Social Media Experience? Social Media Involvement and Its Impact on College Students' Self-Efficacy Perception 259
Ling Fang, Bowling Green State University, USA
Louisa Ha, Bowling Green State University, USA

Chapter 14
What Motivates Online Shoppers to "Like" Brands' Facebook Fan Pages? 279
Mohammad Hatim Abuljadail, Bowling Green State University, USA
Louisa Ha, Bowling Green State University, USA
Fang Wang, Bowling Green State University, USA
Liu Yang, Bowling Green State University, USA

Related References 294

Compilation of References 323

About the Contributors 364

Index 370

Detailed Table of Contents

Preface xvi

Acknowledgment xxi

Introduction xxii

Section 1
The Impact of the Environment on Human Performance

Chapter 1
The Environmental and Technological Factors of Multitasking 1
Lin Lin, University of North Texas, USA
Bill Lipsmeyer, Sikorsky Aircraft Corporation, USA

Research on multitasking, that is, conducting two or more tasks simultaneously or switching quickly between two or more tasks, has focused mostly on a human's capacity to do so inside their brain. However, our daily life experience indicates that our ability to multitask is not only dependent on our brain capacity, but is also related to other factors such as our environments and available resources. Different individuals may have different abilities to multitask due to their expertise, situational awareness, or ability to plan ahead. This chapter discusses the environmental and technological factors of multitasking based on a prior study. The goal is to expand interdisciplinary dialogues and research methodologies to better understand this prevalent phenomenon in our society.

Chapter 2
Blogs' Potentialities in Learning: What Are the Key Variables to Promote Cognitive Empowerment 21
Alessandro Antonietti, Catholic University of the Sacred Heart, Italy
Simona C. S. Caravita, Catholic University of the Sacred Heart, Italy
Barbara Colombo, Catholic University of the Sacred Heart, Italy
Luisa Simonelli, Catholic University of the Sacred Heart, Italy

Growing literature on the impact of new technologies on learning processes suggests that blogs can be effective tools to empower cognition. The chapter describes a line of research which is aimed at exploring the possibility that blogging is related to cognitive skills and can be effectively used in educational settings. In the main study reported in the paper fifty blogs were analysed and classified according to both their content and formal structure. The right (i.e., intuitive-holistic) vs. left (i.e., systematic-analytical) thinking style of blog owners was also assessed, as well as the blog owners' awareness of the psychological processes, activated by the blog that they had devised, both in their own and in other people's mind. Results showed that blog owners are able to use effective communication strategies by differentiating the formal structure of bogs according to the content, but they lack metacognitive awareness about the mental processes activated by the blog. No relation between the blog owner's cognitive style and blog style was found. Implications for the educational use of blogs are discussed. In other two studies such implications were tested in samples of teachers and students. Blogs' potentialities to enhance metacognitive awareness and control in both teachers and students, as well as cooperative learning in students, were critically evaluated. Altogether, results from this research line provide evidence that blogging can be useful to foster cognitive skills in education by empowering the bloggers' awareness of the implied mental processes.

Chapter 3
Assessing the Performance of Decision Support Systems in Military Environment: The 3C Method 45
Jean-Fabrice Lebraty, Lyon 3 University, France
Cécile Godé, Université Lumière Lyon 2, France

This article explores the ability of a decision support system (DSS) to improve the quality of decision making in extreme environment. This DSS is actually based on a networked information system. Academic literature commonly mentions models of fit to explore the relationship between technology and performance, reckoning users' evaluations as a relevant measurement technique for Information System (IS) success. Although effective contributions have been achieved in measurement and exploration of fit, there have been few attempts to investigate the triangulation of fit between "Task-DSS-Decision Maker" under stressful and uncertain circumstances. This article provides new insights regarding the advantages provided by networked IS for making relevant decisions. An original case study has been conducted. It is focused on a networked decision support system called Link 16 that is used during aerial missions. This case study shows that the system improves decision making on an individual basis. Our result suggest the importance of three main fit criteria – Compliance, Complementarity and Conformity – to measure DSS performance under extreme environment and display a preliminary decisional fit model.

Section 2
The Use of Mobile Technology in Human Daily Life

Chapter 4
Giving Up Smoking Using SMS Messages on your Mobile Phone 72
Silvia Cacho-Elizondo, IPADE Business Schoolm, France
Niousha Shahidi, EDC Paris Business School, France
Vesselina Tossan, CNAM, France & EDC Paris Business School, France

The current tendency to use cell phones or other mobile devices for healthcare purposes offers a huge opportunity to improve public health worldwide. In that direction, mobile devices make it easier to offer coaching services through text/video messages, to support individuals trying to break addictions such as smoking. Given that use of such services is still low in France and other countries, it is important to have greater understanding of what leads users to adopt them. Therefore, we propose and validate an explanatory model for the intention to adopt a mobile coaching service to help people to stop smoking. This chapter uses the concepts of vicarious innovativeness, social influence, perceived monetary value, perceived enjoyment, and perceived irritation.

Chapter 5
Smartphones and Self-Broadcasting among College Students in an Age of Social Media.................. 95
Franklin Nii Amankwah Yartey, University of Dubuque, USA
Louisa Ha, Bowling Green State University, USA

This chapter examines the use of smartphones for self-broadcasting via social media among college students. Based on motivation and network externalities theories, our survey of a public university's college students confirmed our hypotheses that network size, years of experience using social media and the time spent on social media positively predict their frequency of self-broadcasting on their smartphones. The results suggest that 85.2% of college students self-broadcast at least once a month by updating their status on SNS and students are likely to self-broadcast within their network. Most students set their profile privacy setting as private or semi-private. But privacy setting does not affect self-broadcast frequency.

Section 3
Information Technologies Characteristics, Improvement and Use

Chapter 6
Effects of Display Characteristics on Presence and Emotional Responses of Game Players 130
Zhiming Wu, Sichuan University, China
Tao Lin, Sichuan University, China
Ningjiu Tang, Sichuan University, China
Shaomei Wu, Sichuan University, China

Large displays are becoming ubiquitous and one important property of large displays is that they afford larger visual angles and physically larger screen sizes. However, there has been little investigation of the effects of large displays on users; furthermore, few studies have employed physiological measures, nor isolated the effects on presence and emotional responses of large visual angle and physically large screen size. This study, then, examines specifically the effects of these two properties of large displays on presence and emotional responses, using physiological as well as subjective methods. The results indicate that the larger visual angle offered by a large display can increase the sense of presence, engagement, and emotional responses when players are playing games. More interestingly, the physically larger sizes offered by a large display seem to also be able to affect these player experiences, even at identical visual angles with small displays.

Chapter 7
Towards Intelligent Window Layout Management: The Role of Mental Map 146
Rui Chen, Sichuan University, China
Tao Lin, Sichuan University, China
Tiantian Xie, Sichuan University, China

Large displays are becoming increasingly pervasive. Larger screen size provides an opportunity for users to see more information simultaneously, but at the cost of managing a larger amount of screen space, which is a great burden declining task performance and user experience. User would do/feel better if this burden could be takeover by the computer itself employing techniques that automate the management of screen space. Some studies on automatic window management have been carried out with some success. However, they mainly focus on utilization of empty screen space and/or overlap elimination while ignore preservation of the mental map of users, which tends to cause user confusion and disorientation in practical use. In this chapter, an empirical model is proposed to identifying the degree of mental map preservation for a window layout rearrangement. Furthermore, a method combining high-level window importance with a genetic multi-objective optimization algorithm is presented to generate recommended window layouts featuring a tradeoff among several conflicting goals: (1) better usage of screen space, (2) lower degree of window overlaps, and (3) better mental map preservation. Results suggest that the method is capable of generating suitable window layouts for users and takes a key step toward developing an automated windows manager.

Chapter 8
Modeling Human Behavior to Reduce Navigation Time of Menu Items: Menu Item Prediction Based on Markov Chain 162
Tian-Tian Xie, Sichuan University, China
Yuxi Zhu, Sichuan University, China
Tao Lin, Sichuan University, China
Rui Chen, Sichuan University, China

With the increase in the number of menu items and the menu structure complexity, users have to spend more time in locating menu items when using menu-based interfaces. Recently, adaptive menu techniques have been explored to reduce the time and menu item prediction plays a crucial role in the techniques. Unfortunately, there still lacks effective prediction models for menu items. This chapter per the authors explores the potential of three prediction models based on Markov chain in predicting top n menu items with human behavior data while interacting with menus - the users' historical menu item selections. The results show that Weighted Markov Chain using Genetic Algorithm can obtain the highest prediction accuracy and significantly decrease navigation time by 22.6% when N equals 4 as compared to the static counterpart. Two application scenarios of these models on mobile devices and desktop also demonstrated the potentials in daily usage to reduce the time spent to search target menu items.

Section 4
Relationship between Information Systems and Networks

Chapter 9
Developing a Successful Facebook Fan Page Based on Costumers' Needs 189
Aikaterini Manthiou, NEOMA Business School, France
Lanlung Luke Chiang, Yuan Ze University, Taiwan
Liang Rebecca Tang, Iowa State University, USA

In this chapter, we determine possible needs that customers are fulfilling by participating in the Facebook fan pages. Understanding the needs that customers aim to pursue is the priority of marketing communications. Therefore, the chapter analyzes a theoretical approach to identify and respond to the needs that consumers seek to satisfy on fan pages. The customer needs are evaluated from functional, hedonic, social, and psychological perspectives and all four of the need components make significant contributions to the overall need on fan pages. Businesses in the food service industry are advised to build an emotional bond with consumers, enhance two-way communication, provide useful information and encourage entertainment on their fan pages. Managers of brands that operate fan pages can be guided by our research with regards to deciding which characteristics or content consumers find important in the these pages.

Chapter 10
Using High-Frequency Interaction Events to Automatically Classify Cognitive Load 210
Tao Lin, Sichuan University, China
Zhiming Wu, Sichuan University, China
Yu Chen, Sichuan University for Nationalities, China

There is still a challenge of creating an evaluation method which can not only unobtrusively collect data without supplement equipment but also objectively, quantitatively and real-time evaluate cognitive load of users based the data. The study explores the possibility of using the features extracted from high-frequency interaction (HFI) events to evaluate cognitive load to respond the challenge. Specifically, back-propagation neural networks, along with two feature selection methods (nBset and SFS), were used as the classifier and it was able to use a set of features to differentiate three cognitive load levels with an accuracy of 74.27%. The main contributions of the research are: (1) knowledge about what detailed features may be predictive of cognitive load changes; (2) demonstrating the potential of using the HFI features in discriminating different cognitive load when suitable classifier and features are adopted.

Chapter 11
How to Become a Cybercriminal? An Explanation of Cybercrime Diffusion 229
Jean-Loup Richet, University of Nantes, France

The main purpose of this chapter is to illustrate a landscape of current literature in cybercrime taking into consideration diffusion of innovation theories and economic theory of competition. In this chapter, a narrative review of the literature was carried out, facilitators leading to cybercrime were explored and explained the diffusion of Cybercriminals' best practices. Cybercrime is compatible with young adults lifestyle (familiarity) and requires little knowledge. Moreover, barriers to entry related to costs (psychological, financial), risks and investments are low. This review provides a snapshot and reference base for academics and practitioners with an interest in cybercrime while contributing to a cumulative culture which is desired in the field. This chapter provides insights into barriers to entry into cybercrime and the facilitators of cybercrime.

Chapter 12
From Relationship to Information: A Study of Twitter and Facebook Usage in Terms of Social Network Size among College Students 241
Chen Yang, Bowling Green State University, USA
Louisa Ha, Bowling Green State University, USA
Gi Woong Yun, Bowling Green State University, USA
Lanming Chen, Bowling Green State University, USA

By measuring social media users' online activities in terms of information broadcasting, information seeking and relational maintenance, this chapter aimed at investigating how college students' SNS usage patterns may affect their online social network sizes. Statistics in this chapter suggested that more information seeking leads to more Twitter followings while more information broadcasting results in a larger size of Twitter followers. The frequency of relationship management, however, did not predict students' social network size on either Twitter or Facebook. Implications of the study were discussed.

Chapter 13
Do College Students Benefit from Their Social Media Experience? Social Media Involvement and Its Impact on College Students' Self-Efficacy Perception 259
Ling Fang, Bowling Green State University, USA
Louisa Ha, Bowling Green State University, USA

As young people are increasingly dependent on Social Networking sites (SNS) to socialize, seek information, and self-broadcast, their SNS consumption has been found to be associated with social capital and social support in a positive way especially among individuals with low psychological assets. This exploratory study investigated SNS involvement in relation to college students' perceived self-efficacy change afterwards based on the social cognitive theory and literature review on social media effects studies. Undergraduate students (N = 395) in a Midwest U.S. public university participated in a web survey in September 2012. Results indicated a positive potential of involving in SNS activities to strengthen users' self-efficacy. Research data also point to the mediating roles of social support and social learning on SNS involvement and self-efficacy, especially among low self-efficacy college students with homogenous SNS networks. Theoretical and practical implications are discussed.

Chapter 14
What Motivates Online Shoppers to "Like" Brands' Facebook Fan Pages? 279
Mohammad Hatim Abuljadail, Bowling Green State University, USA
Louisa Ha, Bowling Green State University, USA
Fang Wang, Bowling Green State University, USA
Liu Yang, Bowling Green State University, USA

Brands' Facebook fan pages have been frequently used as a marketing tool to reach to more individuals; however, Facebook users' motivations to participate in those fan pages are still unclear. This paper investigates the motives that stimulate online shoppers to "like" brands' Facebook fan pages. This study is interested in knowing if online shoppers' proclivity to "like" brands' Facebook fan pages is based on their hedonic or utilitarian motivations. The authors propose a model based on hedonic and utilitarian motivations and uses and gratifications theoretical framework. An online survey was conducted among college students who shopped online in Northwest Ohio (N=198). The findings show that utilitarian motivations have positive significant relationships with "liking" brands' Facebook fan pages.

Related References 294

Compilation of References 323

About the Contributors 364

Index 370

Preface

This publication, *Human Behavior, Psychology, and Social Interaction in the Digital Era*, comprises some of the best articles published in the International Journal of Technology and Human Interaction in 2013. It is our purpose to offer the reader the most up to date research and discussions providing an overview of the trends and advancements in this area.

IJTHI

The first issue of the *International Journal of Technology and Human Interaction* appeared in 2005 due to the increasing research being done in the area where technology and human encounter. Several studies came to the conclusion that the success or even the failure of the implementation of technologies were not due to the technology itself but to interactions between the technology and the user. Furthermore, the problem of technology and human interaction covers all the fields and aspects of our lives, such as education, profession, private, leisure time, just to mention a few.

Taking this into consideration, the journal provides a platform for leading research that addresses issues of human and technology interaction in all domains. The journal aims therefore at publishing interdisciplinary research, including aspects from a wide variety of disciplines. These disciplines may range from more technical ones such as computer science, engineering or information systems to non-technical descriptions of technology and human interaction from the point of view of sociology, psychology, education, communication, management, marketing or even philosophy. The journal also aspires to provide a publication outlet for research questions and approaches that are novel and may find it difficult to be published in established journals following a rigid and exclusive structure. It is open to all research paradigms, being they empirical or conceptual, but requires that they are accessible and reflected. We also encourage the submission of high quality syntheses across research in different specialties that are interesting and comprehensible to all members of the information systems community and related disciplines.

The journal is open to several topics that may include (but are not limited to) the following:

- Adoption of online learning
- All issues related to the interaction of technology and humans, either individually or socially
- Anthropological consequences of technology use
- Ethical aspects of particular technologies (e.g., e-teaching, ERP, etc.)
- Experiential learning though the use of technology in organizations

- HCI design for trust development
- Influence of gender on the adoption and use of technology
- Interaction and conversion between technologies and their impact on society
- Intersection of humanities and sciences and its impact on technology use
- Normative questions of the development and use of technology
- Online Education
- Perceptions and conceptualizations of technology
- Phenomenology of e-government
- Philosophy of Technology
- Questions of computer or information ethics
- Relationship of theory and practice with regards to technology
- Responsibility of artificial agents
- Shaping of e-commerce through law and culture
- Social impact of specific technologies (e.g., biometrics, SCM, PGP, etc.)
- Social shaping of technology and human interaction research
- Technological risks and their human basis
- Technology assessment of software / hardware development
- Technology-Enhanced Learning
- Value of intellectual capital in knowledge management

and all other issues related to the interaction of technology and humans, either individually or socially.

The IJTHI is also attentive to the trends and changes in society and from time to time organizes special issues that cover, discuss and deal with special and particular topics.

Nevertheless the topics covered by the journal are not confined to those mentioned above. Editors are aware that this is a very dynamic field and are open to welcome new topics and discussions.

This Book

One of the most dynamic areas of modern development is information systems and their use and impact in our daily lives. Each year, month, not to say day, we witness the emergence of new technologies, or new features and characteristics of older technologies that, in some way change the way we communicate, interact, express ourselves, play, and work, just to mention a few. In this book, we present 14 works that illustrate how things are changing in today's world.

Section 1: The Impact of the Environment on Human Performance

In this section the reader will find three chapters. The first one - Chapter 1 - The Environmental and Technological Factors of Multitasking – discusses the role of the environment when enabling or constraining the use of information systems in human performance, more specifically, in multitasking. Multitasking can be defined as conducting two or more tasks simultaneously or switching quickly between two or more tasks. So far research has focused mainly on a human's capacity to do so inside one's brain. Nevertheless, research also identified that our ability to multitask depends also on other factors such as environments and available resources. To sum up, this chapter discusses the environmental and technological factors of multitasking and the goal is to expand interdisciplinary dialogues and research methodologies to better understand this prevalent phenomenon in our society.

Chapter 2 - Blogs' Potentialities in Learning: What Are the Key Variables to Promote Cognitive Empowerment? – deals with the role of blogs and their potentiality to promote cognition. Literature suggests that blogs can be effective tools to empower cognition. This chapter describes a line of research which is aimed at exploring the possibility that blogging is related to cognitive skills and can be effectively used in the educational setting. The authors have analysed fifty blogs and classified them according to both their content and formal structure. The right (i.e. intuitive-holistic) vs. left (i.e. systematic-analytical) thinking style of blog owners was also assessed, as well as the blog owners' awareness of the psychological processes activated by the blog that they had devised, both in their own and in other people's minds. Results show that blog owners are able to use effective communication strategies by differentiating the formal structure of bogs according to the content, but they lack metacognitive awareness about the mental processes activated by the blog.

The last chapter in section one - Chapter 3 - Assessing the Performance of Decision Support Systems in Military Environment: The 3C Method - explores the ability of a decision support system (DSS) to improve the quality of decision making in an extreme environment. Academic literature commonly mentions models of fit to explore the relationship between technology and performance, reckoning users' evaluations as a relevant measurement technique for Information System (IS) success. Although effective contributions have been achieved in measurement and exploration of fit, there have been few attempts to investigate the triangulation of fit between "Task-DSS-Decision Maker" under stressful and uncertain circumstances. This article provides new insights regarding the advantages provided by networked IS for making relevant decisions.

Section 2: The Use of Mobile Technology in Human Daily Life

One of the most evident advances in technology was the use of mobile technology. In fact mobile phones, smartphones are incorporating most of the features we could find in computers, laptops and now also in tablets. Furthermore, these devices are incorporating new and different features or helping us in solving daily problems or improving our life as is described, for instance, in Chapter 4 - Giving up smoking using SMS messages on your mobile phone. The use cell phones or other mobile devices for healthcare purposes offers a huge opportunity to improve public health worldwide. In that regard, mobile devices make it easier to offer coaching services through text/video messages, as well as to support individuals trying to break addictions such as smoking. In this chapter, the authors propose and validate an explanatory model for the intention to adopt a mobile coaching service to help people to stop smoking.

Chapter 5 – Smartphones and Self-broadcasting among College Students in an Age of Social Media. The use of smartphones for self-broadcasting among college students using motivation and network externalities theories is also new. In this chapter, author discusses the social and business implications of using smartphones as a self-promotion medium from an information management perspective.

Section 3: Information Technologies Characteristics, Improvement, and Use

In this section, reader may find a selection of articles dealing with more technical aspects having impact on human behaviour and response.

In Chapter 6 – Effects of Display Characteristics on Presence and Emotional Response of Game Players – authors examines specifically the effects of large displays on presence and emotional responses, using physiological as well as subjective methods. Large displays are becoming ubiquitous and one important property of large displays is that they afford larger visual angles and physically larger screen sizes. However, there has been little investigation of the effects of large displays on users; furthermore, few studies have employed physiological measures, nor isolated the effects on presence and emotional responses to large visual angle and physically large screen size. This study intends to address this gap.

Chapter 7 - Mental Map and Genetic Multi-objective Optimization of Window Layout – discusses also some aspects related with displays. Large displays are becoming increasingly pervasive. Larger screen size provides an opportunity for users to see more information simultaneously, but at the cost of managing a larger amount of screen space, which is a great burden on declining task performance and user experience. Some studies on automatic window management have been carried out with some success. However, they mainly focus on utilization of empty screen space and/or overlap elimination while ignoring preservation of the mental map of users, which tends to cause user confusion and disorientation in practical use. This work presents a method to combine high-level window importance with a genetic multi-objective optimization algorithm to generate recommended window layouts featuring a tradeoff among several conflicting goals.

Finally, Chapter 8 - Menu Item Prediction and Its Application based on Markov Chain – explores the potential of three prediction models based on Markov chains in predicting top menu items based on users' historical menu item clicks. With the increase in number of menu items and menu structure complexity, users have to spend more time in locating menu items when using menu-based interfaces. Recently, adaptive menu techniques have been explored to reduce the time and menu item prediction plays a crucial role in these techniques. Unfortunately, an effective prediction model for menu items is still lacking. In this chapter, readers will find a discussion concerning this concern.

Section 4: Relationship between Information Systems and Networks

The last section contains three papers with networks and social media as common ground.

Social media channels are considered as excellent vehicles to create relationships with customers. One specific way to do this is to create a successful brand fan pages on Facebook. In Chapter 9 - Developing a successful Facebook Fan Pages based on costumers' needs – the authors determine possible needs that customers are fulfilling by participating in the Facebook fan pages. Understanding the needs that customers aim to pursue is a priority for marketing communications.

According to Chapter 10 - Using High-frequency Interaction Events to Classify Automatic Cognitive Load - there is exists a challenge for creating an evaluation method which can not only unobtrusively collect data without supplementary equipment but also objectively, quantitatively and real-time evaluate cognitive load of users based the data. The study explores the possibility of using the features extracted from high-frequency interaction (HFI) events to evaluate cognitive load in response to the challenge.

The main purpose of Chapter 11 - How to Become a Cybercriminal? An Explanation of Cybercrime Diffusion - is to illustrate a landscape of current literature in cybercrime taking into consideration diffusion of innovation theories and economic theory of competition. A narrative review of the literature is featured, facilitators leading to cybercrime are explored and the diffusion of cybercriminals' best practices is explained. Cybercrime is compatible with young adults' lifestyle (familiarity) and requires little knowledge. Moreover, barriers to entry related to costs (psychological, financial), risks and investments are low.

Chapter 12 – From Relationship to Information: a Study of Twitter and Facebook Usage in Terms of Social Networks Size Among College Students – explores how college students' SNS usage patterns may be correlated to the sizes of their Twitter followings and followers and to the number of Facebook friends.

In Chapter 13 – Do College Students Benefit from Their Social Media Experience? Social Media Involvement and Its Impact on College Students' Self-Efficacy Perception – authors discuss the association between social capital and social support. This study investigates SNS involvement in relation to college students' perceived self-efficacy change afterwards in a theoretical framework based on the social cognitive theory and review of literature on social media effects.

Finally, in Chapter 14 – What motivates Online Shoppers to "Like" Brands' Facebook Fan pages? – authors bring some light to the discussion concerning the use of fan pages on Facebook. In this chapter, authors investigate the motives of "liking" brands' Facebook fan pages by developing a model based on hedonic and utilitarian motivations approach and uses and gratifications perspective to help organizations better understand the reasons behind their customers' "liking" their brands' Facebook fan pages and develop their Facebook page content strategies accordingly.

Before closing this preface, I would like to say that these papers represent a small part of what can be seen and read in the *International Journal of Technology and Human Interaction* (IJTHI). They represent the hard work of some of our researchers, a little bit from all over the World. They also represent a big effort to bring to the reader what the most recent and updated is being done in those areas. I hope that you enjoy reading them as much as I enjoyed preparing this compilation for you.

Anabela Mesquita
Polytechnic of Porto, Portugal & Minho University, Portugal

Acknowledgment

As Editor I would like to acknowledge the help of those involved in the collation and review of this book, without whose unstinting support the project could not have been completed. A further special note goes to all staff of IGI Global whose contribution throughout the whole process has been invaluable. Thanks to all those who provided constructive and comprehensive reviews.

Special thanks go to Chia-Wen Tsai for having accepted the invitation to write the foreword of this book.

Thank you all.

Anabela Mesquita
Polytechnic of Porto, Portugal & Minho University, Portugal

Introduction

With the rapid development of Internet technology and wireless network, this generation continues to usher in technological advances (Tsai, Shen & Chiang, 2013). Social networks and related technologies (such as mobile technology, blogs, and smartphones) have continued to gain in importance in modern people's lives and workplaces (Cook & Pachler, 2012). For example, social networking services (SNSs) have become a central, virtually unavoidable medium for social interactions (Gosling, Augustine, Vazire, Holtzman & Gaddis, 2011). Moreover, the rapid diffusion of mobile technology devices such as smartphones and tablets has had a significant impact on society, bringing about important changes in the ways of our world (Nedungadi & Raman, 2012). From an academic perspective, very little is yet known about the phenomena involved in the relationship of SNSs, mobile technology, and social interaction (Merchant, 2011; Pimmer, Linxen & Gröhbiel, 2012). In this regard, Dr. Anabela Mesquita collected 12 chapters and edited this book titled "*Human Behavior, Psychology, and Social Interaction in the Digital Era*" to provide researchers and readers state-of-the-art findings in these fields.

Dr. Mesquita is a professor at the Institute of Administration and Accountancy (ISCAP)/Polytechnic School of Porto (IPP), Portugal. She is also the Editor-in-Chief of *International Journal of Technology and Human Interaction*, which is an academic journal indexed by EI-Compendex. In addition, Dr. Mesquita has published numerous papers in various international journals and conferences. She also serves as one of Organizing Committee Chairs of The Third International Conference on E-Technologies and Business on the Web (EBW2015) that will be held at the ESG Management School, Paris, France in 2015. Moreover, Dr. Mesquita is an Associate Editor of the *Information Resources Management Journal* and referee for the *Journal of Cases of Information Technology*. She has also been evaluator and reviewer for European commission projects.

Focusing on academic work, empirical study, and best practices, Dr. Mesquita's book, *Human Behavior, Psychology, and Social Interaction in the Digital Era*, is definite asset as a resource of knowledge for anyone interested in the fields of information systems and networks, SNSs, mobile technology, multitasking, and cybercrime.

With the research findings, knowledge, experience, creativity and innovative applications of technologies presented in this book, readers may be inspired and find the potential and creative applications. It is expected that researchers, educators, students, IT managers, and government healthcare agencies could benefit from *Human Behavior, Psychology, and Social Interaction in the Digital Era*.

Chia-Wen Tsai
Ming Chuan University, Taiwan

REFERENCES

Cook, J., & Pachler, N. (2012). Online people tagging: Social (mobile) network(ing) services and work-based learning. *British Journal of Educational Technology*, *43*(5), 711–725. doi:10.1111/j.1467-8535.2012.01346.x

Gosling, S. D., Augustine, A. A., Vazire, S., Holtzman, N., & Gaddis, S. (2011). Manifestations of personality in online social networks: Self-reported Facebook-related behaviors and observable profile information. *Cyberpsychology, Behavior, and Social Networking*, *14*(9), 483–488.

Merchant, G. (2011). Unravelling the social network: Theory and research. *Learning, Media and Technology*, *37*(1), 4–19.

Nedungadi, P., & Raman, R. (2012). A new approach to personalization: Integrating e-learning and m-learning. *Educational Technology Research and Development*, *60*(4), 659–678.

Pimmer, C., Linxen, S., & Gröhbiel, U. (2012). Facebook as a learning tool? A case study on the appropriation of social network sites from mobile phones in developing countries. *British Journal of Educational Technology*, *43*(5), 726–738. doi:10.1111/j.1467-8535.2012.01351.x

Tsai, C. W., Shen, P. D., & Chiang, Y. C. (2013). The application of social networking sites (SNSs) in e-learning and online education environments: A review of publications in SSCI-indexed journals from 2004 to 2013. *International Journal of Web-Based Learning and Teaching Technologies*, *8*(3), 18–23. doi:10.4018/ijwltt.2013070102

Section 1

The Impact of the Environment on Human Performance

Chapter 1
The Environmental and Technological Factors of Multitasking

Lin Lin
University of North Texas, USA

Bill Lipsmeyer
Sikorsky Aircraft Corporation, USA

ABSTRACT

Research on multitasking, that is, conducting two or more tasks simultaneously or switching quickly between two or more tasks, has focused mostly on a human's capacity to do so inside their brain. However, our daily life experience indicates that our ability to multitask is not only dependent on our brain capacity, but is also related to other factors such as our environments and available resources. Different individuals may have different abilities to multitask due to their expertise, situational awareness, or ability to plan ahead. This chapter discusses the environmental and technological factors of multitasking based on a prior study. The goal is to expand interdisciplinary dialogues and research methodologies to better understand this prevalent phenomenon in our society.

INTRODUCTION

Multitasking has been around for as long as humans have had competing needs, for instance, doing housework while watching a baby. The demands of post-modern lifestyles and the work force have increased pressure on people to do more things in less time, resulting in more of an impetus to multitask or do several things in the same span of time. Interruptions, which require at least some multitasking to be handled, are commonplace for many jobs including managers, physicians, administrative assistants, and small office workers (Fleishman, Costanza, & Marshall-Mies, 1999). Given its commonality in our daily lives and jobs, not surprisingly, multitasking has attracted interest of basic and applied research.

DOI: 10.4018/978-1-4666-8450-8.ch001

The rapid development of technologies has continued to raise our expectation to do things faster. For instance, we eat, talk with friends, make phone calls, or do housework while watching TV. When we work on a computer, we switch immediately to doing something different if the computer is rebooting or buffering. We have changed our habits of reading, searching, and obtaining information from print media such as books or magazines. Increasingly, we surf the Internet for information, follow the hyperlinks from one link to another, one site to another, and click on the graphics and videos as soon as they call on our attention (Zumback, 2006). These are just some examples of common multitasking behaviors and activities. We do not wish to waste even one second of our precious time, so we try to do things concurrently or switch quickly from one task to another!

This current chapter focuses on the environmental and technological factors of people's multitasking behaviors, with the goal to expand interdisciplinary dialogue and research methodologies to better understand this prevalent phenomenon in our society. The chapter will begin by providing an overview of relevant research on multitasking to date. It will then discuss the environmental and technological factors that affect individuals' multitasking behaviors based on themes from a qualitative study. The chapter will conclude by suggesting implications and future research directions.

LITERATURE ON MULTITASKING

In this section, we will provide an overview of research on multitasking through the following lens: 1) limited brain capacity for multitasking; 2) partial attention or rapid switch; 3) cognitive load, automation, and expertise; 4) cultural perceptions of time and individual differences; 5) technological changes; and 6) the perceived new skills, needs and demands.

Limited Brain Capacity for Multitasking

Multitasking has been defined as doing several things at the same time, or as switching quickly between several different tasks (Baddeley, 1996; Meyer & Kieras, 1997). Depending on variations of beliefs in multitasking attitudes and behaviors in different disciplines, different terms have been used to describe and research this phenomenon, including, for example, *dual task, task switching, multitasking,* and *polychronicity* (Baddeley, 1996; Bluedorn, 2002; Hall, 1959; Meyer & Kieras, 1997; Monsell, 2003; Spink & Park, 2005).

Most studies in cognition, psychology, information science, and neuroscience have shown that our ability to engage in concurrent tasks is rather limited (Baddeley, 1996; Broadbent, 1958; Fisch, 2000; Just et al., 2001; Lang, 2001; Spink & Park, 2005). Research shows that multitasking over different types of tasks reduces productivity (Just et al., 2001; Rubinstein, Meyer & Evans, 2001), and that our ability to perform concurrent mental operations is limited by the capacity of the brain's central mechanism (Schweickert & Boggs, 1984). Wickens' Multiple Resource Theory Model (1984) proposed that humans have limited capability in processing information, although several different pools of resources rather than one single source can be tapped. This is because cognitive resources are limited in a finite amount of time and a supply and demand problem may occur when an individual performs two or more tasks that require the same resource in the same time interval.

Both theoretical and empirical studies show that our prefrontal lobes control the executive functions of our brain. Generalized functions include: 1) working memory, 2) self-regulation, 3) internalization of speech, and 4) reconstitution (behavior analysis and synthesis) (Barkley, 1997). According to Barkley (1997), improper development of the self-regulation function is one of the possible causes of Attention Deficit Hyperactivity Disorder (ADHD) with particular emphasis on deficient inhibition development, but this supposition must be taken in the context of the collateral development of all four of the executive functions. In addition, improper development of the four executive functions creates deficiencies on a fifth function, namely motor control-fluency-syntax.

Partial Attention or Rapid Switch

Obviously, multitasking has a lot to do with how we dedicate, focus or divide our attention: does one multitask because one splits one's attention or shifts one's attention between tasks? To what extent does one have control over the attention splits or attention shifts during multitasking? That is, does one multitask because one cannot focus one's attention on a single task for a period time, because one lacks self-discipline, or because one does this intentionally in order to focus or meet certain external or environmental needs at a particular time?

Johnson (2006) describes one kind of multitasking, "It usually involves skimming the surface of the incoming data, picking out the relevant details, and moving on to the next stream. You're paying attention, but only partially. That lets you cast a wider net, but it also runs the risk of keeping you from really studying the fish" (p. 61). Jackson (2008) warned that attention, the key to recapturing our ability to connect, reflect, and relax, is missing in our new world. Ophir, Nass, & Wagner (2009) found that heavy media multitaskers performed worse on task switching than light media multitaskers, possibly due to the fact that the heavy media multitaskers constantly pay attention to all stimuli, resulting in difficulty filtering out irrelevant stimuli in memory (Lin, 2009).

Scholars believe that switching between tasks wastes precious time because the brain is compelled to restart and refocus (Meyer and Kieras, 1997). According to Meyer and Kieras (1997), each time one has this alternation, there is a period in which one will make no progress on either task. The result is that it takes longer to finish any one chore, and that one doesn't do it nearly as well as one would, if one had given it one's full attention. Studies conducted on distracted driving have shown serious consequences when people are distracted or try to multitasking while driving (Just, Keller, & Cynkar, 2008).

Cognitive Load, Automation, and Expertise

Cognitive load plays an important role in both hindering performance and enhancing experience (Ang, Zaphiris, and Mahmood, 2007). Some tasks such as learning new skills may place higher cognitive loads on the brain while other tasks more familiar and automatic, such as typing and driving, may require lower cognitive loads. Poldrack & Foerde (2007) found that people had a harder time learning new things when their brains were distracted by another activity. The functional magnetic resonance imaging (fMRIs) used by researchers showed that learning while distracted or multitasking alters the brain's learning processes and changes the way people learn (Poldrack & Foerde, 2007). Foerde, Knowlton, & Poldrack (2006) found that learning new things is dependent on working memory, while learning based on habit or conditioning, is not as sensitive to working memory.

Tasks can be transferred from a categorization as high cognitive loads to low cognitive loads through repetition (Ang, Zaphiris, & Mahmood, 2007). Just, Keller, and Cynkar (2008) found that novice drivers need all the brain participation they can get for driving, but the level of focus required changes with experience. Over time, the brain re-wires itself. Driving becomes automatic. Scientists call this phenomenon "automaticity" (Just, Keller, & Cynkar, 2008). Automation allows us to do one thing while focusing on something else. In other words, learning to do one task well automatically helps us to multitask. The executive processes of our brains allow us to exert some sort of voluntary control over our behaviors, and help us achieve a goal by ignoring distractions (Weissman, Gopalakrishnan, Hazlett, & Woldorff, 2005). For instance, "if we're performing a task where we want to watch TV and ignore voices that are coming from, say, our children nearby, our frontal region brain may configure the brain to prioritize visual information and dampen down auditory information," said Weissman (NPR, October 2, 2008). Studies have discovered that a non-intrusive background did not affect students' learning performance; and that different background content, intentions or efforts towards the goal had different effects on students' learning performance (Lee, Lin, & Robertson, 2011; Lin, Lee, & Robertson, 2011; Lin, Robertson, & Lee, 2009; Lin, Robertson, & Lee, 2012). In addition, experts are able to flexibly retrieve important aspects of their knowledge with little attentional effort (Bransford et al., 1999).

Cultural Perceptions of Time and Individual Differences

Hall (1959) introduced the concept of polychronicity based on how people in various cultures perceived time. Polychronicity refers to the "extent to which people prefer to engage in two or more tasks or events simultaneously and believe that their preference is the correct way to do things" (Bluedorn, 2001, p. 119). Polychronicity is a continuum, and preferences exist for degrees of engagement. At one extreme is the pattern of focusing on one task at a time, interpreting other potential tasks and events as interruptions. The other extreme is open-ended, involving engagement in several tasks simultaneously (Bluedorn, 2002). According to Hall, people who live in monochronic cultures view time as a linear concept and prefer to complete one task at a time. In contrast, people who live in polychronic cultures view time as cyclical and prefer to engage in more than one task simultaneously. Much has changed since Hall's initial work on cultural perception of time and task. Since it can be seen that this work pre-dated the birth of the Internet, we can rightly assume that the time to complete any given routine task has been greatly compressed.

Bluedorn (2002) summarized earlier research and stated that "polychronic people appear to have more of the following: extraversion, favorable inclination toward change, tolerance of ambiguity, formal education, striving for achievement, impatience and irritability, and frequency of lateness and absenteeism. Those same people appear to have less of the following: conscientiousness or stress. However, Konig, Buhner, and Murling (2005) found that there was no significant correlation between polychronicity and extraversion, nor were they predictors of multitasking performance. Instead, attention, fluid intelligence (i.e., the ability to reason and to solve novel problems), and working memory (i.e., the system of the brain that permits the storage and processing of information needed in the execution of tasks) were the most important predictors of multitasking performance.

Goonetilleke and Luximon (2010) found that there were significant differences in the performance and strategy between monochronic and polychronic individuals in the selective attention test. Monochronic individuals focused their attention on the primary task and achieved higher performance. Polychronic individuals had somewhat better total performance in more than one task under time-constrained conditions. Ishizaka, Marshall and Conte (2010) concluded from their study that employees high in list making should be given prioritized work schedules and tasks and that employees high in achievement strivings may be assigned to a work environment where prioritization of multiple tasks is appropriate as compared to a situation where employees are required to be flexible in deciding which task to focus on from time to time.

Technological Changes

As stated by Hembrooke & Gay, "the ubiquity, pervasiveness and mobility of new technologies encourage a simultaneity of activities that goes beyond anything our culture has heretofore ever known. Indeed, the ability to engage in multiple tasks concurrently seems to be the very essence or core motivation for the development of such technologies" (2003, p. 1). People who have grown up with the convenience of technologies are surrounded by portable media (Oblinger & Oblinger, 2005; Foehr, 2006; Roberts & Foehr, 2008). Such media provide constant connectivity to friends, families, school and work. As a result of the hyper-connectivity, we are layering media in between activities or on top of other activities (Freitas & Griffiths, 2008; Gardner, 2008). Several studies (Foehr, 2006; Roberts & Foehr, 2008; Rideout, Foehr, & Roberts, 2010) reported that children and teens spend an increasing amount of time multitasking, and they manage to pack increasing amounts of media content into the same amount of time each day, for instance, surfing online while watching TV. Nearly one-third of the 8 to18-year-old children and teenagers in the U.S. talk on the phone, instant message, watch TV, listen to music, or surf the Web while they are doing their homework (Foehr, 2006; Roberts & Foehr, 2008).

Each new media adds itself to what we already are, creating both "amputations and extensions" to our senses and bodies, shaping them into a new technical form (McLuhan, 2002; Ong, 2002). McLuhan stated that by continuously embracing technologies, we became automatic devices used to provide error-sensing feedback for the technologies and correct the performance of technologies and that it was this dependency and linkage to technologies that make them an integral part of our lives. As such, whether intentional or not, we have become one large "bio-mechanical" system (McLuhan, 2002; The Matrix, 1999).

Carr (2011) discusses the undeniable re-wiring of the human brain as a function of newly adopted technologies. Tracing back to 8000 BC, Carr discusses the re-wiring of the brain's synaptic connections through such technological changes as movement from oral to written cultures with the introduction of both cuneiform and hieroglyphs, the introduction of the Greek and Roman alphabets, and the eventual adaptation of the printing press. Carr cited the work of developmental psychologist Maryanne Wolf (2008), who made a case for the history of the re-wiring process and gave evidence of the evolutionary states as they related to logographic systems. Today we can extend the trajectory of this evolution to the thousands of software applications that continue to manipulate both text and pictures on our millions of mobile devices.

As greater numbers of software and mobile applications compete for our time, oftentimes our connections with family or friends appear to be on the losing side of the equation. In an environment where information is highly available and digitally pushed to most of us, many people have "the fear of missing out," that is, a fear that if we did not check out what was pushed to us, we might lose some critical information and might suffer some unknown consequence or negative impact on our lives. Our behavior is being dramatically modified by emerging technologies. As Turkle (2011) suggested in her book *Alone Together*, this indicates that our relationship with technology has not yet fully matured.

Perceived New Skills, Needs, and Demands

Some scholars believe that the skills promoted by the convergence of new technologies, including creativity, peer-to-peer learning and multitasking are becoming necessary for success in today's world. Prensky (2001) claimed that younger generations growing up with new media and technologies are comfortable and capable of parallel rather than sequential or linear access to information and are capable of multitasking. The concept of "metatasking" was also suggested as a focused multitasking behavior (Zimmerman, 2007). Metatasking involves simultaneously undertaking any number of tasks that ultimately accomplish one primary objective. Zimmerman (2007) states that as we continue to seek information online and encourage online collaboration, it seems to make sense to encourage a deep focused multitasking, to accomplish as much as we can, and to use these technologies to our advantages. However, claims as the above are far from being documented in research. For instance, Carrier, et al., (2009) compared multitasking skills between younger and older generations and they found that the younger generation exhibited similar mental limitations in multitasking as the older generation did, although members of the younger generation reported more multitasking activities. Hartley, Jonides, and Sylvester's study (2011) showed no difference in managing dual tasks between older and younger adults.

It is obvious that this research area, in particular as multitasking behaviors emerging with new technologies, is ripe for investigation. In addition, much research on multitasking has been done in the lab using quantitative methods, and takes a strongly positivist stance with controlled experiments to identify the optimum distribution of tasks across input and output modality. Clearly, what happens in lab experiments does not often represent a complete picture of what happens in real life. For instance, the media multitaskers in real life may be more internally driven or directed on what to focus and when to switch between the tasks than those who are told what to focus on and when to switch in a lab environment, and as a result, may be better at multitasking (Lin, 2009). Such discrepancies naturally lead to disconnection between research and practice, and reduce the impact of research on practice. Therefore, the goal of this chapter is to look into people's daily experiences so as to better understand the complexity involved in people's daily multitasking activities and to better understand the human capacity for multitasking in natural settings. A qualitative method was incorporated to explore the environmental and technological factors involved in multitasking.

INDIVIDUAL STORIES BEHIND MULTITASKING ACTIVITIES

Methodology

This study used a qualitative approach (Creswell, 1998; Patton, 2001). As researchers, we were interested in what young people do when they multitask. According to van Manen (1990), phenomenology serves as the rationale behind the efforts to understand individuals by entering into their field of perception in order to see life as these individuals see it. In order to ascertain the experiences of multitasking, phenomenology appeared to be an appropriate approach because the researchers were primarily interested in understanding young people's experiences from the perspective of those being studied, by learning directly from those who tell their stories.

The phenomenological method is composed of three research processes: investigation of the phenomena, identification of general themes/essences of the phenomena, and comprehension of essential relationships among these themes (Spiegleberg, 1982; van Manen, 1990). In phenomenology, the researcher first details the individual statements of informants about experiences with the phenomenon before moving to meanings and clusters of meanings. This inductive approach to developing the qualitative narrative shows that the process is one that depends on an emerging design. The purpose of phenomenological inquiry is to discover the essential meaning or existential experience of what it is like to live through a certain experience, although the "meaning or essence of a phenomenon is never simple or one-dimensional. Meaning is multi-dimensional and multi-layered" (van Manen, 1990, p. 78). According to van Manen, "in order to come to grips with the structure of meaning it is helpful to think of the phenomenon described in the text as approachable in terms of meaning units, structure of meaning, or themes" (p. 78).

An important part of this research is devoted to understanding the experiences and discovering the reasons behind multitasking experiences. To find the reasons, it is best to allow the participants to speak for themselves. The following questions led the researchers to start the inquiry: 1) what is involved when one multitasks? 2) why do people multitask? and 3) what role does technology play in multitasking?

The researchers collected stories from 43 undergraduate students majoring in education (ages 21 to 29) at a southern university in the US, and 30 elementary school students (ages 6 to 11) close to the university in the same state. The population selection was decided by availability and relevance (i.e., the younger generations who are relatively more immersed in new media and technologies). Of the 43 undergraduate participants, there was one Caucasian male student. The rest were female students, composed of 40 Caucasian and two African-American female students. All the undergraduate students plan to be K-12 teachers in subjects such as math, science, English, art, and music after they graduate. Of the 30 elementary school children, there were 15 boys and 15 girls.

All the participants contributed their stories to the following questions: "Reflect on an incidence when you were sure that you were multitasking. What were you doing?" In addition, the researchers had discussions with the undergraduate participants about characteristics of multitasking, senses involved in multitasking, and the relationship between multitasking, learning, and technology. The collected stories were initially grouped by activities and contexts, for instance, combining all stories about watching TV while doing homework in one category. Afterwards, the patterns, meanings and themes behind the activities were identified and consolidated. Bracketing experiences enabled the concepts found in the data to be clarified and the categories to emerge. It helped to create an understanding based on the participants' own descriptions.

Findings

Several themes emerged from the study. They are: 1) habit or impulse; 2) need for background noise; 3) conflicting needs over a short period of time; 4) perceived efficiency of multitasking; 5) ubiquitous technologies; 6) individual differences. We will explain these themes with individual comments or stories below.

Habit and Impulse

Ellen, an undergraduate, found herself constantly surfing online, responding to emails, or looking for something to eat while writing a paper. As much as she wanted to focus and complete the paper before doing anything else, Ellen found it almost impossible not to change her attention to other things in the process. She called her behavior impulsive.

Kara also described an impulsive behavior she had when she had to focus on writing a paper to meet a deadline. The more she wanted to focus, the more distracted she became. As soon as she started to work on the paper, she saw many other things she needed to do here and there. A day passed. She found herself writing little; although she did many other things she did not plan to do. Several students echoed Kara's experience. Such an impulsive behavior can be dangerous and life-threatening sometimes. For instance, Carrie described an accident she had while driving:

I remember a multitasking situation where I was driving my car but I got stuck behind a stopped truck. I was going to go around the truck, but a car was on the other side so I couldn't. The truck started moving again so I also started moving. At the same time I looked at the radio to change the song and the truck in front of me had stopped again and I ended up rear-ending him. From now on I just keep my eyes on the road as I change the radio! Lesson learned!

It appears that Carrie has formed a habit of changing radio channels while driving. This may be fine in some occasions; however, in this situation, it caused distraction and interfered with Carrie's response to a sudden urgent situation, resulting in the accident.

Need For Background Noise

Several participants indicated that they needed to multitask in order to focus, particularly when they were doing school work or housework. David, an 8-year-old, said that he would either listen to music or have the TV on so that he wouldn't be bored doing his homework. Dana, an undergraduate says:

I prefer having background noise and an occasional distraction when I'm focusing… I personally find it difficult to focus for extended periods of time on a single task – it becomes monotonous, and I sometimes feel discouraged at a lack of progress.

About half of the participants from both age groups in this study talked about doing homework while watching TV, text messaging, or surfing on the web. This phenomenon is also described in the Kaiser Family Foundation studies, which reported that teens claimed that they cannot tolerate the silence when they do homework, and that they must have something to tune out in order to concentrate (Foehr 2006; Roberts & Foehr, 2008).

It is, however, not clear whether these activities distracted the participants from getting their homework done, or helped them to remain awake and interested in completing their homework. Some researchers point out that performance can be affected by workload being too high or too low (Nachreiner, 1995). Sustained low workload (under-load) can lead to boredom, loss of situational awareness and reduced alertness (Andrade, 2010). It is possible that if an individual is aware that they are losing focus on a task, switching tasks may cause a re-focusing of the brain central processing system. In the study conducted by Andrade (2010), when the primary task was monotonous, the phone operators became bored. A secondary task such as doodling helped them to stay active, reduce daydreaming, and improve memory recall of the primary task (Andrade, 2010). Other studies have also discovered that a non-intrusive background did not affect students' learning performance (Lee, Lin, & Robertson, 2011; Lin, Robertson, & Lee, 2009).

Conflicting Desires or Needs within the Same Period of Time

Often one decides to multitask because one cannot choose which task to do first. Emma, a 10-year old girl said that she had to multitask because she didn't want to miss her favorite show on TV, but she had to finish her homework before she went to bed. The following is a similar story from Anna:

Often, I try to watch TV while I'm doing homework. This usually doesn't work and I end up watching TV; especially if it is a program that I am really interested in. If I don't really care for the program, and someone else is in the room watching it, then I can focus on the homework. I usually tend to focus on whatever sounds more appealing; most of the time TV is more appealing than school work. I also focus on how major the assignment is. If it is something that has to be done now, then I will usually do it. Either way, I cannot accomplish both at the same time. I can't finish my assignment and know everything that happened on a show at the same time.

In the above story, Anna shifted her attention between her dual tasks: doing homework and watching TV. Anna often found herself paying attention to the more appealing – TV. In the following, Inez's story described a similar pattern although her multitasking activity appeared to be pressured by the competing demands at work:

When I am at work I do a lot of multitasking. I am a receptionist, so I have to be answering phones, writing down notes and messages, trying to answer questions, and making sure everyone is helped and greeted when they walk in the door. There are many times when I am helping out a client, and the phone rings. I have to try to help the client with their questions and still manage to answer the phone and try to also help the person on the phone. So I have to switch my focus around throughout the entire time I am at work. One person could need one thing and I have to concentrate on that need, then all of a sudden I have to put my focus on an entirely different task. It can get very stressful dealing with different people's needs and different tasks that are needed to be done. There are many distractions that make it hard to concentrate on any single task, therefore multitasking takes place.

Several undergraduate students told their experiences of working at restaurants or offices where they had to keep several tasks (e.g., customers' orders) in their head while physically doing other things including talking on the phone, working the counter, or refilling coffee cups. The participants did not make any differentiation when it came to holding multiple tasks in queue in their minds and processing them actively. Holding multitasking tasks in mind was perceived as multitasking even when one was not actively carrying out the tasks. This is probably because one experiences an equal amount of mental stress holding tasks in mind as when one tries to conduct a number of tasks physically at the same time.

People who work in this fashion may tune in and out their focus on different things at different times. While they may increase their probability of sensing something from multiple channels, they may not delve deeply enough into any one area unless they make a conscious decision and take action to focus a particular sense on a particular task, often times to its completion. What is unknown, however, is to what extent we are capable of juggling several tasks at the same time, switching them in a short period of time, or holding them in mind for a longer period of time. Or is it simply a reality that any type of professional skill requires a certain level of multitasking?

Perceived Efficiency of Multitasking

Often, what drives people to do multiple things at the same time is the perceived efficiency, effectiveness and productivity. Multitasking is seen as saving time. Edward, an undergraduate, mentioned that he felt it a waste of time in meetings when he could only listen to the speaker without simultaneously doing other things such as surfing online, chatting with friends using Instant Messenger or replying to emails on his computer. Edward's comment was echoed by many, although Edward and other students mentioned that this kind of multitasking behavior was sometimes perceived as inattentive and rude to the speaker and others. Yolanda, an undergraduate, states: "I've had instances where I've had the TV on while studying before. I have also eaten while driving, which is probably not too smart, but I'm always on the go and it saves me time." Barbara, an undergraduate, indicated that she multitasked because she did not need to pay full attention to any of the tasks that she simultaneously carried out:

The times when I have the best multitasking skills is when I do something like watch TV, be in a conversation and do an assignment. I was able to do all of these at once because I only needed to focus on one thing at a time. I could tune out the conversation while writing a paper or focus on what is on the TV while still getting information from all other sources. I need background noise to work most of the time. When I multitask I am able to do all 3 very effectively because my mind can switch tasks quickly and effectively. I also am more used to it since I have been multi-tasking for many years now, probably since the 6th grade.

Helen, an undergraduate, stated:

I believe that I multitask on a daily basis. As a student, mom, and wife I am constantly doing two or more things at once. For example, I cook dinner every night while watching my 10 month old to make sure she will not get hurt crawling around the house. I feed my daughter while eating and talking with my husband during dinner. I use note cards to study while doing the laundry, playing with my little girl, walking the dog, feeding my daughter a bottle, and watching TV. I listen to music while I study. I chat with friends on the phone while checking emails, looking up information, and doing errands outside the home. If it were not for my multitasking abilities I would not be able to get an ounce of work done in any given day.

Zack's story seems more convincing. In schools, we are all asked to take notes while listening to the teacher. Zack, an undergraduate, says: "I feel like I always multitask when I'm trying to take notes and listen to a teacher at the same time. It gets a little hard sometimes; depending on how fast she goes through the notes or how fast she talks."

Taking notes while listening to the teacher is generally accepted as a good practice in schools. Zack's story led us to the concept of doing multiple tasks or involving one's multiple senses for one particular learning goal: to learn the subject matter of the lecture. This possibility, afforded by our technological extensions such as the notebook, the pen, or the computer, seems to create a synergy and affect the outcome or the quality of the goal, which is important to note in the teaching and learning environments. With the rapid integration of technologies in the classrooms, teachers and students are encouraged to use new technologies to pull resources from different channels. Such a practice has brought about many benefits for teaching and learning, while at the same time may have also encouraged multitasking activities (Lin & Bigenho, 2011).

Ubiquitous Technologies

Sixty out of the 73 stories mentioned some form of technology, whether the technology was TV, radio, cell phone, iPod, computer, or other kinds of machines. The following are some examples:

...I will get home; get the internet up as I'm thinking about the order of what to do. At the same time I'm getting paper to print my time sheet and bringing up iTunes. Then lo and behold someone will IM me so I'll start talking to them while pressing print and switching over to check my e-mail. But I'm only able to juggle all of these things because none of them requires my constant attention.

... I work on homework for my online course, have a few instant messenger boxes open, constantly checking email, either listening to my iPod or listening to online radio, and working on a developer program to create 3-D models for a 3-D gaming environment.

I think my multitasking usually involves the computer. We don't have cable at home, but we rent tons of movies from Netflix and the public library. When I am at home I am usually reading blogs or other websites or playing on Neopets on my laptop while watching a movie and chatting on instant messenger or talking to my fiancé if he's home. I have no trouble carrying on a conversation orally with him while typing in a conversation on instant messenger.

As awful as it sounds, I was once driving, talking on the phone, and trying to play my favorite song on my iPod all at once. I got where I intended to go safely, but I couldn't remember at all how I arrived, because I was involved in too many other things besides driving. It's scary to think of what could have happened due to my lack of attention.

Clearly, access to and use of social media, multimedia and technologies provided opportunities to multitask (Rosen, Carrier, & Cheever, 2013; Song, Nam, Lim, & Kim, 2013). The demand of our time each day seems to be expanding as well. Technology has been a boon, but also a bust in some regards. We can easily list the collective number of daily events that soak up our time in undesirable ways, things like junk mail, pop-up windows, telemarketer phone calls, commercials on TV and in movie theaters, to name a few. These assaults on our attention have major negative impacts on us every day including, for example, stealing our precious time through distractions, and by collectively de-sensitizing us to our environment.

Individual Differences

Different individuals have varying degrees of situational awareness and may be affected differently while eating, listening, seeing, texting, and playing. As humans, we develop an almost unconscious sense of which of our five senses should take focusing priority on any given task. For example, one might give the sense of sight priority over hearing or the sense of touch, because we know that the task about to be undertaken is more visually oriented. Another task, such as typing or knitting, might be primary focused on haptic functionality. As it is not easy to do two or more tasks at the same time, certain planning and skills are necessary as articulated by Elizabeth, an undergraduate:

I tend to try to get a head start on things, so in the process of that I try to do more than one thing at a time. Sometimes I'm successful, but other times I find myself overwhelmed, but not defeated. I get it done. Multitasking to me requires skill, time, and patience. I can watch television and do work while talking on the phone. I can cook and wash at the same time, but other than that, I think that I would have to do each task individually. I'm not the best at multitasking, but in certain situations I'm good at it.

Experience tells Elizabeth that she can do more than one thing with some tasks but not with others. Such differences may be due to the nature of the tasks and to her familiarity with the tasks at hand. Not only may one's task priorities and familiarities influence one's ability or decision to multitask, but one's learning style and natural intelligences may also have impacts. Gillian Lynne, who is a choreographer and has created some of the most successful musical theater productions in history, was seen as hopeless and as having a learning disorder in school because she could not concentrate (Robinson, 2006). After seeing a specialist, Gillian was advised to go to a dance school, where she met like-minded people who engaged their entire bodies to think (Beilock, & Goldin-Meadow, 2010; Gardner, 1993; Robinson, 2006). There the kinesthetically-inclined Gillian found her personal formula for success. Personal preferences and learning style differences account for many of our inclinations when it comes to dedicating our time and engagements with our environments.

Gloria, a 10-year-old girl told her multitasking story about playing basketball. She said that she was using her hands to dribble, while trying to protect the ball from getting stolen, looking for an open person under the basket, and weaving in and out trying to get closer to the basket. It is easy to surmise from this story that one must be good at splitting and switching one's attention constantly and rapidly between all the tasks in order to be a good basketball player. Common knowledge tells us that anyone can be a better player through practice. Research has also shown that one can be a better multi-tasker on the trained tasks through repeated practice, although that improved ability does not necessarily transfer when the tasks are changed (Dux, et al., 2009).

Frances, an 8-year-old girl said that she normally looks up her math homework before class and relates the problems for homework to the examples the teacher provides in class. This way she can ask questions about certain problems in the homework not mentioned in the lecture. The fact that Frances was able to plan ahead allowed her to ask additional questions while listening to the teacher, which further advanced her multitasking capability, making her a good student in math. Gus, a 10-year-old boy indicated that he had to multitask when he was playing video games. He had to focus on different aspects of the game and had to press different buttons on the game controller at the same time. Through repeated practice, challenges, and play, Gus had become a good multitasking video game player.

One similarity of Gloria's, Frances' and Gus' stories is that the multitasking skills involved in the tasks – basketball, homework, and video games – can be improved through preparations and repeated practices over time. Rehearsal and planning ahead are skills possessed by chess masters and gifted pianists. Abby, a music teacher indicated that she would try to isolate certain components of taught music and get these components to be "automatic" for her students. Then, as the students learned the next component, they would not need to consciously "think" of the ones now automated. Abby described this practice as doing several things at once, one consciously and one or more subconsciously. This can be seen as one method of reducing cognitive load. And we now know from research on stroke victims, injured football players and military veterans that people can develop strategies to help rehabilitate the brain's executive functions, even if the brain has suffered injuries and/or repeated trauma (Chapman & Kirkland, 2013).

IMPLICATIONS AND FUTURE RESEARCH DIRECTIONS

The purpose of this chapter was to explore the environmental and technological factors related to multitasking. The multitasking experiences shared by the individual participants showed that our multitasking behaviors, activities and abilities are not only constrained by our brain capacity for carrying out two or more tasks concurrently, but also affected by our daily habits, conflicting needs, perceptions, technologies, and individual differences.

In this study, all the participants were asked to tell their stories. As a result, one limitation of the study could be that the participants were only able to speak for their conscious multitasking activities while not being able to speak about the unconscious multitasking activities (i.e., autonomous multitasking activities). However, the 73 stories told by the participants provided rich and diverse multitasking experiences.

As the stories indicate, many students watch TV or use other media when they do their homework. In this case, homework is held as mundane and boring, and hence is hard to execute without other motivational stimulation taking place simultaneously. If this is the case for some students, then the long-held practice of getting students to focus on a lecture, may be an unproductive effort. The stories about playing basketball and video games indicate that many young people are good at multitasking in extra-curricular activities, especially when technology is involved. The mastery and use of technology such as remote control units, cell phones, tablets, game controller devices, digital recording and imaging devices are occurring at progressively earlier ages as children appropriate complex conventions much earlier than some adults, largely due to their sense of play and free time to investigate these functions.

The perceptions and experiences of multitasking were closely related to our attention. Our attention and focus might be dedicated, distracted, switched, or split. These functional attributes can be as a result of our personal preferences and priorities. Each of us has innate and acquired capabilities which come to us through personal experiences. Individual differences in attentional control have a significant impact on student learning in a multimedia instructional environment (Doolittle & Mariano, 2008). Personalization can bring benefits such as increased learning, greater enjoyment, enhanced motivation, and reduced learning time (Kelly, 2008). As educators, we bear the responsibility of looking into the motivation and skills required to help appropriate, motivate and transfer those skills to newer learning environments.

The participants' stories in this study also revealed that our multitasking activities do not all remain in our brains. Our abilities to multitask are not only decided by our innate capacities, but also influenced by our experiences, environments, and technologies. The brain or cognitive ability is important; yet, our interaction with the environment may play an equally important, if not more important role in our ability to multitask. We have come to use the media and numerous technologies as extensions of the human body. The connections of multitasking to attention, preference, priority, ability, technology, and environment indicate that it would be difficult to make a broad claim of someone's ability to multitask without knowing what is involved in a particular multitasking activity. Many of us can handle some multitasking situations more easily while having difficulty or risking our lives multitasking in other situations, as well stated by Salvucci & Taatgen (2008, p.101):

In some situations, multitasking can seem nearly effortless (e.g., walking and talking); for other situations, it can seem extremely difficult if not impossible (e.g., reading and listening to two distinct sentences); for still others, multitasking performance may depend heavily on the individual and/or the environment (e.g., singing while playing an instrument or dialing a phone while driving).

In their most recent book, Salvucci and Taatgen (2011) further presented the concept of "threaded cognition" as a unifying theory of multitasking. They indicated that the key to our multitasking ability is "the ability to take single-task skills and combine them as needed to accomplish a higher-level goal" (p. 7). Therefore, this study leads to further inquiries and investigations such as:

- When does multitasking help attention and when does it impede attention?
- When do new media and technologies enable us and when do they impede us from multitasking?
- To what extent are multitasking skills required in one developing an expertise or professional capacity?
- How does one transfer multitasking abilities in one's daily life, if any, to newer teaching and learning environments?
- How can we incorporate different research paradigms and methodologies to explore multitasking so that we will have both empirical evidence and in-depth understanding of this phenomenon, including its potentials and challenges?

In researching the literature for this and other recent research projects we have noticed that the research on multitasking, when taken as a whole, is rather asymmetrically weighted toward the empirical methodology. Our recommendation is that researchers and scientists might instead welcome more diverse forms of research using various methodologies and conceptual frameworks. In this way issues of power, gender, educational equity, corporate hegemony, and political or technological determinism might also be uncovered in parallel. As humans we are meaning-making machines. A more diverse stanch on this topic should be openly embraced to better understand its impact on teaching, learning, and quality of life in the 21st century and beyond.

CONCLUSION

With our accelerating lifestyles and the constant pressure for greater efficiency from our workforce leaders, multitasking has become an expected skill. We are continuously encouraged to complete as many tasks as possible, rather than give our full attention to perfecting one particular thing in a limited time frame. Multitasking will continue to be a quality of life concern and non-trivial functionality for humans. The developing mantra has become "Perfection is the enemy of good enough!" Business encourages timely forward movement (i.e., just get it good enough to move to the next step) rather than discoveries of deeply focused, highly-extracted knowledge. This study points to the delicate interplay between multitasking performance measured objectively and the individuals' real life experiences over time. It is hoped that this study and future investigations on the topic will help us better understand the environmental and technological factors involved in multitasking behaviors and find solutions to treat the problems.

Note: This current chapter is an enhanced version of an article published at the *International Journal of Technology and Human Interaction* (Lin, 2013) at the invitation of the journal editors. With this update, we have included new literature, discussions, and references.

REFERENCES

Andrade, J. (2010). What does doodling do? *Applied Cognitive Psychology*, *24*(1), 100–106. doi:10.1002/acp.1561

Ang, C. S., Zaphiris, P., & Mahmood, S. (2007). A model of cognitive loads in massively multiplayer online role playing games. *Interacting with Computers*, *19*(2), 167–179. doi:10.1016/j.intcom.2006.08.006

Baddeley, A. D. (1996). Exploring the central executive. *Quarterly Journal of Experimental Psychology*, *49A*(1), 5–28. doi:10.1080/713755608

Barley, R. A. (1997). Behavior inhibition, sustained attention, and executive functions: Constructing a unifying theory of ADHD. *Psychological Bulletin*, *121*(1), 65–94. doi:10.1037/0033-2909.121.1.65 PMID:9000892

Beilock, S. L., & Goldin-Meadow, S. (2010). Gesture changes thought by grounding it in action. *Psychological Science*, *21*(11), 1605–1610. doi:10.1177/0956797610385353 PMID:20889932

Bluedorn, A. C. (2001). Time and organizational culture. In N. M. Ashkanasy, C. P. M. Wilderom, & M. F. Peterson (Eds.), *Handbook of organizational culture and climate*. Thousand Oaks, CA.

Bluedorn, A. C. (2002). *The Human Organization of Time: Temporal Realities and Experience*. Stanford, CA: Stanford University Press.

Bransford, J., Pellegrino, J., & Donovan, M. (1999). *How people learn: Bridging research and practice*. Washington, DC: National Research Council, National Academy Press.

Broadbent, D. (1958). *Perception and communication*. Oxford: Pergamon. doi:10.1037/10037-000

Carr, N. (2011). *The Shallows: What the Internet is doing to our brains*. New York: W.W. Norton Company.

Carrier, L. M., Cheever, N. A., Rosen, L. D., Benitez, S., & Chang, J. (2009). Multitasking across generations: Multitasking choices and difficulty ratings in three generations of Americans. *Computers in Human Behavior*, *25*(2), 483–489. doi:10.1016/j.chb.2008.10.012

Chapman, S. B., & Kirkland, S. (2013). *Make your brain smarter*. New York: Free Press.

Creativity is a key part of the educated mind. (2006). *Edutopia*. Retrieved May 17, 2011 from http://www.edutopia.org/take-chance-let-them-dance

Creswell, J. (1998). *Qualitative inquiry and research design: Choosing among five traditions*. Thousand Oaks, CA: Sage.

Doolittle, P. E., & Mariano, G. J. (2008). Working memory capacity and mobile multimedia learning environments: Individual differences in learning while mobile. *Journal of Educational Multimedia and Hypermedia*, *17*(4), 511–530.

Dux, P. E., Tombu, M. N., Harrison, S., Rogers, B. P., Tong, F., & Marois, R. (2009). Training improves multitasking performance by increasing the speed of information processing in human prefrontal cortex. *Neuron*, *63*(1), 127–138. doi:10.1016/j.neuron.2009.06.005 PMID:19607798

Fisch, S. M. (2000). A capacity model of children's comprehension of educational content on television. *Media Psychology*, *2*(1), 63–91. doi:10.1207/S1532785XMEP0201_4

Fleishman, E. A., Costanza, D. P., & Marshall-Mies, J. (1999). Abilities. In N. G. Peterson, M. D. Mumford, W. C. Borman, P. R. Jeanneret, & E. A. Fleishman (Eds.), *An occupational information system for the 21st century: The development of O*NET* (pp. 175–195). Washington, DC: American Psychological Association. doi:10.1037/10313-010

Foehr, U. G. (2006). *Media multitasking among American Youth: Prevalence, predictors and pairings*. The Henry J. Kaiser Family Foundation.

Foerde, K., Knowlton, B. J., & Poldrack, R. A. (2006). Modulation of competing memory systems by distraction. *Proceedings of the National Academy of Sciences of the United States of America*. Retrieved on May 17, 2011 from http://www.pnas.org/content/103/31/11778.full

Freitas, S., & Griffiths, M. (2008). The convergence of gaming practices with other media forms: What potential for learning? A review of the literature. *Learning, Media and Technology*, *33*(1), 11–20. doi:10.1080/17439880701868796

Gardner, H. (1993). *Multiple intelligences: The theory in practice*. New York: Basic Books.

Gardner, J. S. (2008). *Simultaneous Media Usage: Effects on Attention*. Unpublished doctoral dissertation. Virginia Tech.

Goonetilleke, R. S., & Luximon, Y. (2010). The relationship between monochronicity, polychronicity and individual characteristics. *Behaviour & Information Technology*, *29*(2), 187–198. doi:10.1080/01449290903222697

Gopher, D., Armony, L., & Greenspan, Y. (2000). Switching tasks and attention policies. *Journal of Experimental Psychology. General*, 229–308. PMID:11006903

Hall, E. T. (1959). *The Silent Language*. Garden City, NY: Doubleday.

Hartley, A. A., Jonides, J., & Sylvester, C. (2011). Dual-task processing in younger and older adults: Similarities and differences revealed by fMRI. *Brain and Cognition*, *75*(3), 281–291. doi:10.1016/j.bandc.2011.01.004 PMID:21320741

Hembrooke, H., & Gay, G. (2003). The Laptop and the lecture: The effects of multitasking in learning environments. *Journal of Computing in Higher Education*, *15*(1), 46–64. doi:10.1007/BF02940852

Ishizaka, K., Marshall, S. P., & Conte, J. M. (2001). Individual differences in attentional strategies in multitasking situations. *Human Performance*, *14*(4), 339–358. doi:10.1207/S15327043HUP1404_4

Jackson, M. (2008). *Distracted: The Erosion of Attention and the Coming Dark Age*. Amherst, NY: Prometheus Books.

Johnson, S. (2006). *Everything Bad is Good for You: How Today's Popular Culture Is Actually Making Us Smarter*. Riverhead Books.

Junco, R. (2012). In-class multitasking and academic performance. *Computers in Human Behavior*, *28*(6), 2236–2243. doi:10.1016/j.chb.2012.06.031

Just, M. A., Carpenter, P. A., Keller, T. A., Emery, L., Zajac, H., & Thulborn, K. (2001). Interdependence of non-overlapping cortical systems in dual cognitive tasks. *NeuroImage*, *14*(2), 417–426. doi:10.1006/nimg.2001.0826 PMID:11467915

Just, M. A., Keller, T. A., & Cynkar, J. A. (2008). A decrease in brain activation associated with driving when listening to someone speak. *Brain Research*, *1205*, 70–80. doi:10.1016/j.brainres.2007.12.075 PMID:18353285

Kelly, D. (2008). Adaptive versus learner control in a multiple intelligence learning environment. *Journal of Educational Multimedia and Hypermedia*, *17*(3), 307–336.

Konig, C. J., Buhner, M., & Murling, G. (2005). Working memory, fluid intelligence, and attention are predictors of multitasking performance, but polychronicity and extraversion are not. *Human Performance*, *18*(3), 243–266.

Lang, A. (2001). The limited capacity model of mediated message processing. *Journal of Communication*, 46–70.

Lee, F. J., & Taatgen, N. A. (2002). Multi-tasking as Skill Acquisition. *Proceedings of the twenty-fourth annual conference of the cognitive science society*. pp. 572-577. Mahwah, NJ: Erlbaum.

Lee, J., Lin, L., & Robertson, T. (2011). The impact of media multitasking on learning. *Learning, Media and Technology*, *36*(3), 1–11.

Lin, L. (2009). Breadth-biased versus focused cognitive control in media multitasking behaviors. *Proceedings of the National Academy of Sciences (PNAS) USA 106*: 15521-15522. doi:10.1073/pnas.0908642106

Lin, L., & Bigenho, C. (2011). Note-taking and memory in different media environments. *Computers in the Schools*, *28*(3), 200–216. doi:10.1080/07380569.2011.594989

Lin, L., Lee, J., & Robertson, T. (2011). Reading while watching video: The effect of video content on reading comprehension and media multitasking ability. *Journal of Educational Computing Research*, *45*(2), 183–201. doi:10.2190/EC.45.2.d

Lin, L., Robertson, A., & Lee, J. (2012). The impact of outcome intentions on reading and multitasking performances. *Journal of Educational Technology Development and Exchange*, *5*(1), 77–94.

Lin, L., Robertson, T., & Lee, J. (2009). Reading performances between novices and experts in different media multitasking environments. *Computers in the Schools*, *26*(3), 169–186. doi:10.1080/07380560903095162

Mäntylä, T. (2013). Gender differences in multitasking reflect spatial ability. *Psychological Science*, *24*(4), 514–520. doi:10.1177/0956797612459660 PMID:23462757

McLuhan, M. (2002). Understanding Media: The Extensions of Man (10th, ed). Cambridge, MA: MIT Press.

Meyer, D. E., & Kieras, D. E. (1997). A Computational Theory of Executive Cognitive Processes and Multiple-task Performance: Part 1. Basic Mechanisms. *Psychological Review*, *104*, 229–233. PMID:9009880

Monsell, S. (2003). Task switching. *Trends in Cognitive Sciences*, *7*(3), 134–140. doi:10.1016/S1364-6613(03)00028-7 PMID:12639695

Nachreiner, F. (1995). Standards for ergonomics principles relating to the design of work systems and to mental workload. *Applied Ergonomics*, *26*(4), 259–263. doi:10.1016/0003-6870(95)00029-C PMID:15677027

National Public Radio (NPR). (October 2, 2008). Think You're Multitasking? Think Again. *Morning Edition.* Retrieved on May 17, 2011 from http://www.npr.org/templates/story/story.php?storyId=95256794

Oblinger, D. G., & Oblinger, J. L. (Eds.). (2005). *Educating the Net Generation.* Boulder, CO: EDUCAUSE. Retrieved on May 17, 2011 from http://www.educause.edu/educatingthenetgen

Ong, W. J. (2002). *An Ong Reader: Challenges for Further Inquiry* (T. J. Farrell & P. A. Soukup, Eds.). Cresskill, NJ: Hampton Press Inc.

Ophir, E., Nass, C. I., & Wagner, A. D. (2009). Cognitive control in media multitaskers. *Proceedings of the National Academy of Sciences of the United States of America*, *106*(37), 15583–15587. doi:10.1073/pnas.0903620106 PMID:19706386

Patton, M. (2001). *Qualitative Research and Evaluation Methods* (3rd ed.). London: Sage.

Poldrack, R. A., & Foerde, K. (2007). Category learning and the memory systems debate. *Neuroscience and Biobehavioral Reviews*, *32*(2), 197–205. doi:10.1016/j.neubiorev.2007.07.007 PMID:17869339

Prensky, M. (2001). *Digital Game-based Learning*. The McGraw-Hill Companies.

Rideout, V. J., Foehr, U. G., & Roberts, D. F. (2010). *Generation M^2 Media in the lives of 8- to 18-year-olds.* Menlo Park, CA: Kaiser Family Foundation.

Roberts, D. F., & Foehr, U. G. (2008). Trends in Media Use. In J. Brooks-Gunn & E. Donahue (Ed.). Generation M: Media in the lives of 8-18 year olds. Menlo Park, CA: Kaiser Family Foundation.

Robinson, K. (2006). Take a chance . . . Let them dance: Validating artistic expression.

Rosen, L. D., Carrier, L. M., & Cheever, N. A. (2013). Facebook and texting made me do it: Media-induced task-switching while studying. *Computers in Human Behavior*, *29*(3), 948–958. doi:10.1016/j.chb.2012.12.001

Rubinstein, J. S., Meyer, D. E., & Evans, J. E. (2001). Executive control of cognitive processes in task switching. *Journal of Experimental Psychology. Human Perception and Performance*, *27*(4), 763–797. doi:10.1037/0096-1523.27.4.763 PMID:11518143

Salvucci, D., & Taatgen, N. (2008). Threaded cognition: An integrated theory of concurrent multitasking. *Psychological Review*, *115*(1), 101–130. doi:10.1037/0033-295X.115.1.101 PMID:18211187

Salvucci, D., & Taatgen, N. (2011). *The Multitasking Mind. Oxford Series on Cognitive Models and Architectures*. Oxford: Oxford University Press.

Schweickert, R., & Boggs, G. J. (1984). Models of central capacity and concurrency. *Journal of Mathematical Psychology, 28*(3), 223–281. doi:10.1016/0022-2496(84)90001-4

Song, K., Nam, S., Lim, H. & Kim, J. (2013). Analysis of youngsters' media multitasking behaviors and effect on learning. *International Journal of Multimedia and Ubiquitous Engineering, 8* (4), 191-198.

Spiegleberg, H. (1982). *The Phenomenological Moment: A Historical Introduction* (3rd ed.). The Hague: Martinus Nijhoff. doi:10.1007/978-94-009-7491-3

Spink, A., & Park, M. (2005). Information and non-information multitasking interplay. *The Journal of Documentation, 61*(4), 548–554. doi:10.1108/00220410510607516

The Matrix. (1999). Retrieved at http://www.imdb.com/title/tt0133093/?ref_=nv_sr_1

Turkle, S. (2011). *Alone together: Why we expect more from technology and less from each other*. Perseus Printing, New York: Basic Books.

van Manen, M. (1990). *Researching lived experience: Human science for an action sensitive pedagogy.* Albany: State University of New York Press.

Weissman, D. H., Gopalakrishnan, A., Hazlett, C. J., & Woldorff, M. G. (2005). Dorsal anterior cingulate cortex resolves conflict from distracting stimuli by boosting attention toward relevant events. *Cerebral Cortex, 15*(2), 229–237. doi:10.1093/cercor/bhh125 PMID:15238434

Wickens, C. D. (1984). Processing resources in attention. In R. Parasuraman, J. Beatty, & R. Davies (Eds.), *Varieties of Attention* (pp. 63–101). New York: Wiley.

Wolf, M. (2008). *Proust and the squid: the story and science of the reading brain.* New York: Harper Collins Publishers.

Zimmerman. (2007). Metatasking vs. Multitasking. *Library Journal.* Retrieved on May 17, 2011 from http://www.libraryjournal.com/article/CA6431788.html

Zumback, J. (2006). Cognitive overhead in hypertext learning reexamined: Overcoming the myths. *Journal of Educational Multimedia and Hypermedia, 15*(4), 411–432.

ADDITIONAL READING

Salvucci, D., & Taatgen, N. (2008). Threaded cognition: An integrated theory of concurrent multitasking. *Psychological Review, 115*(1), 101–130. doi:10.1037/0033-295X.115.1.101 PMID:18211187

Salvucci, D., & Taatgen, N. (2011). *The Multitasking Mind. Oxford Series on Cognitive Models and Architectures*. Oxford: Oxford University Press.

KEY TERMS AND DEFINITIONS

Dual Task: A procedure in experimental (neuro) psychology that requires an individual to perform two tasks simultaneously in order to compare performance with single-task conditions (Navon & Gopher, 1979).

Media Multitasking: Media multitasking refers to concurrent use of multiple media (Roberts & Floehr, 2008).

Polychronicity: Polychronicity refers to the "extent to which people prefer to engage in two or more tasks or events simultaneously and believe that their preference is the correct way to do things" (Bluedorn, 2001, p. 119).

Task Switch: Task switching refers to an executive function and a kind of cognitive flexibility that involves the ability to shift attention between one task and another (Monsell, 2003).

Chapter 2
Blogs' Potentialities in Learning:
What Are the Key Variables to Promote Cognitive Empowerment

Alessandro Antonietti
Catholic University of the Sacred Heart, Italy

Barbara Colombo
Catholic University of the Sacred Heart, Italy

Simona C. S. Caravita
Catholic University of the Sacred Heart, Italy

Luisa Simonelli
Catholic University of the Sacred Heart, Italy

ABSTRACT

Growing literature on the impact of new technologies on learning processes suggests that blogs can be effective tools to empower cognition. The chapter describes a line of research which is aimed at exploring the possibility that blogging is related to cognitive skills and can be effectively used in educational settings. In the main study reported in the paper fifty blogs were analysed and classified according to both their content and formal structure. The right (i.e., intuitive-holistic) vs. left (i.e., systematic-analytical) thinking style of blog owners was also assessed, as well as the blog owners' awareness of the psychological processes, activated by the blog that they had devised, both in their own and in other people's mind. Results showed that blog owners are able to use effective communication strategies by differentiating the formal structure of bogs according to the content, but they lack metacognitive awareness about the mental processes activated by the blog. No relation between the blog owner's cognitive style and blog style was found. Implications for the educational use of blogs are discussed. In other two studies such implications were tested in samples of teachers and students. Blogs' potentialities to enhance metacognitive awareness and control in both teachers and students, as well as cooperative learning in students,were critically evaluated. Altogether, results from this research line provide evidence that blogging can be useful to foster cognitive skills in education by empowering the bloggers' awareness of the implied mental processes.

DOI: 10.4018/978-1-4666-8450-8.ch002

INTRODUCTION: BLOGS AS ENVIRONMENTS FOR PERSONALISED LEARNING

A blog (a portmanteau of web log) is a website where entries are written in chronological order and displayed commonly in reverse chronological order. Many blogs provide commentaries or news on a particular subject; other blogs are personal online diaries. When blogging, users can leave comments, provide information to others by posting messages in form of text or multimedia, and build online discussions among bloggers. Thanks to these opportunities, blogs have been shown to provide a good environment for learning, which can be also used to expand in class discussions with all students. Blogs are perceived by learners as useful tools to share their own experiences and to construct knowledge, to improve problem solving and reflection, and to favour the communication of emotions (Chu, Chan &, Tiwari, 2012). Using blogs seems also to promote engagement in collaborative learning (Chu, Chan &, Tiwari, 2012). This happens because students' use of blog is affected by individual differences (related to specific motivation and the level of difficulty attributed to the learning activity), but it does not depend on gender or technical skills (Cakir, 2013). Similarly, the learner's positive perception about blogs as learning environment is independent of discipline background and frequency of use (Chu, Chan &, Tiwari, 2012). Therefore, all students can effectively being engaged in blogging during a learning experience.

A typical blog combines texts and images (sometimes also sounds and music) and includes links to other blogs, web pages and media related to its topic. Therefore, blogs can be conceived as multimedia artefacts and also as hypertexts (Murray & Hourigan, 2008). Such peculiar characteristics make the building and managing of a blog cognitively demanding. The characteristics mentioned above, in fact, "force" people to use specific cognitive strategies whilst constructing and administering their blogs, and this should promote reflection (Blood, 2000; Colombo & Sala, 2011; Wagner, 2003). For example, anyone who builds and manages a blog is asked to make decisions concerning the sections to be included, the number and typologies of the pictures to be uploaded, their layout, the writing style, and so on (Huffaker, 2005). However, such choices should depend not only on the blog's content, but also on the cognitive preferences of the blog owners. Hence the blog structure, as well as any kind of artefact, should mirror the author's cognitive style (Henri, 1992; Xie, Ke, & Sharma, 2010).

Thanks to its intrinsic characteristics, a blog is a user-friendly artefact allowing for high levels of personalization and, due to its specific features, it tends to mirror owner's individual characteristics. In line with this assumption, Guadagno, Okdie and Eno (2008) found that differences in the levels of some personality traits (i.e., openness to new experience and neuroticism) are linked to differences in the likelihood to be a blogger. Likewise, Li and Chignell (2010) provided some evidence that the personality of the blog writer is reflected in the blog style, such in a way that is recognizable by the blog readers too. The way a blog is devised and managed may be also linked to individual cognitive style, that is, the preferred and habitual manner to process information (Antonietti, 2003).

Chapter's Aims

In order to understand the potentiality for cognitive empowerment related to the use of blogs, it is interesting to assess whether blog owners actually personalise the blog they built by arranging the structure of the blog not only according to its content but also in order to match their cognitive style. In addition, it is worth investigating individuals' metacognitive awareness. In other words, to what extent are the blog owners conscious of the mental processes that are activated in other people by the blogs they devised? And, particularly, to what extent are the blog owners aware of the possible match between the features of the blog and the bloggers' individual characteristics?

The studies presented in this chapter addressed these questions. Since blogs are easily handled by almost everyone, they can be viewed as an ecological field for exploring spontaneous multimedia and hypertext information processing. In particular, in this chapter, we first present a short review of the relevant literature on the relations between blogs and cognitive styles and between blogs and metacognition. Then we will describe some studies we carried out to examine two main research questions: 1. To what extent can blogs be seen as artefacts that can reveal personal styles of thinking? and 2. At which level (namely, conscious and intentional or implicit), do relations, if any, between blogs and style emerge?

Blogs and Cognitive Styles

Cognitive Styles

A cognitive style can be defined as the preferred way of thinking (Sternberg, 1997). Such an inclination concerns the peculiar mode of information processing that a person applies across different tasks and which occurs in a variety of behaviours. Cognitive styles are transferable to different domains: people tend to use strategies that belong to their own preferred cognitive styles when facing different kinds of tasks. In short, cognitive styles concern the general tendency to follow thinking strategies of a certain type more frequently than others. They are not abilities, but rather a way of applying abilities. Cognitive styles are also correlated with steady personality traits (Balkis & Isiker, 2005; Fan et al., 2010). For example, a surface learning style (which characterizes students who aim to re-expose notions correctly in a test or exam rather than actually understand them) is associated to the openness to experience (Furnham et al., 2007).

The role of cognitive style has been mainly investigated in instructional contexts. Cook and colleagues (2007) argued that adapting instructions to students' individual style enhances learning, especially in web-based environments. As far as learning from hypermedia is concerned, it has been shown that cognitive styles can be associated to specific browsing patterns (Calcaterra et al., 2005; Fiorina et al., 2007). Cognitive styles are also good predictors of self-efficacy in learning environments (West et al, 2007). Style has been also found to be one of the most important criteria to predict students' preferences for different multimedia learning materials, with students who prefer an intuitive-holistic style using more often animations and simulations and memorizing information better if they employ multimedia materials (Ocepek et al., 2013). Based on data from 272 undergraduate students, Ocepek and colleagues (2013) suggested that information about the students' style should be taken into account in models to recommend different learning multimedia types to different individuals.

Amongst the different cognitive styles, the right/left thinking style (Torrance et al., 1978) is particularly interesting. It is inspired by the theory of brain hemispheric dominance. According to Torrance, we can distinguish between two tendencies. The first one is possibly linked to the right hemisphere, which is associated to the visual-motor code, innovative procedures and intuitive and holistic processing. The second one, allegedly linked to the left hemisphere, is associated to the verbal-abstract code and analytic and sequential procedures. Even though some data supported the biological basis of such a distinction (Fabbri et al., 2007) and specific cognitive functions can be associated to the two brain hemispheres (Martindale, 1999; Springer & Deutsch, 1997), the right/left distinction has to be meant in a metaphoric sense. Right thinkers cannot be considered as people actually "using" prevalently the right hemisphere. They should be considered as individuals showing stylistic profiles that roughly correspond to the features traditionally attributed to that hemisphere (Coffield et al., 2004). Right/left style also includes personality traits (Torrance, 1988): Right-thinkers are imaginative, inventive, enterprising, change-seeking and non-conformist, whereas left-thinkers are realistic, repetitive, settled, planners and conformists. Moreover, cognitive styles associated to hemispheric dominance are related to neuroticism, extraversion, openness and conscientiousness (Zhang, 2006).

The Link Between Blogs and Cognitive Styles

Many characteristics of a blog can be related to the right/left thinking style. For example, bloggers are free to build links between the posts or in the main structure of the blog. Such links can be devised either in a serial or in a holistic-reticular way: The first and second solutions mirror the features of the left and the right style respectively. Moreover, left-thinkers, who are free to personalise the structure of the blog, are expected to choose pre-structured templates, whereas right-thinkers are more innovative and may change the basic structure of the blog. The same may be true as regards the choice of the main topic of the blog: Imaginative topics should be preferred by right-thinkers and informative ones by left-thinkers. Also, the use of pictures may differ: Since the right thinking style includes preferences for the visual code, a highest amount of iconic elements should be included in blogs created by right-thinkers.

Recent literature suggests possible links between the use of blog and specific cognitive strategies (Li & Chignell, 2010; Robertson, 2011; Xie & Sharma, 2011). For instance, Kerawalla and colleagues (2009) found that the presentational style of the blog contents influences users' behaviour, suggesting a relation between the organisation of the blog and the preferred approach to blogs. A comparison between the different blogging behaviour of students on two different courses provided insights into how people could adapt their behaviour across different contexts. They appeared to be able to adjust their blogging behaviour to meet their own needs.

Are people aware of the cognitive peculiarities of blogging? In other words, do they, while building their blogs, use specific cognitive strategies depending on their styles and are they aware of this? The issue of awareness is connected to the topic of metacognition.

Blogs and Metacognition

Metacognition and New Technologies

Metacognition is usually defined as the awareness, knowledge and the control of cognitive processes. The first attempts to define the domain of metacognition were made by Flavell (1979), who proposed a model of metacognition whose key concept was "metacognitive knowledge", which refers to what people perceive and think about how the mind works when engaged in comprehending, memorising and re-elaborating notions.

The role of metacognition in technological environments has been well established (e.g., Colombo & Antonietti, 2013; Cassel, 2002; Cuevas et al., 2004; Cuevas et al., 2002; Kramarski & Mevarech, 2003; Kramarski & Ritkof, 2002). High frequencies of metacognitive-reflection about knowledge, work strategies and personal experiences can be found in blogs used to assist learners (Chan & van Aalst,

2008; Chu et al., 2012). Furthermore, self-regulatory skills are often listed among the abilities that are increasingly regarded as critical to be successful in learning in online education (e.g., Dillon & Greene, 2003). These skills are connected to motivational and behavioural processes but also to metacognitive abilities in planning, monitoring, controlling and evaluating the learning process. In this line of research, in a recent study Lehmann, Hähnlein and Ifenthaler (2013) showed that providing learners with prompts stimulating the use of specific processes of self-regulation in online learning is useful for novice learners. Lastly, individuals who possess higher levels of metacognition (namely, who are more aware of the mental processes they are activating) can use technological tools in a more proper and productive way. Liu, Maddux and Johnson (2004) reported that good achievement is associated with awareness of the actual potentialities of the computer programs that are used to support the learning process.

Metacognition and Blogs

Technological tools can also improve metacognitive skills. Actually, blogs appear to make students more aware of their learning experience (Churchill, 2009; Colombo & Sala, 2011). New technologies and metacognition seem to produce reciprocal cognitive advantages. Blogs, as previously claimed, allow people to express themselves and to become aware of their own psychological features. Furthermore, blogs induce individuals to manage them according to personal characteristics (Deng & Yuen, 2011; Puntambekar & Kolodner, 2005). For example, a person can understand that she/he can express herself/himself better if she/he includes pictures in her/his posts. Thus, blogs can help individuals to be aware of their own psychological profiles since they externalise such a profile, thus permitting them to recognise their features as mirrored by what they created in the blog. Finally, blogs, which are a form of narrative in their own structure and aims, foster reflection and introspection, namely, attitudes and strategies which are usually meant as relevant ways to enhance metacognitive awareness. In fact, it has been argued that the narrative format of reflection and communication can foster metacognitive skills (Antonietti et al., 2014). Blogs help also to promote self-awareness whilst improving communication. As Lee (2009) pointed out, blogs stimulate the building of interpersonal relationships, as well as personal commitment to online contributions, which are vital to successful intercultural exchanges.

Blogs and the Personalization of the Learning Process

As highlighted above, cognitive styles have been successfully explored in educational settings. However, the personalisation of the learning process that cognitive styles imply has to be accompanied by metacognitive awareness in order to improve the learning outcomes. In fact, it is not enough to adapt instructional materials and procedures to the students' learning style: They have to be aware of their own styles so to activate the cognitive resources that individual styles need (Efklides, 2011). Metacognition should be integrated into the cognitive style approach. Bearing in mind that blogs are tools that can be employed in education, it is relevant to consider whether their use is accompanied by an adequate level of metacognitive competence. Thus, it is worth investigating the metacognitive awareness of the blogs' owners. To what extent do people realise the mental processes they activate whilst constructing their blog and those activated by visitors?

An Investigation about Cognitive Style and Metacognition in Blog Construction

Based on the research background previously described, a study (Colombo et al., 2013) was carried out to explore the possibility of applying the concept of "cognitive style" to the use of blogs. To our knowledge, this approach has never been applied in this context. The specific goal was to assess if cognitive coherence could be found in blog construction. Such a coherence is intended both as internal coherence – that is, coherence between the blog's specific contents and the formal aspects – and as coherence between the structure of the log and the owner's cognitive style.

We hypothesised a general category/topic effect, expecting people to make coherent choices about the formal structure of the blog according to the content. We also expected a general coherence between the bloggers' cognitive style and content and the formal structure of their blog. We had no *a priori* predictions about the degree of metacognitive awareness shown by the blog owners concerning the actual strategy used both at a specific level (e.g., writing a post) and at a general level (e.g., organising the blog structure).

Participants were 50 bloggers (28 women), aged between 25 and 52 years (M = 32.5, SD = 2.08). With regards to professional occupations, 12 bloggers were employed in education, 9 worked in commerce, 8 were office employees, 6 were self-employed, 5 were university students, 4 worked in the public health system and 4 were unemployed (2 participants failed to state their occupation). The bloggers participating in the study were first asked to response some questions to establish their level of ICT expertise. None of the participants resulted to be highly competent in any area of computer-supported activities (website construction, multimedia production, software programming, use of word processors, data sheets, and so on). Furthermore, the participants' levels of expertise and frequency of ICT use were comparable to the average values of the general population (Antonietti et al., 2003). All participants had been writing in their blog for at least 7 months.

Possible participants (see below for details concerning blog selection) were first contacted by e-mail. The design of the research was described briefly and the receivers were asked to join the study and to give consent to analyse their blogs. Persons who responded affirmatively were contacted again by sending another e-mail in which they were asked for formal, written permission to analyse their blog. The blogs of those who answered this second email allowing the blog analysis were included in the study. The authors of the blogs were also asked to complete the online questionnaire and interview.

Fifty Italian blogs were analysed. They were selected starting from a beginning range of 98 blogs on the basis that they belonged to the same anchor site, so to guarantee that all blog users involved in the study had the same options and the same constraints when building and administrating their blogs. By browsing the categories listed in the anchor site, the blogs that clearly belonged to one of the selected categories illustrated below were chosen. However, those that appeared to mix different content were excluded. A further selection led us to exclude blogs that had more than one author because it would not be possible to relate the features of the blog to the owner's individual cognitive style. In the final selection, blogs that were not regularly updated were excluded, to ensure that the current features of the analysed blog were actually managed by the person who completed the questionnaires.

The 50 selected blogs were divided according to 5 content categories. Categories were derived from the analysis of the blogs in the Italian blogosphere and compared with blog categories presented in the available literature (Murray & Hourigan, 2008):

1. **Personal Diary:** Blogs that describe the personal everyday life of the owner.
2. **Commentary:** Blogs providing comments on different current news.
3. **Amusement:** Blogs trying to entertain visitors, providing distractions and jokes.
4. **Specific Topic:** Blogs providing specific information on a focused topic, such as cooking, photography, music.
5. **Politic:** Blogs discussing political topics.

The 5 categories were balanced so to obtain 5 subsamples with an equivalent number of blogs (10 for each category) of similar length.

Blogs were further analysed according to different categories concerning their formal structure.

1. **Use of Images:** The number of images and image typology were recorded, according to Mayer's (1993) classification. Images were classified as Decorative (pictures that are not directly relevant to the text), Representative (they portray one element that is mentioned in the text), Organisational (they reflect relations amongst elements in the text), Explanatory (they explain how something – e.g. a system, a process, a mechanism – works).
2. **Relationships between Texts and Images:** The location of the pictures in relation to the corresponding text was taken into account, in order to determine whether Mayer's (2001) spatial contiguity principle (according to which students learn better when corresponding words and pictures are presented near rather that far apart) was or was not followed.
3. **Posts:** Each analysed post was assigned to one of the following 5 categories according to its subject: topic, autobiographical, commentary, question/provocation, reflective.
4. **Main Emotion:** Posts were categorised according to the emotion communicated (joy/happiness, fear/anxiety, sadness/delusion, anger, shame). Two independent judges evaluated the main emotion of each considered post. Ambiguous or controversial posts were discussed. The inter-rater agreement was satisfactory (k = .84).
5. **Self-Indicators:** By using Bruner's (1998) categories, the blogs were classified as containing coping (expressions that clearly indicate action, commitment, resources, appraisal), reflective (the act of reflection and the use of mental verbs) or social (social reference and social localisation) self-indicators.

The owners of the selected blogs completed the SOLAT (Style Of Learning And Thinking) questionnaire in order to match their cognitive style with the "cognitive style" of their blog. The inventory (Torrance, 1988; Torrance et al., 1978) consists of 28 items. In each item, respondents could choose one of two statements, both, or neither. One statement describes an approach that is characteristic of the left mode of thinking and the other a right-thinking approach. Choosing both or neither statements is assumed to correspond to the integrated mode of thinking. Score 1 is assigned to the selected statement, so that three final scores, corresponding to each stylistic profile (left, right and integrated), can be computed. On the basis of such scores, individuals were classified as left-, integrated- or right-thinkers. Acceptable psychometric characteristics of SOLAT have been documented (Albaili, 1993; Torrance, 1988; Torrance et al., 1978). This was true also for the Italian version of SOLAT, which was used in this study (Antonietti et al., 2005).

Participants were asked general questions aimed at exploring metacognitive awareness. The questions were:

1. Why did you begin a blog?
2. Which was your main aim while building your blog?
3. Should you be asked to link your blog to one of the following categories (the same as those considered in our a priori analysis: personal diary, commentary, amusement, specific topic, politic), which one would you choose?
4. How did you choose the graphical elements of your blog?
5. On which elements did you focus when choosing the graphical elements?
6. Did you add any personalised box in your blog? If you answered "yes", please specify which kind of box.
7. What do you like most about your blog?
8. Please, name a blog you especially like and explain why.
9. In your opinion, what do readers thinks about your blog?
10. In your opinion, what do readers like most about your blog?
11. If you had to rate on a five-point scale your satisfaction as a blogger, which score would you assign?

A pilot study was carried out to devise a reliable online version of the metacognitive questionnaire, aiming at collecting answers online and of facilitating the answering procedure. In this pilot study we asked a separate sample of 30 authors of other blogs (belonging to the same categories considered in the study) to respond orally to the aforementioned questions. Responses were recorded and categorised so to devise, for each question, the set of the possible answers that were inserted as multiple choices.

Questions explored the strategies that persons were believed to use/have used in building their blog. Comparing those answers, on the one side, with the actual features of the blogs and, on the other side, with the SOLAT scores, should return indication about the bloggers' metacognitive awareness. To be more precise, the aim of the questions was to use participants' answers regarding their strategies as comparison to the actual strategies (derived from blog analysis) and to the strategies they should use according to their cognitive style.

Data confirmed the first hypothesis and highlighted a general effect depending on the content of the blog, in support of the notion that the authors of the blogs were actually able to differentiate the formal structure according to the content. ANOVAs were computed comparing the scores of different blog categories with reference to image typology (number of images per typology), the relations between text and images (number of images that followed the coherence principle), posts' content (number of posts for each content typology), the main emotion communicated by the posts (the number of posts per each emotion considered) and the use of self-indicators (number of self-indicators per each type). A first difference emerged in the use of Decorative images amongst the blog categories ($F_{4;42}$= 3.00; $p < .05$; η^2= .22). More specifically, a significant difference between Specific Topic and Amusement blogs was found (using Post Hoc Turkey-HSD): in the latter Decorative images were used more frequently. The spatial contiguity principle was also differently observed amongst blog categories ($F_{4;42}$= 3.02; $p < .05$; η^2= .22); in the Amusement blogs the spatial contiguity principle was followed less frequently than in the other kinds of blogs.

As expected, marked differences emerged when considering the content of the posts. Topic posts were more frequent in the Specific Topic blogs; Autobiographical posts were more frequent in Personal and Topic blogs ($F_{4;42}$= 24.72; $p < .001$; η^2= .70); Commentary posts were more frequent in Political and Commentary blogs than in other types of blogs ($F_{4;42}$ = 10.05; $p < .001$; η^2= .49). Reflective posts were more common in Political blogs than in Personal, Topic and Amusement blogs ($F_{4;42}$= 4.59; $p < .001$; η^2= .30); Amusement posts were more common in Amusement blogs than in the other blogs ($F_{4;42}$ = 3.82; $p < .001$; η^2= .27).

Different blog categories appeared also to be associated to different emotions. Positive emotions (joy/happiness) were more evident in Personal, Amusement and Commentary blogs ($F_{4;42}$= 18.71; $p < .001$; η^2= .74). Sadness/delusion were elicited more often in Personal and Commentary blogs ($F_{4;42}$= 13.08; $p < .001$; η^2= .55). Anger was experienced more in Political blogs ($F_{4;42}$= 6.07; $p < .001$; η^2= .37).

It was also found that the use of self-indicators (coping, reflective and social) differed amongst categories. Coping self-indicators were more evident in Personal blogs and less present in Specific Topic blogs ($F_{4;42}$= 26.65; $p < .001$; η^2= .35). Reflective self-indicators occurred more in Personal and Commentary blogs ($F_{4;42}$= 7.30; $p < .001$; η^2= .41). The same was true for social self-indicators ($F_{4;42}$= 14.31; $p < .001$; η^2= .58).

Coherence amongst blog contents and users' aim was tested and it appeared that writers of Political and Topic blogs declared that their main goal was to give people clear and complete information, whereas the others wanted more to entertain their readers (χ^2 (4, N = 46) = 15.99, $p < .005$). No other significant relationship between users' answers and blog categories was found highlighting a lack of metacognitive awareness concerning strategies that were actually effectively used by blog writers.

The second hypothesis was not confirmed: No significant relation between cognitive styles and blog features was found. Scores of the bloggers, classified as right/left/integrated thinkers on the basis of SOLAT scores, were computed with reference to image typology (number of images per typology), relationships between text and images (number of images that followed the coherence principle), posts' content (number of posts for each content typology), main emotion communicated by the posts (number of post for each emotion considered) and use of self-indicators (number of self-indicators for each type). Cross-tabulations between users' cognitive styles (categorised as right, left or integrated) and answers to metacognitive questions were also computed. No significant association emerged.

The study highlights that people act strategically whilst designing their own blogs, differentiating communication strategies and specific contents according to blog general category and adhering to these lines. Yet, they appear not to be aware of this strategic use. Moreover, they do not create their blogs according to their cognitive style. These findings support the notion that people are neither aware of the strategies they actually use, nor of their own cognitive peculiarities that could probably lead them to improve the effectiveness of blog communication.

TOWARD A CONSCIOUS USE OF BLOGS FOR EDUCATIONAL AIMS

As we have seen, blogs are widely popular amongst the new generation and the potentialities of blogs can be used to enhance cognition. For instance, the narrative structure and the multimedia characteristics of blogs can constitute a training field to potentiate different cognitive strategies. Therefore, blogs can be regarded as a motivating activity for students: They can be seen as a mental "gym" to test and exercise cognitive and communication skills. It has been shown that blogs can promote people's interaction and engagement in collaborative learning (Chu et al., 2012). Blogs may also become a pleasant everyday activity that helps students to reflect on their cognitive processes and encourages them to monitor and control their mental strategies in a more effective way, but not if users are left alone in managing them. Yet, the results of the study discussed above show that persons are not aware of their cognitive peculiarities, and hence, fail to rely on them when writing a blog. If individuals are not aware of the mental processes which blogs activate, all the potentialities that these instruments have remain "inert". Higher levels of awareness and more effective use of blogs are needed. Blog use appears to be mainly a spontaneous activity lacking of metacognitive awareness. This leads us to believe that it is naïve to propose the introduction of blogs in instructional and training settings by simply involving learners in reading/using blogs. Instead, attention should be paid to fostering the understanding of the mental processes involved in blog use, so that learners can autonomously regulate their strategies to match their cognitive style. A reciprocal link between the educational use of ICT within a constructivism theoretical framework is documented in literature (Adams, 2011; Wang, 2009). But, referring specifically to blogs, is this cognitively possible?

To answer this question, it may be useful to start from the weakness of educational use of blogs. Actually not all of the studies on blogging yielded enthusiastic results (Deng & Yuen, 2011). For example, in their exploration into the potential of blogs for peer learning and peer support, Hall and Davison (2007) found that, although blogs fostered positive and productive exchanges of peer support, the evidence for reflective conversation was not compelling. Xie, Ke and Sharma (2008) reported similar problems in their study on peer feedback via blogs. Irrespective of the fact that the overall level of student reflective thinking increased over time, peer feedback to each other's blogs did not promote reflective thinking as expected.

Yet, considering the all corpus of researches exploring the effectiveness of blogs, we think that the answer is positive: since the effectiveness of shared communication to enhance cognitive performances has been reported in several researches (e.g., Garza, 2002; Henning, 2003), it is possible to hypothesise that a more cognitive centred use of a blog could be useful, especially in learning or structured contexts. In such a context, blogs become a motivating field to help students to reflect upon their cognitive activities and to be able to monitor and control them in a more effective way. They can also be used in classroom to teach students transferable cognitive abilities and to promote more aware use of their peculiar strategies. For example, a blog in which students and teachers can discuss their learning experiences, share advice, discuss problems, as well as a thematic blog, linked to a specific discipline or topic, can be used to promote active or cooperative learning or to help students to reflect on their naive theories. Alternatively, blogs can be used to support homework, that is, to promote cooperative learning and to motivate students to work transversely on cognitive abilities. Furthermore, blogs can be employed not only to improve social skills amongst students (Goldman et al., 2008) and between students and teachers (Glogowski, 2009), but also to work in a metacognitive way on cognitive styles in an ecological context. We run a few studies aimed at testing these potentialities. Main results are reported in the next paragraphs.

Blogs and Empowerment in School Settings: More Evidence from Research

A first study (Colombo & Sala, 2011) investigated blog's potentialities in a school environment. In this study we mainly focused on possible ways to improve a use of blog, which was more cognitive oriented. The hypothesis was that such a cognitive oriented use of blogs may increase the benefits from using this narrative and social environment on self regulation learning. The authors were also interested in highlighting the specific areas were blogs could actually make a difference in promoting empowerment, when compared to other communication media. Participants were thirty-one kindergarten and primary school teachers and undergraduates. The sample was divided into four experimental groups, which were using different forms of communication and were characterised by different levels of multimedia use: paper communication group (9), computer communication group (8), email communication group (7) and blog communication group (7).

Participants were tested to assess the level of their metacognitive competences, in order to form homogeneous sub-groups. No differences appeared from this pre-analysis, therefore the participants were assigned randomly to the one of the experimental group. They were asked to plan a computer-based activity to be developed with their students. The paper communication group wrote it in a paper-and-pencil modality. The computer communication group used a word editor program (they were free to use multimedia objects, like clip arts or pictures). The email communication group was asked to communicate by emails. They were free to answer or not to answer to other emails and to add multimedia features to their messages. For the blog communication group an online blog was created and participants were instructed on the tool facilities. The fifth group served as control group: They only attended to the pre- and post-tests. Each sub-group had to write down a report of the activity, using the particular medium of his own group and according to three phases: Planning (aims, steps, materials, etc.), Performance (the actual evolution) and Conclusions (assessment, discussion, perspective). All the written texts produced by the four experimental groups were collected and analysed according to the following three main criteria:

1. **Mental Verbs:** Number of mental verbs, which reveal the level and the quality of reflexivity and metacognitive competence, used by teachers and by children.
2. **Language Use:** The style of communication, referring to content, medium, social/emotional context, used by the sub-groups
3. **Metacognitive Strategies:** Number and typologies of strategies used by teachers and by children.

Figure 1. Percentages of the use of mental verbs in the overall sample of the first study (Colombo & Sala, 2011)

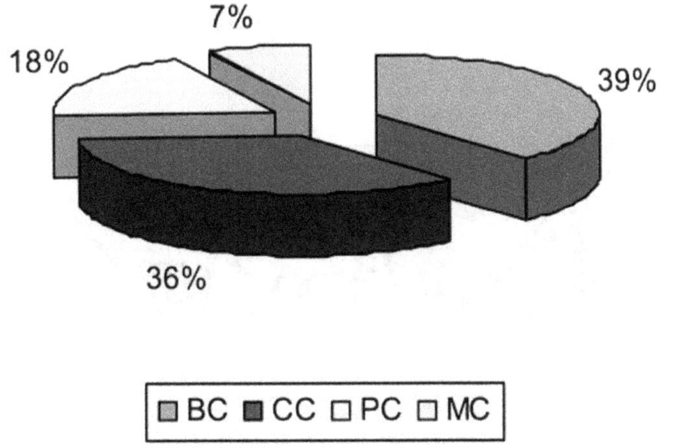

Figure 2. Percentages of the use of mental verbs in the teacher sub-sample of the first study (Colombo & Sala, 2011)

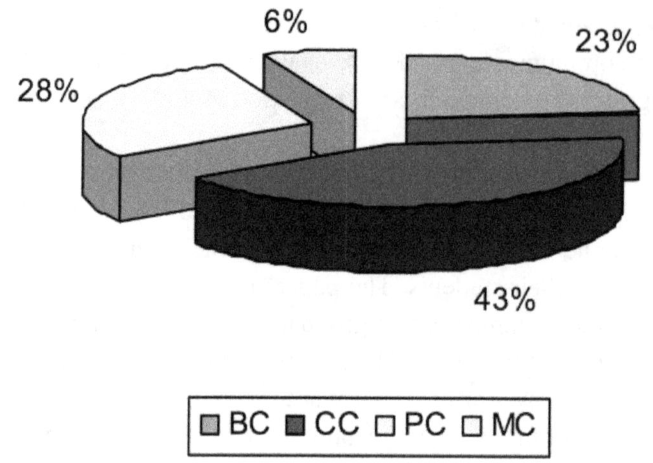

To ensure the homogeneity of the examined contributions, given the different lengths of the teachers' writings, the number of mental verbs used in the first two pages of the three research steps for each sub-group have been computed (Figures 1, 2, and 3).

Number and typologies of strategies used by teachers and by children have been computed for each sub-group and percentages (referring to teachers and children) have been compared (Figures 4 anid 5).

Results (derived by text analysis) highlighted an actual influence of the medium blog in promoting a different cognitive and metacognitive activation of teachers who used this tool. The analysis of communicational styles pointed out a differentiation among the three considered styles only within the Blog communication group (Figure 6): 64% of participants' remarks were related to the content, 26% to a socio-emotional context and 10% to the medium. This differentiation among contents was not found analysing the contents of other groups' communication. Within the blog group, therefore, individuals focused not only on the specific task but also linked more with other participants and also reflected on the medium itself.

Figure 3. Percentages of the use of mental verbs in the children sub-sample of the first study (Colombo & Sala, 2011)

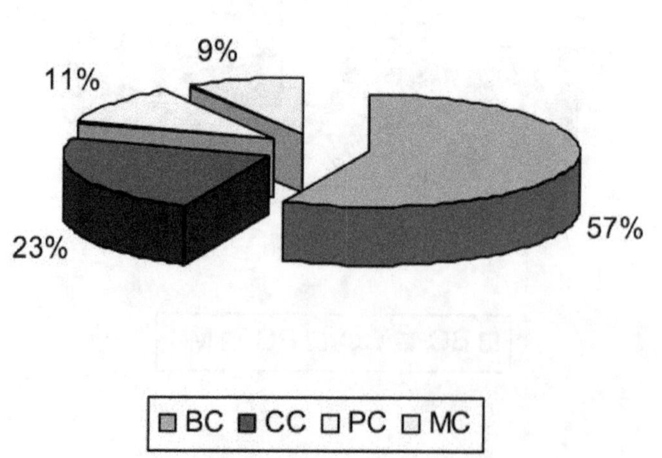

Figure 4. Percentages of strategies used by the teacher sub-sample of the first study; Note. BC: Blog communication group; MC: E-mail communication group; PC: Paper communication group; CC: Computer communication group
(Colombo & Sala, 2011)

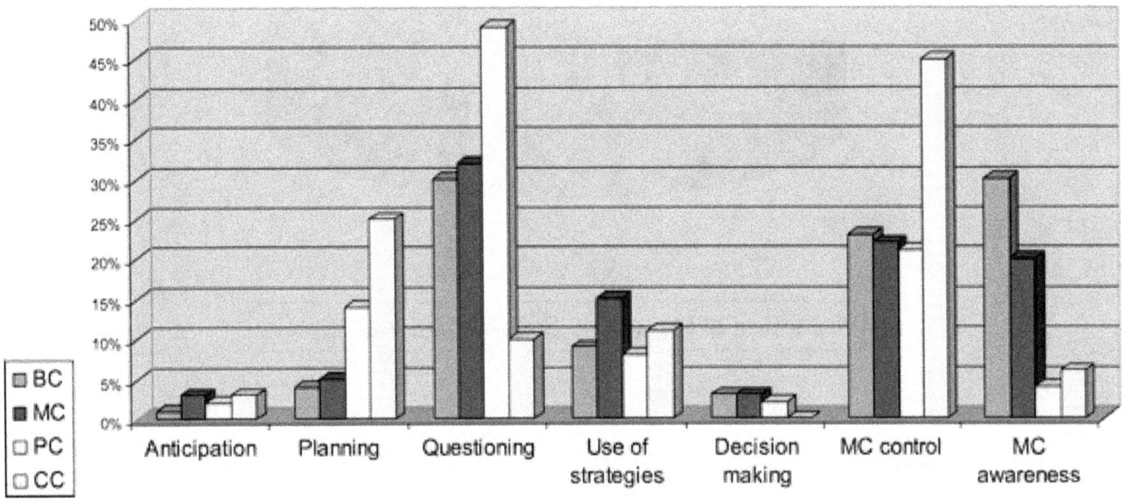

The other sub-groups, instead, focused only on contents closely related to the task. To be more precise, the blog seemed to promote more metacognitive awareness, whereas teachers working with the paper communication group seemed to spend more time questioning. Moreover, the blog seemed to promote metacognitive awareness and control in children, and Blog and computer use were also the sub-groups where the highest use of strategies in children was reported (Blog Communication Group: 33%; Computer Communication: 45%).

Figure 5. Percentages of strategies used by the children sub-sample of the first study; Note. BC: Blog communication group; MC: E-mail communication group; PC: Paper communication group; CC: Computer communication group
(Colombo & Sala, 2011)

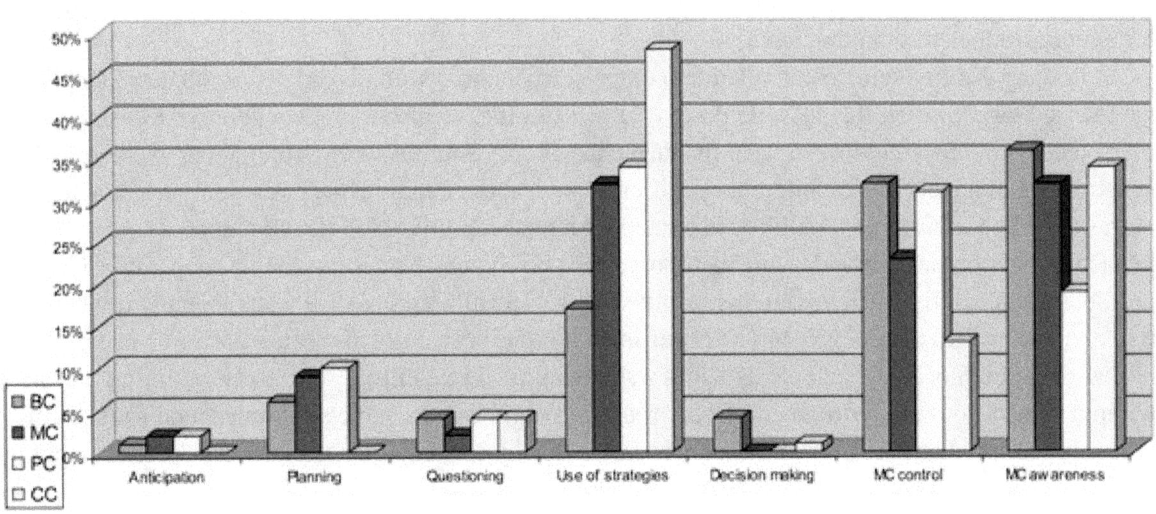

Figure 6. Percentages of themes of the language use in the Blog communication group; first study; Note. BC: Blog communication group
(Colombo & Sala, 2011)

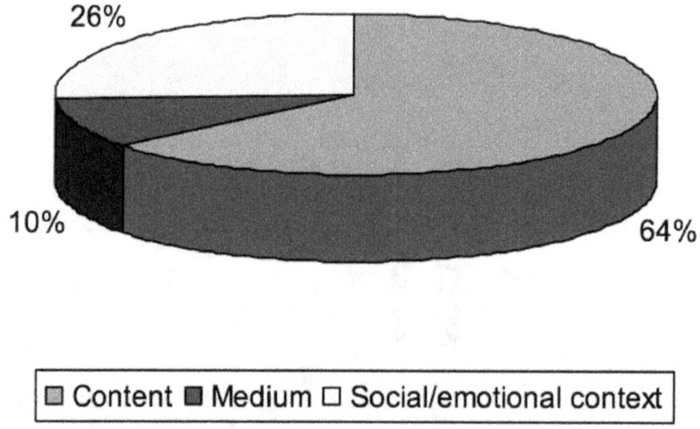

Considering the Use of mental verbs in the different texts, differences among sub-groups have been found. On the whole the Blog group and the Computer group were the groups using more mental verbs (Blog communication group: 39%; Computer communication group: 36%). When the use of mental verbs was considered separately in teachers and children, it is clear that in the blog group more mental verbs were used by children (57%), whereas in the computer communication group were teachers to use them prevalently (43%). The other sub-groups used far less mental verbs.

Communicating through blogs also seemed to promote more metacognitive skills in both teachers and children. Specifically, when considering teachers' use of metacognitive strategies, the Blog communication have been found to promote more metacognitive awareness (28%), while teachers of the Paper Communication group have been shown to spend more time questioning (48%). Moreover, the Blog communication seemed to promote metacognitive awareness (35%) and control (31%) in children.

Results of this study highlight that the use of different media can promote different levels of mentalization and different strategic attitudes. More in detail, blogs, with their common and open space, seem to promote a more social, personal, fluent and informal communication style. Furthermore, in the Blog communication group teachers resulted to be able to balance the use of metacognitive prompts in order to promote the use of metacognition itself among children (while the opposite was true, for example, for simple computer communication).

Following this line of research related to empowerment, in a second study (Colombo & Sala, 2011) we focused on exploring the possible positive effect of blogs in promoting cooperative learning in primary school children, and investigated the influence of online narrative environment on the cooperative construction of novels by the children. Specifically we had three main objectives: analysing whether the narrative online environment building is related with metacognitive skill development; 2) assessing the metacognitive competence and trying to improve it; 3) verifying whether the on line cooperative learning promote metacognition. The researchers worked with 113 children (age 7-9 years) equally divided into an experimental and control group. The experimental group was asked to create a narration in collaboration with children of other schools by using a shared blog. The control group wrote a similar narration without using an online environment. Both groups basically had the same assignment and worked cooperatively. The production of the narration was facilitated by dividing the task into four steps (beginning, crisis, development, solution). In the experimental group, each phase was, in turn, initiated by a class, continued in the second and finished the last one. To be more specific, when children were asked to write the beginning of the story, the first class started the narration and asked the second class to continue the story. After adding events or details to the first part of the narration, the second class would ask the third class to go on working at the narration. The third class would finish the beginning part. Then children were asked to write about the crisis. For this second step the second class would start the narration and ask the third one to continue it. The first class would end the narration related to the crisis section, and so on. Children's work was enriched (both for the experimental and control group) by a metacognitive activity during which students were asked to answer a series of reflective questions related to their work. The same metacognitive prompts were proposed to the control group after each phase.

Before starting the narration, authors run a pre-test designed to assess the metacognitive competence, naive theories on cooperation and groups and cooperative skills (baseline). The same test was repeated at the end of the project (post-test) to assess the effectiveness of the intervention. Metacognitive self-assessment questionnaires and tests of cooperative work (through the composition of original drawings from a defined number of geometric shapes and respecting the basic rules for group role management), with individual and group assessment were used.

Concerning the narration building, researchers analysed the texts produced both for the experimental and the control group. Focusing on the blog, they evaluated the presence of self-indicators (Bruner, 1998): coping (action, commitment, resources, evaluation and consistency), reflective indicators (quality, reflexivity, mental verbs), social-self (social referencing, and location). Concerning the narration, they considered the same self-indicators. In addition, they evaluated the narrative complexity taking into account the length of the tale and the presence and type of morality indicators.

Data highlighted a higher use of coping strategies in blog users (42% of the total presence of self-indicators) and more cognitively complex narrations. In the experimental group the length of the narration (t_{12}= 13.44; $p < .001$) and its level of complexity (t_{12}= 8.40; $p < .001$) were much higher than in the control group. Moreover, in the post-test the experimental group was found to exhibit a better use of procedural strategies, whereas the control group tended to non-sequential paths to solve tasks (χ^2 (5,N = 99)= 16.92; $p < .01$). Focusing on metacognition, students in the experimental group became more aware of the role of each individual within the group (t_{48} = -2.09; $p < .05$), as required by the principle of the positive interdependence of cooperative learning (Nam & Zeller, 2011; Sarfo & Elen, 2011). They also evaluated more realistically the effectiveness of group cooperation (t_{48} = 2.06; $p < .05$). Contrarily to the control group, they were actually able to identify specific and realistic difficulties at some stage of the procedure. These data showed that students who have been blogging were able to reflect on their work and to implement specific metacognitive strategies.

When considering the results from both the two studies, it is also interesting to mark the main differences that emerged between adults and children in the self-regulated environments, as a response to blog use. Among both teachers and children the researchers fond that using blogs favour an improvement of the metacognition. In adults blogs especially improved the metacognitive awareness, whereas in children they especially improved the metacognitive control skill. Furthermore, there was some evidence that the adults are more open to the social aspects which are promoted by the Web 2.0, while the children showed to be especially empowered in their coping strategies by using blogs. From a different point of view it seems that adults run the risk of idealizing social networks and their potential, including cognitive benefits, while Children seem to become more realistic after working in a Web 2.0 environment, also improving their skills of assessing objectively their competence in cooperation tasks. Therefore, it will be worthwhile focusing on the metacognitive control when planning empowerment projects for adults, and on metacognitive awareness when building empowerment projects for children.

The studies we described in this paragraph support the hypothesis that blogs can improve the learning experience at school nevertheless, some research data also indicate that blogs can be useful in empowering the learning also in informal settings. In the study we present in the next paragraph we explored this possibility.

Blogs and Empowerment in Informal Settings: A Study

In a research project on adults (Colombo & Sala, 2011) we investigated whether blogs can promote the metacognitive awareness and foster more effective problem solving strategies in persons working in setting of non-formal learning. Our research hypothesis was that using blog besides promoting more metacognitive awareness and control (and, therefore, more adequate strategies) can also favour more cooperation among the members of a learning group. Also the outcome of cooperative activities may benefit by using blogs. Based on this assumption we hypothesised that mystery stories written using cooperative strategies promoted by using blogs should be better than the ones written in less cooperative environments. Therefore, we tested empowerment differences by using distinct technologies and through different kinds of online cooperation. Participants in the study were 106 adults (mean age: 40.80, SD = 12.75; Male = 56%; Female = 46%), who were randomly split into 4 experimental groups. Each group had the assignment to write a mystery based on given characters and had fifteen days to complete the task. Specific ties to turn a "simple" creative task in a problem solving task were given to the participants. Specifically the subjects were asked to respect specific temporal bounds, to use given characters, and to write a humoristic but realistic plot.

To perform the assigned task each group had to cooperate by using different forms of communication in different kinds of virtual environment, in order to test the effects of types of technological communication on participants' metacognitive awareness and problem solving strategies. The Blog Group was asked to solve the problem cooperatively on a blog devised for the research. A second group, namely the Forum Group, was asked to solve the problem cooperatively using a web forum (without he narrative and multimedia aspects which are peculiar of blogs). The Mail Group was asked to solve the problem cooperatively exchanging mails through a mailing list, while participants in the Control Group were requested to solve the problem individually and to send the solution to the researchers by mail.

Before and after the task participants had to fill the Metacognitive Awareness Inventory (Schraw & Dennison, 1994). Then, after writing the mystery story, participants were also invited to rate their work by considering both their cognitive and practical performances on a 5 point scale by answering open questions.

Meanwhile, mystery stories were rated from 5 external judges (adults with BA degree in different majors, fond of mystery stories) on 5 dimensions (humour, coherence, skilfulness, originality, overall evaluation) on a seven-point scale. The interactions of the participants in the three experimental groups were analysed according to Metacognitive and strategic aspects, including anticipation, planning, monitoring, decision making, use of strategies, and Cooperative aspects, relating to questions, suggestions, answers, proposal acceptances, agreement/disagreement, aim interdependence, task interdependence, role and socio-emotional interdependences.

Table 1. Differences in participants' self-evaluation

Dimensions	F(df)	p
Time managing	4.12 (3,21)	<.05
Engagement	5.75 (3,21)	<.01
Result	45.82 (3,21)	<.001
Plot	5.95 (3,21)	<.01
Coherence	7.18 (3,21)	<.01
Humour	10.61 (3,21)	<.001
Narration	20.61 (3,21)	<.001

When considering the cognitive variables which were assessed we found that in the external evaluation of the narration the blog users ($F(3,16) = 3.25$; $p < .05$) showed higher levels of internal coherence than participants in the other groups. Furthermore, when considering the results from the analysis of the interactions, the participants task interdependence was more present in the Blog group than in the other groups (χ^2 (20) = 46.60; $p < .001$). These results support the hypothesis that interacting in blogs in order to complete cooperative tasks can promote both the cooperation among group members and the use of strategies. The blog structure resulted especially influential on the interdependence among the blog users who were cooperating to complete the group task.

With regards to the metacognitive facets which can be influenced by using blogs assessed by means of the Metacognitive Awareness Inventory), the scores of the Blog users, as compared to the scores of the participants in the other groups, were significantly higher in the subscales regarding Information management strategies ($t(6) = 2.07$; $p < .05$) and conditional knowledge ($t(6) = 2.4$; $p < .05$). These results indicate that writing cooperatively in a blog can actually promote the development of metacognition. Using blogs appeared to be influential especially on the aspects linked to the regulation of cognition. In the self-rating scores, bloggers were also the group reporting the best self-evaluation scores on the all assessed strategic and cognitive dimensions (see Table 1).

Another noteworthy finding that emerged from this study was that, when comparing the self-evaluations with the external evaluations of the performance, participants tended to underestimate their product when the cooperation was not good, and that happened for participants in the Forum and the Mail group, but to overestimate their performance when they perceived good cooperation, that is what happened for the blog users. These tendencies emerged also if participants appeared to be aware of the efficacy of their cooperation, such as forum users who were aware of the failure of their team.

In summary, also in this line of research on the effects of blogs on cognition, blogs resulted to be a form of communication that can favour the cognitive and the practical performance of users in a cooperative setting more than other forms of on-line communication. This is another evidence of the high potentialities of blogs for learning and cognition, stressing the value of using blogs to improve cognitive and metacognitive skills in both formal (such as the school) and informal settings.

CONCLUSION

The studies presented in the previous paragraphs suggest that, even if blogs are usually managed and visited as a leisure activity, without a proper metacognitive awareness, they can actually be used for educational aims.

Findings in the fields of distance education suggested that using technological tools to enhance communication should promote more appropriate cognitive processes (e.g., Deaudelin & Richer, 1999; Kramarski & Zeichner, 2001). More specifically, according to Murray and Hourigan (2008), blogs may provide teachers with an efficient and cost-effective tool that, if employed appropriately, presents learners with alternative means of (self-) expression in the online medium. These writing spaces are adaptable to both individual and group project-based work and consequently offer many possibilities, from a cognitive point of view, in terms of task creation and design. Likewise, Minocha and Roberts (2008) highlighted that blogs can create an added value and can generate a positive experience to engender students' satisfaction and retention, so appearing to be extremely useful to promote learning. Furthermore, Olander (2008) focused on blogs' potentiality in enhancing not only writing skills, but also more general cognitive skills like reading and thinking. Blogs can also enhance problem-solving strategies in learning environment (Te Wang, Huang et al., 2008). However, as recent views in the educational field suggest, individualisation according to cognitive styles and metacognition are important ingredient of the instructional process, which allows a technological tool to be effective (Antonietti & Colombo, 2008).

In this line of research, we aimed to explore the cognitive dimensions underlying blogs. Results from the first study supported the idea that people act strategically while writing a blog and differentiate the communication strategies and contents according to blog general category by keeping coherent to these lines. Nevertheless, our data also showed that people are fully aware neither of this strategic use of blogs nor of their own cognitive features. The main message that we can derive is that bloggers need to enhance their cognitive awareness in order to improve the effectiveness of their blog communication. Programs that aim to join the use of blogs with the enhancement of the awareness about person's cognitive peculiarities may be, therefore, effective in empowering both learning activities and blog communication. In such a frame, blogs have also the value of being a tool that can motivate students to engage in (especially collaborative) learning (Chu et al., 2012).

Data from the other two studies we presented showed that using blogs empowers metacognitive skills in adults and the effectiveness of collaborative learning among children. It also improved the quality of narration among bloggers. The narrative and conversational structure of blogs is the specific feature that can be usefully considered for the cognitive empowerment and deserves attention in research and teaching practice. Indeed, learning itself seems to be facilitated when words are in conversational style (Mayer et al., 2004), hence adding a potentiating effect of narration to the more general multimedia effect already owned by blogs. Narratives and shared modalities of online communication can also be seen as forms of "distributed" cognition where knowledge is embedded in situated discourses (Henning, 2003) and can facilitate the creation of virtual communities (Garza, 2002). Also, the narrative structure of blogs, along with its motivational facet, helps considering these multimedia tools as extremely promising for the cognitive empowerment in learning and communication.

However, another reflection suggested by the results of the studies presented in this chapter is that the mere use of technology is not sufficient to promote cognitive and meta-cognitive benefits. Blogs and the other Web 2.0 environments have an optimal structure in terms of flexibility of multimedia and cognitive structure, and they actually seem to promote a spontaneous use of appropriate cognitive strategies. Nevertheless, those strategies are not accompanied by any metacognitive skills, and an appropriate support at a metacognitive level is needed to improve the effectiveness of the self-regulated learning. Also, to increase the effectiveness of such a path of empowerment, the narrative role of the organization of content is relevant, and this feature is especially allowed and facilitated by blogs.

In summary, the studies we described in this chapter, among others, clearly indicate that blogs are actually useful tools for learning, but learners and bloggers need specific programs aimed at enhancing their awareness of the mental processes involved in the blog using, so to obtain the maximum cognitive benefit from this activity.

REFERENCES

Adams, P. (2011). ICT and pedagogy: Opportunities missed? *Education*, *39*(1), 21–33.

Albaili, M. A. (1993). Inferred hemispheric thinking style, gender, and academic major among United Arab Emirates college students. *Perceptual and Motor Skills*, *76*(3), 971–977. doi:10.2466/pms.1993.76.3.971 PMID:8321616

Antonietti, A. (2003). Cognitive styles assessment. In R. Fernandez-Ballesteros (Ed.), *Encyclopaedia of psychological assessment* (Vol. 1, pp. 248–253). London: Sage. doi:10.4135/9780857025753.n54

Antonietti, A., Calcaterra, A., Colombo, B., & Giorgetti, M. (2003). *Attorno al computer. Strumenti e ricerche su nuove tecnologie e apprendimento* [Around the computer. Instruments and studies about the relation between new technologies and learning]. Rome: Carocci.

Antonietti, A., & Colombo, B. (2008). Computer-supported learning tools: A bi-circular bi-directional framework. *New Ideas in Psychology*, *26*(1), 120–142. doi:10.1016/j.newideapsych.2007.08.001

Antonietti, A., Confalonieri, E., & Marchetti, A. (Eds.). (2014). *Metarepresentation and narrative in educational settings: A cultural framework for reflective thinking*. New York: Cambridge University Press.

Antonietti, A., Fabio, R. A., Boari, G., & Bonanomi, A. (2005). Il questionario "Style of Learning and Thinking" (SOLAT): Dati psicometrici per una validazione e standardizzazione della versione italiana [The questionnaire "Style of Learning and Thinking" (SOLAT): Psychometric measures to validate and standardise the Italian version]. *TPM. Testing Psicometria Metodologia*, *12*, 299–316.

Balkis, M., & Isiker, G. B. (2005). The relationship between thinking styles and personality types. *Social Behavior and Personality*, *33*(3), 283–294. doi:10.2224/sbp.2005.33.3.283

Blood, R. (2000). *Weblogs: A history and perspective*. Retrieved from http://rebeccablood.net/essays/weblog_hystory.html

Bruner, J. S. (1998). Narrative and metanarrative in the construction of Self. In M. Ferrari & R. J. Sternberg (Eds.), *Self awareness: Its nature and development* (pp. 308–331). New York: Guilford Press.

Cakir, H. (2013). Use of blogs in pre-service teacher education to improve student engagement. *Computers & Education*, *68*, 244–252. doi:10.1016/j.compedu.2013.05.013

Calcaterra, A., Antonietti, A., & Underwood, J. (2005). Cognitive style, hypermedia navigation and learning. *Computers & Education*, *44*(4), 441–457. doi:10.1016/j.compedu.2004.04.007

Cassell, J. (2002). "We have these rules inside": The effect of exercising voice in a children's online forum. In S. L. Calvert, A. B. Jordan, & R. R. Cocking (Eds.), *Children in the digital age: Influences of electronic media on development*. Westport, CT: Praeger.

Chan, C. K. K., & van Aalst, J. (2008). Collaborative inquiry and knowledge building in networked multimedia environments. In J. Voogt, & G. Knezek (Eds.), International handbook of information technology in primary and secondary education, 299–316. Dordrecht, The Netherlands: Springer. doi:10.1007/978-0-387-73315-9_18

Chu, S. K. W., Chan, C. K. K., & Tiwari, A. F. Y. (2012). Using blogs to support learning during internship. *Computers & Education*, *58*(3), 989–1000. doi:10.1016/j.compedu.2011.08.027

Churchill, D. (2009). Educational applications of web 2.0: Using blogs to support teaching and learning. *British Journal of Educational Technology*, *40*(1), 179–183. doi:10.1111/j.1467-8535.2008.00865.x

Coffield, F., Moseley, D., Hall, E., & Ecclestone, K. (2004). *Learning styles and pedagogy in post-16 learning. A systematic and critical review*. London, England: Learning & Skills Research Centre.

Colombo, B., & Antonietti, A. (2013). Naïve theories about multimedia learning: A study on primary school textbooks. *Frontiers in Psychology*, *4*(450), 1–19. doi:10.3389/fpsyg.2013.00450 PMID:23382719

Colombo, B., Antonietti, A., Caravita, S., & Sala, R. (2013). Blog content and structure, cognitive style and metacognition. *International Journal of Technology and Human Interaction*, *9*(3), 1–17. doi:10.4018/jthi.2013070101

Colombo, B., & Sala, R. (2011). Blogosphere: cooperative and strategic aspects of self-regulated learning in formal and informal settings. In A. Bartolomé, P. Bergamin, D. Persico, K. Steffens, & J. Underwood (Eds.), *Self-regulated learning in technology enhanced learning environments: Problems and promises, 154- 161*. Aachen: Shaker Verlag.

Cook, D. A., Gelula, M. H., Dupras, D. M., & Schwartz, A. (2007). Instructional methods and cognitive and learning styles in web-based learning: Report of two randomised trials. *Medical Education*, *41*(9), 897–905. doi:10.1111/j.1365-2923.2007.02822.x PMID:17727530

Cuevas, H. M., Fiore, S. M., Bowers, C. A., & Salas, E. (2004). Fostering constructive cognitive and metacognitive activity in computer-based complex task training environments. *Computers in Human Behavior*, *20*(2), 225–241. doi:10.1016/j.chb.2003.10.016

Cuevas, H. M., Fiore, S. M., & Oser, R. L. (2002). Scaffolding cognitive and metacognitive processes in low verbal ability learners: Use of diagrams in computer-based training environments. *Instructional Science*, *30*(6), 433–464. doi:10.1023/A:1020516301541

Deaudelin, C., & Richer, A. (1999). *A learning conversation approach integrating email: Its experiment to support the student learning process at college level*. Paper presented at the 8th EARLI Conference, Goteborg, Sweden.

Deng, L., & Yuen, A. H. K. (2011). Towards a framework for educational affordances of blogs. *Computers & Education*, *56*(2), 441–451. doi:10.1016/j.compedu.2010.09.005

Dillon, C., & Greene, B. (2003). Learner differences in distance learning: Finding differences that matter. In M. G. Moore & W. G. Anderson (Eds.), Handbook of distance education, 235–244.

Efklides, A. (2011). Interactions of metacognition with motivation and affect in self-regulated learning: The MASRL model. *Educational Psychologist*, *46*(1), 6–25. doi:10.1080/00461520.2011.538645

Fabbri, M., Antonietti, A., Giorgetti, M., Tonetti, L., & Natale, V. (2007). Circadian typology and style of thinking differences. *Learning and Individual Differences*, *17*(2), 175–180. doi:10.1016/j.lindif.2007.05.002

Fan, W., Zhang, L., & Watkins, D. (2010). Incremental validity of thinking styles in predicting academic achievements: An experimental study in hypermedia learning environments. *Educational Psychology*, *30*(5), 605–623. doi:10.1080/01443410.2010.496899

Fiorina, L., Antonietti, A., Colombo, B., & Bartolomeo, A. (2007). Thinking style, browsing primes and hypermedia navigation. *Computers & Education*, *49*(3), 916–941. doi:10.1016/j.compedu.2005.12.005

Flavell, J. H. (1979). Metacognition and cognitive monitoring: A new area of cognitive-developmental inquiry. *The American Psychologist*, *34*(10), 906–911. doi:10.1037/0003-066X.34.10.906

Furnham, A., Christopher, A. N., Garwood, J., & Martin, G. N. (2007). Approaches to learning and the acquisition of general knowledge. *Personality and Individual Differences*, *43*(6), 1563–1571. doi:10.1016/j.paid.2007.04.013

Garza, G. (2002). The internet, narrative, and subjectivity. *Journal of Constructivist Psychology*, *15*(3), 185–203. doi:10.1080/10720530290100433

Glogowski, K. (2009). Tracing the emergence of a blogging/writing community: Critical transformations in a grade eight classroom. *Dissertation Abstracts International* Section A, 69.

Goldman, R. H., Cohen, A. P., & Sheahan, F. (2008). Using seminar blogs to enhance student participation and learning in public health school classes. *American Journal of Public Health*, *98*(9), 1658–1663. doi:10.2105/AJPH.2008.133694 PMID:18633075

Granberg, C. (2010). Social software for reflective dialogue: Questions about reflection and dialogue in student teachers' blogs. *Technology, Pedagogy and Education*, *19*(3), 345–360. doi:10.1080/1475939X.2010.513766

Guadagno, R. E., Okdie, B. M., & Eno, C. A. (2008). Who blogs? Personality predictors of blogging. *Computers in Human Behavior*, *24*(5), 1993–2004. doi:10.1016/j.chb.2007.09.001

Hall, H., & Davison, B. (2007). Social software as support in hybrid learning environments: The value of the blog as a tool for reflective learning and peer support. *Library & Information Science Research*, *29*(2), 163–187. doi:10.1016/j.lisr.2007.04.007

Henning, E. (2003). "I click, therefore I am (not)": Is cognition 'distributed' or is it 'contained' in borderless e-learning programmes? *International Journal of Training and Development*, *7*(4), 303–317. doi:10.1046/j.1360-3736.2003.00188.x

Henri, F. (1992). Computer conferencing and content analysis. In A. R. Kaye (Ed.), *Collaborative learning through computer conferencing: The Najaden papers* (pp. 115–136). New York: Springer. doi:10.1007/978-3-642-77684-7_8

Huffaker, D. (2005). The educated blogger: Using weblogs to promote literacy in the classroom. *AACE Journal*, *13*(2), 91–98.

Kerawalla, L., Minocha, S., Kirkup, G., & Conole, G. (2009). An empirically grounded framework to guide blogging in higher education. *Journal of Computer Assisted Learning*, *25*(1), 31–42. doi:10.1111/j.1365-2729.2008.00286.x

Kramarski, B., & Mevarech, Z. R. (2003). Enhancing mathematical reasoning in the classroom: Effects of cooperative learning and metacognitive training. *American Educational Research Journal*, *40*(1), 281–310. doi:10.3102/00028312040001281

Kramarski, B., & Ritkof, R. (2002). The effects of metacognition and email conversation on learning graphing. *Journal of Computer Assisted Learning*, *18*(1), 33–43. doi:10.1046/j.0266-4909.2001.00205.x

Kramarski, B., & Zeichner, O. (2001). Using technology to enhance mathematical reasoning: Effects of feedback and self regulation learning. *Educational Media International*, *38*(2-3), 77–82. doi:10.1080/09523980110041458

Lee, L. (2009). Promoting intercultural exchanges with blogs and podcasting: A study of Spanish-American tele-collaboration. *Computer Assisted Language Learning*, *22*(5), 425–443. doi:10.1080/09588220903345184

Lehmann, T., Hähnlein, I., & Ifenthaler, D. (2014). Cognitive, metacognitive and motivational perspectives on preflection in self-regulated online learning. *Computers in Human Behavior*, *32*, 313–323. doi:10.1016/j.chb.2013.07.051

Li, J., & Chignell, M. (2010). Birds of a feather: How personality influences blog writing and reading. *International Journal of Human-Computer Studies*, *68*(9), 589–602. doi:10.1016/j.ijhcs.2010.04.001

Liu, L., Maddux, C., & Johnson, L. (2004). Computer attitude and achievement: Is time an intermediate variable? *Journal of Technology and Teacher Education*, *12*(4), 593–607.

Martindale, C. (1999). Biological bases of creativity. In R. J. Sternberg (Ed.), Handbook of creativity (137–152). Cambridge, MA: Cambridge University Press.

Mayer, R. E. (1993). Illustrations that instruct. In R. Glaser (Ed.), Advances in instructional psychology, 4, 253–284. Hillsdale, NJ: Lawrence Erlbaum Associates.

Mayer, R. E. (2001). *Multimedia learning*. Cambridge, MA: Cambridge University Press. doi:10.1017/CBO9781139164603

Mayer, R. E., Fennell, S., Farmer, L., & Campbell, J. (2004). A personalization effect in multimedia learning: Students learn better when words are in conversational style rather than formal style. *Journal of Educational Psychology*, *96*(2), 389–395. doi:10.1037/0022-0663.96.2.389

Minocha, S., & Roberts, D. (2008). Social, usability, and pedagogical factors influencing students' learning experiences with wikis and blogs. *Pragmatics & Cognition*, *16*(2), 272–306.

Murray, L., & Hourigan, T. (2008). Blogs for specific purposes: Expressivist or socio-cognitivist approach? *Journal of Eurocall*, *20*, 82–97.

Nam, C., & Zellner, R. D. (2011). The relative effects of positive interdependence and group processing on student achievement and attitude in online cooperative learning. *Computers & Education*, *56*(3), 680–688. doi:10.1016/j.compedu.2010.10.010

Ocepek, U., Bosnić, Z., Šerbec, N. I., & Rugelj, J. (2013). Exploring the relation between learning style models and preferred multimedia types. *Computers & Education*, *69*, 343–355. doi:10.1016/j.compedu.2013.07.029

Olander, M. V. (2008). Painting the voice: Weblogs and writing instruction in the high school classroom. ProQuest Information & Learning). *Dissertation Abstracts International. A, The Humanities and Social Sciences*, *68*, 4265–4265.

Puntambekar, S., & Kolodner, J. L. (2005). Toward implementing distributed scaffolding: Helping students learn science from design. *Journal of Research in Science Teaching*, *42*(2), 185–217. doi:10.1002/tea.20048

Robertson, J. (2011). The educational affordances of blogs for self-directed learning. *Computers & Education*, *57*(2), 1628–1644. doi:10.1016/j.compedu.2011.03.003

Sarfo, F., & Elen, J. (2011). Investigating the impact of positive resource interdependence and individual accountability on students' academic performance in cooperative learning. *Electronic Journal of Research in Educational Psychology*, *9*(1), 73–93.

Springer, S. P., & Deutsch, G. (1997). *Left brain, right brain: Perspectives from cognitive neuroscience*. New York: Freeman.

Sternberg, R. J. (1997). *Thinking styles*. New York: Cambridge University Press. doi:10.1017/CBO9780511584152

Te Wang, K., Huang, Y., Jeng, Y., & Wang, T. (2008). A blog-based dynamic learning map. *Computers & Education*, *51*(1), 262–278. doi:10.1016/j.compedu.2007.06.005

Torrance, E. P. (1988). *Style of Learning and Thinking: Administrator's manual*. Bensenville, IL: Scholastic Testing Service.

Torrance, E. P., McCarthy, B., & Kolesinski, M. T. (1988). *Style of Learning and Thinking*. Bensenville, IL: Scholastic Testing Service.

Torrance, E. P., Reynolds, C. R., Ball, O. E., & Riegel, T. (1978). *Revised norms technical manual for your Style of Learning and Thinking Forms A and B*. Athens, GA: Georgia Studies of Creative Behaviour.

Wagner, C. (2003). Putting another (B)Log on the wire: Publishing learning logs as weblogs. *Journal of Information Systems Education*, *14*(2), 131–132.

Wang, T. (2009). Rethinking teaching with information and communication technologies (ICTs) in architectural education. *Teaching and Teacher Education*, *25*(8), 1132–1140. doi:10.1016/j.tate.2009.04.007

West, C. R., Kahn, J. H., & Nauta, M. M. (2007). Learning styles as predictors of self-efficacy and interest in research: Implications for graduate research training. *Training and Education in Professional Psychology*, *1*(3), 174–183. doi:10.1037/1931-3918.1.3.174

Xie, Y., Ke, F., & Sharma, P. (2008). The effect of peer feedback for blogging on college students' reflective learning processes. *The Internet and Higher Education*, *11*(1), 18–25. doi:10.1016/j.iheduc.2007.11.001

Xie, Y., Ke, F., & Sharma, P. (2010). The effects of peer-interaction styles in team blogs on student's cognitive thinking and blog participation. *Journal of Educational Computing Research*, *42*(4), 459–479. doi:10.2190/EC.42.4.f

Xie, Y., & Sharma, P. (2011). Exploring evidence of reflective thinking in student artifacts of blogging-mapping tool: A design-based research approach. *Instructional Science*, *39*(5), 695–719. doi:10.1007/s11251-010-9149-y

Zhang, L. (2006). Relationships between hemispheric dominance cognitive styles and big five personality traits. *Chinese Journal of Clinical Psychology*, *14*, 354–356.

KEY TERMS AND DEFINITIONS

Blog: A website where entries are written in chronological order and displayed commonly in reverse chronological order. Many blogs provide commentaries or news on a particular subject; other blogs are personal online diaries.

Cognition: Knowledge and to all high-levels processes related to knowledge, such as perception, memory, reasoning, decision making, intuition.

Cognitive Style: A preferred way of thinking. The peculiar mode of information processing applied by a person across different tasks, and which occurs in a variety of behaviours.

Communication: The process of sharing of information, ideas, knowledge and feelings by two or more people who share the same purpose. It may happen through signs, symbols and gestures, which constitutes different form of language.

Media Psychology: A behavioral science that combines the knowledge of psychology and media applications to promote the assessment and evaluation of technology, interfaces, usability, and content, and to find new ways of integrating technology into education, media literacy education.

Metacognition: Usually defined as the awareness, knowledge and the control of cognitive processes.

Multimedia Learning: A specific form of learning that is fostered by different sources of information (e.g., text and images) being presented simultaneously. It is supposed to promote understanding and memorization.

Chapter 3
Assessing the Performance of Decision Support Systems in Military Environment:
The 3C Method

Jean-Fabrice Lebraty
Lyon 3 University, France

Cécile Godé
Université Lumière Lyon 2, France

ABSTRACT

This article explores the ability of a decision support system (DSS) to improve the quality of decision making in extreme environment. This DSS is actually based on a networked information system. Academic literature commonly mentions models of fit to explore the relationship between technology and performance, reckoning users' evaluations as a relevant measurement technique for Information System (IS) success. Although effective contributions have been achieved in measurement and exploration of fit, there have been few attempts to investigate the triangulation of fit between "Task-DSS-Decision Maker" under stressful and uncertain circumstances. This article provides new insights regarding the advantages provided by networked IS for making relevant decisions. An original case study has been conducted. It is focused on a networked decision support system called Link 16 that is used during aerial missions. This case study shows that the system improves decision making on an individual basis. Our result suggest the importance of three main fit criteria – Compliance, Complementarity and Conformity – to measure DSS performance under extreme environment and display a preliminary decisional fit model.

DOI: 10.4018/978-1-4666-8450-8.ch003

INTRODUCTION

At the time of writing, French fighter pilots are deployed in Central African Republic, Mali and Iraq to perform risky missions. As the mass media often reminds us, such air operations are quite dangerous and pilots have to make decisions under risky and stressful conditions. To achieve their goals, some of them are equipped with a specific decision support system, called Link 16. It is a tactical data exchange network which specification is part of the family of Tactical Data Links. With Link 16, users exchange their data and tactical picture in near real time.

This article explores the capacity of a decision support system (DSS) based upon a networked Information System (IS) to improve the quality of decision making (J. F. Lebraty & Godé-Sanchez, 2010). More precisely, the research question can be stated as follows: How does a decision support system based upon a networked information system contribute to improve performance of individual decision making in extreme environment? To address our research question, we based our analysis on the "utilization-focus" stream which aims at examining users' attitudes and perception to measure the way technology affects performance in organizations (Thompson, Higgins, & Howell, 1991). Drawing on such a research stream, some contributions argue that performance impact can be evaluated in investigating task-technology fit, that is when "*a technology provides features and support that "fit" the requirement of the task*" (Goodhue, 1995, p. 214). Models of fit commonly reckons users' evaluations as a relevant measurement technique for information system success (Goodhue, Klein, & March, 2000). Indeed, objective measures – trough metrics – of IS performance are particularly complex to achieve and users' evaluations are considered as an appropriate surrogate (Gelderman, 2002).

Although significant contributions have been achieved in exploration of task-technology fit and measurement of fitness, there have been few published attempts to investigate the trade-off in fitness between "Task-DSS-Decision Maker". Instead, literature chiefly prioritizes the unique task and DSS relationship over the triangulation (Zigurs & Khazanchi, 2008). This article intends to investigate both the relationships between task-DSS and DSS-decision makers in order to provide some new and valuable insights to assess the performance of decision support system based upon a networked IS. We build our analysis from extreme situations, suggesting that when individual decisions are made in such circumstances, the systems' ability to provide decision makers with an appropriate representation of the problem becomes central. Moreover, due to paroxysmal effects, we consider that exploring the way people exploit network information systems to ultimately make decisions in extreme situations allows revealing some specific usages and improvements that could be much more difficult to examine in classical work situations.

This article explores users' evaluations of decision quality when they use a networked IS, in comparison with decision quality without using such a system. Such a benchmarking design is built on a qualitative study funded by the French Ministry of Defense, which especially highlighted the issue of acquiring new capabilities with regards to the multi-role fighter aircraft Rafale outfitted with the Link 16 system. It has been the opportunity to collect pilots and navigators' perceptions on decision making performance with or without this networked system.

This article proceeds as follows. First, we review prior literature on fit models and delineates the notions of DSS based upon a networked IS and extreme situations. Then, we present our method and depict our case study. We discuss the case findings and display three new criteria to advance understanding of fitness. Finally we conclude in highlighting some theoretical and managerial implications of our analysis.

LITERATURE REVIEW

Fit Models

Two models of fit are commonly mentioned and used to explore technology-to-performance questions: the task-technology fit and the cognitive fit models. First theorised by Goodhue (1995), task-technology fit mainly focuses on the relationship between task and technology, and its impact on performance. In that view, technology is considered as a means to allowing an individual to carry out a task, which is reckoned as the action carried out by that individual. As Zack (2007, p. 1671) states: "*the notion that technology should fit the task has become an accepted approach to evaluating the performance impacts of information technology*". Measurement of performance is based on user evaluations of differences in fit before and after technology implementation. According to Goodhue (1995, p. 1828), user evaluations are "*elicited beliefs or attitudes about something and [can be] used to measure many different 'somethings' about systems*". Consequently, user evaluations must be thought of as a measurement technique (rather than as a single theoretical construct) which allows identifying some theoretical perspectives to examine IS success.

Many contributions rely on task-technology fit model to analyse the interplay between decision making performance and IS. For instance, Todd & Benbasat (2000) investigate the relationship between a decisional task and a decision support system to evaluate its impact on decision performance. The model they display is one of the first which clearly focuses on decision making performance. Few years later, Jarupathirun & Zahedi (2007) explore the influence of perceptual factors on the success of web-based spatial decision support systems. Using perceived task-technology fit and perceived goal commitment as major constructs, the authors consider that spatial capabilities and self-efficacy could impact perceived task-technology fit, which in turn influences the success and perceived benefits of a web-based spatial decision support system (both in terms of users' satisfaction and perceived decision quality and efficiency). More recently, Junglas et *al.* (2009) build their studies on the healthcare industry upon the task-technology perspective. Combining seven notions of fit and four human drives into two original research models, the authors examined the determinants of decision to utilise or not utilise mobile technologies. Finaly, Parkes (2013) questions how TTF affects DSS users' attitude and performance. She addresses the research question in deconstructing the task–technology–individual fit model into three two-way interactions: task complexity, task expertise and decisional guidance design.

Although their interest, task-technology fit models have however some well-known limitations, mostly due to under-estimation of individual cognition influence on technology-to-performance questions (Vankatesh & Davis, 2000; Yuan, Archer, Connelly, & Zheng, 2010). As Vessey (1991, p. 220) states, it is necessary to go beyond task-technology fit model and view "*problem solving as an outcome of the relationship between problem representation and problem solving task*". The cognitive fit model suggested by Vessey (1991) states that relationship between information and cognition is an essential component to understand the effect of a decision support system on decision performance. It asserts that decision quality highly depends on the ability of technology to provide decision maker with the appropriate problem representation (Baker, Jones, & Burkman, 2009).

Refining this first approach, Shaft & Vessey (2006) display an "extended cognitive fit model" which details the question of problem representation. The authors distinguish the internal and the external representations of the problem domain to provide a more complete understanding of the relationship between software comprehension and modification. Drawing on Zhang & Norman (1994), internal representation is defined as pre-existing knowledge related to the problem category. External representation is viewed as the formalised image of the problem perceived by a decision maker.

In considering the investigation of internal and external representations as critical to understand technology-to-performance issues, the extended cognitive fit model deepens prior fit models, leading to micro-analysis of fitness and completing studies mainly focused on accounting data and insufficiently on cognitive interactions between IS and users (Dilla, Janvrin, & Raschke, 2010). This article relies on such refinements to address the research question, suggesting that when individual decisions are made in extreme situations, the systems' ability to provide decision makers with an appropriate representation of the problem becomes central. However, few contributions address the trade-offs in fitness between task-technology-and people, chiefly prioritizing the unique task-technology relationship over the triangulation (Zigurs & Khazanchi, 2008). In this article, we investigate both the relationships between task-DSS and DSS-decision makers. Moreover, we build our research on extreme situations. In effect, due to paroxysmal effects, we consider that exploring the way people exploit network information systems to ultimately make decisions in extreme situations allows revealing some specific usages and improvements that could be much more difficult to examine in classical work situations.

Decision Support Systems

DSS is an old concept before being a real tool. We can say that the 70's is the real start for DSS with the work of H.A. Simon, A. Newell and even H. Mintzberg (H Mintzberg, Raisinghani, & Théorêt, 1976; A Newell & Shaw, 1972; Allen Newell & Simon, 1976; Simon, 1960). This term DSS was coined in 1971 by two US researchers (Gorry & Scott Morton, 1971). The idea was to improve the concept of Management Information Systems by including notions of managerial activities developed by Anthony (Anthony, 1965) and H.A. Simon's approaches to decision. As L. Eom (Eom, Lee, & Kim, 1998; Eom, 2004) shows, authors as (Alter, 1977, 2004) have a huge impact on DSS concepts. We can say that at least two types of DSS may be used: data oriented and model oriented. 40 years after, this taxonomy is still valid. D. Arnott and G. Pervan (Arnott & Pervan, 2005, 2008) propose several types of DSS. First, they explain that Personal DSS are the oldest form of DSS developed for independent managers. These DSS may be used in a standalone mode or linked to a data warehouse. Second, Group Support Systems address all systems that allow decision makers to work together. We may find here communications systems (Robey, 1983). We used to talk about groupware (DeSanctis & Gallupe, 1987; Gallupe & DeSanctis, 1988) but nowadays we prefer the term of "collaborative solutions" (Eric, 2009). The third type of DSS is named: "Negotiation Support Systems". Close to GSS (Recio-García, Quijano, & Díaz-Agudo, 2013), these systems are designed to helps people to reach an agreement and solve conflicts (Guo & Lim, 2007). The fourth category used to be called "Executive Information systems" (Watson, Rainer Jr, & Koh, 1991) and are more famous now under the name of "Business Intelligence" (Chen, Chiang, & Storey, 2012; Işık, Jones, & Sidorova, 2013). These systems may be defined as useful dashboards for managers. Other categories may be taken into account as Artificial Intelligence based DSS (Dahal, Almejalli, & Hossain, 2013; Turban, 1995) or Creative Support Systems (Massetti, 1996). In this paper, we will talk about another type of DSS based upon networked information. That kind of DSS has two main features:

- Visualize all elements of interest (as for example friends, foes, terrain features etc.) in layers form;
- Represent links between these elements (as for example the hierarchical unit from which individual elements belong);

The use of a DSS based upon a networked information system is generally mandate or highly recommended. As Chae and Poole mentioned (Chae & Poole, 2005) "Mandates are typically not just simple directives or orders, but also have descriptions of what to do or not to do, accompanied by regulations and documents describing how the mandate must be put into effect. Even when a mandate is framed as a simple directive, attempts to follow it generally generate more detailed policies and procedures that are attached to and become part of the mandate." (p. 150). Even mandatory systems are well framed, mandates as all rules, are subject to interpretation by users. This adds a layer of complexity to these systems.

Networked Information System and Decision Support System in Extreme Environments

This paper examines the use of a DSS when teams operate in extreme environments. Extreme environment is characterized by simultaneously highly levels of change, uncertainty and danger (Aubry, Lièvre, & Hobbs, 2010). The change occurs in a rapid and discontinuous tempo (Faraj & Xiao, 2006). These features point to the time-speed pressure handled by team members as their interactions unfold as well as to differences in the pace of change (Eisenhardt & Graebner, 2007). The uncertainty represents the probability of occurrence of change as well as its time of occurrence and practical details. Cunha et al (2006) consider that uncertainty in extreme environment can mainly root in either expected issues and unexpected causes (i.e. emergence of creeping troubles in team although expected issues are apparently well managed), or unexpected issues and expected causes (i.e. surprising outcomes stemming from pre-identified causes). Performing in extreme environment leads teams to confront to unforeseen situations, that are those that were considered unlikely, or foreseeable but unexpected and which cause surprise because they were overlooked (Weick & Sutcliffe, 2007). Finally, extreme environment entails a high level of danger, which refers to the impact of change. For actors facing high stakes and time pressure (Klein, 1998), the consequences of errors may be tragic, from symbolic (e.g. loss of leadership) to vital danger (e.g. fatal injuries in the course of action).

A vast and established body of literature has investigated decision making in extreme circumstances, especially under uncertainty. For example, some of them suggest techniques of debiasing actors' judgments (Hodgkinson, Bown, Maule, & Glaister, 1999) and simplification of decision-making processes (Schwenk, 1984). In this article, we rely on the naturalistic decision making approach (Klein, 1998; Lipshitz, Klein, & Orasanu, 2001), commonly used to examine decision making under changing and time-pressure environments (for instance, healthcare, fire departments, U.S Navy operations). The naturalistic – or intuitive – perspective states that decision makers do not choose among various alternatives to make decision but refer to typical ways of responding. Taking advantage of their expertise and past-cases experience, they are engaged into a recognition primed process, which enables them to provide rapid and appropriate solutions. Thus, individual decision making will depend on the nature of the decision maker: is he a novice or an expert? (Randel & Pugh, 1996). Facing a complex problem, novice will have to follow a method and have to consider alternatives than to pick the one that satisfied him the most (Simon, Dantzig, & Hogarth, 1987). An expert will first recognize the situation as typical then apply a course of action which he has tested previously. Naturalistic Decision Making is not a normative method but quite the reverse: it is how real experts behave in real settings. So this approach aims to help decision makers to unveil their cognitive biases just to be sure that they have a global situation awareness and that they don't be subjected to a framing effect (Tversky & Kahneman, 1986).

The naturalist model leads to concentrate on situational understanding – situational awareness – as the most important stage of decision making process. As such, Information Systems (IS) which provide decision makers with accurate representation of the situation they deal with play a valuable role. This is the case of networked IS. Providing generic databases and communications devices access facilities to the network users, a networked information system enables greater distribution and ubiquity of information. Allowing refining work and knowledge integration of distributed tasks, such a networked IS helps to compensate for human cognitive limitations and aims at producing better decision (Collins, Ketter, & Gini, 2010). Decision support systems based upon a networked IS enables gathering of large amount of information from integrated applications in order to provide users with (1) visualisation of all elements of interest in a layered form and (2) representation of links between these elements. In that way, the accurate representation of the addressed problem provided by the DSS enables users to make decisions from an updated and complete picture of the environment. They can design their individual process of decision on the connected components of the environment (human, physical, technological, etc.), benefiting from a collective picture of up-to-date interactions displayed by the system.

RESEARCH METHOD

To investigate our research question, we selected an extreme single-case study (Eisenhardt & Graebner, 2007). More precisely, we focused on fighter pilots and/or navigators' decision making in a warfare context, using a decision support system based upon a networked IS called Link 16. Link 16 fits out many NATO fighter aircraft, including the French Rafale currently deployed on various operational theatres.

This case study stood as a pertinent field at least for two reasons: first, aerial war missions are achieved under an important clock-speed pressure (in the order of a few seconds) and changing circumstances (ground to air missile threat, for instance). Risks are prominent since lives can be lost or injured (collateral damages or friendly fire, for example) if poor decisions are made. Uncertainty is high, decision makers constantly facing new conditions and making real time decisions. In that way, pilots and navigators deal with extreme environment. Second, in such circumstances, Link 16 can be viewed as critical for supporting decision making during the whole mission, providing pilots with accurate and up-dated tactical information to make rapid and appropriate decisions.

Context Setting and Data Collection

French Air Force uses many aircrafts. Since 2001, a new approach is developed: an omnirole aircraft fighter. Rafale is a twin-engine, canard delta-wing, multirole fighter aircraft designed and built by Dassault Aviation. This aircraft has a high level of agility and may perform simultaneously air supremacy, interdiction, reconnaissance, and even airborne nuclear deterrent missions. The Rafale is different from other current European fighters in that it is almost entirely built by one country. This aircraft requires an adapted computerized system in order to provide an accurate image of the situation for pilots. Implemented into Rafale aircraft at the beginning of 2006, Link 16 is a wireless Decision Support System, supported by a highly secured networked IS. It provides users with a range of services such as com-

Figure 1. The link-16 network

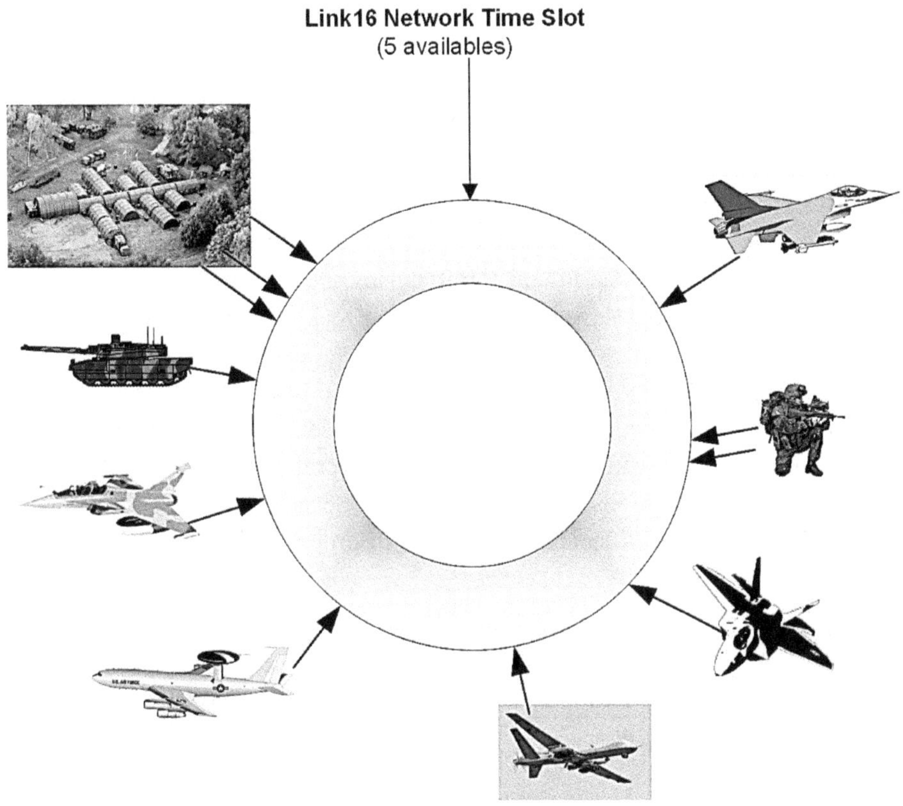

munication devices enabling users to quickly share information on short notice (a kind of instant messaging), and updatable database delivering a real-time picture of the tactical situation. For instance, a fighter aircraft equipped with Link 16 receives on its display screen information concerning friendly and enemy airplanes' positions, counter-battery sites, and location of ground forces. This information comes from other neighbouring fighters, the airborne control system aircraft AWACS (Airborne Warning and Control System), navy ships, and Special Forces units deployed on the battlefield (Figure 1). On the battlefield, Link 16 is used to achieve both air strike and air defence (French airspace control and protection) missions.

In order to consider the benefits of Link 16 on individual decision performance, we encouraged pilots and navigators to compare their way of operating when equipped with the Link 16 system and when equipped only with voice communications. We have not observed significant differences of perception between pilots and navigators. Indeed, they are teammates in theatre, briefing, flying and debriefing together all along their detachment. Even if their activities in flight are specific, they use Link 16 to reach the same goal: achieve the mission. In that way, they noticed and took advantage of the same improvements.

We organised data gathering over a three-year period, from September 2006 to November 2009, in the course of which we visited Rafale squadrons two times (in January 2008 and November 2009). To achieve triangulation, we collected data from multiple sources of information, including individual and collective interviews, field observations, and archival records such as reports from the field, doctrinal documentation, and studies funded by the Department of Defence and the French Ministry of Defence.

Fourteen semi-structured interviews were conducted with pilots and navigators of Rafale, one with a Rafale experimental test pilot and one with an AWACS mission commander used to command and control the battle-space with Link 16. Each interview lasted on average one hour and a half and was tape-recorded and transcribed, except for one of them. In this case, two team members recorded the discussion in writing verbatim and compared their records afterwards. We followed an interview guide focused on users' evaluations of differences in aerial operations conduct with and without Link 16: first, we questioned the Link 16 usages in achieving air strike missions in Afghanistan and/or French airspace control and protection. In doing so, we wanted to understand how the system works and what pilots and navigators concretely do with it, and thanks to it, during the course of action. The second question concerned users' perceived values of Link 16 on decision making and, broadly, on mission achievement. During interviews, we encouraged narration in order to grasp Link 16 effects on decision making under evolving, uncertain and risky conditions. We also re-used interview transcripts of two Rafale pilots (Dubey & Moricot, 2006). These secondary data have been collected by Dubey & Moricot at the very beginning of link 16 implementation into the Rafale aircraft. They provided us with "historical" information on pilots' first perceptions and usages of the DSS and allowed us to deepen our analysis and interpretation of users' evaluations (Kelder, 2005). In order to yield our benchmarking approach

Table 1. Description of interview participants and goals

Quality	Quantity	Period	Goals
Rafale Pilots (secondary analysis)	2	Dubey & Moricot, 2006	Collecting "historical" information on pilots perceptions of Link 16 at the very beginning of its implementation into the Rafale Aircraft
Mirage-5 pilots	6	Sept. 2007	Understanding decision making process without Link 16
Mirage 2000D pilots	3	Oct. 2007	Understanding decision making process without Link 16
Mirage 2000D navigators	3	Oct. 2007	Understanding decision making process without Link 16
Rafale pilots	9	Jan. 2008 – Nov. 2009	Collecting information on users' perceived value of Link 16 concerning decision making process in warfare missions
Rafale navigators	5	Jan. 2008 – Nov. 2009	Collecting information on users' perceived value of Link 16 concerning decision making process in warfare missions
Rafale test pilot	1	Sept. 2008	Collecting information on users' perceived value of Link 16 concerning decision making process in warfare missions
AWACS mission commander	1	June 2008	Collecting information on user's perceived value of Link 16 in commanding and controlling the battle-space
High-ranked officers - Link 16 doctrine	3	Sept. 2009	Understanding the Link 16 doctrine at the institutional level
Total	**33**	**From 2006 to 2009**	

and to better understand the nature of pilots operational work without Link 16, we also interviewed 6 pilots of Mirage-5 (air defence aircraft) and 6 pilots and navigators of Mirage 2000D (strike aircraft) to understand the way they make decision and achieve missions without Link 16. Most of the Rafale pilots come from these squadrons. Interviews have also been the opportunity to thoroughly understand to which extent Link 16 introduced changes in work practices. Finally, we collected data concerning the French doctrinal vision of Link 16, in studying internal archives and interviewing two high-ranking French Air force officers assigned to think the near-future evolutions of Link 16. Table 1 describes the interview participants and goals.

Further, we had the opportunity to observe how Link 16 runs by attending an air operation simulation. We also gathered French after-action reports from the field (Afghanistan and Libya) and NATO training exercises, as well as American institutional monographs and US Air Force after-action reports from Afghanistan and Iraq. They concerned pilots' reviews related to the impact of Link 16 on mission improvement (Table 2).

DATA ANALYSIS

All of these data collection methods allowed us to gather an important amount of multifaceted data. In order to examine the effect of Link 16 on individual decision performance, we had to document and analyse pilots and navigators' evaluations related to the ways Link 16 has modified the process of mission achievement and has affected decision making with regard to their previous way of doing. We investigated the "*continuous interplay between data and concepts*" (Van Maanen, Sorensen, & T., 2007, p. 1149), exploiting our data to question established theory and move on towards new knowledge and some kind of theoretical understanding. Seeking to give sense from our data, we combined three different analysis strategies – narrative, grounded theory and synthetic – as described by Langley (1999). We have recourse to narrative strategy to organize our data and clarify sequences across levels of analysis. Grounded theory (Strauss & Corbin, 2008) involved the systematic comparison between our data in order to gradually elaborate and refine our categories. Finally, synthetic strategy allowed us to reflect upon our data analysis, and construct a model from identified regularities.

Table 2. Collected data (observations and archival records)

	Qualitative Details	**Goals**	**Connections with Interviews**
Shadowing observations	Air operation simulation using Link 16	Observing Link 16 symbol, tactical information displaying, etc. Observing Link 16 usages *in situ* Collecting information on user's perceived value of Link 16 concerning decision making process	AWACS mission commander Rafale pilots and navigators Rafale test pilot
Archival records	French and American After-action reviews from Afghanistan, Libya and Iraq	Understanding Link 16 usages in warfare missions Understanding the impact of Link 16 on mission improvement	Rafale pilots and navigators Rafale test pilot

To achieve such a strategies' combination, we first transcribed data collected from tape-recorded and written interviews, and structured our observations and field notes.

From that raw material, we organised our analysis in two main steps: coding and writing monographs. Concerning the coding, analysis of the interviews, observations and field notes was conducted using NVivo9 software. We completed our analysis in a bottom-up perspective to foster the emergence of our categories. Two themes were developed a priori, 'down' from our research questions and design: (1) perceived changes in air combat missions with Link 16 – interview participants told us their own experience in using this system in theatre and (2) their evaluations of its advantages and values in comparison with previous communication devices. From these two themes, we produced emerging themes (new nodes 'in vivo' from the words, sentences and/or paragraphs) which included a more refined categorisation of usages in terms of changes in mission conduct and decision making. During the whole coding process, we followed an iterative and comparative approach in order to identify regularities. Using queries, we produced qualitative matrix displays in order to compare multiple pairs of items and to build an overall pattern of our data. Finally, we generated a table of our coding structure (Appendix). Concerning the second step of data analysis, a monograph was produced for each interview and sent to interviewees. Feedback allowed us to refine our observations and go deeper into our understanding of the effect of Link 16 on individual decision performance.

Case Study Findings

Recalling the research question presented previously, our goal was to gain a broader understanding of the way and the reasons Link 16 contributes to individual decision making performance in extreme environment. This section provides comprehensive insights from the interviews of reporting pilots and navigators, regarding the perceived usefulness of Link 16 when they make decisions in operational situations. The first and general result is that all the interviewees agree that Link 16 has significantly improved the way they achieve air operations in comparison with voice communication: 100% of respondents spontaneously pointed out that Link 16 enhances mission capabilities, especially in increasing communication quality and sharing abilities. These two testimonies are good examples of what we have encountered: "I can use a metaphor: before [Link 16], we used to run in a tunnel with only torchlight to guide us. Now, with Link 16, you'd switch the light on."; "With Link 16, it becomes simple to perform our tasks. Without Link 16, it's hell."

In addition to such general comments, line-by-line coding of the materials revealed that pilots and navigators evaluate Link 16 efficiency in a different way, whether they focus on the system's ability to support air missions or respond to their requirements during the course of action. We assumed that the first issue concerns the interplay between technology and task: users evaluate improvements regarding the way they run aerial warfare missions with Link 16 and are able to identify the basics of improvements. The second issue addresses the relationship between internal and external representations of the problem domain, that is users assess if their expectations of what the problem domain will be matches or not with the form and the nature of data transmitted by Link 16. This double aspect can be considered as a first global result reminding the relevance of the extended cognitive fit model. The following details the perception of the fit under these assumptions.

The Fit between Aerial Combat Missions and Link 16

During interviews, pilots and navigators were very concerned about Link 16's characteristics and its ability to support their task needs during air combat. More specifically, data analysis shed light on one main result: pilots and navigators consider that the ways Link 16 provides information substantially reduces the risk of data misinterpretation.

First, 67% of respondents argue that Link 16 improves management of informational flow and space. Answers to the open-ended question "In comparison with previous information and communication devices, what are the most important changes introduced by Link 16 to conduct aerial combat missions?" provided detailed information on the technological characteristics of Link 16. In comparison with the way they used to conduct air operations without Link 16, pilots and navigators noticed two main technological improvements. They are related to the quality of (1) visual presentation of the tactical situation and (2) intraflight (aircraft-to-aircraft and between aircrafts and AWACS) and extraflight (between aircraft and ground forces) communications. To this extent, the system's characteristics appear to be in line with the pilots' task needs.

Concerning the presentation of the tactical information (Table 3), Rafale pilots and navigators are provided with continuously updated tracks which visually display the precise location and velocity of detected aircraft, counter-battery sites and/or ground forces. Different icons appear on the screen to indicate the nature of the track (friend, enemy or unknown), warning the pilot of potential threats and helping him to change his flight path, for instance. On screen of the Link 16, data characteristics are designed to be easily understandable and processed during the course of action: friendly forces are symbolised by green circles and enemy forces by red triangles. In addition, other coloured symbols appear to provide information related to the origin of detections (Figure 2). This way, pilots and navigators immediately know if tracks are transmitted by AWACS, their own radar systems, or some other sensors. Such extra information also indicates whether a track has been merged or not by the system. Merging operations are critical since they allow pilots and navigators to collect simplified information on their display. As a result, they don't have to manually differentiate between many tracks nor analyse each anymore.

The second technological improvement perceived by interviewees concerns the quality of communication (Table 4) between aircraft and between aircraft and ground forces (only for Close Air Support). Before Link 16, the main communication channel was radio transmission. Pilots and navigators had to continuously listen to voice communication describing air traffic, mentally convert each description into a location and develop an appropriate tactical response. In addition, during Close Air Support missions they had to communicate with the air controller on the ground in order to obtain the precise target location and thoroughly understand the situation. With Link 16, the biggest part of voice transmission is digitised (except the final dialogue between the pilot and the air controller, just before the assault). Tactical data coming from AWACS are automatically transmitted through the system and pilots do not have to spend time and efforts in radio exchanges anymore. Consequently, the risk of ambiguity is significantly reduced not just because people do not communicate "verbally" anymore but also because data characteristics are relevant to evaluate the significance of the situation.

Figure 2. Display of Link-16

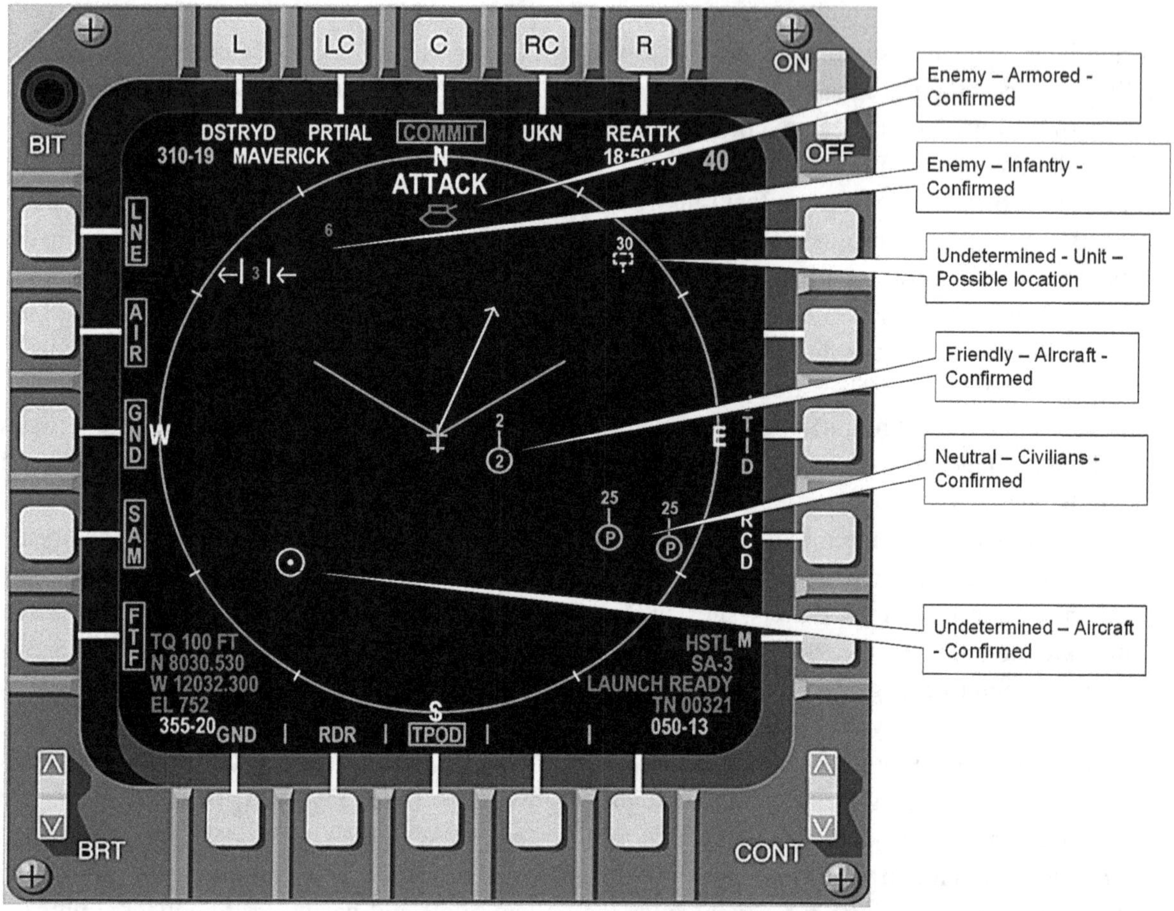

Table 3. Enhancement of the visual presentation

Quality of Visual Presentation of the Tactical Situation		
Proportion of Respondents	**Respondents**	**Example of Verbatim**
75%	A Rafale test pilot	*"What's the key of decision? First of all, it's getting a good representation of the operational situation!"*
	A Rafale pilot	*"With Link 16, you can see UAVs [Unmanned Air Vehicles], for example, you have a little symbol on your screen and other icons tell you "stay aware, there's a counter-battery site over there!" Or visually, graphically show you that the fuel tanker's getting away"*
	A Rafale pilot (secondary analysis)	*"Before Link 16, you had to mentally create the situation by yourself. Now, you don't have to do this work anymore. With Link 16, it's really easy to see everything"*
	A Rafale navigator	*"The implementation of Link 16 is a means to improve considerably information gathering and reduce manual tasks"*

Table 4. Reduction of communication volume

Decrease of Intraflight and Extraflight Communication		
Proportion of Respondents	**Respondents**	**Example of Verbatim**
67%	Two Rafale pilots	*"With Link 16, we quite limit talking"* *"For example, the patrol leader can automatically check the fuel level of his wingmen. No voice communication is required anymore"*
	An AWACS mission commander	*"Link 16 generates an important reduction of voice communication between the AWACS and the aircrafts. As a result, AWACS controllers can concentrate all their attention to the operational context; they became quite more available than before".*

The Fit between Pilots and Representations of the Problem Domain Displayed by Link 16

During interviews, pilots and navigators responses often underlined that services (form and content) provided by Link 16 match adequately with their internal representations of the problem domain. In effect, material coding shows that the usages of Link 16 tend to enhance the fit between decision makers and the system. Pilots and navigators' evaluations of Link 16 clearly indicate that their internal representation of the problem fits with the external representation proposed by the system. In other words, pilots' expectations of what the problem domain will be matches with the form and the nature of data transmitted by Link 16.

Indeed, when they answered to the open ended question: "How do you evaluate the contribution of Link 16 to air combat mission [Close Air Support and airspace protection]?" many of them stressed the reduction of cognitive workload they take advantage of (Table 5).

No mental calculation and compensation are required anymore and they can collect information much faster and accurately. To this extent, the pilots and navigators' perception of mental workload is positive since they feel delivered from information management tasks. When some duties are assigned to the system, they gain time for higher-level decision-making. The results of such compression on time and cognitive workload are extra time saving. This extra time can be used to reckon more alternative courses of action and to make more decisions in a given period of time. Moreover, instead of spending their time to gather and monitor data, they can focus on the essential steps of their mission, which means refining plans and developing sense-making.

Table 5. Reduction of cognitive workload

Decrease of Cognitive Workload		
Proportion of Respondents	**Respondents**	**Example of Verbatim**
75%	Two Rafale pilots	*"Tasks related to information management are pretty reduced now; Link 16 handles much of them"* *"With Link 16, you don't feel overloaded anymore." As a result, "You can dedicate your capacities to tactics"*
	A Rafale test pilot	*"With Link 16, you reduce mental calculation and remain focus on tactics. You're more efficient than before"*

Table 6. Situational awareness improvement

Situational Awareness Improvement		
Proportion of Respondents	**Respondents**	**Example of Verbatim**
67%	A Rafale navigator	*"With Link 16, you can see everything and share everything you see; you have a common visualisation in real time"*
	A Rafale pilot	*"Thank to Link 16, you always know where your friends are, where enemies are, and what kind of tactics you can elaborate in common to protect your friends and eliminate your enemies"*

Some pilots also underline the ability of Link 16 to support what they called "*instinctive fights*." That means that Link 16 has significantly enhanced situational awareness (Table 6); it efficiently supports tactics and mission steering by providing an integrated and real-time interpretation of the tactical situation: pilots are "guided" within the combat environment.

The perception that Link 16 is able to accelerate time and provide combat superiority (Table 7) is shared by many interviewees. It supports pilots in increasing the combat speed, engaging more targets and destroying them. Moreover, Link 16 has considerably modified the way they used to communicate and exchange tactical information. In comparison with the typical analogical voice channel transmissions, Link 16 is viewed as an information multiplier and a quite reliable decision support system.

Our case study clearly demonstrates that Link 16 improves individual perception of decisional performance. The following goes deeper into this first and general result in suggesting new insights to advance fit nature and meanings.

THE 3C METHOD: COMPLIANCE, CONFORMITY, AND COMPLEMENTARITY AS FIT CRITERIA

Case study results enable refining fit concept, especially in entering deeper into fit relationships. In effect, analysis of fitness between task-DSS – aerial missions' characteristics and Link 16 – and DSS-decision makers – pilots' internal representations of the problem domain and representations displayed by Link 16 – reveals the importance of three main criteria to evaluate networked IS-to-performance issues: compliance, conformity and complementarity.

Table 7. Enhancement of decision making performance

Greater Efficiency in Combat Engagement		
Proportion of Respondents	**Respondents**	**Example of Verbatim**
58%	A Rafale pilot	*"Link 16 lets you do something fabulous: increasing combat speed. Actually, you can avoid being seen by others. Your enemies? They still process mental calculation"*
	A Rafale navigator	*"Link 16, it's a revolution! We are in the era of real time and nothing's more valuable than real time when you make a decision!"*
	An AWACS mission commander	*"With Link 16, coordination between aircrafts is easier and faster just because pilots have, or can have, all information they need to achieve their tasks"*

Table 8. The compliance guide

Is there a difference between information provided by the system and the mission you have to perform?
Is there a difference between information provided by the system and the ROEs you have to follow?
Is there a difference between information provided by the system and the SPINs you have to fulfill?

Compliance is usually introduced to examine the relationship between IS and rules of exploitation (Caron, Vanthienen, & Baesens, 2013; Smith, Winchester, Bunker, & Jamieson, 2010; Wallace, Hui, & Cefaratti, 2011). Regarding our case study, Link 16 is in compliance with mandatory guidelines and check-lists used by pilots in the course of action. More precisely, the system is daily designed to be compliant with three major "rules of exploitation": the mission, global rules of engagements (ROE) and special instructions (SPINs). To measure the level of compliance, we asked questions such as displayed in Table 8.

The second criterion is conformity. It is usually introduced by literature on decision support system to complete two kinds of ambitions (Standing, Sims, & Love, 2009): (1) decision maker can be in conformity with the system in totally agreeing with the picture delivered and/or (2) in being socially led to agree. The first perspective examines conformity between users' expectations and the decision process displayed by the system. Mostly based on experimental studies, such a view needs confrontation to real situations to acquire robustness, in particular through qualitative studies (Rowe, 2005). The second perspective of conformity is related to social mechanisms, focusing on the way group members' change their attitude and re-pattern their beliefs through system's usages (Cialdini & Goldstein, 2004). Networked architecture of IS positively affects such individual and social conformity: first, in delivering easily understandable data of the tactical situation, Link 16 significantly enhances conformity between pilots' expectation and decision process displayed by the system; second, in supporting acquisition of a global visualisation of network members' actions, Link 16 allows pilots building up common knowledge and generating coordinated expectations and behaviours. The conformity criterion can be determined by questions such as introduced in Table 9.

Finally, the complementarity criterion concerns the way additional information provided by the system is able to refine user's representations of the problem domain. The complementarity criterion is viewed as the ability of the system to provide an appropriate picture of the situation. Case study findings suggest that the nature of environment fundamentally impacts the complementarity criterion. In extreme situations, decision makers work under time pressure and cannot allocate time to double process the same information. In that way, in reducing cognitive workload, Link 16 allows pilots to save time on computing and calculation tasks and to concentrate on the tactical phase, which is critical during missions. Moreover, Link 16 increases pilots' situational awareness: in line with Endsley (1995), they

Table 9. The conformity guide

Did the information on the screen match your expectations?
Did the system really help you make decisions?
Did you observe maverick's behavior in your team?
After the implementation of this new tool, did you notice any effect on your ability to integrate the team?

Table 10. The complementarity guide

Did you have to manage twice or more the same data coming from different sources?
Did you complete twice or more the same informational task?
Have you ever been disconnected with the ongoing situation?
Did you notice informational shadow area in your system?

obtain a better perception of elements in the environment within a volume of time and space, and they acquire the comprehension of their meaning in current time and in the near future. Table 10 presents some questions to ask in order to measure the complementarity criterion.

During the interviewing process, we had opportunities to deal with the questions of compliance, conformity and complementarity, especially with the first theme of our interview guide (Perceived changes in air strike missions with Link 16, Appendix). Questions regarding these criteria were sorted out and answers were encoded for analysis. In order to encode the answers, we changed them into a yes-no attitude with regards to the chosen criteria. Finally, the criteria measurements for fit are as follows: compliance represents 73%, conformity represents 67% and complementarity represents 58% (Figure 1). As a result, we can consider that Link 16 fits with the users' needs and expectations up to 66%. The following figure displays the way we can illustrate the fit provided by Link16.

According to our observations, if compliance and conformity fit criteria reach a high level. This is not surprising considering quality of conception for such systems. Engineers give a special attention to follow all guidelines and to respect all formal rules. Moreover, extreme situation and discipline contribute to a conformity behavior. The complementarity criterion appears to be set back when actors make decision in extreme situations. This can be explained by the difficulty to create a robust system offering high level situation awareness.

Figure 3. Assessing fit with the 3C method

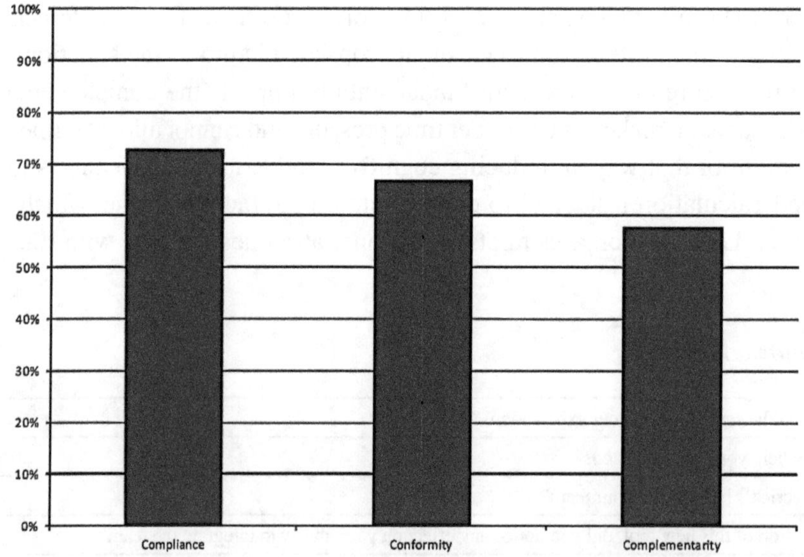

MANAGERIAL IMPLICATIONS: TOWARDS A DECISIONAL FIT MODEL

As we stated at the beginning of this article, extreme environment is characterized by simultaneously three criteria: highly levels of change, uncertainty and danger. Obviously when we talk about extreme situations we think about military cases. But, since 20 years – thanks to globalization – business environment is as tense as the military one. In contemporary organizations, extreme environment analysis is not a trivial topic though: achieving a stable workforce is increasingly difficult, while work in many companies grows complex (especially in the service or creative industries) with tasks intertwined in many and sometimes surprising fluctuant ways, and interdependencies being therefore uncertain or difficult to identify and manage.Let's see how these change, uncertainty and danger criteria fit with current business situation by providing relevant examples. First, regarding the velocity of the evolution of a situation, we can take the example of German economic situation. Stated as robust in 2013 and early 2014, the situation declined suddenly, particularly due to Ukrainian civil war. In few weeks a whole context passed from green to yellow leading many companies to re-assess all their business plans. Second with reference to uncertainty, even if it is a well know criteria in management and economics, its level is higher than ever. Since 1994 and its book and article on the limitation of strategic planning, H. Mintzberg (1994), business managers know that it's better to take into account uncertainty than to try to predict accurately the future. In the case of start-ups, there is a direct link between the capability to predict alternatives and the value of their business plan (Chwolka & Raith, 2012), but as it is very hard to determine, young entrepreneurs have to find other ways to assess the quality of their project. Crowdfunding is a tool designed for this purpose (J.-F. Lebraty & Lobre, 2013). Finally, regarding the notion of danger or high stakes, the example of stock markets seems relevant. On NYSE, the average daily trading value is approximately US$170 billion. In addition, a recent web article on the NASDAQ website was titled "Could Apple Stock Survive A Market Crash?"[1]. So, yes, a majority of modern organizations have to cope with extreme situations and find ways to deal with them. This why since more than 10 years Business Intelligence tools have a huge success (Carneiro Moro, Ribeiro Cortez, & Ferreira Rita, 2014). Simultaneously decision making has known the development of a new theoretical approach named Naturalistic Decision Making (Kahneman & Klein, 2009; Klein, 1998) as we stated in the first part.

New tools, new theory and a context that never happened before push us to propose a new theoretical framework to assess the fit between a decision maker, his tools and his task.

Regarding our advancing on the three fit concepts, we conclude in suggesting an exploratory model, named "Decisional fit model" (Figure 4). Drawing from the extended cognitive fit literature, our model allows analysing the relationships between decision support system based upon a networked IS and individual decision performance. More precisely, it is designed as following:

1. External representation of the problem exclusively comes from an information system or, more specifically, a networked decision support system,
2. Three fit relationships are studied: fit 1 analyzes the interplay between task and networked decision support system, fit 2 focuses on the relationships between decision maker and networked decision support system and fit 3 concerns the link between decision maker and task's characteristics,
3. Unlike extended cognitive fit model, we linked decisional task with two others components: first, the networked decision support system designed to manage task and second, the impact of extreme environment on decision making. In effect, under extreme circumstances, decision must be made quickly. As a result, the mental representation of the problem can simply characterize the decision individual will finally make. This decision is significantly influenced by the levels of fitness.

Figure 4. Decisional fit model

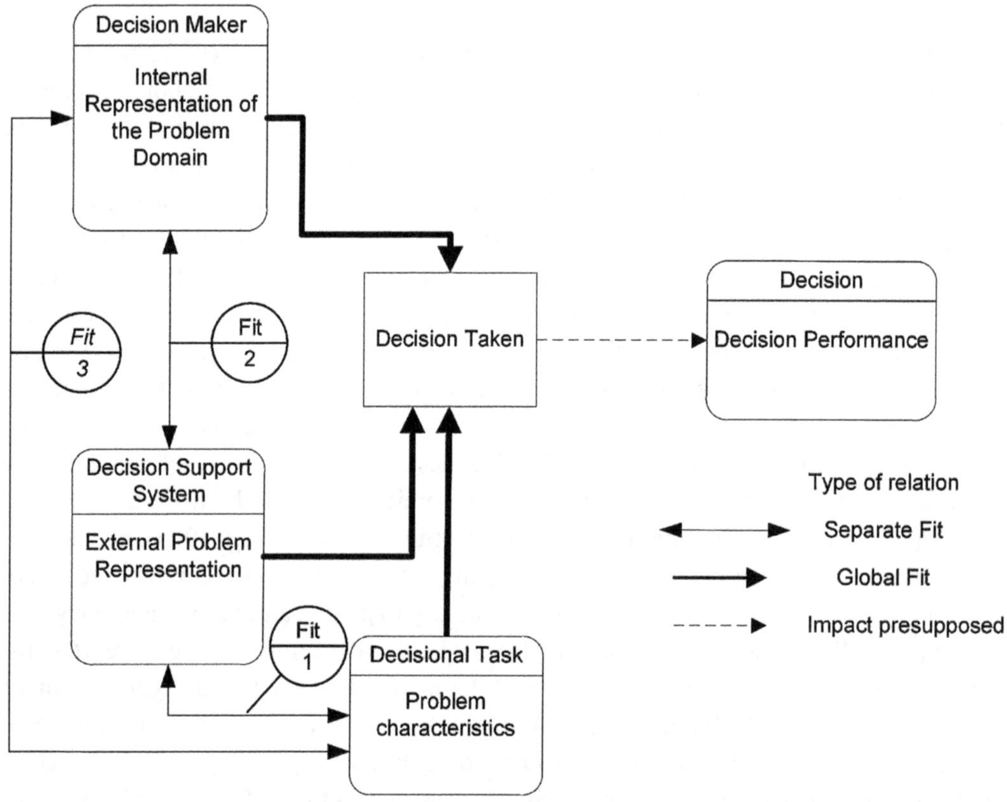

As we can see, three kinds of links are investigated in this model. The first one (classical line) represents the fit relationships. Here, we take into account three relations. First, links between DSS and the decisional tasks have to be examined (Fit1). Then, relations between this DSS and the decision makers must be analyzed (Fit2). We propose to assess these DSS centered relations with our 3 criteria: Compliance, Conformity and Complementarity (3C). Having a high 3C score means that the DSS fits both with the task and the decision makers. Here we did not examine the links between the task and the decision maker (Fit3). Did pilot's beliefs concurs with this specific task or mission? This leads to deal with ethical aspects and may be a sensitive issue in military domain as well as for business concerns. This "Concurrence" criteria could be the forth C.

The second one (line in bold) illustrates the results of the global fit (unification of the three fit relations). Once implemented, this task solution leads to a certain result affecting decision performance. The task solution representation is crucial to make decision. However, we do not examine the relationships between task solution and performance of the decision, assuming that it is positive: Fit – Adapted decision – Decision Performance (dotted line).

This Model is totally adapted to both military and managerial situation. Thus, we started investigation in order to test it in the two following environment. First, we launched a study in two French Universities in order to assess the fit between administrative managers working on business Objects for evaluating the quality of the research performed by academics. Based upon semi structured questionnaires and interviews we will measure the fit level. Depending on this level we will propose ways to improve it by acting both on these three criteria. Second, we will launch a study in the retail bank industry. We want to assess the fit between bank customer managers using Excel specific spreadsheet for deciding to allow credit to students.

CONCLUSION

In this article, we explored the way a Decision Support System based upon networked IS can enhance the decision making performance of individuals in extreme situations. We conducted a case study analysis on the perceived added value of pilots and navigators regarding a new system called Link 16 on air missions. We demonstrated that perceived enhancement of decision performance is drawn on a double fit (task – DSS and DSS – Pilot). Moreover, we provided three criteria contributing to the measurement of overall fit: Compliance, Conformity and Complementarity. This is what we called the 3C method.

In addition, the answer to our initial research question stated as: *"How does a decision support system based upon a networked information system contribute to improve performance of individual decision making in extreme environment?"* is that the specific characteristics of this DSS combined with this extreme environment lead to improve the performance of individual decision making. This improvement results for the enhancing of three fit criteria: Compliance, Conformity and Complementarity.

Our findings have implications for both managerial practices and academic research. Regarding practices, this work will have a direct effect on the training methods that are delivered to French pilots. As we presented in part 5, this study may be useful for business activities.

Regarding the theoretical implications, we contributed to the field of Fit Models by providing new criteria that will have to be tested in different contexts. We provide too a way to assess these criteria and this leads to give a score. This score may be used as a milestone for further improvements. Moreover, by focusing on a special DSS we provided an in-depth insight about the Fit Model by linking the specific characteristics of this system to the fit criteria. We think it's important to drill down into to a detailed view and to take into account characteristics and features even when using a generic model.

Nevertheless, this article suffers from several limits. First of all, it addresses a specific case study and has to be extended in two ways. It has to be tested on other Nation's air forces (e.g. the USA or the UK). Secondly, this research process will have to be conducted in business areas such as the stock market place. Thirdly, we did not study the fit between decision maker and his task. This relation has to be investigated especially in business situation because ethical issues are numerous as financial sector demonstrate. Lastly, a snapshot is interesting but collected data imagery may provide new insights. As a result, this study will have to be repeated periodically in order to provide a dynamic picture of the evolution of fit criteria. For example, we have the intuition that this type of fit could be analysed in terms of coping strategies (Kishita & Shimada, 2011; Lazarus & Folkman, 1984). The occurrence of such strategies could be observed and assessed and thus could lead to a broader definition of fit criteria.

REFERENCES

Alter, S. (1977). A Taxonomy of Decision Support Systems. *Sloan Management Review*, *19*(1), 39–57.

Alter, S. (2004). A work system view of DSS in its fourth decade. *Decision Support Systems, 38*(3), 319–327. doi:10.1016/j.dss.2003.04.001

Anthony, R. N. (1965). *Planning and Control Systems: A framework for analysis.* Cambridge, MA: Harvard University Graduate School of Business Administration.

Arnott, D., & Pervan, G. (2005). A critical analysis of decision support systems research. *Journal of Information Technology, 20*(2), 67–68. doi:10.1057/palgrave.jit.2000035

Arnott, D., & Pervan, G. (2008). Eight key issues for the decision support systems discipline. *Decision Support Systems, 44*(3), 657–672. doi:10.1016/j.dss.2007.09.003

Aubry, M., Lièvre, P., & Hobbs, B. (2010). Project management in extreme environments. *Project Management Journal, 41*(3), 2–4. doi:10.1002/pmj.20185

Baker, J., Jones, D., & Burkman, J. (2009). Using Visual Representations of Data to Enhance Sensemaking in Data Exploration Tasks. *Journal of the Association for Information Systems, 10*(7), 533–559.

Carneiro Moro, S. M., Ribeiro Cortez, P. A., & Ferreira Rita, P. M. R. (2014). Business intelligence in banking: A literature analysis from 2002 to 2013 using text mining and latent Dirichlet allocation. *Expert Systems with Applications, 42*(3), 1314–1324. doi:10.1016/j.eswa.2014.09.024

Caron, F., Vanthienen, J., & Baesens, B. (2013). Comprehensive rule-based compliance checking and risk management with process mining. *Decision Support Systems, 54*(3), 1357–1369. doi:10.1016/j.dss.2012.12.012

Chae, B., & Poole, M. S. (2005). Mandates and technology acceptance: A tale of two enterprise technologies. *The Journal of Strategic Information Systems, 14*(2), 147–166. [REMOVED HYPERLINK FIELD] doi:10.1016/j.jsis.2005.04.001

Chen, H., Chiang, R. H. L., & Storey, V. C. (2012). Business Intelligence and Analytics: From Big Data to Big Impact. *Management Information Systems Quarterly, 36*(4), 1165–1188. Retrieved from http://dblp.uni-trier.de/db/journals/misq/misq36.html#ChenCS12

Chwolka, A., & Raith, M. G. (2012). The value of business planning before start-up — A decision-theoretical perspective. *Journal of Business Venturing, 27*(3), 385–399. doi:10.1016/j.jbusvent.2011.01.002

Cialdini, R. B., & Goldstein, N. J. (2004). Social Influence: Compliance and Conformity. *Annual Review of Psychology, 55*(1), 591–621. doi:10.1146/annurev.psych.55.090902.142015 PMID:14744228

Collins, J., Ketter, W., & Gini, M. (2010). Flexible decision support in dynamic inter-organisational networks. *European Journal of Information Systems, 19*(4), 436–448. doi:10.1057/ejis.2010.24

Cunha, M. P. E., Clegg, S. R., & Kamoche, K. (2006). Surprises in management and organization: Concept, sources and a typology. *British Journal of Management, 17*(4), 317–329. doi:10.1111/j.1467-8551.2005.00470.x

Dahal, K., Almejalli, K., & Hossain, M. A. (2013). Decision support for coordinated road traffic control actions. *Decision Support Systems, 54*(2), 962–975. doi:10.1016/j.dss.2012.10.022

DeSanctis, G., & Gallupe, B. (1987). A foundation for the study of group decision support systems., *33*(12), 1589–1609.

Dilla, W., Janvrin, D. J., & Raschke, R. (2010). Interactive Data Visualization: New Directions for Accounting Information Systems Research. *Journal of Information Systems*, *24*(2), 1–37. doi:10.2308/jis.2010.24.2.1

Dubey, G., & Moricot, C. (2006). *La polyvalence du Rafale ou l'objet total : la relation entre une technologie nouvelle et ses utilisateurs* (Vol. 81). C2SD.

Eisenhardt, K. M., & Graebner, M. E. (2007). Theory Building From Cases: Opportunities and Challenges. *Academy of Management Journal*, *50*(1), 25–32. doi:10.5465/AMJ.2007.24160888

Endsley, M. R. (1995). Toward a theory of situation awareness. *Human Factors*, *37*(1), 32–64. doi:10.1518/001872095779049543

Eom, S. B. (2004). *The Changing Structure of Decision Support Systems Research: An Empirical Investigation through Author Cocitation Mapping*. The 2004 IFIP International Conference on Decision Support Systems, Prato, Tuscany.

Eom, S. B., Lee, S., & Kim, E. B. (1998). A survey of decision support system applications (1988-1994)., *49*(2), 109–120.

Eric, B. (2009). Decisions 2.0: the Power of Collective Intelligence. *MIT Sloan Management Review*, *50*(2), 45. Retrieved from http://proquest.umi.com/pqdlink?did=1625861001&Fmt=7&clientId=9268&RQT=309&VName=PQD

Faraj, S., & Xiao, Y. (2006). Coordination in fast-response organizations. *Management Science*, *52*(8), 1155–1169. doi:10.1287/mnsc.1060.0526

Gallupe, R. B., & DeSanctis, G. (1988). Computer-based support for group problem-finding: An experimental investigation, *12*(2), 277–296.

Gelderman, M. (2002). Task difficulty, task variability and satisfaction with management support systems. *Information & Management*, *39*(7), 593–604. doi:10.1016/S0378-7206(01)00124-0

Goodhue, D. L. (1995). Understanding user evaluations of information systems. *Management Science*, *41*(12), 1827–1844. doi:10.1287/mnsc.41.12.1827

Goodhue, D. L., Klein, B. D., & March, S. T. (2000). User evaluations of IS as surrogates for objective performance. *Information & Management*, *38*(2), 87–101. doi:10.1016/S0378-7206(00)00057-4

Gorry, G. A., & Scott Morton, M. (1971). A Framework for Management Information Systems. *Sloan Management Review*, *12*(1), 55–70.

Guo, X., & Lim, J. (2007). Negotiation support systems and team negotiations: The coalition formation perspective. *Information and Software Technology*, *49*(11-12), 1121–1127. doi:10.1016/j.infsof.2006.11.006

Hodgkinson, G. P., Bown, N. J., Maule, A. J., Glaister, K. W., & Pearman, A. D. (1999). Breaking the frame: An analysis of strategic cognition and decision making under uncertainty. *Strategic Management Journal*, *20*(10), 977–985. doi:10.1002/(SICI)1097-0266(199910)20:10<977::AID-SMJ58>3.0.CO;2-X

Işık, Ö., Jones, M. C., & Sidorova, A. (2013). Business intelligence success: The roles of BI capabilities and decision environments. *Information & Management*, *50*(1), 13–23. doi:10.1016/j.im.2012.12.001

Jarupathirun, S., & Zahedi, F. M. (2007). Exploring the influence of perceptual factors in the success of web-based spatial DSS. *Decision Support Systems*, *43*(3), 933–951. doi:10.1016/j.dss.2005.05.024

Junglas, I., Abraham, C., & Ives, B. (2009). Mobile technology at the frontlines of patient care: Understanding fit and human drives in utilization decisions and performance. *Decision Support Systems*, *46*(3), 634–647. doi:10.1016/j.dss.2008.11.012

Kahneman, D., & Klein, G. (2009). Conditions for intuitive expertise: A failure to disagree. *The American Psychologist*, *64*(6), 515–526. doi:10.1037/a0016755 PMID:19739881

Kelder, J. C. (2005). Using someone else's data: Problems, pragmatics and provisions. *Forum Qualitative Sozial Forschung*, *6*(1), 39.

Kishita, N., & Shimada, H. (2011). Effects of acceptance-based coping on task performance and subjective stress. *Journal of Behavior Therapy and Experimental Psychiatry*, *42*(1), 6–12. doi:10.1016/j.jbtep.2010.08.005 PMID:21074000

Klein, G. (1998). *Sources of Power How People Make Decisions*. Cambridge: MIT Press.

Langley, A. (1999). Strategies for theorizing from process data. *Academy of Management Review*, *24*(4), 691–710.

Lazarus, R. S., & Folkman, S. (1984). *Stress, appraisal, and coping*. New York: Springer.

Lebraty, J. F., & Godé-Sanchez, C. (2010). Information System and Aerial Warfare: Enhancing Decision Making in Extreme Circumstances. In International Council of the Aeronautical Sciences 2010 (pp. 1–10). Nice, France.

Lebraty, J.-F., & Lobre, K. (2013). *Crowdsourcing: One Step Beyond*. Wiley-ISTE. doi:10.1002/9781118760765

Lipshitz, R., Klein, G., & Orasanu, J. (2001). Focus Article. *Taking Stock of Naturalistic Decision Making*, *352*, 331–352.

Massetti, B. (1996). An empirical Examiniation of the Value of Creativity Support Systems on Idea Generation. *Management Information Systems Quarterly*, *20*(1), 83–97. doi:10.2307/249543

Mintzberg, H. (1994). The Fall and Rise of Strategic Planning. *Harvard Business Review*, *72*(1), 107–114. doi:10.1016/0024-6301(94)90173-2

Mintzberg, H., Raisinghani, D., & Théorêt, A. (1976). The structure of un-structured decision processes. *Administrative Science Quarterly*, *21*(2), 246–275. doi:10.2307/2392045

Newell, A., & Shaw, J. C. (1972). The process of creative thinking. In A. Newell & H. A. Simon (Eds.), *Human problem solving* (pp. 144–174). New Jersey, USA: Practice Hall, Englewood Cliff.

Newell, A., & Simon, H. A. (1976). Computer science as empirical inquiry: Symbols and search. *Communications of the ACM*, *19*(3), 113–126. doi:10.1145/360018.360022

Parkes, A. (2013). The effect of task–individual–technology fit on user attitude and performance: An experimental investigation. *Decision Support Systems*, *54*(2), 997–1009. doi:10.1016/j.dss.2012.10.025

Randel, J. M., Pugh, H. L., & Reed, S. K. (1996). Differences in expert and novice situation awareness in naturalistic decision making. *International Journal of Human-Computer Studies*, *45*(5), 579–597. doi:10.1006/ijhc.1996.0068

Recio-García, J. A., Quijano, L., & Díaz-Agudo, B. (2013). Including social factors in an argumentative model for Group Decision Support Systems. *Decision Support Systems*, *56*, 48–55. doi:10.1016/j.dss.2013.05.007

Robey, D. (1983). Cognitive style and DSS design: a comment on Huber's paper., *29*(5), 580–582.

Rowe, F. (2005). Are decision support systems getting people to conform? the impact of work organisation and segmentation on user behaviour in a French bank. *Journal of Information Technology*, *20*(2), 103–116. doi:10.1057/palgrave.jit.2000042

Schwenk, C. R. (1984). Cognitive simplification processes in strategic decision-making. *Strategic Management Journal*, *5*(2), 111–128. doi:10.1002/smj.4250050203

Shaft, T. M., & Vessey, I. (2006). The Role of Cognitive Fit In The Relationship Between Software Comprehension And Modification. *Management Information Systems Quarterly*, *30*(1), 29–55.

Simon, H. A. (1960). *The New Science of Management Decision*. Harper Brothers. doi:10.1037/13978-000

Simon, H. A., Dantzig, A., Hogarth, G. B., Plott, C. R., Raiffa, H., Schelling, T. C., & Winter, S. et al. (1987). Decision Making and Problem Solving. *Interfaces*, *17*(5), 11–32. doi:10.1287/inte.17.5.11

Smith, S., Winchester, D., Bunker, D., & Jamieson, R. (2010). Circuits of Power: A Study of Mandated Compliance to an Information Systems Security De Jure Standard in a Government Organization. *Management Information Systems Quarterly*, *34*(3), 463–486.

Standing, C., Sims, I., & Love, P. (2009). IT non-conformity in institutional environments: E-marketplace adoption in the government sector. *Information & Management*, *46*(2), 138–149. doi:10.1016/j.im.2008.10.001

Strauss, A., & Corbin, J. (2008). *Basics of Qualitative Research Techniques and Procedures for Developing Grounded Theory* (3rd ed.). London: Sage Publication.

Thompson, R. L., Higgins, C. A., & Howell, J. M. (1991). Personal Computing: Towards a Conceptual Model of Utilization. *Management Information Systems Quarterly*, *15*(1), 125–143. doi:10.2307/249443

Todd, P., & Benbasat, I. (2000). The Impact of Information Technology on Decision Making: A cognitive Perspective. In R. Zmud (Ed.), *Framing the Domains of IT Management* (pp. 1–14). Pinaflex.

Turban, E. (1995). *Decision Support and Expert Systems : Management Support Systems*. New Jersey, USA: Prentice Hall PTR.

Tversky, A., & Kahneman, D. (1986). *Rational Choice and the Framing of Decisions*, *59*(4), 251–278.

Van Maanen, J., Sorensen, J., & Mitchell, T. R. (2007). Introduction to special topic forum: The interplay between theory and method. *Academy of Management Review*, *32*(4), 1145–1154. doi:10.5465/AMR.2007.26586080

Vankatesh, V., & Davis, F. (2000). A Theoretical Extension of the Technology Acceptance Model: Four Longitudinal Field Studies. *Management Science*, *46*(2), 186–204. doi:10.1287/mnsc.46.2.186.11926

Vessey, I. (1991). Cognitive fit: A theory based analysis of the graphs versus tables literature. *Decision Sciences*, *22*(2), 219–241. doi:10.1111/j.1540-5915.1991.tb00344.x

Wallace, L., Hui, L., & Cefaratti, M. A. (2011). Information Security and Sarbanes-Oxley Compliance: An Exploratory Study. *Journal of Information Systems*, *25*(1), 185–211. doi:10.2308/jis.2011.25.1.185

Watson, H. J., Rainer, R. K. Jr, & Koh, C. E. (1991). Executive Information Systems: A Framework for Development and a Survey of Current Practices. *Management Information Systems Quarterly*, *15*(1), 13–30. doi:10.2307/249431

Weick, K. E., & Sutcliffe, K. M. (2007). *Managing the unexpected : resilient performance in an age of uncertainty* (2nd ed.). San Francisco: Jossey-Bass.

Yuan, Y., Archer, N., Connelly, C. E., & Zheng, W. (2010). Identifying the ideal fit between mobile work and mobile work support. *Information & Management*, *47*(3), 125–137. doi:10.1016/j.im.2009.12.004

Zack, M. H. (2007). The role of decision support systems in an indeterminate world. *Decision Support Systems*, *43*(4), 1664–1674. doi:10.1016/j.dss.2006.09.003

Zhang, J., & Norman, D. A. (1994). Representations in distributed cognitive tasks. *Cognitive Science*, *18*(1), 87–122. doi:10.1207/s15516709cog1801_3

Zigurs, I., & Khazanchi, D. (2008). From Profiles to Patterns: A New View of Task-Technology Fit. *Information Systems Management*, *25*(1), 8–13. doi:10.1080/10580530701777107

KEY TERMS AND DEFINITIONS

Cognitive Fit: This Fit model explains that if the way a task is displayed is adapted to the decision maker's mental image of this task, then the performance of the decision maker will be high.

Decision Support Systems: an IT system that helps a decision maker. In some cases, such system may totally replace decision makers, in other cases, they assist them. It depends on the nature of the decision problem.

Fit Models: The goal is to understand how to assess the impact of a system on performance. But the notion of performance is hard to define. So, Fit models focus on the components of the systems. If these components are aligned or well adapted all together, it is assumed that performance will be improved.

Managerial Perspective: In management science, it's essential to clearly show how academic results may be applied to real business situations. These managerial implications may be implemented directly by a manager to his own specific context.

Military Applications: IT solutions that are dedicated to extreme situations. These applications may be used in a military context but may be applied to others contexts as high mountains.

Naturalistic Decision Making: Created in the early 90's by Gary Klein, this new approach of decision making focus on how an expert decide in a complex and risky context. This approach is based upon how in natural setting an expert decision maker behave. In that case, decision making is not a choice between options but depends on the early recognition of the situation by the decision maker.

Task Technology Fit: This Fit model explains that if a task is adapted to a technology and to its user, then the performance of this technology will be high.

ENDNOTE

1 On October 2014, 27th - http://www.nasdaq.com/article/could-apple-stock-survive-a-market-crash-cm406390

APPENDIX: THE CODING STRUCTURE

Table 11.

Codes	Sources 18*	Proportion of Respondents
Perceived Changes in Air Combat Missions with Link 16	Total 18	100%
• Mission achievement improvements	18	100%
o In terms of communication	13	75%
▪ Volume of communication	12	67%
▪ Modes of communication	5	33%
o In terms of information management	12	67%
▪ Visualization of the tactical situation	13	75%
▪ Management of dataflow	9	50%
▪ Information screening	1	6%
o In terms of aerial tactics	7	42%
o In terms of workload allocation	5	33%
▪ Between the crew members	4	25%
▪ Between the patrol aircrafts	3	17%
• Enhancement of decision making performance	16	92%
o Reduction of cognitive workload	13	75%
o Situational awareness improvement	12	67%
o Greater efficiency in combat engagements	10	58%
o OODA loop reduction	4	25%
• Changes in work practices	5	33%
New Requirements Introduced by Link 16	Total 18	100%
• Informational issues	9	50%
o Information overload	9	50%
o Data confidentiality	1	8%
o Information ambiguity	1	8%
• Degraded environment	7	42%
• Technical issues	5	33%
o Data merging	5	33%
o Data synchronization	1	6%
o Over-confidence in the system	1	6%

The eighteen coded sources concern: 9 Rafale pilots, 5 Rafale navigators (primary analysis), 2; Rafale pilots (secondary analysis), 1 Rafale test pilot, 1 AWACS mission commander.

Section 2

The Use of Mobile Technology in Human Daily Life

Chapter 4
Giving Up Smoking Using SMS Messages on your Mobile Phone

Silvia Cacho-Elizondo
IPADE Business Schoolm, France

Niousha Shahidi
EDC Paris Business School, France

Vesselina Tossan
CNAM, France & EDC Paris Business School, France

ABSTRACT

The current tendency to use cell phones or other mobile devices for healthcare purposes offers a huge opportunity to improve public health worldwide. In that direction, mobile devices make it easier to offer coaching services through text/video messages, to support individuals trying to break addictions such as smoking. Given that use of such services is still low in France and other countries, it is important to have greater understanding of what leads users to adopt them. Therefore, we propose and validate an explanatory model for the intention to adopt a mobile coaching service to help people to stop smoking. This chapter uses the concepts of vicarious innovativeness, social influence, perceived monetary value, perceived enjoyment, and perceived irritation.

INTRODUCTION

Addictions to tobacco, alcohol, drugs, over-eating, caffeine and pathological gambling are a serious problem for society. Breaking the habit takes enormous willpower, and in many cases the help of therapists or support groups. The rising popularity of smartphones, has led to a dramatic increase in mobile services through apps. The current tendency to use cell phones or other mobile devices for health offers a very interesting opportunity to improve public health worldwide (Stanford Social Innovation Review, 2011). One such service could provide support for giving up smoking. According to the report of PwC (2013), by 2017, mHealth has the potential to save 2.6 billion EUR by helping people quit smoking. This type of service is relatively new in France, hence the relevance of studying the profile of potential adopters.

DOI: 10.4018/978-1-4666-8450-8.ch004

In France, almost 66,000 deaths each year are directly attributable to smoking (5 million in the World), which is the primary cause of avoidable premature death, and the problem of nicotine addiction continues to grow despite efforts to curb it (INPES, 2007). Across the whole French population aged 15-75, the proportion of daily smokers rose from 26.9% to 28.7% between 2005 and 2010, and cigarette sales saw a slight upturn between 2008 and 2009 (from 53.6 billion to 55 billion packets) after dropping significantly between 2001 and 2004 (from 82.5 billion to 54.9 billion, due to substantial increases in the price of tobacco products). However the proportion of smokers that smoke more than ten cigarettes a day is diminishing. The French smokers are usually very young and represent 50% of the smoker people (about 94% of the smokers in EU start smoking before they turn 25). Considering 15 million of French smokers (more than 100 million in the world), more than half would like to stop smoking. Only 750 000 (5% of the smoker people) people stop smoking each year. More than 2 million of smokers used skin patches, nicotine substitute or pharmacological processing in 2010.

Tobacco companies are investing in a new generation of smokeless alternatives to cigarettes as the industry faces growing regulatory threats across the globe. The world's four biggest tobacco companies outside China –Philip Morris International, British American Tobacco, Japan Tobacco International and Imperial Tobacco – are positioning themselves for an increasingly smoke-free future as they seek to entice smokers to non-combustible substitutes such as electronic cigarettes, tobacco vaporizers and nicotine inhalers over the next decade (Wembridge & Thompson, 2012). Nevertheless, the benefits of electronic cigarettes are still controversial. Coaching people to avoid taking a cigarette or an e-cigarette seems to be a better option. According to PcW estimates, out of the 102 million smokers, 48.8 million can potentially use mHealth solutions regularly and 3.9 million smokers could quit smoking successfully. Since January 2010, Health & Human Services (HHS) has invested $5 million dollars to develop its eHealth / mHealth smoking cessation resources aimed at increasing quitting attempts among teens, young adults and adults (Merill, 2011a).

This chapter concerns a mobile coaching service providing support for people trying to stop smoking. The service takes the form of short text messages (SMS or MMS) sent to cell phones to help individuals in a range of situations or anti-smoking activities. The principal objective of this study is to identify drivers fostering the intention to adopt such a service in the young smokers segment in France. The chapter is structured as follows. Firstly, the conceptual framework is presented and after that the model of the intention to adopt the mobile coaching service is introduced. The methodology is then described along with the operationalization of the underlying hypotheses. After reporting the main findings and managerial and social implications, the chapter concludes by considering limitations and avenues for future research.

CONCEPTUAL FRAMEWORK

The effectiveness of a mobile coaching service has already been tested in various countries, including New Zealand where a program to stop smoking was developed and tried out (Whittaker et al., 2008). But such mobile coaching services are relatively little used in France, and this is why they are considered as an innovation for the purposes of this study. Several definitions of an innovation have been proposed. The one used here is by Rogers (1962), who defines an innovation as an idea, practice or object perceived as new by the individual. Diffusion of an innovation is the process by which it is communicated through certain channels over time among the members of a social system (Rogers, 1962). Rogers identifies three factors that explain how an innovation spreads and is adopted: 1) the characteristics of the product or service, 2) the characteristics of consumers and 3) the profiles of different adopter categories through the innovation diffusion process.

Adoption of an innovation can be defined as the initial purchase or repeated purchase of the innovation, depending on the context. For frequent-purchase products, repetition of the purchase is necessary to consider a product adopted, and the threshold of three purchases appears to be an acceptable threshold for judging whether a product has been adopted by the consumer (Cestre, 1996), whereas for durable goods and services, adoption is generally considered to take place from the very first purchase, regardless of regular use or replacement purchases (Le Nagard-Assayag & Manceau, 2011). Gatignon and Robertson (1985) propose a general model of innovation diffusion, taking the conceptual bases proposed by Rogers (1983) (which include the concept of innovation, diffusion over time, influence of interpersonal communication and opinion leaders, the adoption process, the role of innovators and other adopter categories, and the social system in which the diffusion takes place) and adding the influence of marketing campaigns and competitors' actions.

D'Hauteville (1994) develops an attitudinal model of factors that enhance the acceptability of an innovation for consumers that he seeks to explain not only by consumer characteristics but also by product attributes. His empirical studies confirm the results of other studies on innovation, particularly the finding of Ostlund (1974) that perceived attributes play a decisive, more important role in the acceptability of a new product than individual variables, including the specific variable of innovativeness. He also confirms the predominant role of consumer habits in forming attitudes.

Consumer Innovativeness

The concept of consumer innovativeness is not unanimously accepted among researchers (Cestre, 1996; Masson, 2010). Some (Hurt, Joseph & Cook, 1977; Foxall & Haskins, 1986; Venkatraman & Price, 1990) see it as a core personality trait, possessed to varying degrees by all individuals. Midgley and Dowling (1993), Roehrich (1994) and Le Louarn (1997) consider it at the level of all consumer products as a consumer's *generalized unobservable predisposition to purchase new products and brands* rather than sticking to habitual choices and models (Midgley & Dowling, 1993). Others (Goldsmith & Hofacker, 1991; Goldsmith, d'Hauteville & Flynn, 1998) see innovativeness at the level of a product category as a combination of personality traits and attitudes.

For Subin, Mason and Houston(2007), empirical studies paint an inconsistent picture of the relationship between the consumer's innovative predisposition and an innovative behavior (adoption of a new product). They demonstrate that innate innovativeness does not influence adoption behavior *directly*, but *indirectly* through two of the three components of indirect or vicarious innovativeness (defined later), namely modeling and involvement in word of mouth, but not exposure to advertising.

Hirschman (1980) identifies three types of innovativeness: *adoptive innovativeness, indirect innovativeness* or *vicarious innovativeness* and *use innovativeness*. Adoptive innovativeness concerns the actual adoption of a new product. Vicarious innovativeness can be defined as *openness to information on new products of any kind*. This vicarious innovativeness is followed by adoptive innovativeness when consumers actually adopt the product. Finally, use innovativeness involves solving consumer problems with products already available to the consumer. In other words, the consumer invents new uses for an existing product. For the purposes of this study the focus is on the vicarious innovativeness specific to a given field, i.e. the consumer's attitude to new services, namely *mobile coaching services via cell phone*.

Intention to Adopt the Mobile Coaching Service

It is established that intention is a good predictor of actual behavior (Davis, Bagozzi & Warshaw 1989; Ajzen, 1991; Venkatesh & Brown, 2001). Intention is determined by the attitudes to that behavior, and by subjective norms (Ajzen & Fishbein, 1980). Subjective norms relate to the way the subject perceives the opinion of people important to him/her about his/her decision to engage or not engage in a certain behavior. When an information technology-based service is in the early stages of diffusion, as is the case for mobile coaching services to help people stop smoking, the intention to adopt appears a more appropriate object of study than adoptive behavior (Hong & Tam, 2006). This is why we seek to explain the intention to adopt rather than adoptive behavior.

The Technology Acceptance Model (TAM) constructed by Davis (1989) is an adaptation of the theory of Reasoned Action (Fishbein & Ajzen, 1975) designed to model the intention to adopt information systems. Perceived usefulness is defined as the degree to which a person believes that using a particular technology would enhance his or her job performance (Davis, 1989). It determines both the attitude and the intention to use. The perceived ease-of-use is defined as the degree to which a person believes that understanding and using a particular technology would be free from effort (Davis, 1989). This leads the individual towards a perceived usefulness but also to an attitude that directly affects the intention to use (Davis, Bagozzi & Warshaw, 1989, p. 985).

Later, Hong & Tam (2006) showed that the perceived usefulness exercises a direct positive influence on the intention to adopt multi-purpose information services. Information services are devices or instruments that are used to supply users with various types of information: data, video, images (Bergman, 2000). Hong and Tam (2006) define multi-purpose information services as information technology (IT) artifacts that: 1) improve the personal, individual link with the user, 2) offer mobile services and 3) supply a suite of functions for both work and leisure needs. These authors also demonstrate that the perceived monetary value exerts a positive influence on the intention to adopt these services, and that the perceived enjoyment has a positive influence on the intention to adopt. Their study also confirms that the perceived enjoyment has a positive influence on perceived usefulness and perceived ease-of-use, and that social influence has a positive influence on the intention to adopt.

Explanatory Factors for the Intention to Adopt a Mobile Coaching Service and Hypotheses

- **Vicarious Innovativeness:** Vicarious innovativeness towards cell phone coaching services should influence the intention to adopt a specialist service such as help with stopping smoking. Therefore, we propose the following hypothesis:

H1: Vicarious innovativeness towards coaching services exerts a direct, positive influence on the intention to adopt a text-message based mobile coaching service to stop smoking.

Apart from the innovativeness related to the above field, other variables may influence the adoption process. In this study, we consider social influence, perceived monetary value, perceived enjoyment, perceived irritation and demographic profile. Future research could examine the influence of other variables.

- **Social Influence:** This is defined as the measure of a subject's belief of whether significant referents will approve or disapprove of the subject adopting a certain behavior (Ajzen, 1991).This social influence is exerted through messages and signals that help to form perceptions concerning the value of a product or activity (Venkatesh & Brown, 2001). Various studies confirm that peers, superiors and family members all influence consumer behavior and the decisions associated with adoption of products or services (Venkatesh & Brown, 2001; Hong & Tam, 2006).

In the theory of planned behavior, intention, as a determinant of behavior, is conditioned by attitude, subjective norms (or social influence) and perceived behavioral control, which is a kind of perceived ease of adoption (Ajzen, 1991). Mathieson (1991) developed a scale to measure this social influence that was used by Hong & Tam (2006) to demonstrate that social influence has a direct, positive effect on the intention to adopt mobile data services. They define such services as a set of digital data services accessed by a mobile device across a vast geographical zone that can be used to exchange messages, pictures and emails, check flight times, book concert tickets and play games, making no distinction between the services and the devices used to access them. Mobile coaching services, such as support with stopping smoking, are services in which messages of encouragement are sent at random intervals to users, and are therefore a specific category of mobile services. Based on past studies, we propose the following hypothesis:

H2: Social influence (SOCIAL) has a direct, positive effect on the intention to adopt a text-message based mobile coaching service to stop smoking.

- **Perceived Monetary Value:** Most consumers encode prices along a scale and classify products as "expensive" or "cheap" even when they are not familiar with the products concerned. This happens because individuals call on references associated with similar experiences. In general, consumers mentally consider the perceived quality and perceived sacrifice associated with a price. This consideration brings about a perceived monetary value in the consumer's mind, which has an effect on the consumer's intention to adopt a product (Dodds, Monroe, & Grewal, 1991).

When the consumer attaches high importance to the use value, he is willing to make a greater monetary sacrifice. If the consumer perceives the mobile coaching services as having a higher use value than other methods of stopping smoking then he will have the intention to adopt them. Based on these premises, we propose the following hypothesis:

H3: The perceived monetary value (MONEY) exerts a positive influence on the intention to adopt a text-message based mobile coaching service to stop smoking.

- **Perceived Enjoyment:** Seeking pleasurable, fun experiences is a typical personal desire (Rokeach, 1973). The value of enjoyment as regards use of an innovation is conceptualized as the extent to which use of an innovation is agreeable in itself, independently of the expected consequences of its performance. The variables of enjoyment and irritation are rooted in study of the affect and its influence on behavior in general and innovative behavior in particular. Several studies show that the perceived enjoyment value explains adoption of information technologies (Hong & Tam, 2006).

Venkatesh and Morris (2000) and Venkatesh and Brown (2001) argue in their research that perceived enjoyment can have an indirect impact on intention through the perceived ease-of-use. Hong and Tam (2006), meanwhile, suggest that services such as downloadable games, horoscopes or videos can help to pass the time agreeably while a person is waiting for a train. They show that the perceived enjoyment has a positive influence on the intention to adopt mobile data services. In this view, the "fun" aspect of the mobile service will facilitate its adoption and have a positive effect on attitudes. But other researchers such as Chowdhury et al. (2006) find that the influence exerted by perceived enjoyment on the attitude towards mobile advertising is negative and non-significant. To verify this, we propose the following hypothesis:

H4: The perceived enjoyment (ENJOY) exerts a positive influence on the intention to adopt a text-message based mobile coaching service to stop smoking.

- **Perceived Irritation:** Based on the fact that mobile advertising may be perceived by the consumer as invasive, irritating or boring, Chowdhury et al. (2006) propose a scale of perceived irritation, which is not the same thing as the absence of perceived enjoyment. They set out to demonstrate that this construct is negatively associated with the consumer's attitude to mobile advertising.

However, their research leads to the opposite finding, namely that the *influence of irritation on the attitude to mobile adverting is positive but non-significant.* As regards the intention to adopt a mobile coaching service for support while stopping smoking, this leads us to propose the following hypothesis:

H5: Perceived irritation (IRRITA) exerts a negative influence on the intention to adopt a text-message based mobile coaching service to stop smoking.

Considering the conceptual framework developed above, we propose the following model in Figure 1. The following hypotheses compare different groups of individuals:

H6a: The intention to adopt the Mobile Coaching Service differs between individuals who have never tried to stop smoking before and the others.

H6b: The effect of perceived enjoyment is greater in individuals who have never tried to stop smoking before than in others because others know how difficult it may be to stop and not to start again

As this is a model of intention to adopt a technological service, it would appear interesting to compare the model by gender (Venkatesh & Morris, 2000). Gender differences are often addressed in health reports about smokers (Griffith et al, 2012;Briones-Vozmediano et al, 2012).

H7: The impact of vicarious innovativeness on the intention to adopt the mobile coaching service differs according to gender.

Figure 1. Model of adoption of a text-message based mobile coaching service

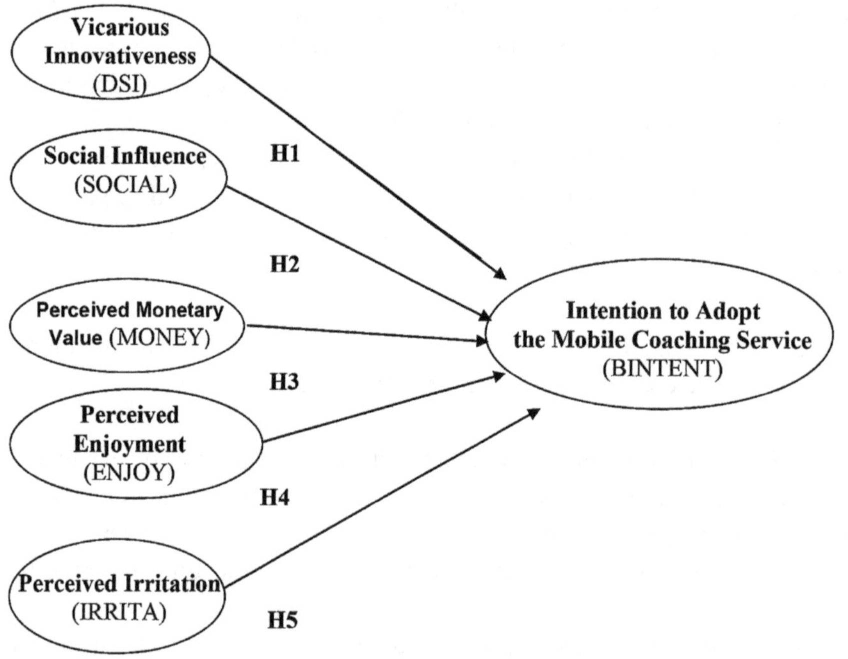

METHODOLOGY

This research originated in a joint project with a French start-up company called *e-Medicis*, which was considering introducing a new service to provide support for people who wanted to stop smoking. The study took the form of a face-to-face survey with a convenience sample of 113 people in several Paris locations. The objectives were to measure vicarious innovativeness as regards mobile services, to measure the intention to adopt the mobile coaching service for stopping smoking, and to validate the hypotheses. The analysis also covered the demographic profile of participants, their smoking behavior (heavy smokers/medium smokers/light smokers/non-smokers).

Below is the analysis of the data collected through the face-to-face survey with smokers. The description of the mobile coaching service tested in the survey is provided in Exhibit 1. The analysis done used the structural equations method. Six 5-point Likert scales from "totally disagree" to "totally agree" were used to measure the variables in the model. The scales are described, together with the authors that inspired them, in Table 1.

- **Intention to Adopt the Mobile Coaching Service** (BINTENT). To measure this construct, we used a French adaptation of the scale proposed by Davis et al (1989), itself adapted from Ajzen and Fishbein (1980). This scale consists of three items: (1) I intend to use this mobile coaching service to stop smoking in the future. (2) I would be prepared to use this mobile coaching service to stop smoking in the future. (3) To stop smoking, I will use this service.
- **Vicarious Innovativeness** (DSI). Goldsmith and Hofacker (1991) distinguish global innovativeness from domain-specific innovativeness that can be applied to a specific category of products or services. They propose a scale to measure consumers' tendency to be among the first to try out

Table 1. Measurement scales

Concept	Label	Items	Adapted From
Intention to Adopt the Mobile Coaching Service	BINTENT	1. I intend to use this mobile coaching service to stop smoking in the future. 2. I would be prepared to use this mobile coaching service to stop smoking in the future. 3. To stop smoking, I will use this service	Davis (1989)
Vicarious Innovativeness	DSI	1. **Involvement:** If I heard a new mobile coaching service was available, I would be interested in using it. 2. **Intention to adopt:** I would consider adopting a new mobile coaching service even if I had not yet heard about it. 3. **Perceived knowledge:** I know more about new mobile coaching services than other people do. 4. **Ease of use:** If I heard that a new mobile coaching service was available and easy to use, I would be interested enough to adopt it. 5. **Need for change:** I adopt a new mobile coaching service because of the advantages it offers me. 6. **Need for cognition:** Before adopting a new mobile coaching service I consider the benefits provided by innovation and its relationship to the status quo. 7. **Price:** I would adopt the mobile coaching service if the price was right (intention to adopt if you have to pay and at what price level, or if the service is free, sponsored by a private or public organization).	Pagani (2007)
Social Influence	SOCIAL	1. People who are important to me would want me to use this mobile coaching service. 2. People who influence my behavior would think I should use this mobile coaching service. 3. People whose opinions I value would prefer me to use this mobile coaching service.	Mathieson (1991)
Perceived Monetary Value	MONEY	1. I expect that this mobile service would have a reasonable price. 2. This mobile service would offer good value for money. 3. I believe that at the right price, this mobile service would be good value.	Dodds, Monroe & Grewal (1991)
Perceived Enjoyment	ENJOY	1. I expect that using this mobile service would be enjoyable. 2. I expect that using this mobile service would be pleasing. 3. I expect that using this mobile service would be entertaining.	Chowdhury, Parvin, Weitenberner & Becker (2006)
Perceived Irritation	IRRITA	1. I feel that this mobile service is irritating. 2. I feel that mobile coaching services are everywhere and, there is no need for another. 3. This mobile service looks annoying.	Chowdhury, Parvin, Weitenberner & Becker (2006)

new products in a specific field. This 5-point scale consists of six items that capture the following constructs: *involvement, usage, intention to adopt, opinion seeking, perceived knowledge* and *need for change*. The scale has been validated for durables and other goods as well as for services.

In 2007, Pagani added two further indicators to this scale. The first, *ease of use*, is an indicator that emerges from the theory of reasoned action and the Technology Acceptance Model (TAM) constructed by Davis (1989). The second, the *need for cognition*, is an indicator drawn from psychological studies. Pagani (2007) uses this scale to measure the vicarious innovativeness regarding third generation (3G) mobile music services, which use technology to generate voice, images and video in any place, network or terminal. Pagani showed in her research that the scale gained reliability when the *opinion seeking* and *usage* items were eliminated. Her final modified scale consists of 6 items measuring involvement, intention to adopt, perceived knowledge, ease of use, need for change and need for cognition. In addition to these items, we include a *price* item (I will adopt a mobile coaching service if the price is right, an item used by Pagani (2007) to test the nomological validity of her scale).

- **Social Influence** (SOCIAL). Following Mathieson (1991), three items were used: (1) People who are important to me would want me to use this mobile coaching service. (2) People who influence my behavior would think I should use this mobile coaching service. (3) People whose opinions I value would prefer me to use this mobile coaching service.
- **Perceived Monetary Value** (MONEY). Following Dodds et al. (1991), three items were used: (1) I expect this mobile service would have a reasonable price. (2) This mobile service would offer good value for money. (3) I believe that at the right price, this mobile service would be good value.
- **Perceived Enjoyment** (ENJOY). We adopt the scale devised by Chowdhury et al. (2006). Four items were used: (1) I expect that using this mobile service would be enjoyable. (2) I expect that using this mobile service would be pleasing. (3) I expect that using this mobile service would be entertaining.
- **Perceived Irritation** (IRRITA). As previously, we use the scale devised by Chowdhury et al. (2006). Three items were used: (1) I feel that this mobile service is irritating. (2) I feel that mobile coaching services are everywhere and, there is no need for another. (3) This mobile service looks annoying.

The Table 1 sums up our measurement scales.

STATISTICAL ANALYSES AND RESULTS

Descriptive Analysis

The sample shows a homogeneous distribution between men (52%) and women (48%). Among the participants, 76% of them are aged between 14 and 21 (N=86). More than 80% of participants (N=92) state that they smoke every day or nearly every day. Smokers can be grouped into three categories according to the number of cigarettes smoked per day: *light smokers* (N=32, 28%), *medium smokers* (N=40, 36%) and *heavy smokers* (N=41, 36%) (see Tables 2 and 3). Of the 113 participants in the study, only 37 stated that they intended to stop smoking in the next 12 months; 76 did not share that intention. However, the average intention to adopt the service does not differ significantly between the two groups, i.e. between people who say they want to stop smoking and the rest (F=2.62 (0)).

Among the people who intend to stop smoking in the next 12 months (N=37), most are Heavy Smokers (more than 10 cigarettes a day) and Light Smokers (fewer than 5 cigarettes a day). Medium Smokers (5 to 9 cigarettes a day) have proportionally less intention of giving up smoking; probably because, according to the arguments collected in the exploratory study they are less worried about their budget and their health, or are under less pressure from their entourage.

Table 2. Classification of smokers by age

Age	Group Size	%
14-17	48	42.5
18-21	38	33.6
22-25	21	18.6
Over 25	6	5.3

Table 3. Classification of smokers by number of cigarettes smoked by day

	Cigarettes per Day	Number	%
Light	< 5 cigarettes	32	28
Medium	5-9 cigarettes	40	36
Heavy	>10 cigarettes	41	36

N = 113 participants.

One of the barriers to adopting the service that emerges from the study is the lack of human contact in the coaching service. This is observed more with individuals over 18 than individuals under 18. The reason may be that the latter who are digital natives and so more used to digital services without human contact. Analysis of the variance shows a significant difference between the averages for the two groups of individuals (F=2.969 and p-value=0.088). Of the 43 individuals who would not use this service because of the lack of human contact, 22 think it could help someone else to stop smoking. This is consistent with extant literature that shows that consumer acceptance of the mobile Internet is linked with a low desire for social contact (Koenigstorfer & Groeppel-Klein, 2012), the mobile Internet being a replacement of social contact (Syed & Nurullah, 2011).

Hypotheses H1 to H7 (except H6a) were tested using the structural equations technique. The maximum likelihood fit function was applied. A two-stage approach was used as recommended by Anderson and Gerbing (1988). First, the measurement instruments for the constructs were assessed by examining the reliability and validity of scales. Then, the relationships were tested.

Scale Reliability and Validity

The reliability of all instruments was tested by the Cronbach's alpha reliability coefficient (see Table 4). All coefficients are acceptable, except those associated with the perceived irritation construct (α=0.56). Consequently, the IRRITA construct was eliminated from the model and hypothesis H5 was not tested. As Table 4 shows, the Jöreskog ρ values are high (except for the IRRITA construct). Each item is better explained by the construct it relates to than by chance. Concerning convergent validity, the influence of relationships (between the measures and their construct) is statistically different from 0. The average extracted variance between a construct and its measures is always above 0.5 (except for the IRRITA construct).

To test discriminant validity, we conducted a Chi squared difference test (by reference to the difference in degrees of freedom). To test the discriminant validity, the analysis recommended by Bagozzi and Yi (1991) relies on a comparison between the χ^2 values of a model that leaves the correlations between the different constructs free, and a model in which the correlations between constructs are fixed at 1. If

Table 4. Scale reliability and validity

	DSI	Social	Money	Enjoy	Irrita	Bintent
Cronbach's alpha	0.86	0.76	0.76	0.83	0.56	0.84
Standardized item loading (> 0.5)						
Item 1	0.64	0.80	0.84	0.76	0.56	0.86
Item 2	0.84	0.84	0.87	0.79	0.72	0.89
Item 3	0.87	0.83	0.74	0.90	0.48	0.85
Item 4	0.72					
Item 5	0.50					
Item 6	0.50					
Item 7	0.76					
Average extracted variance (ρvc>0.5)	0.50	0.53	0.55	0.67	0.35	0.64
Jöreskog ρ (>0.70)	0.86	0.77	0.78	0.86	0.62	0.84

the difference is significant in view of the difference in degree of freedom, it can be concluded that the model tested is better than the constrained model and that the constructs are different. The results for the indicators used in our study are satisfactory *(difference of Chi squared= 22 and difference of ddl =6).*

Estimation of the Model

Having established that the goodness of fit is satisfactory, interpretation of the estimations of linear relationships can begin. Our model has two sub-models one to measure independent variables, one to measure dependent variables and a structural model connecting the latent dependent variables to the latent independent variables. There are several observed independent variables that depend on 4 latent variables. These latent variables are correlated. There are also 3 observed dependent variables (3 items concerning intention) that depend on a latent variable. The indicators are fairly satisfactory (see Table 5). The data thus show a satisfactory goodness of fit for the theoretical model.

This model explains 69% of the variance in intention (R^2). As shown in Table 6, the "perceived enjoyment", "vicarious innovation" and "social influence" variables make a significant contribution to the intention to adopt the mobile coaching service (critical ratio>1.96 or very similar).

Table 5. Goodness of fit

RMSEA (< 0.08)	0.07
NFI	0.77
CFI (> 0.90)	0.90
TLI (> 0.90)	0.82

Table 6. Effects of variables on the intention to adopt the mobile coaching service

Relationships to Test			Estimation	Critical Ratio
Vicarious Innovativeness	→	Intention	0.26	2.32
Social Influence	→	Intention	0.22	1.90
Perceived Monetary Value	→	Intention	-0.17	-1.43
Perceived Enjoyment	→	Intention	0.75	4.73

Consequently, hypotheses H1, H2 and H4 are validated. However, the direct influence of the "perceived monetary value" variable on intention to adopt the mobile coaching service is not significant, and consequently hypothesis H3 is not validated.

The validated model is illustrated in Figure 2.

COMPARISON OF RELATIONSHIPS BETWEEN VARIABLES IN THE SUBSAMPLES

For the first analysis two subsamples are considered:

Sample 1 consists of individuals have already tried to stop smoking (N=56).
Sample 2 consists of individuals who have never tried to stop smoking (N=57).

According to the Table 7, Perceived Monetary Value and Intention to adopt differ between individuals who have never tried to stop smoking before and the others (H6a is validated).

Table 7. Comparison between variables

Variables	Sample 1 Individuals Who Have Already Tried to Stop Smoking (N=56)	Sample 2 Individuals Who Have Never Tried to Stop Smoking (N=57)	F (df=1)	p	Eta^2	Full Sample
Vicarious Innovativeness	2.89 *(1)* .95 *(2)*	2.68 .57	1.23	.269	.01	2.79 .99
Social Influence	2.55 1.31	2.19 1.03	2.73	.101	.02	2.37 1.18
Perceived Monetary Value	3.57 1.23	3.01 1.34	5.23	.024	.05	3.29 1.31
Perceived Enjoyment	2.58 1.16	2.35 1.17	1.07	.303	.01	2.46 1.16
Intention to adopt	2.46 1.30	2.00 1.12	4.17	.043	.04	2.23 1.22

(1) The mean (2) The standard deviation

Figure 2. Result of the model tested

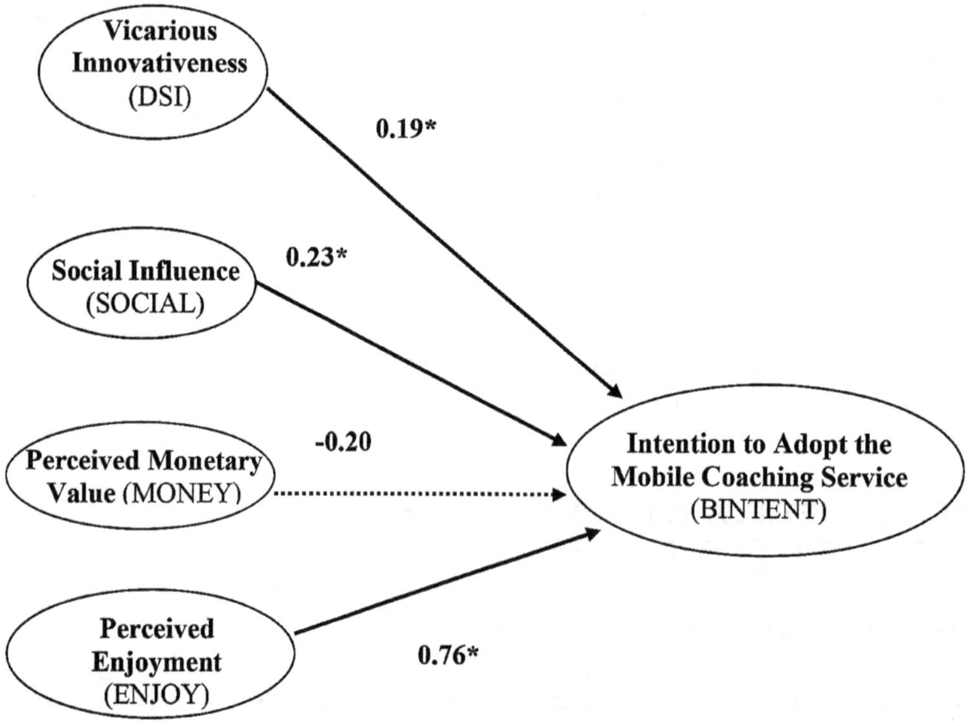

(*): Significant Coefficient (standardized effect), No asterisk: Non-significant

The data are a good fit for the model in each sample (samples 1 and 2:RMSEA=0.068, NFI, CFI and TLI close to 0.90). The percentage of variance in the intention to adopt the mobile service explained by the model is higher for individuals who have never tried to stop smoking (78.6%) than for the others (60.1%).

Table 8 reports the standardized effects of the model's variables on the intention to adopt the mobile coaching service in each of the two subsamples.

For reasons of clarity, only estimations significant at the 0.05 level are shown. Some variables are significant in one sample but not the other. The "perceived monetary value" variable, which was not significant in the full sample, becomes significant in sample 2. Perceived monetary value exerts a nega-

Table 8. Standardized effects on the intention to adopt the mobile coaching service

Variables	Sample 1 Individuals Who Have Already Tried to Stop Smoking (N=56)	Sample 2 Individuals Who Have Never Tried to Stop Smoking (N=57)
Vicarious Innovativeness	0.35	-
Social Influence	-	-
Perceived Monetary Value	-	-0.51
Perceived Enjoyment	0.52	1.01

NB: Only significant coefficients (5%) are reported

Table 9. Standardized effects on the intention to adopt the mobile coaching service

Variables	Females (N=54)	Males (N=59)
Vicarious Innovativeness	0.29	-
Social Influence	-	-
Perceived Monetary Value	-	-
Perceived Enjoyment	0.75	0.82

NB: Only significant coefficients (5%) are reported

tive effect on the intention to adopt the mobile coaching service in individuals who have never tried to stop smoking. The greater the perceived enjoyment, the more consumers who have never tried to stop smoking will have the intention to adopt the mobile coaching service (double the coefficient of the other group). Hypothesis H6b is thus confirmed.

Vicarious innovativeness has a significant effect on the intention to adopt the mobile service in individuals who have already tried to stop smoking. In individuals who have never tried to stop smoking, the intention to adopt the mobile service will depend more on the perceived enjoyment than on other variables.

Otherwise, if we compare individuals who would like to stop smoking (stopped smoking in the past or not) and the others, there is no difference.

The model tested is now estimated for the same population, this time split by gender. In both subsamples, the data show fairly satisfactory goodness of fit for the model. The portion of variance explained is approximately 65%. Table 9 reports the standardized effects of the model's variables on the intention to adopt the mobile coaching service in each of the two gender subsamples.

For reasons of clarity, only estimations significant at the 0.05 level are shown. Vicarious innovativeness has a significant effect on the intention to adopt the mobile service in women but not in men. Validation of our conceptual model is higher in the female sample than the male sample. Hypothesis H7 is validated.

CONTRIBUTIONS, MANAGERIAL IMPLICATIONS, LIMITATIONS, AND AVENUES FOR FUTURE RESEARCH

This study has broad social and managerial implications. The field of mobile health services is providing opportunities to make improvements and reduce costs in the French and wordwide health sector. Worldwide, the mobile health market has an estimated value of $50 billion (Ascari, Bakshi, & Grijpink, 2010). However, attention should not be focused solely on the business opportunities offered by the market, but rather on its social impact. Because these mobile health services can offer continuous "24/7" monitoring and have no geographical barriers, thus, mobile coaching services are open to a larger range of benefits (see Exhibit 2 and 3). With their positioning as accessible services that are complementary to other medical and psychological methods, they also offer a less formal approach to support in the fight to break addictions.

The intention to adopt the text message-based mobile coaching service to help stop smoking was tested with a mostly young target. This is precisely the group with the greatest risk of developing tobacco-related illnesses in the long term, since the lower the age at which people start smoking, the higher their risks of serious tobacco-related illnesses. Yet the study observed that these young smokers do not always intend to stop smoking in the short term. It is important to find innovative ways to motivate them to try and start the process of giving up smoking. With this in mind, mobile services can play an important role because young people are generally heavy users of cell phones and generally find new technologies attractive (Syed & Nurullah, 2011).

Two factors of influence should be taken into consideration: peer influence and the enjoyment generated by this type of remote coaching service, habitually perceived as a game. The more enjoyment consumers get out of using these services, the more they will be prepared to use them in the future. The designers of this type of service need to develop amusing messages to convince early adopters to talk about the effectiveness of the service and how much fun it was to use. Furthermore, if the service is perceived to be at a reasonable price or free because sponsored by a government health organization, the potential user will get more enjoyment out of trying it. The influence exercised by smokers' families and general entourage can also play a decisive role in diffusion of these services by their word of mouth.

The above findings also underline the fact that stimulating curiosity and the smokers' intrinsic interest are more important factors than the perceived monetary value. It would be judicious to ensure that the most favorable techniques and functions are selected, to give the consumer a higher level of perceived enjoyment and intrinsic inter-user motivation.

The negative impact of perceived monetary value for individuals who have never tried to stop smoking before could be explained by the existence of free applications. The smoker may prefer his coaching service for stopping smoking to be free, or provided free by a company or health sector body. However, it is conceivable that when a person really wants to stop smoking, the price to pay will not be a barrier to adoption.

The limitations of this study relate to its small sample size and the overrepresentation of young people, which limits its external validity. Larger-scale studies with samples of different population types (multi-country, different age brackets, different types of smoker) will still be required for closer examination of each segment's needs and preferences. Another possible bias is the low number of participants who had stated a clear intention of stopping smoking in the next 12 months. A much greater effort should have been made to recruit people intending to stop smoking in the near future.

We recommend that an online study should be undertaken with a broader sample comprising subjects of all ages, to determine how the desire (or lack of desire) to stop smoking influences the smoker's perception of the mobile coaching service. Some experience of the service would also be advisable in more advanced stages of the research, once the beta version is available, because the measure of attitude to these new services used in this study is based on respondents' own statements and does not involve them much. A measure that allows closer observation of a behavior could be more effective. It would also be useful to test the perceived enjoyment of this type of service, in order to increase numbers giving up other addictions and/or pay greater attention to other illnesses.

Another interesting avenue to explore is the impact of mobile coaching service in conjunction with e-cigarettes which could grow as greater regulation looms, for example, plain packaging proposals that would remove brands from cigarette packs (Wembridge & Thompson, 2012).

In their research Andrews, Cacho-Elizondo, Drennan and Tossan (2013) showed that perceived usefulness continually proves to be the most important predictor of intentions to use an SMS-assisted smoking intervention. Based on this, health practitioners offering an m-health service must consider its design and delivery to ensure its usefulness to the target group. For Andrew and her colleagues, this could include whether it is used as an additional component to a web-based intervention or a standalone mobile phone-based intervention.

CONCLUSION

This chapter has explored the factors that determine the intention to adopt a text message-based mobile coaching service designed to help people stop smoking. We obtained a fairly robust model in which vicarious innovativeness, social influence and the perceived enjoyment exert a positive influence on the intention to adopt the mobile coaching service. The perceived enjoyment has more influence on the intention to adopt the mobile coaching service in smokers who have never tried to stop smoking before than in smokers who have already tried to stop smoking.

At the methodological level, this research contributes with an improved approach to test the acceptance and potential adoption of new services, not yet available, through the use of a scenario script where the mobile coaching service to quit smoking is described (Exhibit 1).

Kihyun, Gyenung-Min, and Eun Sook (2009) concluded in their research that adoption of an amusing mobile service is a direct reflection of users' way of life insofar as that adoption is in line with socio-economic conditions. In future research, it would be interesting to explore further both the enjoyment aspect of the service and the socio-economic conditions of potential adopters. On this point, male-female differences also require more attention, as our model turned out to be more suited to women. One possible explanation for this could be the level of curiosity associated with this segment, and its usage of mobile applications.

Steenkamp and Gielens (2003) found in their study that the probability of trying an innovation is higher in individuals with a high degree of innovativeness than those with a low degree of innovativeness. These authors compared consumer innovativeness as measured by the attitude scale with innovativeness as measured by affective purchasing behavior for new products, and observe that the results by either measure are equivalent.

Another factor for consideration is the level of involvement. Bloch (1982) and Valette-Florence (1989) noted that a high-involvement individual will have a greater propensity to adopt an innovative behavior than a low-involvement individual.

An additional lesson to be drawn from this study is the discovery of smokers' preference for multimedia communications including images and videos. This trend would certainly help to compensate for the lack of human contact that has often been considered as one of the weaknesses of this type of remote coaching service. A further online study of people from different countries could measure how intending (or not intending) to give up smoking in the short term influences the perception of a mobile coaching service to help people stop smoking. It could also measure the impact of cultural and socio-economic factors, including isolation.

According to Varshney (2014), "Mobile technologies allow information to be made available quickly, but cannot improve the quality of information it is presenting to the patients or healthcare professionals". It remains still true for a mobile phone smoking cessation programme. Therefore, an important challenge is to have a medical approach while coaching people what is not actually verified in the free mobile coaching services.

In addition, it would be interesting to replicate this study with people who want to break other addictive habits such as excessive consumption of medicines, drugs or food or alcohol, or people suffering from health problems such as depression. Support could be provided through prerecorded messages sent by cell phone alone, or in conjunction with other media, for instance as a complement to an internet application, personalized coaching, or treatments such as nicotine patches to stop smoking. In a more advanced stage of research, the service itself could be tried out, to determine the value of using theoretical approaches to examine factors influencing the perceptions and attitudes of participants who use the service.

The future of mobile health or m-health services is promising, it is necessary to continue encouraging further interdisciplinary research to improve them and spread their adoption and wide application in the health sector worldwide. Special focus should be given to developing countries where m-health services or apps could have a significant impact on low-income sensible groups.

ACKNOWLEDGMENT

The authors acknowledge the helpful comments and suggestions of Professor Jean-Pierre Helfer, Dean and Professor at IAE, Paris Panthéon Sorbonne University, President of the Scientific Council at EDC Paris Business School. We would additionally like to thank anonymous referees for their comments, which strengthened this chapter and the support received from our institutions.

REFERENCES

Ajzen, I. (1991). The Theory of Planned Behavior. *Organizational Behavior and Human Decision Processes*, *50*(2), 179–212. doi:10.1016/0749-5978(91)90020-T

Ajzen, I., & Fishbein, M. (1980). *Understanding Attitudes and Predicting Social.* Englewood Cliffs, NJ: Prentice-Hall.

Anderson, J. C., & Gerbing, D. W. (1988). Structural Equation Modeling in Practice: A Review and Recommended Two-step Approach. *Psychological Bulletin*, *103*(3), 411–423. doi:10.1037/0033-2909.103.3.411

Andrews, L., Cacho-Elizondo, S., Drenna, J., & Tossan, V. (2013). Consumer Acceptance of an SMS-assisted Smoking Cessation Intervention: A Multi Country Study. *Journal of Health Marketing Quarterly*, *30*(1), 47–62. doi:10.1080/07359683.2013.758015 PMID:23458481

Ascari, A., Bakshi, A. & Grijpink, F. (2010). mHealth: A New Vision for Healthcare: McKinsey & Company, Inc., and GSMA.

Bagozzi, R. P., & Yi, Y. (1991). Multitrait-Multimethod Matrices in Consumer Research. *JMR, Journal of Marketing Research*, *17*, 426–439.

Bloch, P. H. (1982). Involvement beyond the Purchase Process: Conceptual Issues and Empirical Investigation. *Advances in Consumer Research. Association for Consumer Research (U. S.)*, *9*(1), 413–417.

Briones-Vozmediano, E., Vives-Cases, C., & Peiró-Pérez, R. (2012). Gender Sensitivity in National Health Plans in Latin America and the European Union. *Health Policy (Amsterdam)*, *106*(1), 88–96. doi:10.1016/j.healthpol.2012.03.001 PMID:22465154

Cestre, G. (1996). Diffusion et Innovativité: Définition, Modélisation et Mesure. *Recherche et Applications en Marketing*, *11*(1), 69–88. doi:10.1177/076737019601100105

Chowdhury, H. K., Parvin, N., Weitenberner, C., & Becker, M. (2006). Consumer Attitude toward Mobile Advertising in an Emerging Market: An Empirical Study. *International Journal of Mobile Marketing*, *1*(2).

d'Hauteville, F. (1994). *Un Modèle d'Acceptation du Nouveau Produit par le Consommateur: Cas du Vin allégé en Alcool.* Unpublished Doctoral Dissertation, Université Montpellier II, France.

Davis, F. D. (1989). Perceived Usefulness, Perceived Ease of Use, and User Acceptance of Information Technology. *Management Information Systems Quarterly*, *13*(3), 319–340. doi:10.2307/249008

Davis, F. D., Bagozzi, R. P., & Warshaw, P. R. (1989). User Acceptance of Computer Technology: A Comparison of Two Theoretical Models. *Management Science*, *35*(8), 982–1003. doi:10.1287/mnsc.35.8.982

Dodds, W. B., Monroe, K. B., & Grewal, D. (1991). Effects of Price, Brand, and Store Information on Buyers' Product Evaluations. *JMR, Journal of Marketing Research*, *28*(3), 307–319. doi:10.2307/3172866

Foxall, G., & Haskins, C. G. (1986). Cognitive Style and Consumer Innovativeness: An Empirical Test of Kirton's Adaptation-Innovation Theory in the Context of Food Purchasing. *European Journal of Marketing*, *20*(3/4), 63–80. doi:10.1108/EUM0000000004755

Gatignon, H., & Robertson, T. S. (1985). A propositional Inventory for New Diffusion Research. *The Journal of Consumer Research*, *11*(4), 849–867. doi:10.1086/209021

Goldsmith, R. E., d'Hauteville, F., & Flynn, L. R. (1998). Theory and Measurement of Consumer Innovativeness. *European Journal of Marketing*, *32*(3/4), 340–353. doi:10.1108/03090569810204634

Goldsmith, R. E., & Hofacker, C. H. (1991). Measuring Consumer Innovativeness. *Journal of the Academy of Marketing Science*, *19*(3), 209–221. doi:10.1007/BF02726497

Griffith, D.M., Gunter, K., Watkins D.C. (2012). Measuring Masculinity in Research on Men of Color: Findings and Future Directions. *American Journal of Public Health*, *102*(S2), S187-S194.

Hirshman, E. (1980). Innovativeness, Novelty Seeking and Consumer Creativity. *The Journal of Consumer Research*, *7*(3), 283–295. doi:10.1086/208816

Hong, S., & Tam, K. Y. (2006). Understanding the Adoption of Multipurpose Information Appliances: The Case of Mobile Data Services. *Information Systems Research*, *17*(2), 162–179. doi:10.1287/isre.1060.0088

Hurt, H. T., Joseph, K., & Cook, C. D. (1977). Scales for the Measurement of Innovativeness. *Human Communication Research*, *4*(1), 58–65. doi:10.1111/j.1468-2958.1977.tb00597.x

INPES (French National Health Prevention and Education Institute). (2007), "Le tabac tue un fumeur sur deux. L'industrie du tabac compte sur vous pour les remplacer", *Espace Presse*, retrieved April 4, 2015 from www.inpes.sante.fr/70000/cp/07/cp070820.asp

Kihyun, K., Gyenung-Min, K., & Eun Sook, K. (2009). Measuring the Compatibility Factors in Mobile Entertainment Service Adoption. *Journal of Computer Information Systems*, (Fall): 141–148.

Koenigstorfer, J., & Groeppel-Klein, A. (2012). Consumer Acceptance of the Mobile Internet. *Marketing Letters*, *23*(4), 917–928. doi:10.1007/s11002-012-9206-1

Kotler, P., & Armstrong, G. (2006). *Principles of Marketing* (11th ed.). Pearson Prentice-hall.

Le Louarn, P. (1997). La Tendance à Innover des Consommateurs: Analyse Conceptuelle et Proposition d'une Échelle de Mesure. *Recherche et Applications en Marketing*, *12*(1), 3–19. doi:10.1177/076737019701200101

Le Nagard-Assayag, E., & Manceau, D. (2011). *Le Marketing de l'Innovation, de la Création au Lancement de Nouveaux Produits* (2nd ed.). Paris: Dunod.

Masson, J. (2010). Effets de la Modification d'un Attribut Constitutif d'un Produit Alimentaire sur son Adoption par les Consommateurs: Le cas du Vin à teneur réduite en Alcool. Unpublished Doctoral Dissertation, Montpellier Supagro, Centre International d'Etudes Supérieures en Sciences Agronomiques, France.

Mathieson, K. (1991). Predicting User Intentions: Comparing the Technology Acceptance Model with the Theory of Planned Behavior. *Information Systems Research*, *2*(3), 173–191. doi:10.1287/isre.2.3.173

McArdle, M. (2014). Thank You For E-Smoking, Bloomberg Businessweek, 00077135, October 2, Edition 4366.

Merrill, M. (2011a). HHS mHealth initiatives target smoking cessation, *Healthcare IT News*, retrieved April 4, 2015 from www.healthcareitnews.com/news/hhs-mhealth-initiatives-target-smoking-cessation

Merrill, M. (2011b). Study finds potential in social media tools for smoking cessation, *Healthcare IT News*, retrieved April 4, 2015 from www.healthcareitnews.com/news/study-find-potential-social-media-health-tools-smoking-cessation

Midgley, D. F., & Dowling, G. R. (1993). A Longitudinal Study of Product Form Innovation: The Interaction between Predispositions and Social Messages. *The Journal of Consumer Research*, *19*(4), 611–625. doi:10.1086/209326

Ostlund, L. E. (1974). Perceived innovation attributes as predictors of innovativeness. *The Journal of Consumer Research*, *1*(2), 23–29. doi:10.1086/208587

Pagani, M. (2007). A Vicarious Innovativeness Scale for 3G Mobile Services: Integrating the Domain Specific Innovativeness Scale with Psychological and Rational Indicators. *Technology Analysis and Strategic Management*, *19*(6), 709–728. doi:10.1080/09537320701711207

Poe, M. (2014). 'The issues are complicated' with e-cigarettes, *The Times West Virginian*, March 9.

PwC (PricewaterhouseCoopers) report June 2013, Socio-economic impact of m-health, An assessment report for the European Union, Retrieved November 2014, from http://www.pwc.fr/socio-economic-impact-of-mhealth-an-assessment-report-for-the-european-union.html

Rodgers, A., Corbett, T., Bramley, D., Riddell, T., Wills, M., Lin, R.-B., & Jones, M. (2005). Do u Smoke after txt? Results of a Randomised Trial of Smoking Cessation Using Mobile Phone Text Messaging. *Tobacco Control*, *14*(4), 255–261. doi:10.1136/tc.2005.011577 PMID:16046689

Roehrich, G. (1994). Innovativités Hédoniste et Sociale: Proposition d'une Échelle de Mesure. *Recherche et Applications en Marketing*, *9*(2), 19–42. doi:10.1177/076737019400900202

Rogers, E. M. (Ed.). (1962). *Diffusion of Innovations*. New York: The Free Press.

Rogers, E. M. (Ed.). (1983). *Diffusion of Innovations* (3rd ed.). New York: The Free Press.

Rokeach, M. (1973). *The Nature of Human Values*. New York: The Free Press.

Steenkamp, J.-B. E. M., & Gielens, K. (2003). Consumer and Market Drivers of the Trial Probability of New Consumer Packaged Goods. *The Journal of Consumer Research*, *29*(December), 368–384. doi:10.1086/378615

Subin, I., Mason, C. H., & Houston, M. B. (2007). Does Innate Consumer Innovativeness relate to New Product/ Service Adoption Behavior? The Intervening Role of Social Learning via Vicarious Innovativeness? *Journal of the Academy of Marketing Science*, *35*(1), 63–75. doi:10.1007/s11747-006-0007-z

Syed, S. F., & Nurullah, A. S. (2011). Use of Mobile Phones and the Social Lives of Urban Adolescents: A Review of Literature. *Trends in Information Management*, *7*(1), 1–18.

Thompson, C., & Wembridge, M. (2012). Big Tobacco Push for Cigarette Alternatives. *Financial Times, August* 12.

Valette-Florence, P. (1989). Conceptualisation et Mesure de l'Implication. *Recherche et Applications en Marketing*, *4*(1), 57–78. doi:10.1177/076737018900400104

Varshney, U. (2014). A model for improving quality of decisions in mobile health. *Decision Support Systems*, *62*, 66–77. doi:10.1016/j.dss.2014.03.005

Venkatesh, V., & Brown, S. (2001). A Longitudinal Investigation of Personal Computers in Homes: Adoption Determinants and Emerging Challenges. *Management Information Systems Quarterly*, *25*(1), 71–102. doi:10.2307/3250959

Venkatesh, V., & Morris, M. (2000). Why don't Men ever Stop to Ask for Directions? Gender, Social Influence and their Role in Technology Acceptance and Usage Behavior. *Management Information Systems Quarterly*, *24*(1), 115–139. doi:10.2307/3250981

Venkatraman, M. P., & Price, L. (1990). Differentiating between Cognitive and Sensory Innovativeness: Concepts, Measurement, and Implications. *Journal of Business Research*, *20*(4), 293–315. doi:10.1016/0148-2963(90)90008-2

Wembridge, M., & Thompson, C. (2012), Big Tobacco bets on e-cigarette future, Financial Times, August 12. Retrieved November, 2014, from http://www.ft.com/cms/s/0/cb76997e-e0a5-11e1-b465-00144feab49a.html#ixzz23RJXFAJZ

Whittaker, R., Maddison, R., McRobbie, H., Bullen, C., Denny, S., Dorey, E., & Rodgers, A. et al. (2008). A Multimedia Mobile Phone-Based Youth Smoking Cessation Intervention: Findings from Content Development and Piloting Studies. *Journal of Medical Internet Research*, *10*(5), e49. doi:10.2196/jmir.1007 PMID:19033148

KEY TERMS AND DEFINITIONS

Breaking Addictions: Reduce addictive behaviors like cigarettes, drugs, alcohol adoption of an innovation: use an innovation at least 3 times for consumer goods, buy an industrial item.

Health & Human Services (HHS): "It is the United States government's principal agency for protecting the health of all Americans and providing basic human services".

Mobile Health or mHealth: It considers all the declinations of health care using mobile portable or wireless devices.

Perceived Enjoyment: The fun you may have while playing with an application.

Perceived Irritation: Based on the fact that mobile advertising may be perceived by you as invasive, irritating, or boring.

Perceived Monetary Value: The value you obtain according to the price quality ratio.

Social Influence: Influence of the people who matter for you towards your behavior.

Technology Acceptance Model (TAM): It was constructed by Davis (1989). It is an adaptation of the theory of Reasoned Action (Fishbein & Ajzen, 1975) designed to model the intention to adopt information systems.

Vicarious Innovativeness: It is defined as *openness to information on new products of any kind*. This kind of innovativeness is followed by adoptive innovativeness when consumers actually adopt the product.

APPENDIX

Exhibit 1: Description of the Mobile Coaching Service provided in the Interviews

We are now going to describe the main characteristics of a SMS assisted smoking cessation service which allows you to receive SMS or MMS message on your mobile phone to help you quit smoking:

Every morning you would receive an SMS or MMS message on your mobile phone to encourage you to not smoke all day.

If during the day, you feel that you will break your effort to not smoke, you can send an SMS message to a special mobile phone number to ask for help.

You will receive a supportive SMS message back immediately that would be specifically adapted to your needs.

In the middle or at the end of the day, if you haven't sent any alert messages, you'll receive an SMS or MMS message of congratulations and encouragement.

You would be able to modify the timing of receiving the messages to suit the times of day that are most convenience for you.

Exhibit 2: Smokeless Trends and the Tobacco Industry

According to Thompson and Wembridge (2012):

Tobacco companies are changing their strategic focus. The change in direction comes as the industry faces increased regulatory pressure both in mature markets and emerging economies. For instance, Australia has passed the world's most strident anti-tobacco regulation, dubbed "plain packaging", whereby tobacco products are sold in drab, standardised packs with graphic health warnings replacing brands. Plain packaging has also being considered in the UK and the European Union.

In 2011, Euromonitor estimated that the smokeless tobacco market, which also includes chewing tobacco and snuff, was already worth $14bn of the $664bn world, with cigarettes still accounting for over 90 per cent of the total. The smokeless tobacco category has continued its upward trajectory in the last years. According to Information Resources Inc.(IRI), in the 52 ended Jan 20, 2014,the category rang up $5.16 billion at U.S. c-stores.-

In order to affront all these pressures and new consumption trends, tobacco companies are seeking alternative solutions, Philips Morris plans to launch a cigarette under the Marlboro brand in 2016 where the tobacco is heated rather than burnt creating less smoke and tar. This product innovation is done to attract more health-conscious smokers. In 2011, British American Tobacco set up a new business unit, Nicoventures, which focuses on smoking alternatives. Nicoventures is working towards launching an inhalable pure nicotine product within the next two years or so. The expected goal is that his product will offer a safer alternative to smoking while still meeting smokers' cravings.

Electronic cigarettes (e-cigarettes), still unregulated, are the biggest component of the current market for cigarette substitutes, however in the past, companies have invested in smokeless alternatives with mixed results. As an illustration we can mention the case of the US conglomerate RJR Nabisco who pioneered one of the first smokeless cigarettes in the 80s, called Premier, at an estimated cost of more than $300m, but it sold poorly due to consumer concerns over the taste and smell.

Another issue affecting e-cigarettes is that they have nicotine inside and the effects on health of inhaling nicotine vapor are not completely known. Experts consider that it is better than inhaling nicotine and byproducts from tobacco on fire.

Matt Myers, president of *Campaign for Tobacco-Free Kids* says: "*The concern with e-cigarettes is that it will re-glamorize smoking, the gesture. Therefore, people who used to smoke may be lured back" (Poe, 2014).*

Neutral Packs: Another Recent Trend on Tobacco

The neutral cigarette packs were rapidly accepted by smokers in Australia after their introduction in December 2012 according to a research published by Tobacco Control on November 11th 2014. 28% of them approved this new regulation before it was implemented and they became 49% afterwards.

The same measure is planned to be introduced in France in 2016 to dissuade consumers from cigarettes: all packagings of cigarettes are going to have the same shape, the same size and the same typography without any logo. The brand name will continue to appear in small letters.

This measure combined with a price rise allowed to decrease the number of Australian daily smokers above 14 from 15,1% in 2010 to 12.8% in 2013 according to a research reported by *French journal Libération* on November 11th.

Exhibit 3: Some Examples of Smoking Cessation Mobile Apps

Coaching people to avoid taking a cigarette seems to remain a better option than smokeless alternatives such as the e-cigarette. In that contest, mobile apps emerge as a promising medium to deliver different treatments to promote health, which include smoking cessation. The first research study about a mobile phone smoking cessation program was published by (Rodgers et al., 2005). Nowadays, there are an increasing number of Mobile apps to coach smokers during their smoking cessation efforts offering them different features.

One of these apps is Livestrong **MyQuit Coach**. A related study of social media tools for smoking cessation that surveyed 266 users, released by the University of Southern California's Institute for Communication Technology Management (CTM) and the USC Center for Body Computing (CBC) found that the ability to immediately and continually track cigarette consumption along with encouragement and social support can lead to smoking cessation (Merryl, 2011b).

Another app, **QuitNow**, indicates for how long you haven't smoked, the number of cigarettes you have avoided and the money saved. Based on indicators of the World Health Organization, this app analyses your progresses and you can exchange on the chat with other concerned people.

Other apps available are: **iCoach**, an online digital health coaching platform that helps individuals quit smoking. **SmokefreeTXT** is a mobile text messaging in the United States that propose smokers to send them messages. **Texquit** asks details about the smoker's habits (frequentation, quantity, spending, reasons for quitting, difficulties to resist smoking…). The messages are personalized and sent 4 weeks before and 4 weeks after the quit date (1-5 messages per week).

Chapter 5
Smartphones and Self-Broadcasting among College Students in an Age of Social Media

Franklin Nii Amankwah Yartey
University of Dubuque, USA

Louisa Ha
Bowling Green State University, USA

ABSTRACT

This chapter examines the use of smartphones for self-broadcasting via social media among college students. Based on motivation and network externalities theories, our survey of a public university's college students confirmed our hypotheses that network size, years of experience using social media and the time spent on social media positively predict their frequency of self-broadcasting on their smartphones. The results suggest that 85.2% of college students self-broadcast at least once a month by updating their status on SNS and students are likely to self-broadcast within their network. Most students set their profile privacy setting as private or semi-private. But privacy setting does not affect self-broadcast frequency.

INTRODUCTION

Smartphones and social networking sites (SNS) are globally used and these gadgets are helping to modify the lives of people. These technologies provide access and connections to networks of people with convenience and the ability to share and receive information from a fluid and shifting Internet network (Enteen, 2006). Users are thus able to shrink time and space with these technologies by remotely connecting to friends, family, and business associates. Mobile phones continue to be tools that facilitate social sharing of information (Yang, Wang, & Li, 2013). From an information management standpoint, smartphones have immense self-broadcasting power (Bert, Giacometti, Gualano, & Siliquini, 2014) and SNS sites remain frequent destinations for smartphone users (Randall, 2014).

DOI: 10.4018/978-1-4666-8450-8.ch005

Smartphones as self-broadcasting tools and location-based technologies have been studied by several scholars (Costa, Benevenuto, & Merschmann, 2013; Karikoski & Soikkeli, 2013; Kang & Jung, 2014). Other studies discussed self-presentation on social networks (Hilsen & Helvik, 2014; Michikyan, Subrahmanyam, & Dennis, 2014). Studies have also hypothesized the relationship between Facebook friends and social well-being (Indian & Grieve, 2014; Wang, Kosinski, Stillwell, & Rust, 2014). Past studies have also examined social capital and Facebook use (Ellison, Steinfield, & Lampe, 2007; Jiang & de Bruijn, 2014; Li & Chen, 2014). However, little to no research exists on the role smartphones play in self-broadcasting in relation to the number of friends that one has on Facebook. Scholars have discussed smartphone use (Cacho-Elizondo, Shahidi, & Tossan, 2013) and the motivations behind social network use (Ross, Orr, Sisic, Arseneault, Simmering, & Orr, 2009). Though Tong, Van der Heide, and Langwell (2008) have explored the relationship between the number of friends and interpersonal impressions on Facebook, they did not examine smartphone use in the same context. The Nielsen Company's *Social Media Report* reports the rise of SNS use through smartphones and digital technologies (The Nielsen Company, 2012). Young adults between 18 and 24 years spend 59% of their time using the applications on their phones to browse the Internet (Bryant, 2013). However, owning a smartphone does not mean all applications will be utilized (Verkasalo, Lopez-Nicolas, Molina-Castillo, & Bouwman, 2010).

The present study seeks to fill the gap in research on the relationship between the use of Facebook through smartphones and how network size influences self-broadcasting intensity. The study builds on Wei's 2008 study on motivations for using mobile phones, which encouraged further examination of the mobile phone as a location-based broadcast device. From a technology and human interaction framework, this study offers a new way of conceptualizing smartphones as a self-broadcast medium for college students. The authors propose telephones have evolved from a point-to-point interpersonal communication medium to facilitating SNS broadcasting of information and news. The study employs motivation theory (Deci, 1975) and network externalities theory (Katz & Shapiro, 1985) as theoretical frameworks. Motivation theory explains what motivates an individual to adopt a technology, while network externalities theory determines the multiplier network value users gain from using mobile technology. According to a 2012 Pew survey and a 2012 Google survey, 55% and 39% of those surveyed reported they owned smartphones, respectively. Pew survey results in the same year also indicated that 67% of Internet users use social networking sites (Pew Internet Research, 2012). The following section presents a summary of relevant literature on self-broadcasting, Facebook, smartphones, gender and identity, self-esteem, and privacy online.

Self-Broadcasting

Self-broadcasting is a community communication style in which an individual self-projects her/his identity (Pempek, Yermolayeva, & Calvert, 2009; Underwood, Kerlin, & Farrington-Flint, 2011). Self-broadcasting may entail posting pictures on a SNS such as Facebook, composing and posting status messages, commenting on posts of others on Facebook, or checking into buildings or various places through smartphones.

Advantages of Self-Broadcasting

Amanda Lenhart (2013), senior research specialist at the Pew Internet & American Life Project, said that 84% of 18 to 29 year olds capture images with their phones and 68% record videos on the go. Lenhart also asserted that there are now more opportunities given the technologies at the disposal of young adults between 18 and 29 that enables them to be heard in cyberspace. Various businesses use SNS to share their activities (Greer & Ferguson, 2011), and college students benefit from sharing information through their SNS as well. Cirucci (2013) affirms that teens are the majority of social media users (73% of teens use SNS in the United States). These teens reported that they comment on their friends' SNS walls (86%) and post pictures (83%). Previous studies have suggested that people use SNS to construct profiles to share content (Boyd & Ellison, 2007; Kim, Kim, & Nam 2010). The use of SNS provides many benefits to users, such as convenience in keeping in touch with friends, life satisfaction, and building social capital (Boyd & Ellison, 2007; Johnston, Tanner, Lalla, & Kawalski, 2013; Li & Chen, 2014). People who use SNS like to be accepted by others, which can influence their self-esteem (Barker, 2009). Self-broadcasting allows the user to present a positive identity on Facebook or display intelligence (Mehdizadeh, 2010; Pempek et al., 2009), which may in turn attract "likes." These are important benefits for users looking for affirmations online (Hunt, Atkin & Krisnan, 2012). Other benefits include maintaining offline relationships (Craig & Wright, 2012). Utz, Tanis, and Vermeulen (2012) also add that SNS can enable the reaffirming of existing relationships and the encouragement of collective action. Kim and Lee (2011) suggest that "the number of Facebook friends and positive self-presentation may enhance users' subjective well-being" (p. 359), though this kind of happiness may not be rooted in perceived social support provided by friends on Facebook. However, sincere self-presentation may improve happiness grounded in the social support offered by Facebook friends (Kim & Lee, 2011). Young adults thus seem to need affirmation from friends and family on social networks to boost their self-esteem. Receiving "likes" on Facebook posts may also give them the motivation to continue posting or updating their status on SNS such as Facebook.

Smartphones and Facebook

Smartphones offer users the capability for increased interactivity and interaction, and while cell phones are everywhere, not much is known of the actual use of the applications they contain (LaRue, Mitchell, Terhorst, & Karimi, 2010). Understanding the change in communication patterns due to mobile phone usage helps in the development of innovative applications. *Mashable* (2012), an Internet news blog, defines a smartphone as "a mobile phone built on a mobile operating system, with more advanced computing capability and connectivity than a feature phone". Smartphone users are able to share their location via SNS such as Facebook, which disclose where they are physically within their network (Sutko & de Souza e Silva, 2011). Users are also able to broadcast their location to a large number of their friends.

Facebook is an SNS that facilitates communication and connections between individuals and groups. There are various social plug-ins embedded on Facebook's interface that enhance the user's experience. There are presently 160.7 million Facebook users in the U.S. (Facebook n.d.). The total number of monthly active users on Facebook is now approximately 1.06 billion (Facebook, 2014). According to a Pew Internet Research survey (2013), 63% of Facebook users logged onto the site several times in a day, while only 14% reported that they visit Facebook less than once a week.

Cell Phone Use/Motivating Factors

There have been a number of studies that have examined cell phone use and SNS (LaRue, et al., 2010; Underwood et al., 2011). For example, Fang & Ha (2011) found a positive relationship between SNS and cell phone use among college students. Adolescents use cell phones because of the convenience of social network applications on their smartphones, whereas adults use cellphones for family and security reasons (Context-Based Research Group, 2002; Ling, 2000). Apart from using these devices to stay connected (Wei & Lo, 2006), smartphones are also used for entertainment and seeking information on the Internet, becoming a part of the lives young adults (Baym, Zhang & Lin, 2004). SNS creates a medium for individuals and groups with different interests to facilitate new connections. SNS also helps users maintain relationships in addition to facilitating the sharing of content within various networks (Ha & Hu, 2013). According to Ha & Hu (2013), students with a vast number of friends have high SNS involvement and are more likely to seek and share information within their network. Some studies that used network externalities found that enjoyment, number of friends on a network, and usefulness were the factors for using SNS technology (Lin & Lu, 2011). A 2008 study examining motivations for cell phone use by Wei suggested younger users were more likely to use their cell phones for obtaining entertainment and news. Younger users of SNS are motivated to engage with these technologies because they not only gain social capital, but also develop their online identities.

Self Esteem, Social Media, and GPA

SNS use influences social connectivity, social capital, self-esteem, and identity development (Tazghini & Siedlecki, 2013; Tiggemann & Slater, 2014). Spies and Margolin (2014) found in a study on psychosocial development in adolescents that SNS use is differentially related to adolescent's self-esteem, sociability, and identity development. Greitemeyer, Mügge, and Bollermann (2014) in their study also affirmed that having few responses from Facebook friends threatens self-esteem of Facebook users and affects their perception of having a meaningful life. Research has also suggested that the need for popularity also motivates users to use SNS (Utz et al., 2012). Studies have also suggested that SNS use by college students is negatively related to their GPA (Junco, 2012a; Junco, 2012b). Turner & Croucher's 2014 study on college students' SNS use, however, suggested that use of social and traditional media was not related to student GPA. The authors used survey methods to explore college students' use of various SNS as well as the utilization of traditional media such as television. Based on multiple regression analysis, the results suggested that college students' use of traditional media was a significant predictor of GPA. Rosen, Mark Carrier, and Cheever (2013) asserted that electronic communication is fulfilling. They examined how technological distractions influenced learning. They administered a questionnaire that gauged learning strategies adopted by students. The results suggested that having a positive outlook on technology did not prevent students from focusing while studying. Only those who were shifting their attention to technologies around them were likely to be distracted. Those who accessed Facebook while studying had lower GPAs.

MOTIVATION AND NETWORK EXTERNALITIES

Previous studies employed motivation theory to explain individual behavior for accepting information technology (Lin & Lu, 2011). Intrinsic and extrinsic motivations are central to the acceptance of information technology (Deci, 1975). *Intrinsic motivation* is performing an action due to an interest in that action rather than external factors (Ciampa, 2014; Davis, Bagozzi, & Warsaw, 1992). For example, receiving "likes" on a Facebook page may invoke intrinsic satisfaction. Extrinsic motivation is executing an action due to its discerned value; for example, there is value in monetizing a video on YouTube because one may receive money from clicks and views. The present study thus proposes that the intrinsic benefit of self-expression and the extrinsic benefit of self-promotion constitute an individual's discerned benefits in smartphone usage.

Network externality is the value users receive from a product; this value increases as the number of users and complementary services and products increase (Katz & Shapiro, 1985; Strader, Ramaswani, & Houle, 2007). Once the number of users reaches an ideal number, relative benefit is achieved (Lin & Bhattacherjee, 2008; Wang, Hsu, & Fang, 2004). There are two types of network externalities: direct network externalities and indirect network externalities (Gupta & Mela, 2008; Katz & Shapiro, 1985; Lin & Bhattacherjee, 2008). Direct network externality is "the increase in users of a particular product or service, where a user's benefits increase" (Lin & Lu, 2011, p. 1153). The more friends one has on Facebook the higher the chances of receiving many "likes" on a post. Indirect network externality is "an increased sense of user value from using a product or service; the effect the user obtains from such a product or service increases with the increase of related complementary products" (Lin & Lu, 2011, p. 1153). When a product has network externalities, consumer acceptance is driven by the number of users motivating other users to adopt that product (Wang, Lo, & Fang, 2008). A product with limited number of users is more likely to experience technological lockout (Schilling, 2002). The capability to build community around a product helps to add value to that product through the increase of users (Zhou & Lu, 2011). Mäntymäki and Islam (2014), in a study using network externalities, noted the ubiquity and the rapid increase of virtual worlds. The authors affirmed that network externalities play a crucial role in the adoption of information technologies; however, there is limited research on the role of network externalities in the sustained use of communication technologies. In another study on reasons why people play social games on their smartphones, Wei and Lu (2014) also affirmed that network externalities and individual gratifications immensely influenced the intention to play these games on their phones.

Network externality has received attention in traditional information systems and economics (Dickinger, Arami, & Meyer, 2008); there is, however, limited research on their application to developing mobile services (Zhou & Lu, 2011). This study seeks to fill this gap in the literature. Employing the theories of direct and indirect network externalities, the present study suggests that peer network externality is one of the sources of network externalities. The study focuses on college student status updates because they are an important feature used by young adults to disseminate information within their networks. Hence our first hypothesis is that network size provides the network externalities explaining the frequency of self-broadcasting.

H1: The more friends and followers college students have on Facebook, the more likely they are to self-broadcast using Facebook applications on their smartphones.

In addition, since experience in using social media may affect the skill level of the users and knowledge of the utility value of social media for communication and self-promotion, the study also examines the effect of SNS experience on the intensity of self-broadcast. Hence we hypothesize that SNS experience positively predict frequency of update.

H2: The longer college students have used SNS, the more likely they are to update their status and share content with others using social media on their smartphones.

Another important factor to predict self-broadcasting frequency is the time spent on social media and the frequency of update. The time spent on the medium reflects the importance of the medium to the user. If college students spent a lot time on social media, then it is more likely that they will utilize the medium to communicate to their network or promote themselves.

H3: The more the time college students spend on social media, the more likely they are to update their status using social media on their smartphones.

Mobile Communication SNS, Gender, and Identity Performance

Gender differences have been noted in mobile and SNS use, with females spending more time on their phones (Leung & Wei, 2000), and male profile pictures on SNS stressing status and risk while female profile pictures stressing emotions (Tifferet & Vilnai-Yavetz, 2014). Not only are mobile technologies being used to strengthen bonds between people (Campbell & Russo, 2003; Ling, 2008), but they have also become extensions of identity (Campbell, 2008; Katz & Sugiyama, 2005; Lobet-Maris, 2003). Through SNS and integrated media, people are able to represent themselves online (Boyd, 2008; Hull & Stornaiuolo, 2010). Through self-presentation, mobile phone users create their identities online by using SNS applications (Boyd, 2011). Integrated media thus plays a salient role in helping youth develop a stronger sense of social identity and cohesion through self- representation (Hilsen & Helvik, 2014). The motivation of people to create, share, collaborate, and self-represent themselves has led to the creation Web 2.0 technologies.

Differences in social media use within ethnic groups have been noted by a number of scholars, with Latinos using Facebook less than Caucasians (Hargittai, 2008), while students of Vietnamese, Latino, African-American, and Indian ancestry exhibit an elaborate use of Facebook (Grasmuck, Martin, & Zhao, 2009). A study on the effect of gender, ethnicity, and income on the use of digital communication technologies suggested that Latino and African-American students use cell phones as convenience tools to gain support from a social network (Junco, Merson, & Salter, 2010).

As previous literature indicated the effect of gender and ethnicity on cell phone and social media use, so two variables were added in a regression model to show the potential moderating effect of these variables on self-broadcasting intensity. Though income was not found in previous studies to have any impact on social media use, the relative higher cost of smartphones may show income to play a role in the result. This was included in the model as well. Figure 1 presents a summary of the research model for this study.

Privacy

Social media and smartphones serve as cybernetic nodes, allowing users to feel more connected (Edenius & Rämö, 2011), while enabling businesses to access user personal information to customize their information to target prospective consumers (Scoble & Israel, 2014). The issue of privacy when using mobile phones or other electronic gadgets is salient. According to Scoble and Israel (2014), we are entering an era when much contextual information will be collected and stored by many businesses to help deliver customized information and products to consumers. New technological gadgets now know an individual better than family and friends do. Through contextual data, gadgets are now able to know the context

Figure 1. Research model

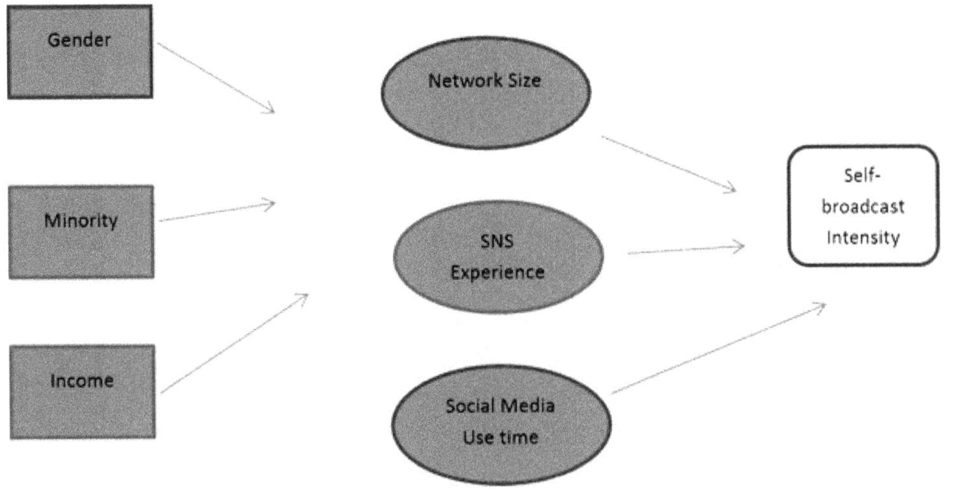

of user online Google searches, favorite restaurants, books, movies, and other private user information. Consumers are able to connect, collaborate, and share information, dislodging marketers from their positions of power and becoming co-creators of media content (Hanna, Rohm, & Crittenden, 2011).

Recent studies have highlighted the importance of privacy on social network cites (Feng & Xie, 2014; McGarry, 2013; Stern & Kumar, 2014). The ability to share and disseminate information to a large audience is riddled with issues of privacy that need to be addressed. Strangers are able to share another's information in cyberspace, which become permanent fixtures of the Internet. Employers are reviewing the Facebook pages of potential employees and making hiring decisions based on what they see. Apart from using default privacy settings, users are improving their privacy but excluding sensitive information on their SNS (Young & Quan-Haase, 2013). According to a Pew survey teens have accepted application downloading, with 58% of all teens having downloaded an app on their tablet or smartphone (2013). Teens are, however, careful about these apps for privacy reasons; they avoid or delete apps that pose a threat to their privacy. Teen girls are most sensitive about location privacy and the kind of information that apps are storing or sending out (Pew Research Internet Project, 2013). According to the Pew report, 51% of teens who use apps avoided them because of privacy issues, and 26% of teens that use apps uninstalled them because they discovered the apps were collecting and sharing personal information they did not want to make public. There were also 46% of teens that turned off location tracking on their apps or cell phones because they were concern about their privacy. More teen girls reported having turned off location tracking (59%) compared to teen boys (37%). With regards to managing privacy on SNS, a Pew report indicated that 48% of all social media users have some problems changing their privacy settings, but 49% reported that it was easy for them (Pew Internet Research Project, 2012). Among young adults between 18 and 29 years old, 57% said it was not difficult to manage their privacy settings (Pew Internet Research Project, 2012). Privacy settings may affect the sense of security but can increase the reach/impact of the posting. Since privacy is an important issue in social media, we pose the following questions on the privacy setting of social media in the study:

RQ1. What are the social media profile privacy settings of college student smartphone users?
RQ2. Does privacy settings affect frequency of self-broadcasting?

RESEARCH METHOD

Data for this study was collected using a self-administered online survey of college students enrolled at a public university in the Midwest United States in 2012. Students were recruited from 32, small-size, general education classes and three large lecture, introductory classes. This ensured a varied range of majors and class standings. As an incentive to participate, students received extra class credit for their participation. The survey took approximately 15 minutes to complete, and respondents answered questions related to their use and opinion of a variety of media. A total of 346 responses from college students were received. A total of 263 respondents (76%) reported owning smartphones. Only those who owned smartphones were included in the study. There were four major variables related to this study: network size (followers and friends on Facebook), self-broadcasting intensity, experience on social media, and time spent on social media.

Measures

- **Self-Broadcast:** Posting an update, sharing a link to other media sites, and posting links from other sites were counted as self-broadcasting activities. The intensity of self-broadcasting was measured by both the frequency of update and the sum of these activities. Respondents were asked to indicate how frequently they update their social network page. Seven items were provided for this question: 1) several times an hour; 2) every several hours; 3) every day; 4) once to several times a week; 5) between once a week and once a month; 6) less than once a month; 7) hardly ever update.
- **Network Size and Experience in Social Media:** Network size was measured by an open-ended question. Respondents were asked the approximate number of "friends" or "followers" they had on the social networking site Facebook.
- **SNS Experience:** In social media was measured by asking respondents the number of years they have used SNS.
- **Network Diversity:** Types of "friends" on a user's Facebook page were reported by respondents. 11 types of friends were identified: a) Parents; b) Spouse or current romantic partner; c) Siblings and relatives (other than parents or spouse); d) Coworkers from a current or previous job; e) Neighbors; f) Classmates in college; g) High school classmates; h) Fellow members of a group; i) People that one never met in person; j) People that one met in person once; k) other.
- **Smartphone Ownership:** The smartphones used by college students were measured by asking respondents what kind of cellular phone they had. Respondents were presented with four options: a) android smartphones; b) iPhones; c) other smartphones; d) regular cellular phones. Respondents who indicated their phones were a, b, or c were counted as smartphone users.
- **Social Media Applications Installed:** Respondents were asked what types of applications were installed on their cell phone or other mobile devices. Respondents were presented with thirteen options to select from: games; news; weather; educational; social media such as Facebook and LinkedIn; location-based social media devices such as Foursquare, Brightkite; media such as TV, radio, magazines; health; business; communication such as e-mail/texting or Skype; dating services; shopping; and other.
- **Social Media Profile Settings**: Respondents were asked how they set their profiles and the social media they used most. Respondents were presented with three options: a) public; b) partially private (friends of friends or your networks can see it); c) private (friends only).

RESULTS

A comparison of the basic demographic characteristics of smartphone owners and non-smartphone owners was conducted before our analysis. We found no significant demographic differences between smartphone and non-smartphone owners among the college student sample, except minority status (see Table 1). Overall gender and income does not affect social network size, only minority status was a significant predictor of network size. Network diversity (types of friends in the network) had no effect on social media update frequency. The findings also suggest that setting a profile public or private or installing social media mobile apps on a smartphone has no impact on update frequency. Most of our respondents set their profile to private. For smartphone users 43% identified themselves as male and 57% female. Non-smartphone users were comprised of 39.5% male and 60.5% female. With regards to the types of phones owned by the students in this sample, 31.8% owned Android smartphones, 39.3% owned IPhones, 4.9% had other smartphones, and 24% owned regular smartphones. Caucasians represented 92.7% of non-smartphone owners and 77.7% of smartphone owners. African Americans represented 4.9% of non-smartphone owners and 13.8% of smartphone owners. There were 2.3% Hispanic smartphone owners and 1.2% Asian smartphone owners, with no Hispanic and Asian non-smartphone owners. Native Americans reported a count of 1.2% non-smartphone owners and 0.8% smartphone owners. Respondents who identified their ethnicity as other reported 1.2% non-smartphone owners and 4.2% smartphone owners. As expected, smartphone users have significantly higher use of social media than non-smartphone owners with 5.4 more hours of use per week.

Table 1. Profile of smartphone owners and non-smartphone owners, college students

		Smart Phone Users (N=263)	**Non-Smart Phone Users (N=83)**
Ethnicity	Caucasian	77.7%	92.7%
	African-American	13.8%	4.9%
	Hispanic	2.3%	0.0%
	Asian	1.2%	0.0%
	Native American	0.8%	1.2%
	Other	4.2%	1.2%
Annual Household Income	Under US $ 30,000	61.3%	58%
	$30,001-$60,000	11.7%	13.6%
	$60,001-$90,000	13.7%	14.8%
	$90,001-$150,000	9.0%	8.6%
	Over $150,000	4.3%	4.9%
Gender	Male	43.0%	39.5%
	Female	57.0%	60.5%
*Mean SNS Use Time * (Significantly different at p< 0.05)*		20.67 hours per week	15.25 hours per week

Note: Missing data was excluded from the analysis.

PREDICTORS OF SELF-BROADCASTING

H1 predicted that network size and self-broadcasting (i.e. updating one's status on Facebook through applications on smartphones) were positively correlated. A multiple regression analysis indicated that network size is positively related to self-broadcasting among college students in terms of frequency of update (*beta* = .19, $p = 0.003$). This hypothesis was supported.

H2 posited that the longer a college student has used SNS, the more likely the student will update their social media on their smartphones. The results supported this hypothesis. The regression analysis showed a significant positive relationship between social media experience and the frequency of status update (*beta* = 24, $p < 0.001$). The hypothesis was supported.

H3 posited that total social media use time and update frequency was positively correlated. The result showed that those who update more frequently were also likely to spend more time in social media (*beta*= .14, $p = 0.03$). H3 is supported. But among the three variables, social media use time is the least important predictor of SNS updates. Hence the three variables: network size, social media experience, and social media use time explained 12.8% of total variance of Facebook update frequency.

Importance of Social Media Applications on Smartphones and Privacy Settings

Our findings also highlight the importance of mobile applications on smartphones. Social media application was the most commonly installed application with 87.3% of smartphone owners installing it. The other popular mobile application type is communication related applications: 67.1% had applications that allowed them to email, text, and Skype. In addition, 51.6% of respondents had installed TV and radio applications on their phones, 48.4% owned magazine related apps and 18.3% had health related apps. 11.5% reported having business applications. There were 3.6% of respondents who had dating apps on their phones (these were the least popular apps), 31.7% had shopping apps, and 4% of respondents had other applications installed on their phones.

In terms of privacy settings, we found that a majority of students set their profile settings as private (60.6%), indicating that they don't want to share their information publicly, only 14.8% of the students set their profile setting as public. Profile settings do not significantly impact the frequency of self-broadcasting (beta=-0.09, n.s.).

DISCUSSION AND IMPLICATIONS

The results show that 85.2% of college students self-broadcast at least once a month by updating their status on SNS. Three quarters of these students have smartphones (76%). This is slightly below the 2013 Pew Internet Research Project (Pew) statistic, which reported that 18 to 24 year-olds have smartphones ownership of 79% (Smith, 2013). Our results suggested students are likely to self-broadcast within their network and the larger the network size, the more likely college students self-broadcast within their network. However, we caution that having a bigger network size does not necessarily mean one will self-broadcast; users may be more cautious in updating their status as their network size increases. Public figures with large network sizes may update less frequently and be more thoughtful in composing status messages since the media and audiences are always monitoring them.

There were slightly more females (57%) than males (43%) among smartphone users with a high number owning iPhones (39.3%). Minority student smartphone users were more likely to have a larger network size, which has implications for media and advertising companies who may consider targeting minority students (African Americans, Hispanics, Asians, Native Americans) with relevant products and services. Additionally there was a significant positive relationship between social media experience and the frequency of status update. So the longer college students have been on social media platforms the more likely they are to update their status. Literature available also suggests women are more active users of cell phones when compared with men (Leung & Wei, 2000). But gender did not affect the relationship between update frequency and network size or social media use. After controlling for gender, our results still indicated a strong relationship between network size and self-broadcasting. Social media use time was not a strong predictor for status updates, but network size and social media experience were more important in our sample.

It was also interesting to note that owners of smartphones had a significantly lower GPA than non-smartphone owners (χ^2=19.67, df=5, $p < 0.01$). Non-smartphone owners are twice more likely to have a 3.5 GPA or above than smartphone owners. Because of the close relationship between social media and smartphone use, these results affirm studies that have also demonstrated that the use of SNS has a negative relation to GPA (Junco, 2012a; Junco, 2012b). However, a 2011 study by Hanson, Drumheller, Mallard, McKee, and Schlegel on cell phone and SNS use suggested that females had a higher GPA, whereas Turner and Croucher's 2014 study on college students' SNS use suggested that use of social and traditional media was not related to student GPA. The present study contradicts Turner and Croucher's results. Though studies have suggested that SNS use influences social connectivity, social capital, self-esteem, and identity development (Johnston et al., 2013; Spies & Margolin, 2014). The lack of self-esteem difference between smartphone and non-smartphone owners in this study is interesting to report. With regards to social media use time, those who updated more frequently were more likely to spend more time on social media. However, this variable was the least important in predicting SNS updates. Hence to predict who the frequent self-broadcasters are, one should focus on the network size and the social media experience, rather than just how much time they spent on social media. There are other important personality variables and offline social capital that may explain the frequency of self-broadcasting that future research should include.

Implications for the results in light of network externalities is important because college students with larger networks have a larger resource of people that they can connect with, making them more motivated to self-broadcast and share information with the people in their network. Literature available suggests that with regards to network externalities the number of college students who have smartphones and/or have Facebook accounts will attract more non-users to the networks of these technologies (Wang et al., 2008). Thus the ability of current users to build interest groups by attracting more people to use an SNS service contributes to adding value to the experience of a product or service (Zhou & Lu, 2011). Those who join deem there are some benefits to be derived from these technologies, since more people are using them. Benefits may include using smartphones/social media to connect with friends and family (Ling, 2004), and sharing opinions about products and services through various SNS networks. For most consumers it is important to have compatible products with those in their network to help facilitate a seamless flow of information through self-broadcasting. For example, in this study 76% of respondents have smartphones and the benefits they derive from interacting with people in their networks may increase as more friends and family own smartphones and join SNS sites. Users with fewer people in their network are less likely to self-broadcast or post links from other news sites. The implications for advertisers, media companies, and news stations is that it is more beneficial to target users with larger social networks who are also smartphone users rather than targeting users who have a smaller number of people in their network. About 21% of students in this study update their status every few hours, and 16.1% update once every day. However, a greater percentage of students update once to several times a week (30.4%), making 18 to 25-year-olds in this sample frequent self-broadcasters. This also makes them an important audience for various marketers.

However, having a longer experience with SNS does not necessarily mean students will post links from other sites using their smartphones as we found very few respondents did so in their social media use. Media companies and advertisers should not expect college students to disseminate their content to other users in their network because updates tend to be more personal in nature. It is salient for advertisers and media companies to know that the inclination to self-broadcast is due more to their own desire to share their own information within their own private network. Since advertisers and media are more interested in a user's interest to spread their content rather than the personal news of their network contacts, such difference in interest need to reconciled in social media marketing. Social media marketers must relate their messages to personal interest of the users. The use of mobile applications by college students may also have implications for advertisers, media, and health companies. Colleges and universities may also benefit in targeting their recruiting efforts by employing the communication channels used by these students by encouraging current students and alumni to share interesting updates or post pictures they had when they were in college. Advertising currently occurs within many mobile applications, but once a user purchases the full application these advertisements may cease. Advertisers, media, health companies, and colleges can explore ways to utilize these Web 2.0 technologies to reach their respective audiences. As health-related applications were installed by 18.3% of college students in this study. Relevant health information can be integrated into these applications, since college students are a generation that feels immune to most health issues. Health organizations may want to especially target individuals with large SNS networks and those who have smartphones that facilitate and give out contextual data (Scoble & Israel, 2014) and spread health information among their friends in the network. The low percentage of students making their profile public shows that students are more aware of protecting their own personal information than previous studies suggested. But the high percentage of students with private profile means they may think social media is more a private medium than public medium and be less cautious in their posting. As we did not examine the content of the posting and update, we only found privacy setting does not have any relationship with update frequency.

The present study has some limitations. There were more females than males in our sample; however, this did not affect the results after controlling for gender. The authors did not examine the attitudes respondents had towards mobile applications and the features they had on their cell phones. Future studies could examine qualitative data that could be compared to this study. Various social networking sites (e.g.Twitter, Instagram, &Tumblr) and social networking business sites (LinkedIn) have interfaces that facilitate various types of content sharing, and some users may prefer staying longer on one SNS than the other. Facebook, for example, has the most users compared to Twitter, Instagram, or Tumblr. Some people may prefer using Facebook over Twitter because Facebook is more saturated with users. Others may prefer Twitter or LinkedIn for various reasons. Each of these social media serves a different broadcast function. LinkedIn is more of a social business tool compared to Facebook, Twitter, or Tumblr, and people use each of these mediums differently depending on their needs and networks. Future studies could compare and contrast self-broadcasting among these social media to help note the differences that exist. We strongly caution against generalizing the results for this study.

ACKNOWLEDGMENT

This research was funded by the Bowling Green State University Research Capacity Enhancement Grant and the Toledo Blade (Toledo, OH).

REFERENCES

Barker, V. (2009). Older adolescents' motivations for social network site use: The influence of gender, group identity, and collective self-esteem. *Cyberpsychology & Behavior*, *12*(2), 209–213. doi:10.1089/cpb.2008.0228 PMID:19250021

Baym, K. N., Zhang, B. Y., & Lin, M. C. (2004). Social interactions across media. *New Media & Society*, *6*(3), 299–318. doi:10.1177/1461444804041438

Bert, F., Giacometti, M., Gualano, M. R., & Siliquini, R. (2014). Smartphones and health promotion: A review of the evidence. *Journal of Medical Systems*, *38*(1), 1–11. doi:10.1007/s10916-013-9995-7 PMID:24346929

Boyd, D. (2008). Why youth love social networking sites: The role of networked publics in teenage social life. In D. Buckingham (Ed.), *Youth, identity, and digital media* (pp. 119–142). Cambridge, MA: MIT Press.

Boyd, D. (2011). Social network sites as networked publics: Affordances, dynamics, and implications. In Z. Papacharissi (Ed.), *A networked self: Identity, community, and culture on social network sites* (pp. 39–58). New York: Routledge.

Boyd, D., & Ellison, N. (2007). Social network sites: Definition, history, and scholarship. *Journal of Computer-Mediated Communication*, *13*(1), 210–230. doi:10.1111/j.1083-6101.2007.00393.x

Bryant, S. (2013). *More than half of the time spent on Internet is via mobile*. Retrieved June 19, 2014 from http://www.ad-ology.com/half-time-spent-internet-via-mobile/

Cacho-Elizondo, S., Shahidi, N., & Tossan, V. (2013). Intention to adopt a text message-based mobile coaching service to help stop smoking: Which explanatory variables? *International Journal of Technology and Human Interaction*, *9*(4), 1–19. doi:10.4018/ijthi.2013100101

Campbell, S. W. (2008). Mobile technology and the body: Apparategeist, fashion, & function. In J. Katz (Ed.), *Handbook of mobile communication studies* (pp. 153–164). Cambridge, MA: MIT Press. doi:10.7551/mitpress/9780262113120.003.0012

Campbell, S. W., & Russo, T. C. (2003). The social construction of mobile telephony: An application of the social influence model to perceptions and uses of mobile phones within personal communication networks. *Communication Monographs*, *70*(4), 317–334. doi:10.1080/0363775032000179124

Ciampa, K. K. (2014). Learning in a mobile age: An investigation of student motivation. *Journal of Computer Assisted Learning*, *30*(1), 82–96. doi:10.1111/jcal.12036

Cirucci, A. M. (2013). First person paparazzi: Why social media should be studied more like video games. *Telematics and Informatics*, *30*(1), 47–59. doi:10.1016/j.tele.2012.03.006

Context-Based Research Group. (2002). *The mobiles: Social evolution in a wireless society.* Retrieved from Context Based Research Group website: http://www.contextresearch.com/context/pdf/theMOBILES.pdf

Costa, H., Benevenuto, F., & Merschmann, C. H. L. (2013). Detecting tip spam in location-based social networks. *Proceedings of the ACM Symposium on Applied Computing, Coimbra, Portugal.* Retrieved from http://homepages.dcc.ufmg.br/~fabricio/download/sac2013.pdf

Craig, E., & Wright, B. K. (2012). Computer-mediated relational development and maintenance on Facebook. *Communication Research Reports*, *29*(2), 119–129. doi:10.1080/08824096.2012.667777

Davis, F. D., Bagozzi, R. P., & Warsaw, P. R. (1992). Extrinsic and intrinsic motivation to use computers in the work place. *Journal of Applied Social Psychology*, *22*(14), 1111–1132. doi:10.1111/j.1559-1816.1992.tb00945.x

Deci, E. L. (1975). *Intrinsic motivation.* New York: Plenum Publishing. doi:10.1007/978-1-4613-4446-9

Dickinger, A., Arami, M., & Meyer, D. (2008). The role of perceived enjoyment and social norm in the adoption of technology with network externalities. *European Journal of Information Systems*, *17*(1), 4–11. doi:10.1057/palgrave.ejis.3000726

Edenius, M., & Rämö, H. (2011). An office on the go: Professional workers, smart phones and the return of place. *International Journal of Technology and Human Interaction*, *7*(1), 37–55. doi:10.4018/jthi.2011010103

Ellison, N. B., Steinfield, C., & Lampe, C. (2007). The benefits of Facebook "friends": Social capital and college students' use of online social network sites. *Journal of Computer-Mediated Communication*, *12*(4), 1143–1168. http://jcmc.indiana.edu/ doi:10.1111/j.1083-6101.2007.00367.x

Enteen, J. (2006). Spatial conceptions of URLs: Tamil Eelam networks on the World Wide Web. *New Media & Society*, *8*(2), 229–249. doi:10.1177/1461444806061944

Facebook reports fourth quarter and full year 2012 results. (2014, May 30). *Facebook.* Retrieved from http://investor.fb.com/releasedetail.cfm?ReleaseID=736911

Facebook statistics by country. (n.d.). *Socialbakers.* Retrieved March 14, 2012, from http://www.socialbakers.com/Facebook-statistics/

Fang, L., & Ha, L. (2011). *Who are the heavy users of social network sites among college students? A study of social network sites and college students.* Paper presented at the Association for Education in Journalism and Mass Communication Annual Convention, St. Louis, MO. Retrieved from http://citation.allacademic.com/meta/p_mla_apa_research_citation/5/2/0/0/4/pages520046/p520046-1.php

Feng, Y., & Xie, W. (2014). Teens' concern for privacy when using social networking sites: An analysis of socialization agents and relationships with privacy-protecting behaviors. *Computers in Human Behavior*, *33*, 153–162. doi:10.1016/j.chb.2014.01.009

Grasmuck, S., Martin, J., & Zhao, S. (2009). Ethno-racial identity displays on Facebook. *Journal of Computer-Mediated Communication*, *15*(1), 158–188. http://jcmc.indiana.edu/ doi:10.1111/j.1083-6101.2009.01498.x

Greer, F. C., & Ferguson, A. D. (2011). Using Twitter for promotion and branding: A content analysis of local television Twitter sites. *Journal of Broadcasting & Electronic Media*, *55*(2), 198–214. doi:10.1080/08838151.2011.570824

Greitemeyer, T., Mügge, D. O., & Bollermann, I. (2014). Having responsive Facebook friends affects the satisfaction of psychological needs more than having many Facebook friends. *Basic and Applied Social Psychology*, *36*(3), 252–258. doi:10.1080/01973533.2014.900619

Gupta, S., & Mela, C. F. (2008, November). What is a free customer worth? *Harvard Business Review*, *86*, 102–109, 138. Retrieved from http://hbr.org/magazine PMID:19009724

Ha, L., & Hu, X. (2013). Social media involvement among college students and general population: Implications to media management. In M. Friedrichsen & W. Mühl-Benninghaus (Eds.), *Handbook of Social Media Management: Value Chain and Business Models in Changing Media Markets*. New York: Springer-Verlag. doi:10.1007/978-3-642-28897-5_44

Hanna, R., Rohm, A., & Crittenden, L. V. (2011). We're all connected: The power of social media ecosystem. *Business Horizons*, *54*(3), 265–273. doi:10.1016/j.bushor.2011.01.007

Hanson, T. L., Drumheller, K., Mallard, J., McKee, C., & Schlegel, P. (2011). Cell phones, text messaging, and Facebook: Competing time demands of today's college students. *College Teaching*, *59*(1), 23–30. doi:10.1080/87567555.2010.489078

Hargittai, E. (2008). The digital reproduction of inequality. In D. Grusky (Ed.), *Social stratification*. Boulder, CO: Westview Press.

Hilsen, A., & Helvik, T. (2014). The construction of self in social medias, such as Facebook. *AI & Society*, *29*(1), 3–10. doi:10.1007/s00146-012-0426-y

Hull, G. A., & Stornaiuolo, A. (2010). Literate arts in a global world: Reframing social networking as cosmopolitan practice. *Journal of Adolescent & Adult Literacy*, *54*(2), 85–97. doi:10.1598/JAAL.54.2.1

Hunt, D., Atkin, D., & Krisnan, A. (2012). The influence of computer-mediated communication apprehension on motives for Facebook use. *Journal of Broadcasting & Electronic Media*, *56*(2), 187–202. doi:10.1080/08838151.2012.678717

Indian, M., & Grieve, R. (2014). When Facebook is easier than face-to-face: Social support derived from Facebook in socially anxious individuals. *Personality and Individual Differences*, *59*, 102–106. doi:10.1016/j.paid.2013.11.016

Jiang, Y., & de Bruijn, O. (2014). Facebook helps: A case study of cross-cultural social networking and social capital. *Information Communication and Society*, *17*(6), 732–749. doi:10.1080/1369118X.2013.830636

Johnston, K., Tanner, M., Lalla, N., & Kawalski, D. (2013). Social capital: The benefit of Facebook 'friends'. *Behaviour & Information Technology*, *32*(1), 24–36. doi:10.1080/0144929X.2010.550063

Junco, R. (2012a). Too much face and not enough books: The relationship between multiple indices of Facebook use and academic performance. *Computers in Human Behavior*, *28*(1), 187–198. doi:10.1016/j.chb.2011.08.026

Junco, R. (2012b). In-class multitasking and academic performance. *Computers in Human Behavior*, *28*(6), 2236–2243. doi:10.1016/j.chb.2012.06.031

Junco, R., Merson, D., & Salter, W. D. (2010). The effect of gender, ethnicity, and income on college students' use of communication technologies. *Cyberpsychology, Behavior, and Social Networking*, *13*(6), 619–627. doi:10.1089/cyber.2009.0357 PMID:21142986

Kang, S., & Jung, J. (2014). Mobile communication for human needs: A comparison of smartphone use between the US and Korea. *Computers in Human Behavior*, *35*, 376–387. doi:10.1016/j.chb.2014.03.024

Karikoski, J., & Soikkeli, T. (2013). Contextual usage patterns in smartphone communication services. *Personal and Ubiquitous Computing*, *17*(3), 491–502. doi:10.1007/s00779-011-0503-0

Katz, J. E., & Sugiyanma, S. (2005). Mobile phones as fashion statements: The co-creation of mobile communication's public meaning. In R. Ling & P. Pederson (Eds.), *Mobile communications: Renegotiation of the social sphere* (pp. 63–81). London: Springer. doi:10.1007/1-84628-248-9_5

Katz, M. L., & Shapiro, C. (1985). Network externalities, competition and compatibility. *The American Economic Review*, *75*, 424–440. Retrieved from http://www.jstor.org

Kim, H. J., Kim, M. S., & Nam, Y. (2010). An analysis of self-construals, motivations, Facebook use, and user satisfaction. *Journal of Human-Computer Interaction*, *26*(11-12), 11–12. doi:10.1080/10447318.2010.516726

Kim, J., & Lee, R. J. E. (2011). The Facebook paths to happiness: Effects on the number of Facebook friends and self-presentation on subjective well-being. *Cyberpsychology, Behavior, and Social Networking*, *14*(6), 359–364. doi:10.1089/cyber.2010.0374 PMID:21117983

LaRue, M. E., Mitchell, M. A., Terhorst, L., & Karimi, A. H. (2010). Assessing mobile phone communication utility preferences in social support network. *Telematics and Informatics*, *27*(4), 363–369. doi:10.1016/j.tele.2010.03.002

Lenhart, A. (2013). *Young adults, mobile phones and social media: Technology and the transition to adulthood*. Retrieved June 10, 2013, from http://www.pewinternet.org/2013/05/07/young-adults-mobile-phones-and-social-media-technology-and-the-transition-to-adulthood/

Leung, L., & Wei, R. (2000). More than just talk on the move: Uses and gratifications of the cellular phone. *Journalism & Mass Communication Quarterly*, *77*(2), 308–320. doi:10.1177/107769900007700206

Li, X., & Chen, W. (2014). Facebook or Renren? A comparative study of social networking site use and social capital among Chinese international students in the United States. *Computers in Human Behavior*, *35*, 116–123. doi:10.1016/j.chb.2014.02.012

Lin, C. P., & Bhattacherjee, A. (2008). Elucidating individual intention to use interactive information technologies: The role of network externalities. *International Journal of Electronic Commerce*, *13*(1), 85–108. doi:10.2753/JEC1086-4415130103

Lin, K. Y., & Lu, H. P. (2011). Why people use social networking sites: An empirical study integrating network externalities and motivation theory. *Computers in Human Behavior*, *27*(3), 1152–1161. doi:10.1016/j.chb.2010.12.009

Ling, R. (2000). "We will be reached": The use of mobile telephony among Norwegian youth. *Information Technology & People*, *13*(2), 102–120. doi:10.1108/09593840010339844

Ling, R. (2004). *The mobile connection: The cell phone's impact on society*. San Francisco: Morgan Kaufmann.

Ling, R. (2008). *New tech, new ties: How mobile communication is reshaping social cohesion*. Cambridge, MA: MIT Press.

Lobet-Marris, C. (2003). Mobile phone tribes: Youth and social identity. In L. Fortunati, J. Katz, & R. Riccini (Eds.), *Mediating the human body: Technology, communication and fashion* (pp. 87–92). Mawah, NJ: Lawrence Erlbaum.

Mäntymäki, M., & Islam, A. (2014). Social virtual world continuance among teens: Uncovering the moderating role of perceived aggregate network exposure. *Behaviour & Information Technology*, *33*(5), 536–547. doi:10.1080/0144929X.2013.872190

Mashable. [Web log post]. (n.d.). *Smartphone*. Retrieved September 14, 2012, from http://mashable.com/follow/topics/smartphone/

McGarry, C. (2013). Control who can find you on Facebook. *PC World*, *31*(11), 178–179.

Mehdizadeh, S. (2010). Self-presentation 2.0: Narcissism and self-esteem on Facebook. *Cyberpsychology, Behavior, and Social Networking*, *13*(4), 357–364. doi:10.1089/cyber.2009.0257 PMID:20712493

Michikyan, M., Subrahmanyam, K., & Dennis, J. (2014). Can you tell who I am? Neuroticism, extraversion, and online self-presentation among young adults. *Computers in Human Behavior*, *33*, 179–183. doi:10.1016/j.chb.2014.01.010

Nielsen Company (The). (2012). *State of the media: Social media report*. Retrieved from The Nielsen Company website: http://blog.nielsen.com/nielsenwire/social/2012/

Pempek, T. A., Yermolayeva, Y. A., & Calvert, S. L. (2009). College students' social networking experiences on Facebook. *Journal of Applied Developmental Psychology*, *30*(3), 227–238. doi:10.1016/j.appdev.2008.12.010

Pew Internet Research Project. (2012). *A comparison of results from surveys by the Pew Research Center and Google Consumer Surveys*. Retrieved June 10, 2014, from http://www.people-press.org/2012/11/07/a-comparison-of-results-from-surveys-by-the-pew-research-center-and-google-consumer-surveys/

Pew Internet Research Project. (2012). *Privacy management on social media sites*. Retrieved June 10, 2014 from http://www.pewinternet.org/2012/02/24/main-findings-12/

Pew Internet Research Project. (2012). *Teens and mobile app privacy*. Retrieved June 10, 2014 from http://www.pewinternet.org/2013/08/22/teens-and-mobile-apps-privacy/

Pew Research Internet Project. (2013). *Frequency of social media use*. Retrieved June 10, 2014 from http://www.pewinternet.org/2013/12/30/frequency-of-social-media-use/

Randall, N. (2014). Let your fingers do the talking: Students' social distance as revealed through cell phone contact lists. *Sociological Spectrum*, *34*(2), 136–142. doi:10.1080/02732173.2014.878608

Rosen, L. D., Mark Carrier, L. L., & Cheever, N. A. (2013). Facebook and texting made me do it: Media-induced task-switching while studying. *Computers in Human Behavior*, *29*(3), 948–958. doi:10.1016/j.chb.2012.12.001

Ross, C., Orr, S. E., Sisic, M., Arseneault, J. M., Simmering, M. G., & Orr, R. R. (2009). Personality and motivations associated with Facebook use. *Computers in Human Behavior*, *25*(2), 578–586. doi:10.1016/j.chb.2008.12.024

Schilling, M. A. (2002). Technology success and failure in winner-take-all markets: The impact of learning orientation, timing and network externalities. *Academy of Management Journal*, *45*(2), 387–398. doi:10.2307/3069353

Scoble, R., & Israel, S. (2013). *Age of context: Mobile, sensors, data and the future of privacy*. Lexington, KY: Patrick Brewster Press.

Smith, A. (2013). *Pew research Internet project*. Retrieved June 5, 2014, from http://www.pewinternet.org/2013/06/05/smartphone-ownership-2013/

Spies Shapiro, L., & Margolin, G. (2014). Growing up wired: Social networking sites and adolescent psychosocial development. *Clinical Child and Family Psychology Review*, *17*(1), 1–18. doi:10.1007/s10567-013-0135-1 PMID:23645343

Stern, T., & Kumar, N. (2014). Improving privacy settings control in online social networks with a wheel interface. *Journal of the Association For Information Science & Technology*, *65*(3), 524–538. doi:10.1002/asi.22994

Strader, T. J., Ramaswani, S. N., & Houle, P. A. (2007). Perceived network externalities and communication technology acceptance. *European Journal of Information Systems*, *16*(1), 54–65. doi:10.1057/palgrave.ejis.3000657

Sutko, D. M., & de Souza e Silva, A. (2011). Location-aware mobile media and urban sociability. *New Media & Society*, *13*(5), 807–823. doi:10.1177/1461444810385202

Tazghini, S., & Siedlecki, K. L. (2013). A mixed method approach to examining Facebook use and its relationship to self-esteem. *Computers in Human Behavior*, *29*(3), 827–832. doi:10.1016/j.chb.2012.11.010

Tifferet, S., & Vilnai-Yavetz, I. (2014). Gender differences in Facebook self-presentation: An international randomized study. *Computers in Human Behavior*, *35*, 388–399. doi:10.1016/j.chb.2014.03.016

Tiggemann, M., & Slater, A. (2014). NetTweens: The internet and body image concerns in preteenage girls. *The Journal of Early Adolescence*, *34*(5), 606–620. doi:10.1177/0272431613501083

Tong, S. T., Van der Heide, B., Langwell, L., & Walther, J. B. (2008). Too much of a good thing? The relationship between number of friends and interpersonal impressions on Facebook. *Journal of Computer-Mediated Communication, 13*(3), 531–549. doi:10.1111/j.1083-6101.2008.00409.x

Turner, J., & Croucher, S. (2014). An examination of the relationships among United States college students' media use habits, need for cognition, and grade point average. *Learning, Media and Technology, 39*(2), 199–214. doi:10.1080/17439884.2013.777349

Underwood, J. D. M., Kerlin, L., & Farrington-Flint, L. (2011). The lies we tell and what they say about us: Using behavioural characteristics to explain Facebook activity. *Computers in Human Behavior, 27*(5), 1621–1626. doi:10.1016/j.chb.2011.01.012

Utz, S., Tanis, M., & Vermeulen, I. (2012). It is all about being popular: The effects of needs for popularity on social network site use. *Cyberpsychology, Behavior, and Social Networking, 15*(1), 37–42. doi:10.1089/cyber.2010.0651 PMID:21988765

Verkasalo, H., Lopez-Nicolas, C., Molina-Castillo, F. J., & Bouwman, H. (2010). Analysis of users and non-users of smartphone applications. *Telematics and Informatics, 27*(3), 242–255. doi:10.1016/j.tele.2009.11.001

Wang, C. C., Hsu, Y., & Fang, W. (2004). Acceptance of technology with network externalities: An empirical study of internet instant messaging services. *Journal of Information Technology Theory and Application, 6*, 15–28. Retrieved from http://aisel.aisnet.org/jitta/

Wang, C. C., Lo, S. K., & Fang, W. (2008). Extending the technology acceptance model to mobile telecommunication innovation: The existence of network externalities. *Journal of Consumer Behaviour, 7*(2), 101–110. doi:10.1002/cb.240

Wang, N. N., Kosinski, M. M., Stillwell, D. D., & Rust, J. J. (2014). Can well-being be measured using Facebook status updates? Validation of Facebook's gross national happiness index. *Social Indicators Research, 115*(1), 483–491. doi:10.1007/s11205-012-9996-9

Wei, P., & Lu, H. (2014). Why do people play mobile social games? An examination of network externalities and of uses and gratifications. *Internet Research, 24*(3), 313–331. doi:10.1108/IntR-04-2013-0082

Wei, R. (2008). Motivations for using the mobile phone for mass communications and entertainment. *Telematics and Informatics, 25*(1), 36–46. doi:10.1016/j.tele.2006.03.001

Wei, R., & Lo, V. H. (2006). Staying connected while on the move: Cell phone use and social connectedness. *New Media & Society, 8*(1), 53–72. doi:10.1177/1461444806059870

Yang, Y., Wang, X., & Li, L. (2013). Use mobile devices to wirelessly operate computers. *International Journal of Technology and Human Interaction, 9*(1), 64–67. doi:10.4018/jthi.2013010105

Young, A., & Quan-Haase, A. (2013). Privacy protection strategies on Facebook. *Information Communication and Society, 16*(4), 479–500. doi:10.1080/1369118X.2013.777757

Zhou, T., & Lu, Y. (2011). Examining mobile instant messaging user loyalty from the perspectives of network externalities and flow experience. *Computers in Human Behavior, 27*(2), 883–889. doi:10.1016/j.chb.2010.11.013

ADDITIONAL READING

Baek, K., Holten, A., Harp, D., & Yaschur, C. (2011). The links that bind: Uncovering novel motivations for linking on Facebook. *Computers in Human Behavior, 27*(6), 2243–2248. doi:10.1016/j.chb.2011.07.003

Baek, K., Holten, A., Harp, D., & Yaschur, C. (2011). The links that bind: Uncovering novel motivations for linking on Facebook. *Computers in Human Behavior, 27*(6), 2243–2248. doi:10.1016/j.chb.2011.07.003

Bakke, E. (2010). A model and measure of mobile communication competence. *Human Communication Research, 36*(3), 348–371. doi:10.1111/j.1468-2958.2010.01379.x

Bakke, E. (2010). A model and measure of mobile communication competence. *Human Communication Research, 36*(3), 348–371. doi:10.1111/j.1468-2958.2010.01379.x

Barak, O., & Pavelski, J. (2013). The worst social media blunders of 2013. Retrieved June 10, 2014, from http://nypost.com/2013/12/31/worst-social-media-blunders-of-2013/#

Barak, O., & Pavelski, J. (2013). The worst social media blunders of 2013. Retrieved June 10, 2014, from http://nypost.com/2013/12/31/worst-social-media-blunders-of-2013/#

Bloustien, G. (2007). "Wigging people out": Youth music practice and mediated communities. *Journal of Community & Applied Social Psychology, 17*(6), 446–462. doi:10.1002/casp.938

Bloustien, G. (2007). "Wigging people out": Youth music practice and mediated communities. *Journal of Community & Applied Social Psychology, 17*(6), 446–462. doi:10.1002/casp.938

Chang, C., & Heo, J. (2014). Visiting theories that predict college students' self-disclosure on Facebook. *Computers in Human Behavior, 30*, 79–86. doi:10.1016/j.chb.2013.07.059

Chang, C., & Heo, J. (2014). Visiting theories that predict college students' self-disclosure on Facebook. *Computers in Human Behavior, 30*, 79–86. doi:10.1016/j.chb.2013.07.059

Chen, T., Harper, S., & Yesilada, Y. (2011). How do people use their mobile phones?: A field study of small device users. *International Journal of Mobile Human Computer Interaction, 3*(1), 37–54. doi:10.4018/jmhci.2011010103

Chen, T., Harper, S., & Yesilada, Y. (2011). How do people use their mobile phones?: A field study of small device users. *International Journal of Mobile Human Computer Interaction, 3*(1), 37–54. doi:10.4018/jmhci.2011010103

Chen, W., & Lee, K. (2013). Sharing, liking, commenting, and distressed? The pathway between Facebook interaction and psychological distress. *Cyberpsychology, Behavior, and Social Networking, 16*(10), 728–734. doi:10.1089/cyber.2012.0272 PMID:23745614

Chen, W., & Lee, K. (2013). Sharing, liking, commenting, and distressed? The pathway between Facebook interaction and psychological distress. *Cyberpsychology, Behavior, and Social Networking, 16*(10), 728–734. doi:10.1089/cyber.2012.0272 PMID:23745614

Chiou, W., Chen, S., & Liao, D. (2014). Does Facebook promote self-interest? Enactment of indiscriminate one-to-many communication on online social networking sites decreases prosocial behavior. *Cyberpsychology, Behavior, and Social Networking*, *17*(2), 68–73. doi:10.1089/cyber.2013.0035 PMID:24102570

Chiou, W., Chen, S., & Liao, D. (2014). Does Facebook promote self-interest? Enactment of indiscriminate one-to-many communication on online social networking sites decreases prosocial behavior. *Cyberpsychology, Behavior, and Social Networking*, *17*(2), 68–73. doi:10.1089/cyber.2013.0035 PMID:24102570

Cliff, L., Ellison, N., & Steinfield, C. (2006). A face (book) in the crowd: Social searching vs. social browsing. *Proceedings of the 2006 20th Anniversary Conference on Computer Supported Cooperative Work, New York*, 167-170. doi:10.1145/1180875.1180901

Cliff, L., Ellison, N., & Steinfield, C. (2006). A face (book) in the crowd: Social searching vs. social browsing. *Proceedings of the 2006 20th Anniversary Conference on Computer Supported Cooperative Work, New York*, 167-170. doi:10.1145/1180875.1180901

Colapinto, C., & Benecchi, E. (2014). The presentation of celebrity personas in everyday Twittering: Managing online reputations throughout a communication crisis. *Media Culture & Society*, *36*(2), 219–233. doi:10.1177/0163443714526550

Colapinto, C., & Benecchi, E. (2014). The presentation of celebrity personas in everyday Twittering: Managing online reputations throughout a communication crisis. *Media Culture & Society*, *36*(2), 219–233. doi:10.1177/0163443714526550

Cousins, C. K., & Robey, D. (2005). Human agency in a wireless world: Patterns of technology use in nomadic computing environments. *Information and Organization*, *15*(2), 151–180. doi:10.1016/j.infoandorg.2005.02.008

Cousins, C. K., & Robey, D. (2005). Human agency in a wireless world: Patterns of technology use in nomadic computing environments. *Information and Organization*, *15*(2), 151–180. doi:10.1016/j.infoandorg.2005.02.008

DiMicco, J. M., & Millen, D. R. (2007). Identity management: Multiple presentations of self in Facebook. *Proceedings of the 2007 International ACM Conference on Supporting Group Work, Sanibel Island, FL*, 383-386. doi:10.1145/1316624.1316682

DiMicco, J. M., & Millen, D. R. (2007). Identity management: Multiple presentations of self in Facebook. *Proceedings of the 2007 International ACM Conference on Supporting Group Work, Sanibel Island, FL*, 383-386. doi:10.1145/1316624.1316682

Donath, J., & Boyd, D. (2004). Public displays of connection. *BT Technology Journal*, *22*(4), 71–82. doi:10.1023/B:BTTJ.0000047585.06264.cc

Donath, J., & Boyd, D. (2004). Public displays of connection. *BT Technology Journal*, *22*(4), 71–82. doi:10.1023/B:BTTJ.0000047585.06264.cc

Dowdall, C. (2009). Impressions, improvisations and compositions: Reframing children's text production in social network sites. *Literacy*, *43*(2), 91–99. doi:10.1111/j.1741-4369.2009.00521.x

Dowdall, C. (2009). Impressions, improvisations and compositions: Reframing children's text production in social network sites. *Literacy*, *43*(2), 91–99. doi:10.1111/j.1741-4369.2009.00521.x

Ellison, N. B., Heino, R., & Gibbs, J. (2006). Managing impressions online: Self presentation processes in the online dating environment. *Journal of Computer-Mediated Communication*, *11*(2), 415–441. http://jcmc.indiana.edu/ doi:10.1111/j.1083-6101.2006.00020.x

Ellison, N. B., Heino, R., & Gibbs, J. (2006). Managing impressions online: Self presentation processes in the online dating environment. *Journal of Computer-Mediated Communication*, *11*(2), 415–441. http://jcmc.indiana.edu/ doi:10.1111/j.1083-6101.2006.00020.x

Fifield, N. (2014, May 30). Right to be forgotten means "old rules of the Internet no longer apply." *The Telegraph*. Retrieved from http://www.telegraph.co.uk/technology/google/10866517/Right-to- be-forgotten-means-old-rules-of-the-Internet-no-longer-apply.html

Fifield, N. (2014, May 30). Right to be forgotten means "old rules of the Internet no longer apply." *The Telegraph*. Retrieved from http://www.telegraph.co.uk/technology/google/10866517/Right-to- be-forgotten-means-old-rules-of-the-Internet-no-longer-apply.html

Fu, H., Lin, P., Yue, H., Huang, G., & Lee, C. (2014). Group mobility management for large- scale machine-to-machine mobile networking. *IEEE Transactions on Vehicular Technology*, *63*(3), 1296–1305. doi:10.1109/TVT.2013.2284596

Fu, H., Lin, P., Yue, H., Huang, G., & Lee, C. (2014). Group mobility management for large- scale machine-to-machine mobile networking. *IEEE Transactions on Vehicular Technology*, *63*(3), 1296–1305. doi:10.1109/TVT.2013.2284596

Garcia, D., & Sikström, S. (2014). The dark side of Facebook: Semantic representations of status updates predict the Dark Triad of personality. *Personality and Individual Differences*, *67*, 92–96. doi:10.1016/j.paid.2013.10.001

Garcia, D., & Sikström, S. (2014). The dark side of Facebook: Semantic representations of status updates predict the Dark Triad of personality. *Personality and Individual Differences*, *67*, 92–96. doi:10.1016/j.paid.2013.10.001

Geser, H. (2004). Towards a sociological theory of the mobile phone. *Sociology in Switzerland: Sociology of the Mobile Phone*. Retrieved from http://socio.ch/mobile/index_mobile.htm

Geser, H. (2004). Towards a sociological theory of the mobile phone. *Sociology in Switzerland: Sociology of the Mobile Phone*. Retrieved from http://socio.ch/mobile/index_mobile.htm

Greenhow, C., & Robelia, B. (2009). Informal learning and identity formation in online social networks. *Learning, Media and Technology*, *34*(2), 119–140. doi:10.1080/17439880902923580

Greenhow, C., & Robelia, B. (2009). Informal learning and identity formation in online social networks. *Learning, Media and Technology*, *34*(2), 119–140. doi:10.1080/17439880902923580

Greenhow, C., & Robelia, B. (2009b). Old communication, new literacies: Social network sites as social learning resources. *Journal of Computer-Mediated Communication*, *14*(4), 1130–1161. http://jcmc.indiana.edu/ doi:10.1111/j.1083-6101.2009.01484.x

Greenhow, C., & Robelia, B. (2009b). Old communication, new literacies: Social network sites as social learning resources. *Journal of Computer-Mediated Communication, 14*(4), 1130–1161. http://jcmc.indiana.edu/ doi:10.1111/j.1083-6101.2009.01484.x

Harter, S. (1999). *The construction of the self: A developmental perspective.* New York: Guilford Press.

Harter, S. (1999). *The construction of the self: A developmental perspective.* New York: Guilford Press.

Hjorth, L. (2009). Web U2: Emerging online communities and gendered intimacy in the Asia-Pacific region. *Knowledge, Technology & Policy, 22*(2), 117–124. doi:10.1007/s12130-009-9077-9

Hjorth, L. (2009). Web U2: Emerging online communities and gendered intimacy in the Asia-Pacific region. *Knowledge, Technology & Policy, 22*(2), 117–124. doi:10.1007/s12130-009-9077-9

Humphreys, L. (2008). Mobile social networks and social practice: A case study of dodgeball. *Journal of Computer-Mediated Communication, 13*(1), 341–360. http://jcmc.indiana.edu/ doi:10.1111/j.1083-6101.2007.00399.x

Humphreys, L. (2008). Mobile social networks and social practice: A case study of dodgeball. *Journal of Computer-Mediated Communication, 13*(1), 341–360. http://jcmc.indiana.edu/ doi:10.1111/j.1083-6101.2007.00399.x

Humphreys, L. (2011). Who's watching whom? A study of interactive technology and surveillance. *Journal of Communication, 61*(4), 575–595. doi:10.1111/j.1460-2466.2011.01570.x

Humphreys, L. (2011). Who's watching whom? A study of interactive technology and surveillance. *Journal of Communication, 61*(4), 575–595. doi:10.1111/j.1460-2466.2011.01570.x

Jean-Richard, V., Crump, L., Daugla, D. M., Hattendorf, J., Schelling, E., & Zinsstag, J. (2014). The use of mobile phones for demographic surveillance of mobile pastoralists and their animals in Chad: Proof of principle. *Global Health Action, 7*.

Jean-Richard, V., Crump, L., Daugla, D. M., Hattendorf, J., Schelling, E., & Zinsstag, J. (2014). The use of mobile phones for demographic surveillance of mobile pastoralists and their animals in Chad: Proof of principle. *Global Health Action, 7*.

Johnston, B. (2009, March 31). Social networking and Craigslist top mobile destinations, says Openwave. *Wireless and Mobile News.* Retrieved from http://wirelessandmobilenews.com

Johnston, B. (2009, March 31). Social networking and Craigslist top mobile destinations, says Openwave. *Wireless and Mobile News.* Retrieved from http://wirelessandmobilenews.com

Junco, R., & Cotten, S. R. (2012). No A 4 U: The relationship between multitasking and academic performance. *Computers & Education, 59*(2), 505–514. doi:10.1016/j.compedu.2011.12.023

Junco, R., & Cotten, S. R. (2012). No A 4 U: The relationship between multitasking and academic performance. *Computers & Education, 59*(2), 505–514. doi:10.1016/j.compedu.2011.12.023

Kaba, B., & Osei-Bryson, K. M. (2012). An empirical investigation of external factors influencing mobile technology use in Canada: A preliminary study. *International Journal of Technology and Human Interaction, 8*(1), 1–14. doi:10.4018/jthi.2012040101

Kaba, B., & Osei-Bryson, K. M. (2012). An empirical investigation of external factors influencing mobile technology use in Canada: A preliminary study. *International Journal of Technology and Human Interaction*, *8*(1), 1–14. doi:10.4018/jthi.2012040101

Kaba, B., & Osei-Bryson, K. M. (2012). An empirical investigation of external factors influencing mobile technology use in Canada: A preliminary study. *International Journal of Technology and Human Interaction*, *8*(1), 1–14. doi:10.4018/jthi.2012040101

Kaba, B., & Osei-Bryson, K. M. (2012). An empirical investigation of external factors influencing mobile technology use in Canada: A preliminary study. *International Journal of Technology and Human Interaction*, *8*(1), 1–14. doi:10.4018/jthi.2012040101

Kakihara, M., Sorensen, C., & Wiberg, M. (2002, May). *Fluid interaction in mobile work practices.* Paper presented at the meeting of The First Global Mobile Roundtable, Tokyo, Japan. Retrieved from http://mobility.lse.ac.uk/download/kakiharasorensenwiberg2002.pdf

Kakihara, M., Sorensen, C., & Wiberg, M. (2002, May). *Fluid interaction in mobile work practices.* Paper presented at the meeting of The First Global Mobile Roundtable, Tokyo, Japan. Retrieved from http://mobility.lse.ac.uk/download/kakiharasorensenwiberg2002.pdf

Kapidzic, S. (2013). Narcissism as a predictor of motivations behind Facebook profile picture selection. *Cyberpsychology, Behavior, and Social Networking*, *16*(1), 14–19. doi:10.1089/cyber.2012.0143 PMID:23249240

Kapidzic, S. (2013). Narcissism as a predictor of motivations behind Facebook profile picture selection. *Cyberpsychology, Behavior, and Social Networking*, *16*(1), 14–19. doi:10.1089/cyber.2012.0143 PMID:23249240

Karpinski, A. C., Kirschner, P. A., Ozer, I., Mellott, J. A., & Ochwo, P. (2013). An exploration of social networking site use, multitasking, and academic performance among United States and European university students. *Computers in Human Behavior*, *29*(3), 1182–1192. doi:10.1016/j.chb.2012.10.011

Karpinski, A. C., Kirschner, P. A., Ozer, I., Mellott, J. A., & Ochwo, P. (2013). An exploration of social networking site use, multitasking, and academic performance among United States and European university students. *Computers in Human Behavior*, *29*(3), 1182–1192. doi:10.1016/j.chb.2012.10.011

Katz, J. E., & Aakhus, M. (Eds.). (2002). *Perpetual contact: Mobile communication, private talk, public performance*. New York: Cambridge University Press. doi:10.1017/CBO9780511489471

Katz, J. E., & Aakhus, M. (Eds.). (2002). *Perpetual contact: Mobile communication, private talk, public performance*. New York: Cambridge University Press. doi:10.1017/CBO9780511489471

Kenski, K., & Stroud, N. J. (2006). Connections between internet use and political efficacy, knowledge, and participation. *Journal of Broadcasting & Electronic Media*, *50*(2), 173–192. doi:10.1207/s15506878jobem5002_1

Kenski, K., & Stroud, N. J. (2006). Connections between internet use and political efficacy, knowledge, and participation. *Journal of Broadcasting & Electronic Media*, *50*(2), 173–192. doi:10.1207/s15506878jobem5002_1

Khattab, I., & Love, S. (2008). Mobile phone use across cultures: A comparison between the United Kingdom and Sudan. *International Journal of Technology and Human Interaction*, *4*(2), 35–52. doi:10.4018/jthi.2008040103

Khattab, I., & Love, S. (2008). Mobile phone use across cultures: A comparison between the United Kingdom and Sudan. *International Journal of Technology and Human Interaction*, *4*(2), 35–52. doi:10.4018/jthi.2008040103

Kirschner, P. A., & Karpinski, A. C. (2010). Facebook® and academic performance. *Computers in Human Behavior*, *26*(6), 1237–1245. doi:10.1016/j.chb.2010.03.024

Kirschner, P. A., & Karpinski, A. C. (2010). Facebook® and academic performance. *Computers in Human Behavior*, *26*(6), 1237–1245. doi:10.1016/j.chb.2010.03.024

Lampe, C., Ellison, N. B., & Steinfield, C. (2007). A familiar face(book): Profile elements as signals in an online social network. *Proceedings of the SIGCHI Conference on Human Factors in Computing Systems, San Jose, CA*. doi:10.1145/1240624.1240695

Lampe, C., Ellison, N. B., & Steinfield, C. (2007). A familiar face(book): Profile elements as signals in an online social network. *Proceedings of the SIGCHI Conference on Human Factors in Computing Systems, San Jose, CA*. doi:10.1145/1240624.1240695

Lampe, C., Ellison, N. B., & Steinfield, C. (2008). Changes in the use and perception of Facebook. *Proceedings of the 2008 ACM Conference on Computer Supported Cooperative Work, New York*. doi:10.1145/1460563.1460675

Lampe, C., Ellison, N. B., & Steinfield, C. (2008). Changes in the use and perception of Facebook. *Proceedings of the 2008 ACM Conference on Computer Supported Cooperative Work, New York*. doi:10.1145/1460563.1460675

Li, N., & Chen, G. (2009). Analysis of a location-based social network. *Proceedings of the 2009 International Conference on Computational Science and Engineering, Washington, D.C.*, *4*, 263-270. doi:10.1109/CSE.2009.98

Li, N., & Chen, G. (2009). Analysis of a location-based social network. *Proceedings of the 2009 International Conference on Computational Science and Engineering, Washington, D.C.*, *4*, 263-270. doi:10.1109/CSE.2009.98

Lu, Y., Deng, Z., & Wang, B. (2010). Exploring factors affecting Chinese consumers' usage of short message service for personal communication. *Information Systems Journal*, *20*(2), 183–208. doi:10.1111/j.1365-2575.2008.00312.x

Lu, Y., Deng, Z., & Wang, B. (2010). Exploring factors affecting Chinese consumers' usage of short message service for personal communication. *Information Systems Journal*, *20*(2), 183–208. doi:10.1111/j.1365-2575.2008.00312.x

Mainsah, H. (2011). "I could well have said I was Norwegian but nobody would believe me": Ethnic minority youths' self-representation on social network sites. *European Journal of Cultural Studies*, *14*(2), 179–193. doi:10.1177/1367549410391926

Mainsah, H. (2011). "I could well have said I was Norwegian but nobody would believe me": Ethnic minority youths' self-representation on social network sites. *European Journal of Cultural Studies*, *14*(2), 179–193. doi:10.1177/1367549410391926

Marwick, E. A., & Boyd, D. (2010). I tweet honestly, I tweet passionately: Twitter users, context, collapse and the imagined audience. *New Media & Society*, *13*(1), 114–133. doi:10.1177/1461444810365313

Marwick, E. A., & Boyd, D. (2010). I tweet honestly, I tweet passionately: Twitter users, context, collapse and the imagined audience. *New Media & Society*, *13*(1), 114–133. doi:10.1177/1461444810365313

Marwick, E. A., & Boyd, D. (2011). To see and be seen: Celebrity practice on Twitter. *Convergence (London)*, *17*(2), 139–158. doi:10.1177/1354856510394539

Marwick, E. A., & Boyd, D. (2011). To see and be seen: Celebrity practice on Twitter. *Convergence (London)*, *17*(2), 139–158. doi:10.1177/1354856510394539

Maximilien, M. E. (2008). Mobile mashups: Thoughts, directions, and challenges. *The IEEE International Conference on Semantic Computing*, 597-600. doi:10.1109/ICSC.2008.100

Maximilien, M. E. (2008). Mobile mashups: Thoughts, directions, and challenges. *The IEEE International Conference on Semantic Computing*, 597-600. doi:10.1109/ICSC.2008.100

Pai, P., & Arnott, D. C. (2013). User adoption of social networking sites: Eliciting uses and gratifications through a means–end approach. *Computers in Human Behavior*, *29*(3), 1039–1053. doi:10.1016/j.chb.2012.06.025

Pai, P., & Arnott, D. C. (2013). User adoption of social networking sites: Eliciting uses and gratifications through a means–end approach. *Computers in Human Behavior*, *29*(3), 1039–1053. doi:10.1016/j.chb.2012.06.025

Papachararissi, Z. (2011). Conclusion: A networked self. In Z. Papacharissi (Ed.), *A networked self: Identity, community, and culture on social network sites* (pp. 304–318). New York: Routledge.

Papachararissi, Z. (2011). Conclusion: A networked self. In Z. Papacharissi (Ed.), *A networked self: Identity, community, and culture on social network sites* (pp. 304–318). New York: Routledge.

Repo, P., Hyvonen, K., Pantzar, M., & Timonen, P. (2006). Inventing use for a novel mobile service. *International Journal of Technology and Human Interaction*, *2*(2), 49–64. doi:10.4018/jthi.2006040103

Repo, P., Hyvonen, K., Pantzar, M., & Timonen, P. (2006). Inventing use for a novel mobile service. *International Journal of Technology and Human Interaction*, *2*(2), 49–64. doi:10.4018/jthi.2006040103

Rohn, U. (2014). Social networking sites across cultures and countries: Proximity and network effects. *Qualitative Research Reports in Communication*, *14*(1), 28–34. doi:10.1080/17459435.2013.835339

Rohn, U. (2014). Social networking sites across cultures and countries: Proximity and network effects. *Qualitative Research Reports in Communication*, *14*(1), 28–34. doi:10.1080/17459435.2013.835339

Salimkhan, G., Manago, A., & Greenfield, P. (2010). The construction of the virtual self on MySpace. *Cyberpsychology. Journal of Psychological Research on Cyberspace*, *4*(1), 1–19. Retrieved from http://www.cyberpsychology.eu/index.php

Salimkhan, G., Manago, A., & Greenfield, P. (2010). The construction of the virtual self on MySpace. *Cyberpsychology. Journal of Psychological Research on Cyberspace*, *4*(1), 1–19. Retrieved from http://www.cyberpsychology.eu/index.php

Seifert, J., Dobbelstein, D., Schmidt, D., Holleis, P., & Rukzio, E. (2014). From the private into the public: Privacy-respecting mobile interaction techniques for sharing data on surfaces. *Personal and Ubiquitous Computing*, 1–14.

Seifert, J., Dobbelstein, D., Schmidt, D., Holleis, P., & Rukzio, E. (2014). From the private into the public: Privacy-respecting mobile interaction techniques for sharing data on surfaces. *Personal and Ubiquitous Computing*, 1–14.

Shah, D. V., Kwak, N., & Holbert, R. L. (2001). "Connecting" and "disconnecting" with civic life: Patterns of internet use and the production of social capital. *Political Communication*, *18*, 141–162. doi:10.1080/105846001750322952

Shah, D. V., Kwak, N., & Holbert, R. L. (2001). "Connecting" and "disconnecting" with civic life: Patterns of internet use and the production of social capital. *Political Communication*, *18*, 141–162. doi:10.1080/105846001750322952

Shu, W., & Chuang, H. Y. (2011). Why people share knowledge in virtual communities. *Social Behavior and Personality*, *39*(5), 671–690. doi:10.2224/sbp.2011.39.5.671

Shu, W., & Chuang, H. Y. (2011). Why people share knowledge in virtual communities. *Social Behavior and Personality*, *39*(5), 671–690. doi:10.2224/sbp.2011.39.5.671

Shuter, R., & Chattopadhyay, S. (2014). A cross-national study of cultural values and contextual norms of mobile phone activity. *Journal of Multicultural Discourses*, *9*(1), 61–70. doi:10.1080/17447143.2013.859262

Shuter, R., & Chattopadhyay, S. (2014). A cross-national study of cultural values and contextual norms of mobile phone activity. *Journal of Multicultural Discourses*, *9*(1), 61–70. doi:10.1080/17447143.2013.859262

Sorenson, S. B., Rui, S., Jingwen, Z., & Jia, X. (2014). Self-presentation on the Web: Agencies serving abused and assaulted women. *American Journal of Public Health*, *104*(4), 702–707. doi:10.2105/AJPH.2013.301629 PMID:24524489

Sorenson, S. B., Rui, S., Jingwen, Z., & Jia, X. (2014). Self-presentation on the Web: Agencies serving abused and assaulted women. *American Journal of Public Health*, *104*(4), 702–707. doi:10.2105/AJPH.2013.301629 PMID:24524489

Valenzuela, S., Park, N., & Kee, K. F. (2009). Is there social capital in a social networking site?: Facebook use and college students' life satisfaction, trust, and participation. *Journal of Computer-Mediated Communication*, *14*(4), 875–901. doi:10.1111/j.1083-6101.2009.01474.x

Valenzuela, S., Park, N., & Kee, K. F. (2009). Is there social capital in a social networking site?: Facebook use and college students' life satisfaction, trust, and participation. *Journal of Computer-Mediated Communication*, *14*(4), 875–901. doi:10.1111/j.1083-6101.2009.01474.x

Wang, S., & Stefanone, M. A. (2013). Showing off? Human mobility and the interplay of traits, self-disclosure, and Facebook check-ins. *Social Science Computer Review*, *31*(4), 437–457. doi:10.1177/0894439313481424

Wang, S., & Stefanone, M. A. (2013). Showing off? Human mobility and the interplay of traits, self-disclosure, and Facebook check-ins. *Social Science Computer Review*, *31*(4), 437–457. doi:10.1177/0894439313481424

Wei, R., Lo, V., Xu, X., Chen, Y., & Zhang, G. (2014). Predicting mobile news use among college students: The role of press freedom in four Asian cities. *New Media & Society*, *16*(4), 637–654. doi:10.1177/1461444813487963

Wei, R., Lo, V., Xu, X., Chen, Y., & Zhang, G. (2014). Predicting mobile news use among college students: The role of press freedom in four Asian cities. *New Media & Society*, *16*(4), 637–654. doi:10.1177/1461444813487963

Wiredu, G. (2007). User appropriation of mobile technologies: Motives, conditions and design properties. *Information and Organization*, *17*(2), 110–129. doi:10.1016/j.infoandorg.2007.03.002

Wiredu, G. (2007). User appropriation of mobile technologies: Motives, conditions and design properties. *Information and Organization*, *17*(2), 110–129. doi:10.1016/j.infoandorg.2007.03.002

Yee, N., Bailenson, J. N., & Ducheneaut, N. (2009). The proteus effect: Implications of transformed digital self-representation on online and offline behavior. *Communication Research*, *36*(2), 285–312. doi:10.1177/0093650208330254

Yee, N., Bailenson, J. N., & Ducheneaut, N. (2009). The proteus effect: Implications of transformed digital self-representation on online and offline behavior. *Communication Research*, *36*(2), 285–312. doi:10.1177/0093650208330254

Young, K. (2011). Social ties, social networks and Facebook experience. *International Journal of Emerging Technologies and Society*, *9*, 20–34. Retrieved from http://www.academia.edu/922467/Social_Ties_Social_Networks_and_the_Facebook_Experience

Young, K. (2011). Social ties, social networks and Facebook experience. *International Journal of Emerging Technologies and Society*, *9*, 20–34. Retrieved from http://www.academia.edu/922467/Social_Ties_Social_Networks_and_the_Facebook_Experience

KEY TERMS AND DEFINITIONS

Identity: The tangible and intangible layered strands of a person that evolve contextually.

Network Externality: The value users receive from a product or service by linking with a large number of people.

Self-Broadcasting: The dynamic process of creating and disseminating narratives and images using smartphone applications.

Self-Esteem: The value that one attaches to herself or himself.

Smartphone: A cellular phone that is able to connect to the World Wide Web and able to give contextual information through installed applications.

Social Plugin: An application that facilitates real-time World Wide Web connectivity and sharing.

Web 2.0: Social technologies that facilitate creation of content, sharing, and interactivity on the World Wide Web.

APPENDIX

Questionnaire: SNS Involvement, Experience in Social Media

22. **What kind of cellular phones do you have?**
    a. Android Smartphones
    b. iPhones
    c. Other smartphones
    d. Regular cellular phones
29. **What types of mobile applications do you have on your cell phone/other mobile devices? (Check all that apply)**
    Games
    News
    Weather
    Educational
    Social media such as Facebook, LinkedIn
    Location-based social media devices such as Foursquare, Brightkite
    Media such as TV, radio, magazines
    Health
    Business
    Communication such as e-mail/texting or Skype
    Dating services
    Shopping
    Other (please specify) ______________________
40. **How many years have you been a member of social networking sites, such as Facebook (Write 0 if you are not a social network site user and skip to Q. 51)?**
    ____
42. **How do you set your profile on the social network site you use most often?**
    a. public
    b. partially private (friends of friends or your networks can see it)
    c. private (friends only)
43. **How frequently do you update your social network page?**
    a. Several times an hour
    b. Every several hours
    c. Everyday
    d. Once to several times a week
    e. Between once a week and once a month
    f. Less than once a month
    g. Hardly ever update
45. **Approximately how many "Friends" or "Followers" do you have on the main social network sites that you use? Write 0 if you don't use that social network site.**
    Facebook _______
    Twitter _______
    Other social media sites _______

46. **Please check the type of people you have on your Facebook friends list in following categories (Check all that apply).**
    a. Parents
    b. Spouse or current romantic partner
    c. Siblings and relatives (Other than parents or spouse)
    d. Co-workers from your current or previous job
    e. Neighbors
    f. Classmates in college
    g. People you went to high school with
    h. Members of a group you belong to
    i. People you have never met in person
    j. People you have met in person only one time
    k. Other
48. **How frequent do you do anything below on your social network sites?**

Table 2.

	Almost Daily	**Several Times a Week**	**Once a Week**	**Once a Month or Less**	**Never**
a. Post news content from other news media	○	○	○	○	○
b. Link to other media sites	○	○	○	○	○
c. Read news posted on the site	○	○	○	○	○
d. Post pictures taken by myself or people I know	○	○	○	○	○
e. Post Pictures from other sites	○	○	○	○	○
f. Post videos made by myself or people I know	○	○	○	○	○
g. Post Videos from other sites	○	○	○	○	○
h. Stay in touch with family members, including all communications with them on SNS	○	○	○	○	○
i. Stay in touch with close friends, including all communications with them on SNS	○	○	○	○	○
j. Connect with old friends that you lost touch with	○	○	○	○	○
k. Make new friends	○	○	○	○	○
l. Connect with other people who share your hobbies or interests	○	○	○	○	○
m. Find potential romantic partners or people you might like to date.	○	○	○	○	○
n.Read comments or posts by celebrities, politicians or athletes	○	○	○	○	○
o. Post comments to, or share something a friend has posted	○	○	○	○	○
p. Send instant messages to or chat with a friend through the social networking site	○	○	○	○	○
q. Tag people in posts, photos or videos	○	○	○	○	○
r. Play game	○	○	○	○	○
s. Find some useful shopping information, such as coupon or deal information.	○	○	○	○	○
t. Post product review/comments	○	○	○	○	○
u. Learn something useful from others	○	○	○	○	○
v. Receive encouraging messages from others	○	○	○	○	○

53. **What is your disposable personal income after taxes per month (including allowances)?**
    a. Under $500
    b. $500 - $1,500
    c. $1,501 - $3,000
    d. Over $3,000
54. **What was your total household income before taxes for the most recent calendar year (January through December)?**
    a. Under US$ 30,000
    b. $30,001-$60,000
    c. $60,001-$90,000
    d. $90,001-$150,000
    e. Over $150,000
60. **Please indicate your GPA.**
    a. Higher than 3.5
    b. Between 3 and 3.5
    c. Between 2.5 and 3
    d. Between 2 and 2.5
    e. Between 1.5 and 2
    f. Between 1 and 1.5
    g. Lower than 1
78. **Please check the categories that best represent yourself for each statement below.**

Table 3.

	Not at All True	Barely True	Moderately True	Exactly True
a. I can always manage to solve difficult problems if I try hard enough	○	○	○	○
b. If someone opposes me, I can find means and ways to get what I want	○	○	○	○
c. It is easy for me to stick to my aims and accomplish my goals	○	○	○	○
d. I am confident that I could deal efficiently with unexpected events	○	○	○	○
e. Thanks to my resourcefulness, I know how to handle unforeseen situations	○	○	○	○
f. I can solve most problems if I invest the necessary effort	○	○	○	○
g. I can remain calm when facing difficulties because I can rely on my coping abilities	○	○	○	○
h. When I am confronted with a problem, I can usually find several solutions	○	○	○	○
i. If I am in a bind, I can usually think of something to do	○	○	○	○
j. No matter what comes my way, I'm usually able to handle it	○	○	○	○

79. **How do you agree to the following statements?**

Table 4.

	Strongly Disagree	2	3	4	Strongly Agree
a. I feel that I am a person of worth, at least on an equal plane with others	O	O	O	O	O
b. I feel that I have a number of good qualities	O	O	O	O	O
c. All in all, I am inclined to feel that I am a failure	O	O	O	O	O
d. I am able to do things as well as most other people	O	O	O	O	O
e. I feel I do not have much to be proud of	O	O	O	O	O
f. I take a positive attitude toward myself	O	O	O	O	O
g. On the whole, I am satisfied with myself	O	O	O	O	O
h. Other people look to me as a source of conversation topics	O	O	O	O	O
i. I am likely to share with others information that I have read in the news	O	O	O	O	O
j. I enjoy discussing a range of topics with others	O	O	O	O	O
k. Other people appreciate the contributions I make in conversations	O	O	O	O	O

80. **Do you agree to the following statements about your attitudes toward academic study?**

Table 5.

	Strongly Disagree	2	3	4	Strongly Agree
a. I believe that I have the ability to complete college	O	O	O	O	O
b. I believe it is possible for me to make good grades	O	O	O	O	O
c. I doubt that I can make the effort to finish college	O	O	O	O	O
d. I am determined to do what it will take in order to succeed with my goals	O	O	O	O	O
e. I question why I need a degree for the career I want to pursue	O	O	O	O	O

81. **How would you evaluate your level of self-regulatory capability?**
    a. High
    b. Somewhat high
    c. Neither high nor low
    d. Somewhat low
    e. Low

83. **How do you think about the following statements regarding social trust?**

Table 6.

	Never	2	3	4	All of the Time
a. Generally speaking, would you say that people can be trusted	○	○	○	○	○
b. People try to take advantage of you if they got the chance	○	○	○	○	○
c. People try to be fair	○	○	○	○	○
d. You can't be too careful in dealing with people	○	○	○	○	○
e. People try to be helpful	○	○	○	○	○
f. People are just looking out for themselves	○	○	○	○	○

84. **How do you agree to the following statements regarding life satisfaction?**

Table 7.

	Strongly Disagree	2	3	4	Strongly Agree
a. In most ways my life is close to my ideal	○	○	○	○	○
b. The conditions of my life are excellent	○	○	○	○	○
c. I am satisfied with my life	○	○	○	○	○
d. So far I have gotten the important things I want in life	○	○	○	○	○
e. If I could live my time over, I would change almost nothing	○	○	○	○	○

Section 3

Information Technologies Characteristics, Improvement and Use

Chapter 6
Effects of Display Characteristics on Presence and Emotional Responses of Game Players

Zhiming Wu
Sichuan University, China

Ningjiu Tang
Sichuan University, China

Tao Lin
Sichuan University, China

Shaomei Wu
Sichuan University, China

ABSTRACT

Large displays are becoming ubiquitous and one important property of large displays is that they afford larger visual angles and physically larger screen sizes. However, there has been little investigation of the effects of large displays on users; furthermore, few studies have employed physiological measures, nor isolated the effects on presence and emotional responses of large visual angle and physically large screen size. This study, then, examines specifically the effects of these two properties of large displays on presence and emotional responses, using physiological as well as subjective methods. The results indicate that the larger visual angle offered by a large display can increase the sense of presence, engagement, and emotional responses when players are playing games. More interestingly, the physically larger sizes offered by a large display seem to also be able to affect these player experiences, even at identical visual angles with small displays.

INTRODUCTION

Recent advances in technology have made large displays ever more ubiquitous. There are various areas in which the use of a large display can be appealing or even critical, such as visual surveillance, air traffic control, design, visual analytics, presentations, collaboration work, and entertainment. Large displays possess two unique and potentially important properties: larger visual angle and physically larger screen size. However, the two characteristics of large displays are often ignored by HCI (human-computer inter-

DOI: 10.4018/978-1-4666-8450-8.ch006

action) researchers. Research on user response to media information is usually performed using screens of average size; the results are then generalized, perhaps unwittingly, to all screen sizes. This neglects the important possibility that different display characteristics may significantly affect the psychological importance of the information displayed (Ravaja, Turpeinen, Saari, Puttonen, & Keltikangas-Jarvinen, 2008; Liu, C., Chapuis, O., Beaudouin-Lafon, M., Lecolinet, E., & Mackay, W. E., 2014; Andrews, C., & North, C., 2013). Also, interaction design philosophies based on average screen size uniformly applied to newer displays may also lead to unintended consequences because screen size could augment or diminish expected effects (Ravaja et al., 2008). Recent studies (e.g., (Elliott & Hearst, 2002; Martinez, Jhala, & Yannakakis, 2009; Mary et al., 2006; D. S. Tan, Gergle, Scupelli, & Pausch, 2004; D. Tan, Gergle, Scupelli, & Pausch, 2006; Olsen, B. I., Laeng, B., Kristiansen, K. A., & Hartvigsen, G., 2013; Ichino, J., Isoda, K., Hanai, A., & Ueda, T. 2013)) have already begun to isolate certain interesting characteristics of large display, but most of them have assumed that larger display fill a greater percentage of the viewer's visual field, and physical size is often confused with visual angle. In fact, although researchers have studied the effects of display characteristics such as visual angle, resolution, brightness, contrast and color, little has been done to systematically isolate the effects of physical sizes and visual angles on user experiences.

The long-term goal of our research aims to isolate and study characteristics unique to large displays so that we form a theoretical basis for understanding how different display characteristics affect user experience. We believe that taking a bottom-up approach and understanding each of these fundamental characteristics in isolation rather than taking a top-down approach and studying how one display technology differs from another will be much more productive in the long term. Such an approach will allow us to build a general theory that explains effects induced by various display technologies simply by recombining our understanding of display characteristics and then studying their specific interactions(D. S. Tan, 2004).

The present study focuses specifically on game players; their presence, engagement, perceived challenge, emotional arousal, and emotional valence are measured. Based on previous empirical studies and theory from psychophysiology, we evaluate the following two main experimental hypotheses using subjective and physiological measures.

H1: Larger visual angles offered by large displays will induce greater presence, engagement, emotional arousal, and emotional valence as compared to small displays when game players are playing a 3-D game.

H2: Physically larger screen size offered by large displays will induce greater presence, engagement, emotional arousal, and emotional valence, even at identical visual angles with small displays, when game players are playing a 3-D game.

BACKGROUND

Effects of Large Display on Users

Robertson et al. (Mazaeva, Ntuen, & Lebby, 2001) document a series of user studies demonstrating productivity benefits from using large displays with an eye toward novel software applications that might better support the way information workers multitask between their projects and applications. These studies showed a significant performance benefit as well as a satisfaction preference for large displays.

In addition to productivity benefits, Czerwinski et al. (Czerwinski et al., 2003) also document results showing that larger displays lead to improved recognition memory and peripheral awareness. Tan et al. (D. S. Tan, Czerwinski, & Robertson, 2003) reported a series of studies demonstrating the advantages of large displays on 3D navigation in virtual worlds. They suggested that while large displays increase performance for all users on average, females improved so much that the normal advantage male users have over females in virtual 3D navigation disappears when using large displays. These studies revealed that the wider fields of view provided by large displays lead to increased ability to process optical flow cues during navigation, cues that females are more reliant upon than males. Thus, large displays actually serve to eliminate a gender bias, at least for the set of tasks tested. In separate studies, Tan et al. (D. S. Tan et al., 2004; D.S. Tan, Gergle, Scupelli, & Pausch, 2003) also found that large displays provide for a more immersive experience when performing spatial tasks, such as navigating 3D environments. This immersion allows the user to better perform navigation in 3D, in addition to building better cognitive maps of the virtual world.

Researches have been carried out to explore the relationship between design factors and users' experience. For example, Ragan et al. (Ragan, Endert, Bowman, & Quek, 2012) explores how spatial layout, interactivity, and persistent visibility affect learning working with large displays; the effect of interior bezels of tiled-monitor large displays on visual search, tunnel steering, and target selection is studied by Bi et al.(Bi, Bae, & Balakrishnan, 2010). Researchers also explored the relationship between screen size and viewer enjoyment using different forms of media content. Lombard et al. (Lombard, Reich, Grabe, Bracken, & Ditton, 2000) reported that viewers experienced a greater sense of enjoyment when watching rapid point of view movements on large screens. Similarly, Ohtani and Mitshuhashi (Olesen, Yannakakis, & Hallam, 2008) reported that users preferred larger (20", 42", and 70") television displays when watching dramatic content.

Although many researchers have explored the cognitive benefits of large displays, most of them did not isolate the effects of physical size and visual angle. Usually, participants were required to sit in front of large and small displays with identical viewing distance when completing experimental tasks. Furthermore, almost all experiments in previous studies used passive viewing (e.g., viewing pictures and watching film segments) as the experimental task. It seems arbitrary to generalize findings from passive viewing tasks to highly interactive tasks, such as gaming, before investigating the validity of such generalization. In addition, there still lacks a clear understanding of how larger screen size and visual angle offered by large display affect presence, engagement and emotional responses of game players although large displays have become prevalent in game environments.

Game Player Experience

Presence has recently been identified as a potentially important variable in video game research that may affect use and a variety of outcomes of exposure, ranging from enjoyment to aggression (Bracken & Skalski, 2009; Lee & Peng, 2006; Tamborini & Skalski, 2006). Presence is often generally defined as user's subjective sensation of "being there" in a scene depicted by a medium but there is still a lack of universal definition. In the research field of media presentation, Lessiter's definition (Lessiter, Freeman, Keogh, & Davidoff, 2001) has been widely accepted and suggests that presence includes four sub-dimensions: sense of physical space, engagement, ecological validity and negative effects. *Sense of physical space* (also referred to as "spatial presence") refers to users' sense of being located within a spatially contiguous physical environment. *Engagement* refers to users' involvement and interest in the content of the display environment, and their general enjoyment of the media experience. *Ecological validity* relates to the believability and realism of the content and naturalness and solidity of the environment. *Negative effects* refer to uncomfortable feelings such as tiredness and simulator sickness (see (Lessiter et al., 2001) for details). An early study has reported the effects of visual angles of display and room illumination on the sensation of presence during normal television viewing and find that the two factors "arrival" and "departure" have very different relationships to viewer memory and attitude change, but they did not find evidence that visual angle and room illumination affected the sensation of presence (Kim & Biocca, 1997). In the field of game research, some studies have suggested that presence is important for game player experience and regarded it as a part of player experience evaluation (IJsselsteijn, Poels, & de Kort, 2008; Tamborini & Skalski, 2006).

In addition, previous studies (Qin, Wei, & Zeng, 2012; Witmer & Singer, 1998) have suggested that presence has important emotional consequences. People differ in their ability to experience presence, and these differences may be associated with emotional responses to the mediated environment. In this study, we are also interested in the effects of different display parameters on emotional responses. According to dimensional emotion theory, emotions can be located in a two-dimensional space, as coordinates of valence and arousal (Pedersen, Togelius, & Yannakakis, 2009). The emotional arousal dimension indicates the level of activation as regards the emotional experience, and ranges from very excited or energized at one extreme to very calm or sleepy at the other. Emotional valence describes the degree to which an affective experience is negative (unpleasant) or positive (pleasant).

Our pilot study(Lin, Hu, Imamiya, & Omata, 2006) has investigated the effects of the different viewing conditions on emotional arousal of game players using subjective and physiological measures and show interesting results. In the current study, we extend the previous results in the following three aspects: (a) the effects of different viewing conditions on emotional valence are investigated besides emotional arousal; (b) the previous study mainly focuses on the average physiological responses during a long-term game session and the current study further examines the effects of different viewing conditions on physiological responses during short-term game events; and (c) presence of game players is examined using a multi-dimension questionnaire.

EXPERIMENT

Participants

Twenty male and four female university students aged 18 to 30 participated in the experiment in return for JPY500. Before the experiment, all participants were required to complete a background questionnaire about their experience with large displays and with the game, along with personal information such as sex and handedness. Twenty of the 24 subjects were experienced with the game, while the other 4 subjects were somewhat experienced or completely inexperienced. In addition, all participants reported that they were somewhat experienced or completely inexperienced with large displays.

Tasks

The experimental tasks were performed in an HCI laboratory. A 50" large display (*Sanyo SlimView*) and a 17" standard desktop display (*Dell Ultrasharp*) were used (see Figure 1 (a)). We chose a simple game session from a popular 3D video game called Super Mario 64 (NINTENDO®) as the experimental task.

Figure 1. [(a) 50" large display and 17" desktop monitor, and (b) screen shot of the game task]

The game task was to run all regulated paths, defeat a monster, and obtain two red stars. A player of average skill requires about 6 minutes completing the game task. Figure 1 (b) shows a screen shot of the task.

The game was played on a NINTENDO 64 gaming console under three viewing conditions: (a) playing on the large display at 80° horizontal visual angles; (b) playing on the large display at 20° horizontal visual angles; and (c) playing on the small display at 80° horizontal visual angles (see Figure 2). The players logically could play the game on the small screen at a 20° visual angle but the viewing distance is so far that subjects may not clearly see the content of screen. This task condition was thus ignored.

Figure 2. [Task conditions: (a) playing on the large display at 80° visual angles, (b) at 20° visual angles, and (c) playing on the small display at 80° visual angles.]

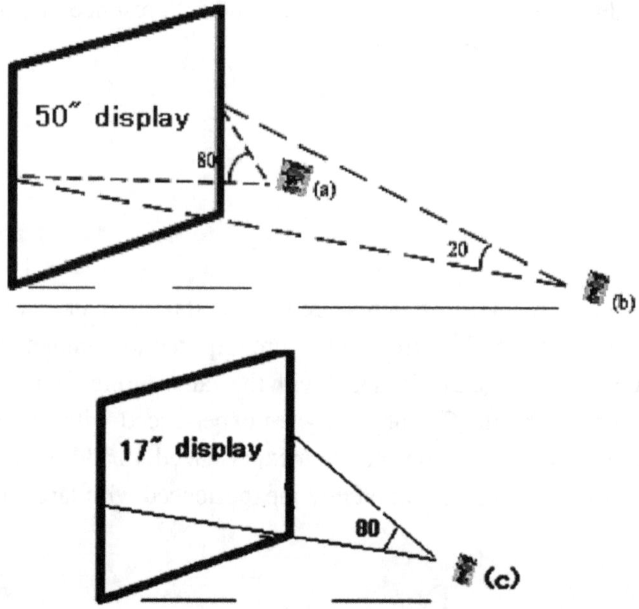

The order of presentation of these conditions was balanced. Participants are required to play the game task in each viewing condition as quickly and as well as possible. Task completion time and game score were recorded as task performance index. Other factors such as resolution (1024 x 768), refresh rate (60Hz), color, brightness and contrast were held as constant as possible between displays. Since the game task is very simple for participants, it is reasonable to expect that there would be no difference in physiological response caused by task performance differences across task conditions. Sound effects were also switched off because they could influence physiological reaction. These measures ensure that any differences observed within task conditions can be attributed to changes in the display conditions, and not to changes in the background music or to task performance.

Experimental Procedure

The experiment was divided into four phases: a welcome phase, a practice phase, a game phase, and a debriefing phase. During the welcome phase, all participants were required to sign a consent form with a detailed description of the experiment, its duration, and its research purpose. Each participant also filled out a background questionnaire. During the practice phase, instructions were read to each participant describing the game task rules, and each was given a brief tutorial on how to complete the task. Participants were then allowed to practice for about 10 minutes on the large display to reduce any short-term physiological responses brought about by the novelty of the large display. Participants then completed the game session under each specified condition. After each session, participants had about 10 minutes to rest and to complete questionnaires rating their experience with the game session. During the course of play, participants were neither encouraged nor discouraged from talking. At the end of game play, participants discussed their impressions of the experiment.

Evaluation Methods for Player Experiences

It has been generally agreed that emotional experiences comprise three components: subjective experience, expressive behavior and physiological responses (e.g., physiological arousal) (Ravaja, 2004). In the study, two kinds of measures (i.e., subjective and physiological measures) were adopted to evaluate experiences of game players.

Subjective Measures

Sense of presence was measured after each condition with the ITC-Sense of Presence Inventory (ITC-SOPI), a 44-item self-reporting instrument(Lessiter et al., 2001). The ITC-SOPI identifies four presence-related elements: (a) spatial presence (19 items, e.g., "I had a sense of being in the game scenes," "I felt I was visiting the game world"); (b) engagement (13 items, e.g., "I felt involved in the game environment," "My experience was intense"); (c) ecological-validity/naturalness (5 items, e.g., "The content of the game seemed believable to me," "The game environment seemed natural"); and (d) negative effects (6 items, e.g., "I felt dizzy," "I felt nauseous"). This study employs only 37 items addressing the first three elements which are most relevant for the current study. The wording of some of the items was slightly altered to adapt the instrument specifically to video games. Each of the items was rated on a 9-point scale, ranging from 1 (strongly disagree) to 9 (strongly agree).

Participants rated their emotional valence and arousal to each of the conditions using 9-point pictorial scales. The valence scale consists of 5 graphical depictions of human faces in expressions ranging from a severe frown (most negative) to a broad smile (most positive). Similarly, for arousal ratings, there are 5 graphical characters varying from a state of low visceral agitation to one of high visceral agitation. Emotional arousal and valence are scored on these two scales, which resemble Lang's Self- Assessment Manikin (Lang, 1980).

Physiological Measures

Physiological measuring has begun to be used to evaluate user experience in HCI evaluations. Physiological response is involuntary and thus is regarded as an objective evaluation technology. In recent several game studies (Mandryk & Atkins, 2007; Qin et al., 2012), physiological measures such as Galvanic skin response (GSR) and heart rate (HR) have shown the potential of evaluating presence and emotional responses of game players. In the study, we chose GSR and HR to index arousal, and also measured HR to index engagement for attention-requiring tasks based on previous literature.

GSR is a measure of skin conductivity. As humans experience arousal, their sympathetic nervous system is activated, resulting in increased sweat gland activity and skin conductance. GSR has also frequently been used as a measure of arousal (Pedersen et al., 2009; Ravaja, 2004). Arousal is also accompanied by increased sympathetic nervous system (SNS) activity which causes the heart to speed up; for example, a recent study showed that heart rate appears to index primarily emotional arousal during video game playing (Johnston, Anastasiades, & Wood, 1990). However, heart activity is innervated also by the parasympathetic nervous system (PNS). Increased cardiac parasympathetic activity causes the heart to slow down and is associated with information intake and attention. Therefore, heart rate is also used to measure attention (e.g., (Reeves, Lang, Kim, & Tatar, 1999)). Although the dual innervations of the heart activity may entail interpretative difficulties, we assume that heart rate has the potential to index both emotional arousal and engagement in certain contexts.

The GSR signal at 64 Hz and the EKG (electrocardiograph) signal at 128 Hz were collected during game play using Biofeedback System (ProComp Infiniti™ System and CardioproTM$^{2.0}$ Software from Thought Technology Ltd.). GSR sensors were placed on the two left fingers and EKG electrodes were placed in the standard configuration of two electrodes on the chest and one on the abdomen. HR was computed from EKG signals. Before computing HR, we inspected the EKG signals and corrected some erroneous samples where two beats were counted in a single sampling interval or only one beat was counted across two sampling intervals.

RESULTS

Results from subjective data analysis are described first, followed by the results from physiological data analysis. Correlations between subjective and physiological measures are then presented. In this study, we define significance at the 5% level, i.e., $P < 0.05$. Differences between the various experimental conditions were analyzed using a Paired-Samples T-Test, and patterns of differences between individuals were analyzed using non-parametric statistical techniques (Wilcoxon Signed Ranks Test). Analysis of correlations among measures was performed using Pearson Correlation.

Table 1. Mean subjective response to each task condition. Identifying strongly with the experience is reflected by a higher mean. Significant differences are shaded in gray. P values are the level of significance

	80° vs. 20° Visual Angles on the Large Display			50" vs. 17" Sizes at the 80° Visual Angle		
	Mean (80°)	Mean (20°)	P	Mean(50")	Mean(17")	P
Spatial Presence	5.15	4.25	.001	5.15	4.66	.045
Engagement	6.81	5.94	.001	6.81	6.56	.031
Ecological-Validity	5.59	5.67	.073	5.59	5.51	.062
Emotional Arousal	6.95	6.21	.001	6.95	6.22	.035
Emotional Valence	7.11	6.21	.027	7.11	6.54	.058

Subjective Evaluation Results

The chi-squared statistic was used to determine whether subjective responses were influenced by order of presentation of experimental condition and task performance; there were no significant effects on any of the subjective evaluations. Statistical analysis indicates that spatial presence and engagement was significantly higher when participants played on the large display at the large visual angle (80°) than when playing at the small angle (20°) (see Table 1). Also, the large visual

angle was found to influence participant emotional response: participants playing on the large display at a large visual angle (80°) reported more emotional arousal and more positive (larger values) emotional valence than when playing on the large display at a small visual angle (20°) (see Table 1). However, there was no significant difference in Ecological-Validity/naturalness between the two visual angles.

We also subjectively evaluated spatial presence, engagement, emotional arousal, valence and Ecological-Validity/naturalness between play on the large screen (50") and the small one (17") at an identical visual angle (80°). The results indicated that the physically larger screen size led to significantly higher spatial presence, engagement, and emotional arousal as compared to the physically smaller screen. No significant differences were found in ecological-validity/naturalness and emotional valence (see Table 1). Notably, the large visual angle produced greater positive evaluation of emotional valence, but the physically large screen size did not.

In addition, as a general attitude, when subjects were asked, given a choice, which condition would they choose to play game, all subjects reported that they would choose to play the game at the 80° visual angles.

Physiological Evaluation Results

A one-way ANOVA analysis and post-hoc comparisons were used to examine the differences in task performance (game score, task completion time, etc.) among three task conditions and did not find statistically significant differences. A Paired-Samples T-Test showed that the mean GSR was significantly higher when playing on the large display at the large visual angle (80°), as compared to the small visual angle (20°) (mean difference = 0.409 μm, $t_{23} = 6.46$, $P < 0.001$). This pattern was consistent for 22 of 24 participants, a significant trend ($Z = -4.086$, $P < 0.001$) (see Figure 3). Notably, subjects 13 and 18 experienced lower GSR at the large visual angle (80°) than at the small visual angle (20°).

Figure 3. [The differences in mean GSR between playing on the large display at 80o and 20o visual angles. Mean GSR is higher when playing at the 80o visual angle as compared with playing at the 20o visual angle.]

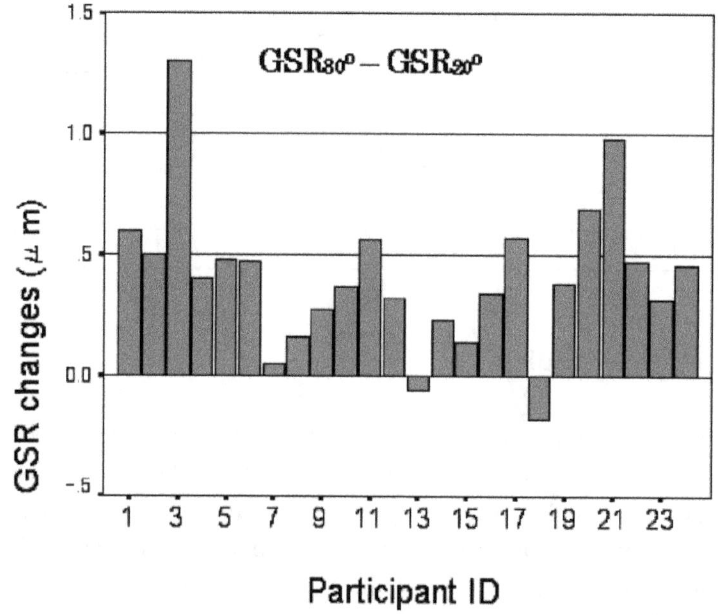

We further examined the subjects' task performance and playing process by reviewing the video tape. The game content and task performance of all subjects except subject 13 and 18 did not apparently change at the large and small visual angles. Subject 13 and 18 were found to encounter some accident difficulties (e.g. being attacked by unexpected bombs and carelessly falling from a mountain road) when playing at the small visual angle. These difficulties may cause the game tasks to suddenly become more arousing and resulted in higher GSR, which are in line with Ravaja's study (Ravaja et al., 2008) . In the post-interviews, the two subjects also reported that they feel more tension when meeting accident difficulties due to their misplays. Thus, we assume that the unexpected changes of game content of subject 13 and 18 diminished the effects of the large visual angle on GSR changes.

In addition, we investigated the difference in GSR resulting from play on the large and small screens at the identical 80° visual angles. Although the large screen (50") showed a larger increase in mean GSR than the small one (17"), these results were not statistically significant (mean difference = 0.18 μm, $t_{23} = 6.34$, $P > 0.05$).

Increased cardiac sympathetic activity is related to emotional arousal and causes increased HR, whereas increased cardiac parasympathetic activity is related to information intake and attention engagement and causes HR to decrease. To reduce the interpretative difficulties entailed by the dual innervations, we examined HR changes, respectively, in response to short-term arousing and attention-requiring events, rather than mean comparison. A typical arousing event and a typical attention-requiring event were chosen. The arousing event consists of defeating a monster (about 25 seconds), and the attention-requiring event consists of crossing a slippery and narrow foot bridge to get a red star, then returning (about 30 seconds). HR for the 10 seconds preceding these events was computed as the baseline. We defined the change in HR during an event versus its baseline as △HR ($\triangle HR = HR_{event} - HR_{baseline}$).

All 24 participants succeeded in completing the two events under all task conditions (24 x 3 task conditions = 72 events). The data analysis shows that all 72 arousing events caused, to some extent, increases in HR versus their baseline HR; more importantly, participants experienced a significantly greater acceleration in mean HR in response to arousing events when playing on the large display at the large visual angle than when playing at the small visual angle (mean $\triangle HR_{\text{large visual angle}}$ = 1.12 bmp vs. mean $\triangle HR_{\text{small visual angle}}$ = 0.86 bmp, t_{23} =6.85, $P < 0.05$). Thus, we conclude that the large visual angle has caused higher emotional arousal and thus resulted in greater acceleration in HR, as compared to the small visual angle. However, we did not find significant differences in HR changes versus baseline between playing on the physically large and the physically small size screens at an identical visual angle.

In addition, 70 of 72 attention-requiring events caused decreases in HR versus their baselines. (One participant experienced HR increases, rather than decreases, when playing under conditions (a) and (b). The participant encountered unexpected difficulties and nearly fell from the narrow bridge. The attention-requiring event in a sense may have become an arousing event, thus causing HR increases. We excluded this participant.) We then compared the other 23 participants' HR changes. The results show that attention-requiring events under the condition of playing on the large display at the large visual angle caused significantly greater deceleration in mean HR versus baseline HR than playing at the small visual angle (mean $\triangle HR_{\text{large visual angle}}$ = -0.98 bmp vs. mean $\triangle HR_{\text{small visual angle}}$ = -0.71 bmp, t_{22} =-4.97, $P < 0.05$). Therefore, we assume that the large visual angle has caused higher engagement and thus resulted in greater deceleration in HR compared to the small visual angle. We did not find significant differences in HR changes in response to attention-requiring events for play on the large vs. small screens at an identical visual angle.

Correlations between Subjective Data and Physiological Response

The subjective and physiological results show that the larger visual angle has significant effects on both player GSR response and self-reported presence and emotional response. We are interested in whether and how the GSR changes caused by large and small visual angles correlate with subjective rating changes in emotional arousal, spatial presence and engagement caused by the two visual angles

We could not directly compare the means of time-series GSR data to the results from the subjective questionnaires. Physiological response has very large individual differences, and individual baselines must be taken into account. Thus, to permit comparison of the time-series GSR data with one-time subjective data, we normalized the measures, transforming them into dimensionless numbers between −1 and +1. For each individual, the difference between the large and small visual angles was divided by the maximum range of that individual's response. The time-series GSR data were normalized using the following formula (1):

$$GSR_{normalized} = \frac{MeanL - MeanS}{MAX\{PeakL - MinL, PeakS - MinS\}} \tag{1}$$

where L refers to playing at the large visual angle, and S refers to playing at the small visual angle. The corresponding normalizing equation (2) for the one-time subjective data is:

$$Subjective_{normalized} = \frac{L - S}{8} \tag{2}$$

Figure 4. [Normalized GSR is correlated with normalized emotional arousal (R = 0.923, P < 0.001).]

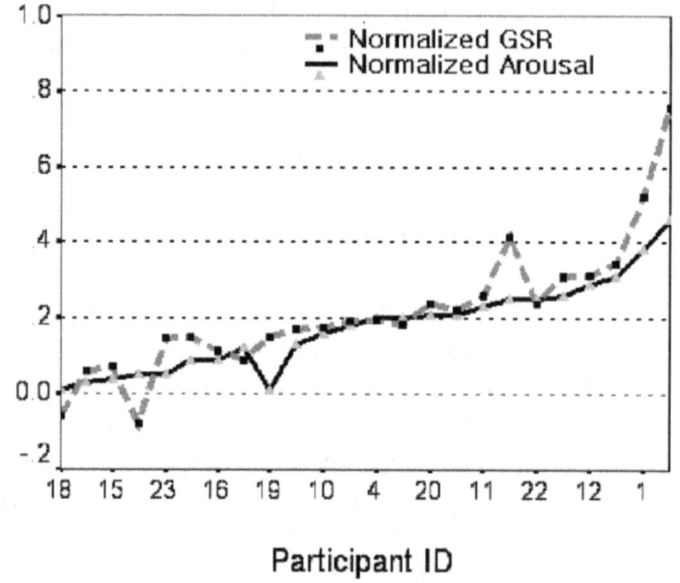

These normalized measures were then correlated across all individuals using the Pearson correlation.

Using Pearson's coefficient, we analyzed all participants' data and found that normalized GSR was positively correlated with normalized arousal rating ($R = 0.923$, $P < 0.001$) (see Figure 4). The positive correlation of normalized emotional arousal and GSR suggests that the amount by which subjective ratings of arousal increase when playing at the large visual angle is proportional to the amount to that GSR increases during that condition. In addition, normalized GSR is positively correlated with normalized spatial presence ($R = 0.734$, $P < 0.001$) and normalized engagement ($R = 0.681$, $P < 0.05$). This suggests that the amount by which GSR increased when playing at a large visual angle is proportional to the increase in subjective spatial presence and engagement ratings.

DISCUSSION

The experiment described in this paper demonstrates that physical size and visual angle are two important factors to consider when designing display systems. To our knowledge, this is the first study that isolates the effects of the large visual angles and physically large screen sizes offered by large displays on presence and emotional responses.

The experimental results show that there are significant increases in physiological arousal indexed by GSR changes when playing at the larger visual angle, and that the GSR changes between the different visual angles can positively and significantly correlate with the changes of subjective ratings in emotional arousal, spatial presence and engagement; in addition, HR shows significantly larger acceleration and deceleration, respectively, in response to the increase in physiological arousal and engagement when playing on the large display at the large visual angle versus the small visual angle. These results support our hypothesis 1 (**H1**) and suggest that larger visual angles offered by large display can improve game players' sense of spatial presence and engagement when playing a 3-D game; furthermore, large visual angles seems to be an emotional "amplifier", which also could cause more intense emotional responses when playing the game.

More interestingly, when playing the game even at a visual angle identical with small screen sizes, the results show that physically large screen size significantly increases the subjective sense of spatial presence, engagement and emotional arousal, and also seems to induce different trends in physiological responses (HR and GSR), although they do not reach statistical significance. One explanation for the lack of statistical significance may be psychophysiological testing is perhaps not as robust as HCI evaluators might like it to be. The considerable individual differences in physiological responses would have contributed to the lack of statistical significance and this problem might be reduced by using greater numbers of participants or under more tightly controlled conditions. These experimental results support our hypothesis 2 (**H2**) suggest that physically large screen size offered by large display, even at identical visual angles with small display, is also an important display characteristic that can significantly affect user experience and must be considered in designing our future display systems. One explanation for the effects of physically larger size is that they may induce players to adopt different cognitive strategies.

In addition, we normalized GSR and subjective rating data and found that there are significant correlations between GSR changes and subjective ratings of emotional arousal, presence and engagement. The correlations show that the subjective and physiological components seem to be able to confirm each other, and thus provide interesting evidence that subjective ratings correlate with physiological responses.

In summary, the larger visual angles show significant effects on player experience as measured by physiological response and subjective rating, as we expected; furthermore, physically large screen size may also be an important display characteristics affecting user experience. This study has practical implication for HCI design—both visual angle and physical screen size can augment or diminish expected effects, contributing to a redefinition of the requirements for what constitutes effective and desirable HCI when designing display environments.

FUTURE WORK

The long term goal of our project is to building a general theory that explains effects induced by recombining various display characteristics so that user can craft the interaction environments that best support user experience. The present study takes an initial step toward the goal and additional work should be required. There might be an interesting point in angle-presence or size-presence curve as it is usually assumed to represent the optimal visual angle or size at which user will be immersed. Future work is required to determine the shape of the curve as one increase display size (or visual angle) from a traditional desktop display to a large wall-size display and beyond. It would be interesting to see if presence change is an abrupt shift that happens when a certain size or visual angle is achieved or if it is more continuous across a series of sizes or angles. This would allow us to gain a deeper understanding of display characteristics and how it relates to immersion and presence.

We also must fully examine the interaction between physical size and visual angle. Maintaining constant visual angles was merely a means of isolating and studying physical display size as an interesting display characteristic. The study does not suggest that large displays should be intentionally used at equivalent visual angles to their smaller counterparts. Instead, we need clearly understand the interactions between display size and visual angles (or other factors) so that we can design display systems that make optimal use of large displays.

There may be other factors such as gender, age, habit, eyesight and interactivity that may affect user experience, but we do not have clear understanding of how each of these contributes to the effects and how interact with each other. We believe that this could form an entire research agenda, and would add not only our understanding of large displays, but of the human visual and perceptual systems in general.

Facial EMG is the most studied index of emotional valence (Ravaja, 2004). The facial EMG was initially chose to evaluate players' emotional valence but EMG signal collection need place tiny sensors on the subjects' face and most of them report strong uncomfortable feelings and may distract their attention. The collection of facial EMG signals thus was given up in the present study and the physiological difference indexing emotional valence (e.g. facial EMG) among different display parameters will be considered in future step. In addition, physiological response has very large individual differences and the analysis method of physiological response also needs further exploration.

ACKNOWLEDGMENT

This study was sponsored by the Scientific Research Foundation for the Returned Overseas Chinese Scholars, State Education Ministry and by the Grants-in-Aid for Scientific Research of the Japan Society for the Promotion of Science (No. 19500106).

REFERENCES

Andrews, C., & North, C. (2013). The Impact of Physical Navigation on Spatial Organization for Sensemaking. *Visualization and Computer Graphics. IEEE Transactions on, 19*(12), 2207–2216.

Bi, X., Bae, S.-H., & Balakrishnan, R. (2010). Effects of interior bezels of tiled-monitor large displays on visual search, tunnel steering, and target selection. Paper presented at the Computer Human Interaction. doi:10.1145/1753326.1753337

Bracken, C. C., & Skalski, P. (2009). Telepresence and Video Games: The Impact of Image Quality. *PsychNology Journal, 7*(1), 101–112.

Czerwinski, M., Smith, G., Regan, T., Meyers, B., Robertson, G., & Starkweather, G. (2003). Toward characterizing the productivity benefits of very large displays. *Proc. Interact, 2003*, 9–16.

Elliott, A. M. E., & Hearst, M. A. (2002). A comparison of the affordances of a digital desk and tablet for architectural image tasks. *International Journal of Human-Computer Studies, 56*(2), 173–197. doi:10.1006/ijhc.2001.0520

Ichino, J., Isoda, K., Hanai, A., & Ueda, T. (2013, April). Effects of the display angle in museums on user's cognition, behavior, and subjective responses. In *Proceedings of the SIGCHI Conference on Human Factors in Computing Systems* (pp. 2979-2988). ACM. doi:10.1145/2470654.2481413

IJsselsteijn, W., Poels, K., & de Kort, Y. A. W. (2008). The Game Experience Questionnaire: Development of a self-report measure to assess player experiences of digital games. *Eindhoven: TU Eindhoven. FUGA Deliverable D3.3. Technical Report.*

Johnston, D. W., Anastasiades, P., & Wood, C. (1990). The relationship between cardiovascular responses in the laboratory and in the field. *Psychophysiology, 27*(1), 34–44. doi:10.1111/j.1469-8986.1990.tb02175.x PMID:2339186

Kim, T., & Biocca, F. (1997). Telepresence via television: Two dimensions of telepresence may have different connections to memory and persuasion. *Journal of Computer-Mediated Communication, 3*(2), 1–22.

Lang, P. J. (1980). Behavioral treatment and bio-behavioral assessment: Computer applications. *Technology in mental health care delivery systems*, 119 -137.

Lee, K. M., & Peng, W. (2006). What Do We Know About Social and Psychological Effects of Computer Games? A Comprehensive Review of Current Literature. In P. Vorderer & J. Bryant (Eds.), *Playing video games. Motives, responses, and consequences* (pp. 325–346). Mahwah, NJ: Lawrence Erlbaum Associates.

Lessiter, J., Freeman, J., Keogh, E., & Davidoff, J. (2001). A Cross-Media Presence Questionnaire: The ITC-Sense of Presence Inventory. *Presence (Cambridge, Mass.), 10*(3), 282–297. doi:10.1162/105474601300343612

Lin, T., Hu, W. H., Imamiya, A., & Omata, M. (2006). Large display size enhances user experience in 3D games. *Smart Graphics. Proceedings, 4073*, 257–262.

Liu, C., Chapuis, O., Beaudouin-Lafon, M., Lecolinet, E., & Mackay, W. E. (2014). Effects of display size and navigation type on a classification task. In *Proceedings of the 32nd annual ACM conference on Human factors in computing systems* (pp. 4147-4156). ACM. doi:10.1145/2556288.2557020

Lombard, M., Reich, R. D., Grabe, M. E., Bracken, C. C., & Ditton, T. B. (2000). Presence and Television. *Human Communication Research, 26*(1), 75–98.

Mandryk, R. L., & Atkins, M. S. (2007). A fuzzy physiological approach for continuously modeling emotion during interaction with play technologies. *International Journal of Human-Computer Studies, 65*(4), 329–347. doi:10.1016/j.ijhcs.2006.11.011

Martinez, H. P., Jhala, A., & Yannakakis, G. N. (2009, 10-12 Sept. 2009). *Analyzing the impact of camera viewpoint on player psychophysiology.* Paper presented at the Affective Computing and Intelligent Interaction and Workshops, 2009. ACII 2009. 3rd International Conference on.

Mary, C., George, R., Brian, M., Greg, S., Daniel, R., & Desney, T. (2006). *Large display research overview.* Paper presented at the CHI '06 extended abstracts on Human factors in computing systems, Montreal, quebec, Canada.

Mazaeva, N., Ntuen, C., & Lebby, G. (2001, 25-28 July 2001). *Self-Organizing Map (SOM) model for mental workload classification.* Paper presented at the IFSA World Congress and 20th NAFIPS International Conference, 2001. Joint 9th.

Olesen, J. K., Yannakakis, G. N., & Hallam, J. (2008, 15-18 Dec. 2008). *Real-time challenge balance in an RTS game using rtNEAT.* Paper presented at the Computational Intelligence and Games, 2008. CIG '08. IEEE Symposium On.

Olsen, B. I., Laeng, B., Kristiansen, K. A., & Hartvigsen, G. (2013). Size does matter: Women mentally rotate large objects faster than men. *Scandinavian Journal of Psychology*, *54*(3), 196–204. doi:10.1111/sjop.12043 PMID:23448540

Pedersen, C., Togelius, J., & Yannakakis, G. N. (2009). *Modeling player experience in super mario bros*. In Computational Intelligence and Games, 2009. CIG 2009. IEEE Symposium on (pp. 132-139). IEEE. doi:10.1109/CIG.2009.5286482

Qin, Han-shu, Wei, Yan, & Zeng, Shao-hua. (2012). Parameters optimization for SVM classification based on adaptive NGA. *International Journal of Advancements in Computing Technology, 4*(18), 490-497. doi: 18.5810.4156/ijact.vol4.issue

Ragan, E. D., Endert, A., Bowman, D. A., & Quek, F. (2012). *How spatial layout, interactivity, and persistent visibility affect learning with large displays*. In *Proceedings of the International Working Conference on Advanced Visual Interfaces* (pp. 91-98). ACM. doi:10.1145/2254556.2254576

Ravaja, N. (2004). Contributions of Psychophysiology to Media Research: Review and Recommendations. *Media Psychology*, *6*(2), 193–235. doi:10.1207/s1532785xmep0602_4

Ravaja, N., Turpeinen, M., Saari, T., Puttonen, S., & Keltikangas-Jarvinen, L. (2008). The psychophysiology of James Bond: Phasic emotional responses to violent video game events. *Emotion (Washington, D.C.)*, *8*(1), 114–120. doi:10.1037/1528-3542.8.1.114 PMID:18266521

Reeves, B., Lang, A., Kim, E. Y., & Tatar, D. (1999). The effects of screen size and message content on attention and arousal. *Media Psychology*, *1*(1), 49–67. doi:10.1207/s1532785xmep0101_4

Tamborini, R., & Skalski, P. (2006). The role of presence in the experience of electronic games. In P. Vorderer & J. Bryant (Eds.), *Playing video games: Motives, responses, and consequences* (pp. 225–240). Lawrence Erlbaum Associates.

Tan, D., Gergle, D., Scupelli, P., & Pausch, R. (2006). Physically large displays improve performance on spatial tasks. *ACM Trans. Comput.-Hum. Interact., 13*(1), 71-99. doi: http://doi.acm.org/10.1145/1143518.1143521

Tan, D. S. (2004). *Exploiting the Cognitive and Social Benefits of Physically Large Displays (thesis)*. Carnegie Mellon University.

Tan, D. S., Czerwinski, M., & Robertson, G. (2003). *Women go with the (optical) flow*. In *Proceedings of the SIGCHI conference on Human factors in computing systems*, 209-215.

Tan, D. S., Gergle, D., Scupelli, P., & Pausch, R. (2003). With similar visual angles, larger displays improve spatial performance. In *Proceedings of the SIGCHI conference on Human factors in computing systems* (pp. 217-224). ACM. doi:10.1145/642611.642650

Tan, D. S., Gergle, D., Scupelli, P., & Pausch, R. (2004). Physically large displays improve path integration in 3D virtual navigation tasks. *In Proceedings of the CHI 2004 Conference on Human Factors in Computing Systems*, 439-446. doi:10.1145/985692.985748

Witmer, B. G., & Singer, M. J. (1998). Measuring presence in virtual environments- A Presence questionnaire. *Presence (Cambridge, Mass.)*, *7*(3), 225–240. doi:10.1162/105474698565686

ADDITIONAL READING

Bennett, K. B., & Flach, J. M. (2011). *Display and Interface Design*. CRC Press Inc.

Books, H. (2011). *Articles on Digital Display Connectors, Including: Digital Visual Interface, Serial Digital Interface, Apple Display Connector, Hdmi, Digital Flat Pan*. Hephaestus Books.

Cadiz, J. J., Gupta, Anoop, & Jancke, Gavin. (2001). Sideshow: Providing Peripheral Awareness of Important Information.

Habelski, S. (2008). *A Design Study in 3D User Interface for Large Interactive Displays*. VDM Verlag.

Lyon, D. (2007). *Surveillance Studies: An Overview*. Cambridge: Polity Press.

Medina, J. J. (2008). *The biology of recognition memory*. Psychiatric Times.

Myers, R. L. (2002). *Display Interfaces: Fundamentals and Standards*. Wiley Series in Display Technology. doi:10.1002/0470855762

Norman, K. A., & O'Reilly, R. C. (2003). Modeling hippocampal and neocotical contributions to recognition memory: A complementary learning systems approach. *Psychological Review*, *110*(4), 611–646. doi:10.1037/0033-295X.110.4.611 PMID:14599236

KEY TERMS AND DEFINITIONS

Air Traffic Control: Is a service provided by ground-based controllers who direct aircraft on the ground and through controlled airspace, and can provide advisory services to aircraft in non-controlled airspace.

Collaboration Work: Is the collective work of two or more individuals where the work is undertaken with a sense of shared purpose and direction, that is attentive and responsive to the environment.

Immersion into Virtual Reality: Is a perception of being physically present in a non-physical world.

Multitasking: Originated in the computer engineering industry. It refers to the ability of a microprocessor to apparently process several tasks simultaneously.

Peripheral Awareness: Is the strategy takes advantage of human's innate ability to stay aware of things in our periphery. It works by filling users' peripheral attention with information such that it envelops them without distracting them.

Recognition Memory: Is a subcategory of declarative memory. Essentially, recognition memory is the ability to recognize previously encountered events, objects, or people. When the previously experienced event is reexperienced, this environmental content is matched to stored memory representations, eliciting matching signals.

Surveillance: Is the monitoring of the behavior, activities, or other changing information, usually of people for the purpose of influencing, managing, directing, or protecting them.

User Experience (UX): Involves a person's behaviors, attitudes, and emotions about using a particular product, system or service.

Visual Analytics: Is an outgrowth of the fields of information visualization and scientific visualization that focuses on analytical reasoning facilitated by interactive visual interfaces.

Visual Angle: Is the angle a viewed object subtends at the eye, usually stated in degrees of arc. It also is called the object's angular size.

Chapter 7
Towards Intelligent Window Layout Management:
The Role of Mental Map

Rui Chen
Sichuan University, China

Tao Lin
Sichuan University, China

Tiantian Xie
Sichuan University, China

ABSTRACT

Large displays are becoming increasingly pervasive. Larger screen size provides an opportunity for users to see more information simultaneously, but at the cost of managing a larger amount of screen space, which is a great burden declining task performance and user experience. User would do/feel better if this burden could be takeover by the computer itself employing techniques that automate the management of screen space. Some studies on automatic window management have been carried out with some success. However, they mainly focus on utilization of empty screen space and/or overlap elimination while ignore preservation of the mental map of users, which tends to cause user confusion and disorientation in practical use. In this chapter, an empirical model is proposed to identifying the degree of mental map preservation for a window layout rearrangement. Furthermore, a method combining high-level window importance with a genetic multi-objective optimization algorithm is presented to generate recommended window layouts featuring a tradeoff among several conflicting goals: (1) better usage of screen space, (2) lower degree of window overlaps, and (3) better mental map preservation. Results suggest that the method is capable of generating suitable window layouts for users and takes a key step toward developing an automated windows manager.

DOI: 10.4018/978-1-4666-8450-8.ch007

INTRODUCTION

The last decade has witnessed an unprecedented growth of information, display size and multi-task interfaces. The larger screen size provides an opportunity for users to see more information in more windows at the same time, but at the cost of managing a larger amount of screen space. For example, when using some information-intensive software such as visual analytics and stock market quotation analysis, users often need frequent window operations (i.e., *add, move, delete* and *resize*) to obtain the desired information. These operations are not relevant to the task and, as a result, drag users out of their task domain. Users' mental resources consumed by window operations make no contribution to the task performing. Additionally, when a window operation is completed, a recall process is necessary before users can return into their task context and proceed. This is quite similar to process scheduling performed by computer operating systems. Consequently, such methods of manually managing windows not only induce users' attention to be interrupted, thereby decreasing their overall task performance (Bly & Rosenberg, 1986; Bury, Davies, & Darnell, 1985; Davies, Bury, & Darnell, 1985; Johnson-Laird, 1985; Kandogan & Shneiderman, 1997; Sandberg-Diment, 1984), but also dissipate valuable mental resources. Research on automatic window layout optimization aiming to reduce the amount of needed window operations throughout the process of task performing is thus of great importance.

Improving the interaction with multi-window interfaces has been a challenge for HCI and interface researchers and some efforts have been made (Beaudouin-Lafon, 2001; Hutchings & Stasko, 2002a, 2002b, 2004; Trivedi, Lai, & Zhang, 2001). They can generally be split into two sub-categories: (1) automatically optimizing window layout, in which interfaces focus on helping users effectively utilize display space usually by maximizing visual scope and minimizing the overlap between windows, and (2) developing more efficient window operations, in which interfaces focus on considering whether the standard window operations could be improved or expanded to better help users efficiently operate the windows in the context of multi-tasks. This study lies in the former sub-category and, specifically, focuses on the integration of automatic optimization of window layout and preservation of users' mental map as well as their balance.

Some researchers (Funke, Neal, & Paul, 1993; Lüders, Ernst, & Stille, 1995; Trivedi, et al., 2001) have explored the method of automatic window layouts with some success. Nevertheless, they have rarely been adopted in practical multi-window interfaces, since most studies focus their attention on ways to better exploit empty display space and ignore the fact that the destruction of user's mental map of window layout induced by window deployment adjustment tends to cause users confusion and frustration. The term mental map refers to the structural cognitive information a user creates internally by observing the layout of the graph (Purchase, Hoggan, & Görg, 2007). This internal cognitive structure represents the user's underlying understanding of the information. Several studies (Coleman & Parker, 1996; Misue, Eades, Lai, & Sugiyama, 1995; Purchase, 1998; Purchase, et al., 2007; Purchase & Samra, 2008) have suggested that it is important for dynamic graph layout to preserve consistent mental map of users, otherwise confusion may result. For this reason, a transformation between two layouts needs to ensure that the successor resembles the previous one in order to preserve the mental map (Pilgrim, 2007). A recent study (Archambault & Purchase, 2013) reveals that users can orient themselves in dynamic data significantly easier, no matter how many targets they are tracking, with the help of mental map. Consequently, mental map preservation is useful to prevent users from getting lost in a working environment requires seeking assorted objects over time, especially when the number of which is large. Three factors had been identified in (Freire & Rodríguez, 2006) which affect mental map preservation in interactive dynamic graphs: (1) the predictability of navigational actions, (2) the degree of change from one view to the next, and (3) the traceability of changes once they occur. However, applying these guidelines to the domain of automated optimization of window layout appears intuitively viable as well. User confusion and disorientation could be prevented if they can foresee some of the outcome of a rearrangement of a window layout. The rearrangement itself must also be controlled to a reasonable extent to preserve users' existing perception of the layout, otherwise user frustration would be caused, suggesting that searching for the most suitable degree of reform is crucial. What's more, changes made to a window layout should exhibit how they are emerging to users so that they can be efficiently integrated into users' mental map. As a window layout can also be treated as a graph consisting of rectangles, the aforementioned suggestions are also appropriate for the domain of automatic window layout. However, how to integrate mental map preservation into traditional methods of automatic window layout is still a great challenge.

To reduce user's manual adjustment to a window layout, the following three aspects may be considered: making use of empty screen space, minimizing overlap between windows and preservation of mental map. Better utilization of blank screen space and less overlap lead to neater window layouts with larger windows, which are capable of showing larger amount of information to users at the same time and, therefore, greatly reduce the probability that users manually adjust the layout. By means of preservation of the mental map, users could easily find the target window without the need for tuning the layout, which would in turn potentially trigger a chain of layout adjustments making a mess of the whole layout. Besides, the extra time cost as well as the cognitive load of both visual search and mental map reconstruction could also be eliminated. Users would not be interrupted and are kept in their task domain, thereby achieving higher overall task performance. However, the three aspects are in conflict, among which a compromise has to be achieved. Multi-objective optimization algorithms are commonly used to obtain the tradeoff among several conflicting optimization objectives, ensuring that performance improvement of a certain objective would not cause a performance decrease of any other objectives (Coello, Lamont, & Van Veldhuizen, 2007). For this reason, they are very appropriate choices for the problem domain.

This chapter presents a genetic multi-objective optimization algorithm to automatically optimizing window layout while featuring preservation of users' mental map. This algorithm is based on a specifically designed window layout representation suitable for evolutionary search, and attempts to achieve a compromise among (1) maximizing utilization of empty screen space, (2) minimizing overlap between windows and (3) maximizing the degree of mental map preservation. These objectives are in conflict with each other. Results demonstrate that the method is capable of generating suitable window layouts for users and takes a key step toward developing an adaptive, intelligent windows manager.

RELATED WORK

There exists a large body of research exploring window management methods, which allow users to arrange multiple windows on the screen (for history and review, see (Myers, 1988)). Within this body of work, most researchers have explored ways to better exploit empty display space and minimize the overlap of windows. A genetic algorithm was proposed to remove the overlap from the windows and to compact the layout of the windows (Trivedi, et al., 2001). *QuickSpace* operations focus on using empty space to grow one window while maintaining the visible contents of other windows, aiming to help users maintain awareness primarily on single-monitor systems (Hutchings & Stasko, 2002b). Non-overlapping

Dragging helps keep windows visible by automatically moving recently occluded windows to empty space (Bell & Feiner, 2000). Both of these systems could be enhanced by shrinking operations that assist in creating empty display space. The CUBRICON Intelligent Window Manager (*CIWM)* (Funke, et al., 1993) automates every window management operation, deciding the position, size, visibility, and depth order for every window as well as when to automatically close windows. The above techniques have generally not been adopted since inferring desired window placement and size, and indeed existence could cause users considerable confusion and frustration because they could greatly or even completely rearrange the layout and thus destroy users' mental map of the layout (Misue, et al., 1995).

Scheme Constraints Window Manager (*SCWM*) (Badros, Nichols, & Borning, 2000), as an improved strategy for *CIWM*, is rooted in a constraint-based system with the aim of balancing automated window management with user control by allowing the user to group and interact with windows by setting constraints (relationships). While this is an innovative and interesting system, there are two potential pitfalls in a relationship-based format. The first is user forgetfulness: if a user makes a relationship and subsequently forgets that the relationship exists, or makes a relationship that inadvertently breaks another relationship, the consequences of a single action could be confusing and overwhelming; furthermore, relationships must be defined per session (machine and window instance) and since user desktops can change many times per session, this may be a time-consuming process.

Some new window operations have also been presented to improve the interaction with multi-window applications. Tan, Meyers, and Czerwinski (2004) developed a novel interaction technique that allows users to replicate arbitrary regions of existing windows into independent windows called *WinCuts*. Hutchings and Stasko (2002a) presented a set of new operations called *shove*, *expand* and *jostle* for managing screen real estate. Benefiting from these operations, windows could acquire more desktop space. Beaudouin-Lafon (2001) developed peeling and rotating operations with the aim of letting users access different windows more easily by treating them more as physical pieces of paper and less as rigid rectangles. Hutchings and Stasko (2004) presented a window shrinking operation that specifically intends to help users display a window's relevant information.

Some studies focused particularly on evolving the behavior of window switching to alleviate the loss of user performance induced by such indispensable operations. One of these is the *SCOTZ* (Tak, Scarr, Gutwin, & Cockburn, 2011), who provides a modal interface in which all open windows as well as the desktop window are organized as thumbnails grouped by applications. Different layouts of the interface, featuring enhanced predictability (i.e., stability) of window location and accessibility (i.e., size morphing) of recently or frequently visited windows respectively, are generated using tree-map algorithms (Shneiderman, 1992). *SCOTZ* outperformed Microsoft Windows taskbar and Alt+Tab in user preference as well as mental cost according to the test results. However, more effort may be put into how to weaken the requirement of window switching so as to maintain users' working flow.

Waldner, Steinberger, Grasset, and Schmalstieg (2011) put emphasis on making use of attributes extracted from windows' content and managing windows in a compositing way. Their system derives windows' importance maps from a saliency-based visual attention model employing image processing techniques. Based on this information, spatial arrangement is carrying out to move important regions of occluded windows to little-importance desktop areas. Furthermore, vital contents of occluded windows are made visible within the non-important areas (free space) of overlay windows which are rendered transparently. Mental map preservation has been simply ignored may be a potential weakness of this system. Users' mental map could be destroyed if automatic spatial adjustments take place, especially at the time when users are concentrating their attention.

The mental map concept of graph drawing has triggered a large amount of researches on its value of applications in various fields after it was first proposed in (Eades, Wei, Misue, & Sugiyama, 1991). The most studied applications of mental map lie in the field of information visualization, or more precisely, dynamic graph drawing. Along with the dynamic variation of a graph, the layout of the graph has to be adjusted to reflect the changes of the relationships between elements (e.g., nodes and links) in the graph. Manually updating a graph layout is time-consuming and attention-intensive so that it is usually done by a layout algorithm automatically. Such layout algorithms have to be carefully designed aiming at not only releasing users from the burden of manual layout adjustments, but also ensuring that users find and understand valuable information in the updated graph at a minimal cost. However, when a layout adjustment takes place, considerable re-arrangement of the graph elements may destroy users' formed mental map of the original graph, which leads to an additional load of the rebuilding process of the mental map and thus has a negative impact on users' overall task performance. Consequently, preservation of users' mental map becomes a necessary concern for layout algorithm designers. The question that how to integrate mental map preservation into common layout optimization methods should be answered.

There is quite a large body of literature digging into automatic graph layout methods combining mental map preservation. The pioneer study (Eades et al., 1991) which emphasized the necessity of preserving users' mental map of a diagram argued that mental map would not be disturbed at all if the "shape" of the diagram is preserved. It proposed several mathematical models for mental map based on the geometric attributes of the diagram in order to make formally specifying success or otherwise of re-arrangement algorithms for diagrams possible. In this study, mental map was modeled as the orthogonal ordering, cluster and topology of the elements a diagram contained. More specifically, the orthogonal ordering model indicated the up, down, left and right relationship between the elements should not be changed; the cluster model suggested that the elements which are close together should stay close together; the topology model recommended to preserve the "insideness" relationship and alike between the elements. Henceforward, the mental map concept has been widely applied to various areas involving stabilized visualization of evolving data.

A later work (Misue, Eades, Lai, & Sugiyama, 1995) illustrated some methods aiming at ensuring node disjointness and providing overall and detailed views while preserving orthogonal ordering and (to some extent) topology. Storey, Fracchia, and Müller (1999) proposed a simple but flexible layout adjustment algorithm called *SHriMP* based on a fisheye lens paradigm. It is capable of customizing a detailed view by specifying a focus point in the graph while preserving the mental map of the layout by preserving the orthogonality, proximity and topology of the graph elements. Mashima, Kobourov, and Hu (2012) employed the advantage of the geographic map metaphor to visualize large-scale dynamic relational data and considered mental map preservation as fixing relative node positions. A similar study (Hu, Kobourov, & Veeramoni, 2012) proposed an approach for visualizing multiple relationships defined on the same dataset also using a geographic map metaphor. Their method ensures preservation of mental map by maintaining not only node stability but also cluster stability and color stability. Angelini, Cortese, Di Battista, and Patrignani (2013) defined two operations, namely *flip* and *skip*, to describe natural transformation of a graph embedding that preserve the mental map. The degree of mental map preservation is depicted by the number of the operations needed to accomplish a topological morphing, the less, the better. Considering the studies of (Liu, Hu, North, & Shen, 2013) and (Gansner, Hu, & North, 2013) on visualization of streaming text data, mental map preservation is achieved by means of controlling the difference of node positions between consecutive time steps. Although with dissimilar implementation, the work of (Sallaberry, Muelder, & Ma, 2013) and (Xu, Kliger, & Hero, III, 2013) also utilize node motion regulating between consecutive time steps of dynamic graphs for mental map preserving. Some other researches incorporated mental map preservation mechanism into heuristic algorithms. For instance, Lin, Lee, and Yen (2011) presented a simulated-annealing based graph drawing algorithm taking into account mental map preservation. When a slight modification is made to the original graph, their algorithm preserves the drawing contours of the graph and the relative positions of nodes so as to preserve the mental map. However, the applicability of this algorithm is limited to small-size graphs due to the limitations inherited from the simulated annealing scheme. Note that keeping the changes made to node positions under control is a frequently adopted paradigm to preserve mental map in the mentioned studies. Besides minimizing node motion, there are another seven layout principles for mental map preservation identified in the graph drawing literature, which were described in detail in (Bergman, 2010).

Automatic window layout is to some extent a simplified particular case (with no edges, and the role of nodes is played by windows) of dynamic graph layout problem. However, mental-map-preserving featured window layout evolvement has been rarely examined. In this chapter, we incorporate a fitness function defining the degree of window movement into a multi-objective genetic algorithm for automatic window layout optimization. The detail of the design of the algorithm will be discussed in the following sections.

ASPECTS AFFECTING WINDOW LAYOUT

Window's Importance

In our design, window's importance refers to the degree of window's relatedness to user's task. Importance of a window is evaluated by the proportion of its overall active time to the total time taken to get the task done, and it is a changing attribute during user's task. More important window desires larger screen space and should not be covered by less important ones. This feature is used as the weight coefficient in our algorithm.

Utilization of Empty Screen Space

As is limited, the valuable screen space should be fully exploited to exhibit more information to users. That is to say, all the available screen space needs to be utilized for rendering windows. By doing so, windows in the layout would less possibly overlap so that users could be less forced to manually adjust the layout. Consequently, larger windows better achieve the goal as they occupy larger screen space. The degree to which a layout makes use of screen space can be easily represented by the proportion of the overall occupied screen space by all of its component windows to all available screen space, as illustrated with the area surrounded by bold lines in Figure 1, which is formed by four windows (depicted as numbered rectangles).

Amount of Overlap among Windows

Although larger windows accomplish higher screen space utilization, they overlap each other with a significantly higher probability within a fixed screen space. Overlaps need to be minimized as they induce difficulty in recognizing and accessing information, which is the main factor that compels users

Figure 1. Representation of the degree of screen space utilization

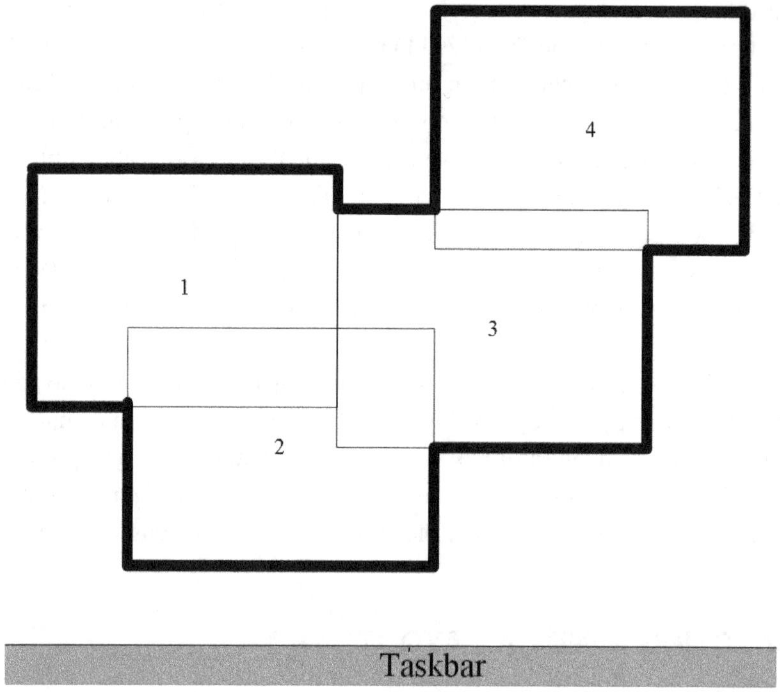

to perform manual layout adjustment operations. For minimizing overlaps, windows may be assigned a smaller size and moved to other screen positions so as not to intersect with others. The degree of overlaps within a layout can be represented by the proportion of the total amount of lost screen space of all involving windows due to overlaps. Taking Figure *1* as an example, the four windows should have occupied more screen space if they were arranged free of overlap. However, overlap induces their loss of screen space, the proportion of which indicates the degree of overlap of the layout.

Note that the above mentioned two factors are conflicting. One of them desires larger windows while the other prefers smaller ones. A compromise between the two has to be obtained to generate suitable layouts.

Preservation of Mental Map

Changes have to be introduced to generate a better layout. However, destruction of user's mental map maybe potentially takes place along with the changes. It is noteworthy that mental map preservation tends to keep layout unchanged while alteration of layout is definitely indispensable to generate a better-arranged layout. Hence, the tradeoff between layout change and preservation of mental map should be considered. The modifications made to a layout have to be controlled in such a level that tradeoff between layout optimization and mental map preservation can be reached.

We present a method to evaluate the degree of mental map preservation during layout optimization based on empirical observation and analysis. The content of user's mental map of a window layout is considered as connections between certain information and screen position of its corresponding holder window. Thus, we assume that mental map preservation can be achieved through consistency preservation of these connections.

Figure 2. A layout composed of six numbered windows and its corresponding mental map built by user containing imprecise screen position of each window

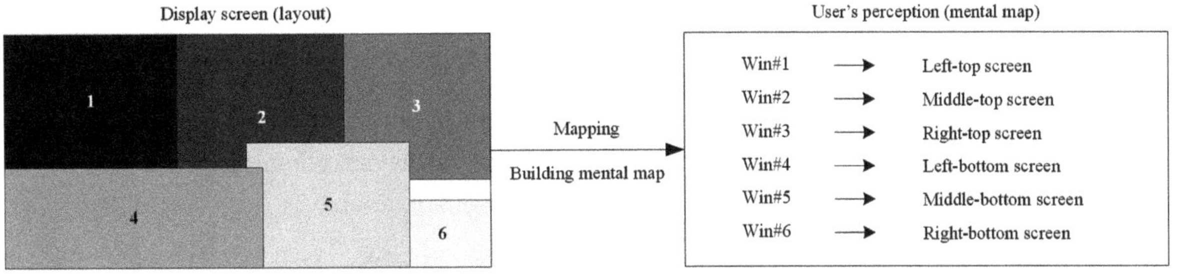

Most of user's mental resource is spent on the task at hand, which produces the imprecise nature of the connections due to lack of available mental resource (e.g., users can never memorize precise screen coordinates of windows, while they get an impression of rough screen position instead). Figure 2 gives an intuitive illustration of this notion. Accordingly, a window's layout can be amended to some extent while achieving preservation of connection consistency at the same time.

Intersection of an optimized window and its corresponding original window, as the black area shown in Figure 3, implicates the degree of connection consistency the optimized window preserves. The proportion of the intersection to the original window quantitatively gives the level of preservation of connection consistency or, namely, mental map preservation with respect to our assumption.

Focus Area

Users tend to manage different screen area in different ways. As we observed, user prefers to place the most task-related window onto certain screen area which serves as the focus area of the user. Continuing to work using optimized layout for a relatively long time without altering it supports the inference that the suggested layout is appropriate for the current task. User's focus area is learnt by our system at this time, where the most task-related window is to be placed in later optimization runs.

Figure 3. Intersection of optimized window and its corresponding original window

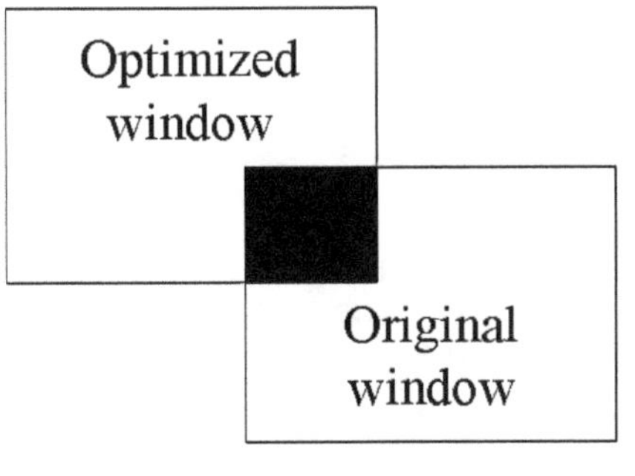

MULTI-OBJECTIVE OPTIMIZATION (MOO)

Multi-objective optimization problem was defined by Osyczka (1985) as the problem of finding a vector of decision variables which satisfies constraints and optimizes a vector function whose elements represent the objective functions. These functions form a mathematical description of performance criteria which are usually in conflict with each other. Hence, the term optimize means finding such a solution which would give the values of all the objective functions acceptable to the decision maker.

Formally, a general MOO problem is defined as maximizing (or minimizing) $F(x) = \left(f_1(x),\ldots,f_k(x)\right)$ subject to $g_i(x) \geq 0$, $i = \left\{1,\ldots,m\right\}$ and $h_j\left(x\right) = 0$, $j = \{1,\ldots,p\}$, $x \in \Omega$. An MOO problem solution maximizes (or minimizes) the components of a vector $F(x)$ where x is an n-dimensional decision variable vector $x = \left\{x_1,\ldots,x_n\right\}$ from some universe Ω. It is noted that $g_i(x) \geq 0$ and $h_j(x) = 0$ represent constraints that must be fulfilled while maximizing (or minimizing) $F(x)$ and Ω contains all possible x that can be used to satisfy an evaluation of $F(x)$ (Coello, et al., 2007).

MOO methods are capable of finding a tradeoff among conflicting optimization objectives, hence exactly fit the problem we are dealing with. We employ a genetic MOO algorithm, featuring an Aggregate Objective Function (AOF) (or Linear Aggregating Function) (Coello, et al., 2007) as its fitness function, to generate suitable layout considering all the described factors.

AUTOMATIC WINDOW LAYOUT BASED ON GENETIC MOO ALGORITHM

Each chromosome is representative of a layout of the open windows, which is a one-dimensional array of genes. Each gene is a tuple as (x, y, w, h) consisting of the screen position (x and y, i.e., screen coordinates of the left-top vertex) and size (w and h, i.e., width and height) of a window. A layout is determined by the positions and sizes of windows and its appropriateness is determined by fitness function of the algorithm.

Taking the affecting factors we suggested into account, four objective functions are formed, as listed below.

$$f_1 = \sum_{i=1}^{n} \frac{I_i}{I_{\max}} \cdot A(V_i) \Big/ (A(S) - A(T))$$

$$f_2 = \sum_{i=1}^{n} \frac{I_i}{I_{\max}} \cdot A(V_i) \Big/ \sum_{i=1}^{n} A(W_i)$$

$$f_3 = \sum_{i=1}^{n} \frac{I_i}{I_{\max}} \cdot A(W_i \cap W_i') \Big/ A(W_i)$$

$$f_4 = A(F \cap W_{core}) \big/ A(F)$$

Table 1. 1Symbol explanation

n	Number of open windows in a layout	T	Taskbar
I_i	Importance of window i	W_i	Window i
$I_{\max}$	Importance value of the most important window in a layout	W'_i	Corresponding Optimized window of window i
$A(x)$	This function calculates the occupied screen area of object x	W_{core}	The most important window
V_i	Visible parts of window i	F	Focus area
S	Whole screen		

For explanation of involving symbols in the functions, see Table 1.

The four objective functions evaluate a layout's degree of (1) utilization of empty screen space, (2) non-overlapping, (3) mental map preservation and (4) overlap between the most important window and the learnt focus area respectively, whose aggregation gives the final fitness function $F = f_4 \cdot \sum_{k=1}^{m} w_k \cdot f_k$, where $\sum_{k=1}^{m} w_k = 1$ and $m = 3$. f_4 is treated as the weight of the final fitness as it does not conflict with others. Before the focus area is learnt, f_4 is simply ignored.

Emulation tests verify the ability of our algorithm to generate suitable layouts and Figure 4 demonstrates one of them. Rectangles with different colors represent windows with different importance. The black rectangle represents the most important window while the white one indicates the worst important window. An acceptable layout can be produced by our algorithm after 300 (more or less) generations of evolution. Normally, another 400 or 500 generations of evolution are needed to find an optimal approximation and further evolutions gain little reward. These emulation tests are applied with 14-inch and 19-inch (common size of laptop and PC displays) widescreen monitors.

Note that more important windows are assigned larger screen space and each optimized window covers part of its original window (see the initial layout for reference) as expressions of mental map preservation.

CONCLUSION AND FUTURE WORK

We present a genetic MOO algorithm based windows layout optimization method in this chapter. Our algorithm employs three mutually restricted optimization objectives as: (1) maximizing utilization of empty screen space; (2) minimizing windows overlap; (3) preserving user's mental map, and searches the optimal compromise solution among them to generate an optimized layout that fits user's cognition. By doing so, the number of involved manual windows layout adjustment operations is reduced so that user can spend more valuable mental resources on the task at hand and achieve higher overall task performance.

Figure 4. The process that our early emulation prototype system generates an optimized layout

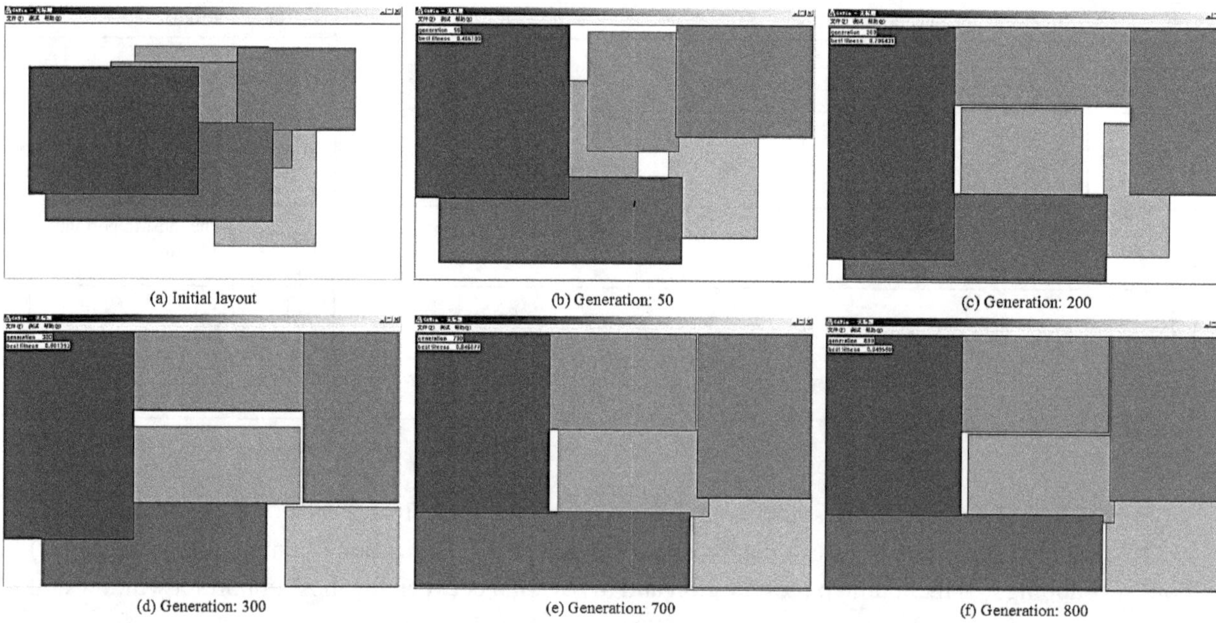

(a) Initial layout (b) Generation: 50 (c) Generation: 200

(d) Generation: 300 (e) Generation: 700 (f) Generation: 800

However, some windows could be too small to work with as larger number of open windows involved in the layout. The problem can be mitigated by lowering the weight of f_2 or simply using a larger display. One should realize that to arrange large number of windows within a fixed-size screen while maintaining their readability and accessibility at the same time is simply not applicable. As a result, there comes the demand for a screen space recycling mechanism to take screen space from unimportant windows so as to offer more to important ones, or even do not render the unimportant windows at all. A main concern when applying similar strategies is keeping accessibility and/or identifiability of the trivial windows because it is still possible that users would recall them in the future. Some of the test results achieve relatively high fitness while the corresponding layouts are not suitable for practical use without user adjustment. Some windows in the layout are too flat or too thin to expose user-interested contents, and only their menu bars and/or tool bars could be shown. User has to perform a few window operations in order to access desired information, which tends to break the mental map. A possible solution to the problem is to incorporate constraints to shape windows (to prevent them from being too flat or too thin, i.e., to keep them in a low aspect ratio). However, it may be more promising to move critical content (e.g., the content around the current position of caret or cursor) to the center of the optimized window so that users can instantly continue their task without layout adjustment.

Besides, there may be two aspects of major improvement of this study that should be considered in the future: (1) porting the algorithm to portable devices (e.g., smart phones and tablets) with smaller screens, which are increasingly used for dealing with complex tasks; (2) exploring how users memorize and identify a specific window among a group of windows within a task context and what implications it gives to the design of screen space recycle mechanism.

Modern portable devices employ touch screens as interactive devices while PCs or laptops adopt keyboards and mouses, which implies that a user would get totally different experience interacting with portable devices. Thus, porting the algorithm to portable devices would possibly demand fundamental studies on the way that user interacts with such devices. Screen space recycle mechanism seems to be necessary for a better arrangement of window layout, especially for devices with small screens. It is unreasonable to assign a fixed amount of screen space to a currently idle window. A better practice might be offering screen space of the idle windows to the active one. Before performing the screen space rearrangement, however, it should be ensured that users could easily identify and retrieve a certain idle window when needed. The answer to how users memorize and identify a window serves as a basis for achieving the assurance.

Our algorithm is currently more suitable for the layout optimization in a single-application task involving multiple windows (e.g., visual analytics) due to the lack of an importance-driven screen space recycle mechanism. Windows are usually kept open and unchanged for the purpose of real-time reference (in order to grasp global information) in this type of tasks, revealing little dependence on an automated screen space recycle system. Nevertheless, this kind of mechanism would be particularly helpful in a multi-application task (e.g., a writing task using Word, Internet Browser, a translator and a reference manager) in which the user usually works with a primary window most of the time and thus real-time reference is not frequent. In this situation, screen space occupied by the reference windows should be automatically and smoothly transferred to the primary window while ensuring an acceptable level of cost when they are recalled.

Implementation of the full system is in progress and, when done, the system will be deployed on our test workstations. User tests will be then performed to evaluate the practical effect of the system on releasing the load introduced by manual window layout management and, thus, improving user productivity. Feedbacks from subjects will be collected at the same time for future refining. For the sake of quicker implementation and verification, we employ a simple AOF method. After some performance optimization of our algorithm, other MOO algorithms, such as algorithms based on the concept of Pareto optimal like NPGA, SPEA and so on, will be incorporated into our model and experiments will be conducted to find out the one that fits our model best.

REFERENCES

Angelini, P., Cortese, P. F., Di Battista, G., & Patrignani, M. (2013). Topological morphing of planar graphs. *Theoretical Computer Science*, *514*, 2–20. doi:10.1016/j.tcs.2013.08.018

Archambault, D., & Purchase, H. C. (2013). Mental map preservation helps user orientation in dynamic graphs. In W. Didimo & M. Patrignani (Eds.), Lecture Notes in Computer Science: Vol. 7704. *Graph Drawing* (pp. 475–486). Berlin, Germany: Springer Berlin Heidelberg. doi:10.1007/978-3-642-36763-2_42

Badros, G. J., Nichols, J., & Borning, A. (2000). SCWM: an intelligent constraint-enabled window manager. In Proceedings of the AAAI Spring Symposium on Smart Graphics (pp. 76–83). Palo Alto, CA: AAAI Press; Retrieved from http://scwm.mit.edu

Beaudouin-Lafon, M. (2001). Novel interaction techniques for overlapping windows. In *Proceedings of the 14th Annual ACM Symposium on User Interface Software and Technology* (pp. 153–154). New York, NY: ACM. doi:10.1145/502348.502371

Bell, B. A., & Feiner, S. K. (2000). Dynamic space management for user interfaces. In AckermanM. EdwardsK. (Eds.), *Proceedings of the 13th annual ACM Symposium on User Interface Software and Technology* (pp. 239–248). New York, NY: ACM. doi:10.1145/354401.354790

Bergman, D. S. (2010). *Mental map preservation principles and related measurements for quantitative evaluation of force-directed graph layout algorithm behaviour* (Master's thesis). School of Interactive Arts and Technology, Simon Fraser University. Retrieved from http://summit.sfu.ca/item/10032

Bly, S. A., & Rosenberg, J. K. (1986). A comparison of tiled and overlapping windows. In *Proceedings of the SIGCHI Conference on Human Factors in Computing Systems* (pp. 101–106). New York, NY: ACM. doi:10.1145/22627.22356

Bury, K. F., Davies, S. E., & Darnell, M. J. (1985). *Window management: a review of issues and some results from user testing* (Technical Report No. HFC-53).

Coello, C. A. C., Lamont, G. B., & Van Veldhuizen, D. A. (2007). Evolutionary Algorithms for Solving Multi-Objective Problems (2nd ed.). Genetic and Evolutionary Computation. New York, NY: Springer.

Coleman, M. K., & Parker, D. S. (1996). Aesthetics-based graph layout for human consumption. *Software, Practice & Experience*, *26*(12), 1415–1438. doi:10.1002/(SICI)1097-024X(199612)26:12<1415::AID-SPE69>3.0.CO;2-P

Davies, S. E., Bury, K. F., & Darnell, M. J. (1985). An experimental comparison of a windowed vs a non-windowed operating system environment. In *Proceedings of the Human Factors and Ergonomics Society Annual Meeting* (Vol. 29, No. 3, pp. 250–254). Santa Monica, CA: Human Factors and Ergonomics Society. doi:10.1177/154193128502900310

Eades, P., Wei, L., Misue, K., & Sugiyama, K. (1991). *Preserving the mental map of a diagram* (Technical Report No. IIAS-RR-91-16E). Retrieved from http://www.cs.tsukuba.ac.jp/~misue/publications/techreport/iias-rr-91-16e.pdf

Freire, M., & Rodríguez, P. (2006). Preserving the mental map in interactive graph interfaces. In CelentanoA. (Ed.), *Proceedings of the Working Conference on Advanced Visual Interfaces* (pp. 270–273). New York, NY: ACM. doi:10.1145/1133265.1133319

Funke, D. J., Neal, J. G., & Paul, R. D. (1993). An approach to intelligent automated window management. *International Journal of Man-Machine Studies*, *38*(6), 949–983. doi:10.1006/imms.1993.1044

Gansner, E. R., Hu, Y., & North, S. (2013). Visualizing streaming text data with dynamic graphs and maps. In W. Didimo & M. Patrignani (Eds.), Lecture Notes in Computer Science: Vol. 7704. *Graph Drawing* (pp. 439–450). Berlin, Germany: Springer Berlin Heidelberg. doi:10.1007/978-3-642-36763-2_39

Hu, Y., Kobourov, S. G., & Veeramoni, S. (2012). Embedding, clustering and coloring for dynamic maps. In *Proceedings of the Pacific Visualization Symposium (PacificVis)* (pp. 33–40). Piscataway, NJ: IEEE. doi:10.1109/PacificVis.2012.6183571

Hutchings, D. R., & Stasko, J. (2002a). *New operations for display space management and window management* (Technical Report No. GIT-GVU-02-18). Retrieved from College of Computing / GVU Center, Georgia Institute of Technology website: https://smartech.gatech.edu/handle/1853/3270

Hutchings, D. R., & Stasko, J. (2002b). QuickSpace: new operations for the desktop metaphor. In L. Terveen & D. Wixon (Eds.), *CHI '02 Extended Abstracts on Human Factors in Computing Systems* (pp. 802–803). New York, NY: ACM. doi:10.1145/506443.506605

Hutchings, D. R., & Stasko, J. T. (2004). Shrinking window operations for expanding display space. In CostabileM. F. (Ed.), *Proceedings of the Working Conference on Advanced Visual Interfaces* (pp. 350–353). New York, NY: ACM. doi:10.1145/989863.989922

Johnson-Laird, A. (1985). They look good in demos, but windows are a real pain. *Software News*, *42*, 36–37.

Kandogan, E., & Shneiderman, B. (1997). Elastic windows: evaluation of multi-window operations. In PembertonS. (Ed.), *Proceedings of the ACM SIGCHI Conference on Human Factors in Computing Systems* (pp. 250–257). New York, NY: ACM. doi:10.1145/258549.258720

Lin, C.-C., Lee, Y.-Y., & Yen, H.-C. (2011). Mental map preserving graph drawing using simulated annealing. *Information Sciences*, *181*(19), 4253–4272. doi:10.1016/j.ins.2011.06.005

Liu, X., Hu, Y., North, S., & Shen, H.-W. (2013). CompactMap: A mental map preserving visual interface for streaming text data. In *Proceedings of the IEEE International Conference on Big Data* (pp. 48–55). Piscataway, NJ: IEEE. doi:10.1109/BigData.2013.6691714

Lüders, P., Ernst, R., & Stille, S. (1995). An approach to automatic display layout using combinatorial optimization algorithms. *Software Focus*, *25*(11), 1183–1202.

Mashima, D., Kobourov, S., & Hu, Y. (2012). Visualizing Dynamic Data with Maps. *IEEE Transactions on Visualization and Computer Graphics*, *18*(9), 1424–1437. doi:10.1109/TVCG.2011.288 PMID:22184261

Misue, K., Eades, P., Lai, W., & Sugiyama, K. (1995). Layout adjustment and the mental map. *Journal of Visual Languages and Computing*, *6*(2), 183–210. doi:10.1006/jvlc.1995.1010

Myers, B. A. (1988). A taxonomy of window manager user interfaces. *IEEE Computer Graphics and Applications*, *8*(5), 65–84. doi:10.1109/38.7762

Osyczka, A. (1985). Multicriteria optimization for engineering design. *Design Optimization*, 193-227.

Pilgrim, J. von. (2007). Mental map and model driven development. *Electronic Communications of the EASST, 7*.

Purchase, H. C. (1998). Performance of layout algorithms: Comprehension, not computation. *Journal of Visual Languages and Computing*, *9*(6), 647–657. doi:10.1006/jvlc.1998.0093

Purchase, H. C., Hoggan, E., & Görg, C. (2007). How important is the "mental map"? – an empirical investigation of a dynamic graph layout algorithm. In M. Kaufmann & D. Wagner (Eds.), Lecture Notes in Computer Science: Vol. 4372. *Graph Drawing* (pp. 184–195). Berlin, Germany: Springer Berlin Heidelberg. doi:10.1007/978-3-540-70904-6_19

Purchase, H. C., & Samra, A. (2008). Extremes are better: investigating mental map preservation in dynamic graphs. In G. Stapleton, J. Howse, & J. Lee (Eds.), Lecture Notes in Computer Science: Vol. 5223. *Diagrammatic representation and inference* (pp. 60–73). Berlin, Germany: Springer Berlin Heidelberg. doi:10.1007/978-3-540-87730-1_9

Sallaberry, A., Muelder, C., & Ma, K.-L. (2013). Clustering, Visualizing, and Navigating for Large Dynamic Graphs. In W. Didimo & M. Patrignani (Eds.), Lecture Notes in Computer Science: Vol. 7704. *Graph Drawing* (pp. 487–498). Berlin, Germany: Springer Berlin Heidelberg. doi:10.1007/978-3-642-36763-2_43

Sandberg-Diment, E. (1984). Value of windowing is questioned. *The New York Times*, p. 35.

Shneiderman, B. (1992). Tree visualization with tree-maps: 2-d space-filling approach. *ACM Transactions on Graphics*, *11*(1), 92–99. doi:10.1145/102377.115768

Storey, M.-A. D., Fracchia, F. D., & Müller, H. A. (1999). Customizing a Fisheye View Algorithm to Preserve the Mental Map. *Journal of Visual Languages and Computing*, *10*(3), 245–267. doi:10.1006/jvlc.1999.0124

Tak, S., Scarr, J., Gutwin, C., & Cockburn, A. (2011). Supporting window switching with spatially consistent thumbnail zones: design and evaluation. In P. Campos, N. Graham, J. Jorge, N. Nunes, P. Palanque, & M. Winckler (Eds.), Lecture Notes in Computer Science: Vol. 6946. *Human-Computer Interaction – INTERACT 2011* (pp. 331–347). Berlin, Germany: Springer Berlin Heidelberg.

Tan, D. S., Meyers, B., & Czerwinski, M. (2004). WinCuts: manipulating arbitrary window regions for more effective use of screen space. In E. Dykstra-Erickson & M. Tscheligi (Eds.), *CHI '04 Extended Abstracts on Human Factors in Computing Systems* (pp. 1525–1528). New York, NY: ACM. doi:10.1145/985921.986106

Trivedi, N., Lai, W., & Zhang, Z. (2001). Optimizing windows layout by applying a genetic algorithm. In *Proceedings of the IEEE Congress on Evolutionary Computation* (Vol. 1, pp. 431–435). Piscataway, NJ: IEEE. doi:10.1109/CEC.2001.934423

Waldner, M., Steinberger, M., Grasset, R., & Schmalstieg, D. (2011). Importance-driven compositing window management. In TanD.FitzpatrickG.GutwinC.BegoleB.KelloggW. A. (Eds.), *Proceedings of the SIGCHI Conference on Human Factors in Computing Systems* (pp. 959–968). New York, NY: ACM.

Xu, K. S., Kliger, M., & Hero, A. O. III. (2013). A regularized graph layout framework for dynamic network visualization. *Data Mining and Knowledge Discovery*, *27*(1), 84–116. doi:10.1007/s10618-012-0286-6

KEY TERMS AND DEFINITIONS

Focus Area: The screen region a user habitually watches at first glance.

Genetic Algorithm: A heuristic search algorithm inspired by the theory of biological evolution, which is typically applied to searching a huge space for the optimal approximation of the real optimum.

Mental Map: An internal cognitive representation within each individual's mind, which mirrors that individual's knowledge about different locations.

Multi-Objective Optimization: A class of optimization technique which is good at finding an optimal compromise among several mutually conflicting objectives.

Screen Space Recycling: An automated mechanism designed for the purpose of transferring the screen space occupied by unimportant or idle windows to the important or active windows, which attempts to more effectively allocate the finite screen space.

Window Importance: An abstract concept which depicts the relevance between a window and the user's task.

Window Layout: The formation of assorted open windows which are requisites for accomplishing the user's task.

Chapter 8
Modeling Human Behavior to Reduce Navigation Time of Menu Items:
Menu Item Prediction Based on Markov Chain

Tian-Tian Xie
Sichuan University, China

Tao Lin
Sichuan University, China

Yuxi Zhu
Sichuan University, China

Rui Chen
Sichuan University, China

ABSTRACT

With the increase in the number of menu items and the menu structure complexity, users have to spend more time in locating menu items when using menu-based interfaces. Recently, adaptive menu techniques have been explored to reduce the time and menu item prediction plays a crucial role in the techniques. Unfortunately, there still lacks effective prediction models for menu items. This chapter per the authors explores the potential of three prediction models based on Markov chain in predicting top n menu items with human behavior data while interacting with menus - the users' historical menu item selections. The results show that Weighted Markov Chain using Genetic Algorithm can obtain the highest prediction accuracy and significantly decrease navigation time by 22.6% when N equals 4 as compared to the static counterpart. Two application scenarios of these models on mobile devices and desktop also demonstrated the potentials in daily usage to reduce the time spent to search target menu items.

INTRODUCTION

With the increase in menu items and the menu structure complexity, users have to spend more time and efforts in locating menu items when using WIMP (windows, icons, menus, pointer) interfaces, and this tends to cause the decrease in task performance and extra mental load. How to decrease the visual search

DOI: 10.4018/978-1-4666-8450-8.ch008

time of menu items and improve task performance has been an open problem in the HCI (human-computer interaction) field. Although some research efforts have been made to solve the problem, most of them focused on the optimization of static menu structure such as menu item grouping (E. Lee & MacGregor, 1985; Norman, 1991), layout (Cockburn, Gutwin, & Greenberg, 2007) and format (Parkinson, Sisson, & Snowberry, 1985; Shih & Goonetilleke, 1998). These approaches can decrease the navigation time and improve task performance to some extent, but their efficacy is evidently decreasing with the increase of menu structure complexity and the number of menu items.

A few more efforts have also been made to explore adaptive menu technologies in response to the challenge. Adaptive menu technologies usually rely on three approaches: spatial, graphical or ephemeral. Spatial approaches (L. Findlater & McGrenere, 2008; Mitchell & Shneiderman, 1989) re-organize menu items to reduce navigation time and, to a lesser degree, to aid visual search. For example, an adaptive split menu moves or copies the most frequently and/or recently used items to the top of the menu for easier access (K. Z. Gajos, Czerwinski, Tan, & Weld, 2006). The main drawback of this approach is that spatial consistency cannot be maintained (Cockburn et al., 2007). For graphical approaches, they reduce navigation time through changing the background of predicted items (K. Gajos et al., 2005; K. Z. Gajos et al., 2006; Tsandilas & Schraefel, 2005).

As an alternative to spatial and graphical adaptation, Leah Findlater, Moffatt, McGrenere, and Dawson (2009) proposed the use of a temporal dimension and introduced *ephemeral adaptation* as a new adaptive menu technique to reduce navigation time. Ephemeral adaptive approaches use a combination of abrupt and gradual onset to provide initial adaptive support, which then gradually fades away; The goal is to draw the user's attention to a subset of adaptively predicted items, thus, reducing visual search time. To demonstrate the benefit of ephemeral adaptation, they further compared the effects of two predefined conditions of prediction accuracy (50% VS. 79%). The results show that when the accuracy with which the adaptive algorithm predicts the users' needs is relatively high (79%), ephemeral adaptation significantly offers better performance and user satisfaction benefits over traditional static menus, and a performance benefit over a graphical adaptation technique, whereas it does not, when adaptive accuracy is low (50%). It suggests that obtaining high prediction accuracy of menu items is crucial to the success of adaptive menus. However, there still lacks an effective prediction method of menu items based on historical menu click data.

Conventional prediction models are merely based on frequencies, *Most Frequently Used* (MFU) or time, such as *Most Recently Used* (MRU), or both including *Split Recency and Frequency*(SR&F) (Leah Findlater & McGrenere, 2004) and *Combined Recency and Frequency*(CRF) (D. Lee et al., 1999). However, the accuracy of these methods are usually low.

More sophisticated models utilized context information (e.g., at a bus station or in an office). Kamisaka, Muramatsu, Yokoyama, and Iwamoto (2009) utilized a wrapper algorithm to select features from time and position attributes for a Naive Bayes classifier (NB) to recommend applications for mobile devices. This NB model was better than conventional algorithms, zero-attribute rule (ZeroR), which is a simple and light-weight MFU-based method. However, they didn't find location attributes useful to the item order. Kurihara, Moriyama, and Numao (2013) proposed another context-aware approach, event frequency - inverse context frequency (EF-ICF), based on term frequency - inverse document frequency (TF-IDF) to create a prediction model, where the contexts (i.e., location, time and both) correspond to document and application events to terms. They then utilized a split menu to combine the frequency based ranking list and the context based ranking list. Their results showed an improvement on the NB model proposed by Kamisaka et al. (2009). However, the model EF-ICF was just a supplement of frequency based method, namely only a small part of the menu items was set according to this algorithm while other items were set by MFU model.

A recently proposed prediction model, AccessRank (Fitchett & Cockburn, 2012), combined a simple Markov chain with time and frequency, and resulted in better performance in accuracy compared to time and frequency models. A further study into Markov chain might brought more benefits.

Markov chain models have been widely applied in other areas such as weather forecast (Gabriel & Neumann, 1962), and user navigation pattern (Borges & Levene, 2000) with great success. All menu items of a system in nature constitute a discrete finite space and each menu clicking corresponds to a transition between two states stimulated by human behavior. Making full use of the clicking behavior might be able to help us reduce the navigation time. Thus, Markov chain was applied to predict menu items in our study with menu clicking data. Specifically, we first explored three approaches based on Markov chain (i.e., Absolute Distribution Markov Chain, Probability Summation Markov Chain and Weighted Markov Chain based on Genetic Algorithm), along with Top-N techniques, and compared them with previous prediction methods from diverse areas of computer science, including MRU, MFU, SR&F,CRF and AccessRank. Secondly, we compared the visual search time and user preferences between the static and highlighting menu display based on a prediction method selected to verify that the selected prediction method was effective. At last we demonstrated two actual application scenarios of prediction models based on Markov chain to envisage its application.

Our results showed that Weighted Markov Chain based on Genetic Algorithm could obtain the highest prediction accuracy and significantly decrease navigation time by 22.6% when N = 4 as compared to the static counterpart. And models proposed in this study utilizing clicking behavior can be generalized to reduce menu locating time in daily usage on different platforms.

MARKOV CHAIN MODELS FOR MENU ITEM PREDICTION

We mainly investigated three prediction models based on Markov chain: Absolute Distribution Markov Chain Prediction (ADMCP), Probability Summation Markov Chain Prediction (SPMCP) and Genetic Algorithm based Weighted Markov Chain Prediction (WMCP-GA) (Xia & Zhu, 2010).

Absolute Distribution Markov Chain Prediction (ADMCP)

For a sequence of random states, the transition probability matrix and initial distribution are used to calculate the absolute distribution of future states. The steps to construct and apply k-step ADMCP are below:

1. Identify the state space and the corresponding state of each time interval, $E=\{E_1,E_2,\cdots,E_m\}$ where m is the state space size.
2. Markov property test.
3. Calculate the k-step transition frequency matrix.

4. Calculate the k-step transition probability matrix $\mathbf{P}^{(k)} = (p_{ij}^{(k)})_{1\le i,j\le m}$. The matrix defines the state transition rules, where $\mathbf{P}^{(k)}$ is the probability of state E_i after a lag of k steps that transits to state E_j, $E_i, E_j \in E$.
5. Make predictions based on the initial distribution and transition probability matrix. If we take the Lth period as the initial period, and E_i as the initial state, then the initial distribution is $\mathbf{P}(L) = (0,\cdots,0,1,0,\cdots,0)$, where $\mathbf{P}(L)$ is a normal vector and the ith component is 1. Therefore the state of the distribution of the $(L+k)\mathrm{th}$ period can be obtained via:

$$\mathbf{P}(L+k) = \frac{\mathbf{P}(L)\mathbf{P}^{(k)}}{\|\mathbf{P}(L)\mathbf{P}^{(k)}\|} = (p^{(k)}(E_1), p^{(k)}(E_2), \cdots, p^{(k)}(E_m)) \tag{1}$$

where $\|\cdot\|$ is the 1-norm of a vector. Then, the predicted state E_j has:

$$E_j = \underset{E_s \in E, 1\le s\le m}{\arg\max}\, p^{(k)}(E_s) \tag{2}$$

Probability Summation Markov Chain Prediction (SPMCP)

It utilizes various step transition probability matrices and multi-period initial distribution summation to make predictions. As more historical data are concerned, this model should have higher prediction accuracy. The method requires at least 2 steps summation to perform predictions, while ADMCP is a special situation of SPMCP where only 1 step is used. The modeling process is the same as ADMCP, but needs all the matrices from 1-step to k-step, $\mathbf{P}^{(1)}, \mathbf{P}^{(2)}, \cdots, \mathbf{P}^{(k)}$, instead. Based on the initial distributions, $\mathbf{P}(L), \mathbf{P}(L-1), \cdots, \mathbf{P}(L-k+1)$, of the [illegible] periods, apply the formula below.

$$\mathbf{P}(L+1) = \frac{\sum_{i=1}^{k} \mathbf{P}(L-i+1)\mathbf{P}^{(i)}}{\|\sum_{i=1}^{k} \mathbf{P}(L-i+1)\mathbf{P}^{(i)}\|} \tag{3}$$

Then, we can get the $(L+1)\mathrm{th}$ period state distribution and the predicted state E_j according to formula (2).

Genetic Algorithm Based Weighted Markov Chain Prediction (WMCP-GA)

The basic idea of WMCP-GA is to obtain the optimal weights via a genetic algorithm for weighted Markov chain (WMC). Compared to SPMCP, WMCP considers not only historical data but also the importance of each historical period. The calculation process between WMCP and SPMCP only varies when computing $\mathbf{P}(L+1)$.

$$\mathbf{P}(L+1) = \frac{\sum_{i=1}^{k} w_i \mathbf{P}(L-i+1)\mathbf{P}^{(i)}}{\|\sum_{i=1}^{k} w_i \mathbf{P}(L-i+1)\mathbf{P}^{(i)}\|} \tag{4}$$

w_i is the ith component of the weight vector of the k-step transition probability matrix. The determination of the weight vector is a combinatorial optimization problem. To solve this problem, a simple genetic algorithm was used in this study. The weights of WMCP are evolved by selection, crossover and mutation operators for a certain generation, and are evaluated by prediction accuracy (i.e., fitness value). The process is detailed in the first experiment.

When Top-N is added, the predicted set of the $(L+k)\mathrm{th}$ period is S, $S \subseteq E$, $|S|=n$, $n=1, 2,\ldots, m$, $\forall E_i \in S, \forall E_j \in C_E S$:

$$p^{(k)}(E_i) > p^{(k)}(E_j) \tag{5}$$

where $1 \leq i \leq n; 1 \leq j \leq m-n$. By default, Top-N is set to Top-1 (N is the same to n when referred).

Evaluation for Prediction Algorithms

Herlocker, Konstan, Terveen, and Riedl (2004) recommended six measures for evaluating prediction algorithms of a recommender system. We chose three of the six measures (i.e., accuracy, coverage and user evaluation) as the metrics for algorithm evaluation. In addition, modeling time was adopted in the study.

For each algorithm, prediction accuracy is computed as the average of the percentage of correct predictions for each task sequence. Prediction coverage is computed as the number of menu items for which predictions can be formed as a percentage of the number of menu items in all task sequences. Modeling time is recorded from the start of modeling on a training set to the end of testing on a test set. These three measures are calculated with ten-fold cross validation (Kohavi, 1995). For user evaluation, we seek *explicit (ask)* (Herlocker et al., 2004) via questionnaires to obtain users' subjective feedbacks about preferences, and *implicit (observe)* (Herlocker et al., 2004) via logs to analyze the difference of the visual search time for each tested algorithm.

EXPERIMENTS

The main goal of our experiment was to examine the efficacy of various algorithms in menu item prediction. A drawing software, AyuDraw, was used as the experimental software platform, and users' click data of menu items were collected when they were completing the given drawing tasks. We first examined various prediction algorithms, along with Top-N techniques, and compared them with previous prediction methods listed in the introduction, and then analyzed the visual search time and user preferences between the static and highlighting menu display based on a selected prediction method.

Experiment One: Menu Item Prediction Modeling

We first collected sequential click data of menu items, and then applied three prediction methods (ADMCP, SPMCP, WMCP-GA) with different parameter configurations on these data. In addition, the three methods were evaluated in terms of prediction accuracy, coverage, and modeling time, and compared with previous methods (MRU, MFU, SR&F, CRF and AccessRank) in terms of prediction accuracy.

Platform

We selected an open source PC paint software, AyuDraw, as the platform, which can be easily run on PC and Tablet PC. There are only 13 menu items related to painting in the software, and they all are located under the "Tool" menu. These menu items are: Select, Eraser, Zoom, Pen, Brush, Text, Font, Line, Circle, Rectangle, RoundRect, Undo and Redo. Figure 1 shows the overview of these 13 items in the software. The state space of menu predictions is composed of these 13 menu items, while other menu items (such as "open" and "save") irrelevant to painting are ignored for simplicity. Those non-menu items related to color and graphic model option (hollow or solid, line weight, eraser size, etc.) are not included as well. Each letter from A to M represents a corresponding state within those 13 states. Then the state space is

$$E = \{A, B, C, D, E, F, G, H, I, J, K, L, M\}.$$

A 2.2 GHz AMD Phenom II N970 laptop with 4 GB of RAM and Microsoft Windows 7 was used to construct these models for measure calculation (e.g., accuracy, coverage, and modeling time).

Figure 1. The AyuDraw platform that shows the 13 menu items related to painting task under "Tool" menu

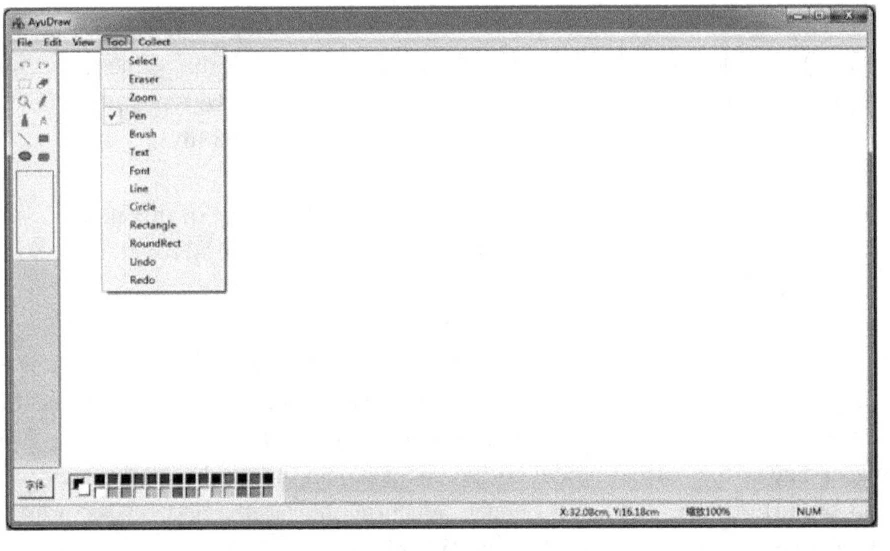

Task

For each drawing task, participants should imitate one out of ten picture templates. Moreover, participants shall only use right-click popup menu to select menu items, which are the same to those under "Tool" menu. Those ten templates are pictures of gate, desk, dormitory, stars, mountains, forests, pastoral, pencil, soccer field and portrait.

Participants

Thirty-two participants (eight females) participated in the experiment. Their ages range from 20 to 32 and all of them have user experience for paint software of such kind.

Procedure

The platform was sent to participants with instructions about the software and given tasks in detail. They were asked to get familiar with the software before starting the experimental tasks, and should complete the given tasks in two weeks. Each completion of tasks produced a state sequence of menu items selected (e.g., *BCDADLBE*) and a picture. As there were noises inside each sequence, a preprocess was used to filter two kinds of those noises below:

1. Unconscious menu selection. Participants draw nothing after a selection of a menu item. This is often caused by error selection. They are filtered out while collecting data on the client side.
2. Repetitive menu selection. While under a menu state, participants reselect the same menu item again. As menu functions related to drawing except 'undo' and 'redo' are consistent (namely, no need to reselect the same menu item to trigger the same function), only repetitive 'undo' and 'redo' selections remain in the sequence.

Then, chi-square test was carried out on those preprocessed data to guarantee that they could be used in Markov chain prediction modeling (we only tested on 1-step Transition Frequency Matrix of the 13 menu states). In order to obtain the prediction accuracy and modeling time for each model within different configurations of Top-N and steps, we constructed those three types of model with different configurations below (Each modeling procedure ran in a single thread only):

3. ADMCP. We construct Top-N × Step (6×10) = 60 configurations of this model.
4. SPMCP. As SPMCP requires at least two steps, we constructed Top-N × Step (6×9) = 54 configurations of this model.
5. WMCP-GA. WMCP-GA also requires at least two steps, we construct Top-N × Step (6 × 9) = 54 configurations of this model. Genetic Algorithm on evolving weights: Each weight is encoded as an 8-bit binary number, the population size is 20, and the default generation number is 100. The fitness function is that after normalization of each weight vector, ten-fold cross validation is used to calculate the accuracy of the current model on the data set, and the produced accuracy is then set to be the fitness of the current individual. For each selection process, the best fitness individual is always selected, while others run through a Roulette Wheel. Single point crossover operator (crossover rate = .80) and single point mutation (mutation rate = .01) are used. For this model, we take another 10 typical test with Top-3 to study the characteristic of weights with the highest prediction accuracy.

We calculated the prediction coverage of Top-1 ADMCP from step-1 to step-10 as the coverage of all three models.

Then, in order to study the impact of generation number on prediction accuracy and modeling time for WMCP-GA, we added another two typical tests, where the second and the third tests set the generation number to 50 and 500. We also set Top-N to Top-3, because predictions with Top-3 have been largely used in many menu interaction studies (Cockburn et al., 2007; Leah Findlater et al., 2009; Sears & Shneiderman, 1994; Tsandilas & Schraefel, 2007).

At last, we compared those predictive algorithms introduced in introduction with these three Markov chain based menu prediction methods from Top-1 to Top-10. As AccessRank is dependent on both Markov and CRF, we also integrated these three Markov chain prediction methods into AccessRank.

Results and Analyses

The data collection continued for two weeks. We had collected 55 drawing tasks from those 32 participants. The total menu item number was 1389, the shortest sequence length was 3, the longest was 62, and the average was 25.

The preprocess filtered out 4.39% of the original items, and 1328 menu items remained. The frequencies of these menu items followed Zipf distribution (R^2 = .84); The data collected was valid (Cockburn et al., 2007). After the calculation, the chi-square test showed that: $\chi^2 = 1362.25$ ($p < .01$). Therefore, the Markov property was satisfied, and we could use methods based on Markov chain to predict.

For each Top-N from 1 to 6, Figure 2 shows the relationship between steps and the prediction accuracy of each model. Table 1 shows the characteristics of the prediction accuracy under certain Top-N and prediction model type. Table 2 shows the best weights of WMCP-GA from Top-1 to Top-6. Figure 3 shows the prediction coverage of ADMCP under Top-1. The impact of the generation number of WMCP-GA on the prediction accuracy and the prediction modeling time under Top-3 shows in Table 3. The comparison results of those algorithms are shown in Figure 4.

Table 1. Under certain GA configurations, the statistics of the prediction accuracy with different Top-N of these three prediction models

Top-N	ADMCP				SPMCP				WMCP-GA			
	Accuracy (%)			BS	Accuracy (%)			BS	Accuracy (%)			BS
	M	*SD*	*Max*		*M*	*SD*	*Max*		*M*	*SD*	*Max*	
1	21.25	3.90	26.84	2	27.66	0.91	29.70	3	30.82	1.37	31.94	5
2	33.51	3.64	39.43	1	41.45	2.25	44.79	4	46.63	1.03	47.76	5
3	44.67	4.43	50.78	4	52.68	2.23	55.31	2	56.74	0.65	57.60	8
4	54.53	4.60	63.97	1	62.45	2.35	66.20	2	67.88	1.39	69.51	6
5	63..67	3.74	72.14	1	71.21	1.91	74.96	2	76.37	0.75	77.35	7
6	72.07	3.81	81.59	1	78.09	1.75	80.84	2	83.30	0.59	84.35	7

Note. BS = best step, which leads to the highest prediction accuracy for a specific model under a certain Top-N.

Figure 2. The relationship between the prediction accuracy of three model types and steps under different Top-N

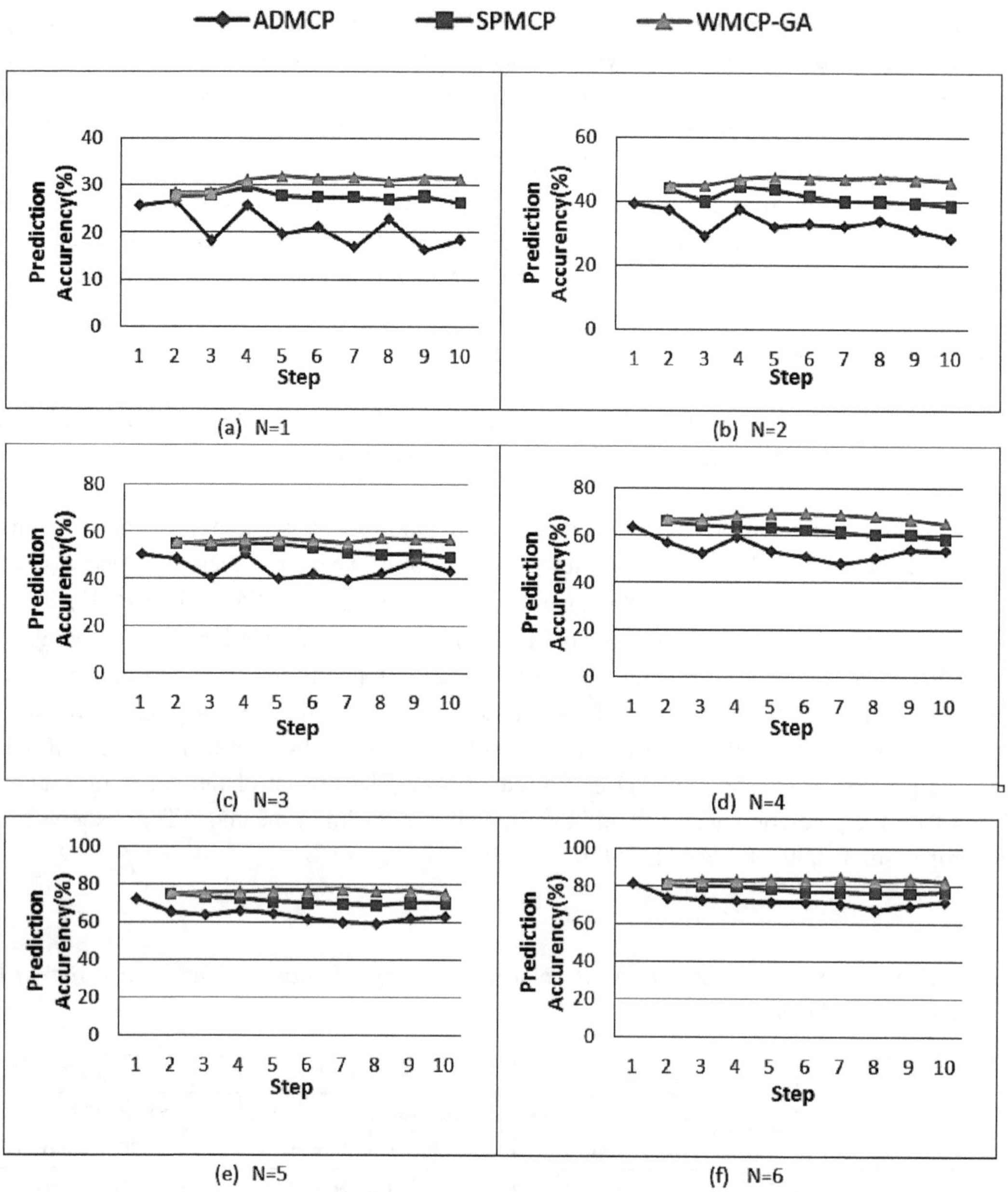

Prediction Accuracy

The factors that directly affect Prediction Accuracy are Top-N and model type. For each model type, the accuracy also depends on the steps used of Markov chain. Moreover, the accuracy of WMCP-GA is also affected by various parameter settings of GA (to avoid the experiment being a GA research, we only consider the influence of generation number on prediction accuracy). We follow through four steps to study the relationship between accuracy and each factor:

Table 2. Optimal weight vectors for WMCP-GA with the initial population size set to 20 and generation number to 100

Top-N	Steps	Optimal Weights							
		1	2	3	4	5	6	7	8
Top-1	5	0.37	0.17	0.05	0.37	0.04			
Top-2	5	0.45	0.13	0	0.28	0.13			
Top-3	8	0.30	0.35	0.04	0.25	0.04	0.00	0.00	0.00
Top-4	6	0.53	0.27	0.02	0.13	0.06	0.00		
Top-5	7	0.54	0.20	0.03	0.14	0.02	0.06	0.00	
Top-6	7	0.58	0.15	0	0.25	0	0.02	0.00	

Table 3. The influence of generation number on the prediction accuracy and the modeling time for WMCP-GA (n=3).

Steps	50 Generations		100 Generations		500 Generations	
	Accuracy (%)	Time(s)	Accuracy (%)	Time(s)	Accuracy (%)	Time(s)
2	55.81	5	55.91	9	55.90	45
3	55.97	5	56.16	11	55.99	52
4	56.40	6	56.98	12	57.01	60
5	57.31	7	57.26	13	58.06	64
6	56.27	7	57.11	14	57.69	70
7	55.67	8	57.00	15	57.25	74
8	56.18	8	56.41	16	57.92	78
9	55.38	8	57.01	16	57.71	81
10	53.39	9	56.76	18	57.32	84

Note. As the modeling time is calculated with ten-fold cross validation, all the time recorded is ten times of the time period from the start of modeling on a training set to the end of testing on a test set.

Figure 3. Prediction coverage rate of three model types under different steps

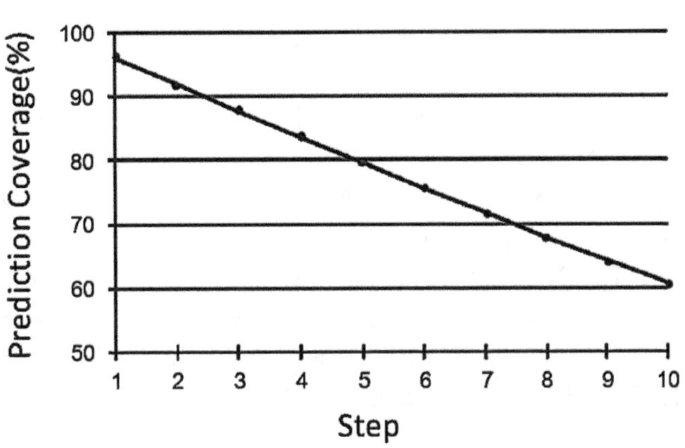

Figure 4. The Prediction accuracy of these algorithms on menu interaction datasets from Top-1 to Top-10

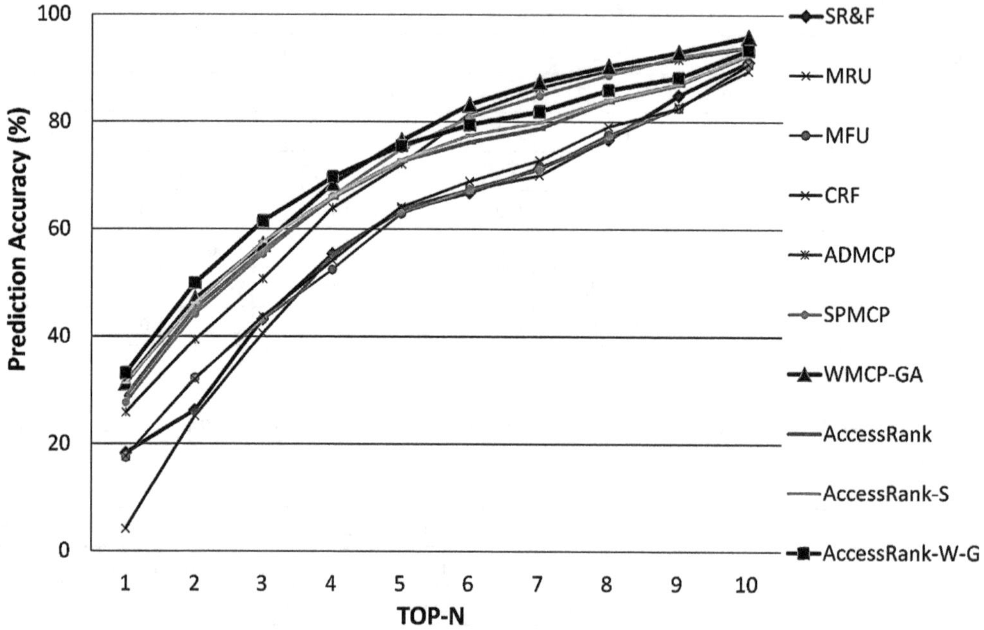

1. When parameter settings of GA are fixed, Bivariate correlation (Pearson) in SPSS is used to analyze the correlation between Top-N, model type and accuracy. For those with two-side significance (p < .05), ANOVA will be used to further analyze the relationship.
2. As Markov chain prediction relies on the steps used, we also try to find the relationship between steps and prediction accuracy when other parameters are regarded as constant.
3. For a specific Top-N, the best settings of step will be found out according to accuracy.
4. In order to guarantee the configuration of GA parameters is within proper range, the experiment only checks the impact of generation number on the accuracy under different step when N is 3.

- The prediction accuracy is significantly correlated to Top-N ($p < .001$) and model type ($p < .001$). The prediction accuracy of each model is significantly different, and the accuracy sort from high to low is WMCP-GA, SPMCP and ADMCP. When *n* increases, Top-N model increases the prediction accuracy of the interaction model significantly (*Sig.* < .001).

When considering model types only, Independent-Samples T (T') test shows: ADMCP ($M = 48.28$, $SD = 17.91$), SPMCP ($M = 55.59$, $SD = 17.51$), WMCP-GA ($M = 60.28$, $SD = 18.05$). Compared to ADMCP, SPMCP ($F_{59, 53} = .038$, *Sig.* = .030), WMCP-GA ($F_{59, 53} = .020$, *Sig.* = .001), and between SPMCP and WMCP-GA, we have $F_{53, 53} = .111$, *Sig.* = .173.

When only considering Top-N of ANOVA analysis. LSD (heterogeneity of variance, *Sig.* = .435) was used to compare the different impacts on accuracy between each Top-N. Each pair is significantly different (*Sig.* < .001). The average prediction accuracy of the interaction model from Top-1 to Top-6 is 26.39%, 40.28%, 51.12%, 61.37%, 70.18% and 77.61%.

- When the step increases, there is a trend that the prediction accuracy of each Markov chain prediction model decreases (Figure 2). The trend varies from high to low is ADMCP, SPMCP and WMCP-GA. For each Top-N, the *SD* of the prediction accuracy of each prediction type of the interaction model exists: ADMCP > SPMCP > WMCP-GA. Namely, the influence of the prediction accuracy stability on each prediction type by steps follows the same relationship (Table 1).
- When step is 1, ADMCP can reach the highest prediction accuracy; For SPMCP, 2 steps offer the highest prediction accuracy. But when N is low, the steps that offer the best accuracy vary for both methods. For WMCP-GA, the best step varies according to N (Table 1). Under certain Top-N, the best weights are listed in Table 2. In which, it suggests that the weights of step 1, step 2 and step 4 are the main components if we set the threshold of weight to 0.10.
- The steps used in WMCP-GA to offer the highest prediction accuracy are 5 steps under three different generation numbers (Table 3). After comparing the accuracy under different settings of generation number using ANOVA, the LSD result shows, the average accuracy between 100 and 500 configurations does increase significantly (*Sig.* = .001).

Prediction Coverage

The coverage of each Markov chain based model only depends on the menu sequence itself for a specific task, and on the steps used for each prediction. When a task is specified and the max number of steps used is fixed, the coverage of each model is the same. Thus, the simplest prediction model, ADMCP, is used to find the coverage of these three models.

Figure 3 shows the prediction coverage of Top-1 ADMCP under different steps, which represents the coverage of three model types. The coverage decreases according to steps. When step is 1, the coverage rate reaches the highest, 95.86%; And when step is 10, the coverage rate reaches the lowest, 60.55%. As the average length of a task sequence is 24, when more steps are used, the coverage drops significantly.

Modeling Time

The factors that affect modeling time are the same to those of prediction accuracy. We follow through two steps to study the relationship between modeling time and those factors.

1. When parameter settings of GA are fixed, Bivariate correlation (Pearson) in SPSS is used to analyze the correlation among Top-N, model type and modeling time. For those with two-side significance ($p < .05$), ANOVA will be used to further analyze the relationship.
2. For a specific Top-N, we analyze the impact of generation number on the modeling time of WMCP-GA under different steps.

- The modeling time varies significantly among those three prediction models ($p < .001$). When considering the impact of model types only, the independent sample t-test shows that the average modeling time of ADMCP is 2.73 ms (*SD* = .52), it increases to 7.17 ms (*SD* = 1.40) for SPMCP and 14746.31ms (*SD* = 3232.23) for WMCP-GA. When compared to ADMCP, the difference from SPMCP ($F_{53, 59}$ = 40.60, *Sig.* = .000) and WMCP-GA ($F_{53, 59}$ = 139.61, *Sig.* = .000) are significant. When compared to SPMCP, WMCP-GA ($F_{59, 59}$ = 125.455, *Sig.* = .000) varies significantly.

Table 4. The Configurations of 10 compared algorithms for menu prediction

Algorithms	Configurations
MRU	
MFU	
SR&F	select $(n+1)/2$ item(s) with MRU, and the rest with MFU.
CRF	$p = 2$ and $\lambda = 0.1$ (Fitchett & Cockburn, 2012)
ADMCP	step = 1 (the third result of prediction accuracy)
SPMCP	step = 2 (the third result of prediction accuracy)
WMCP-GA	step = 4, generation = 100, population = 20, crossover rate = .80, mutation rate = .01 (the third result of prediction accuracy suggests 4-step is the best)
AccessRank	$p = 2$, $\lambda = 0.1$, $\alpha = 1.65$, $\delta = 0$, and $w_{t_n} = 1$ (as timing data is not collected, we set w_{t_n} to 1)
AccessRank-S	$p = 2$, $\lambda = 0.1$, $\alpha = 1.65$, $\delta = 0$, $w_{t_n} = 1$, and step = 2
AccessRank-W-G	$p = 2$, $\lambda = 0.1$, $\alpha = 1.65$, $\delta = 0$, $w_{t_n} = 1$, step = 4, generation = 100, population = 20, crossover rate = .80 and mutation rate = .01

Note. AccessRank-S = AccessRank-SPMCP, where SPMCP is integrated into AccessRank. AccessRank-W-G = AccessRank-WMCP-GA, where WMCP-GA is integrated into AccessRank.

- The modeling time of WMCP-GA related to the generation number and the step is approximately linear (Table 3).

Comparisons with Previous Algorithms

Algorithms compared are listed in Table 4, those configurations are selected according to analyses above and other researches.

When $n \leq 4$, AccessRank-W-G has the highest accuracy (Figure 4). As suggested, highlighting is useful when the minimal accuracy level is higher than 60% (Antwarg, Lavie, Rokach, Shapira, & Meyer, 2012). When $n = 3$, AccessRank-W-G first reaches the accuracy of 61.45%, followed by WMCP-GA (57.12%). When $n > 4$, WMCP-GA performs the best. When $n = 5$, WMCP-GA has the accuracy of 76.53%, followed by AccessRank-W-G (75.59%).

Discussion

The result is discussed from aspects of measures of the three prediction models and the comparison of previous algorithms:

- ADMCP: ADMCP offers the lowest prediction accuracy under the same condition. When step is 1, the prediction accuracy reaches its maximum. However, when $n \leq 3$, the best step is different. As the least amount of history needed to predict is one step, ADMCP can reach the highest prediction coverage of 95.86%. Moreover, the modeling time is short (M = 2.73ms), each menu interaction can update the prediction model rapidly. So, this model can be used not only when global learning is needed at run time, but also when per user and per item learning is needed.
- SPMCP: The prediction accuracy of SPMCP is between that of ADMCP and WMCP-GA. 2 steps probability summation can provide the highest accuracy. However when $n \leq 2$, the best step also varies. The least requirement of step is 2 for SPMCP, so the max prediction coverage is 91.71%. The modeling time is short (M = 7.17ms), and the property of modeling time is the same to ADMCP.
- WMCP-GA: The prediction accuracy of WMCP-GA is the highest. Although the best step is different under each Top-N, we can infer from Table 2 that step 1, step 2 and step 4 influenced the prediction accuracy most. As WMCP-GA also requires 2 steps at least, the maximum prediction coverage is 91.71%. The time to train (update) this model is the longest (M = 14.7s, 100 generations). If CPUs do not improve significantly, this model cannot be used when real-time per item learning is needed. But it can be used when per person and global learning is needed.

The Comparison shows that, when $n \leq 5$, current existing methods surveyed cannot reach the accuracy of 80% for this application. And when compared with those three methods, MRU, MFU, SR&F and CRF are not suitable for menu prediction. But the combination of CRF and Markov chain based prediction methods (AccessRank) can improve accuracy when n is low.

Experiment Two: Interaction Efficiency Verification

We integrated a specific prediction model into the original platform based on the analyses of experiment one. Then we compared the visual search time and user preferences between the traditional and the highlighting menu interaction to verify the hypotheses:

1. The visual search time of menu items with an adequate prediction technique is shorter than that of the static menu item.
2. Participants prefer the highlighting menu display to the static menu display.

Once the hypotheses are verified, we make conclusions that the prediction method can get a satisfied menu prediction result. Thus, the selected prediction method is an effective prediction method for menu item prediction.

Selection of Prediction Model

In the experiment, we used five conditions below to select a proper prediction model:

1. Select prediction models with the highest prediction accuracy under the same configuration. (From the result of experiment one, WMCP-GA is selected).

Table 5. Prediction accuracy and coverage rate (%) of 4-step WMCP-GA from Top 3 to 5

Top-N	Prediction Accuracy	Prediction Coverage
3	57.12	83.51
4	68.55	83.51
5	76.53	83.51

2. The accuracy of prediction models must be higher than 50% (Because low prediction accuracy will prevent users from searching for the target item).
3. Only consider the prediction methods with $n \leq 5$ (As the goal of decreasing visual search time, the actually shown number of prediction items shall be small. Otherwise, a second level visual search time might increase. As the total number of menu items is 13, we decided that *n* should be less than half).
4. The prediction coverage must be higher than 80% (This indicates the utilization rate of prediction).
5. Utilize histories of the last four item selections suggested by experiment one.

Under these conditions, the selected prediction model was 4-step WMCP-GA with Top 3 ~ 5. The prediction accuracy and coverage of the prediction model is shown in Table 5. As this study focuses on Markov chain only, we haven't selected 4-step AccessRank-WMCP-GA, though its accuracy is higher when *n* is less than 5.

Figure 5. Top-3 menu items highlighted by adding "Stars" as prefix and suffix. The last selected menu item was Pen, which triggered this prediction.

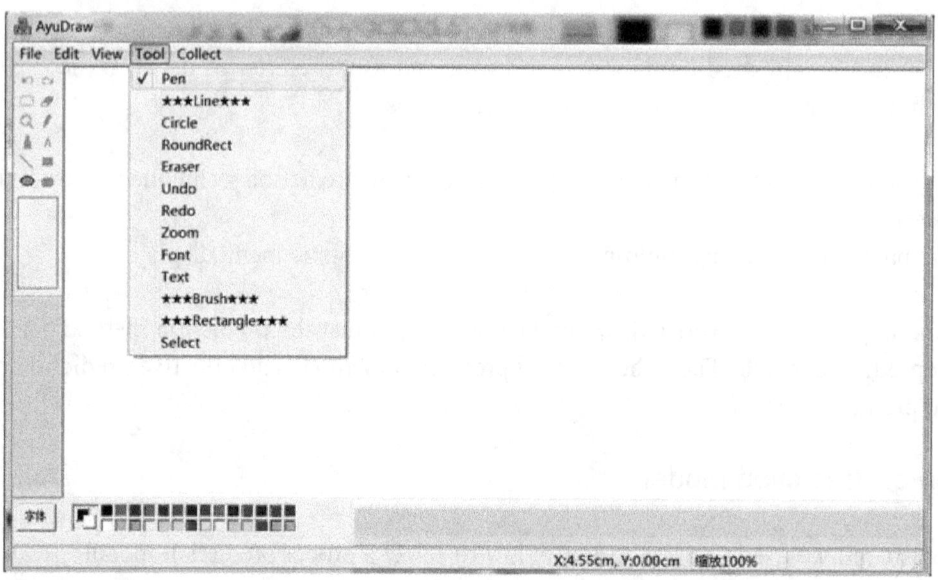

Platform

We had modified three parts of the platform used in experiment one:

1. Integrating prediction models (A well trained 4-step WMCP-GA model was integrated into the software. The last four menu item states were used as the input to calculate current possible items. Moreover, the predicted items were highlighted by adding "Stars" as prefix and suffix. Figure 5 shows the display result of a prediction under Top-3).
2. Collecting visual search time (We recorded the period between the time a user opened up a right-click popup menu and had finished selecting a menu item as the visual search time for the menu item).
3. Varying the order of menu item sequences under "Tool" menu (As for each participant, they will perform one task under 4 different configurations in the experiment, the variation is used to eliminate the effect of memory. The order is random generated previously).

Task Design

The design of the task content is:

1. Each participant chooses one picture out of five different templates to imitate (In this way we can clearly define the task as well as maintain the diversity of tasks performed).
2. Each imitation is carried out under 4 different versions of the software (The purpose is to find out under which Top-N, the model performs the best. These four versions of the painting software are the platform with a default static menu, the platform with the Top-3 4-step WMCP-GA model integrated, the platform with the Top-4 4-step WMCP-GA model integrated, and the platform with the Top-5 4-step WMCP-GA model integrated).
3. We undertake two phases to verify the hypotheses (Firstly, we conduct a survey on user preferences, which provides better experience when users have completed the entire tasks. Those four options are: (a) static menu is better; (b) menu with Top-3 is better; (c) menu with Top-4 is better; (d) menu with Top-5 is better; (e) neither. Secondly, the collected visual search time from the four modified platforms is compared to each other).

Results and Analyses

Ten participants, nine males, took part in this experiment. Their ages range from 20 to 30. They all are familiar with painting software. The data gathered from these ten participants covered all five types of templates. There were 40 pictures in total and 628 sequence elements (Each element records a menu item selection, a pairwise visual search time and a hit flag). The longest task sequence contained 46 elements, while the shortest contained seven. The average length of the interaction sequence was 16.

The subjective questionnaire result is shown in Table 6. Nine participants thought that platforms with prediction offered better menu interaction efficiency than the static counterpart, only one participant chose neither. After analyzing the sequence, we find out that the participant who could not decide which was better had lower prediction coverage (42.86%) as his interaction sequence was quite short (seven). In those platforms with prediction, 4 participants preferred Top-3 and another 4 preferred Top-4, while

Table 6. Survey results of subjective questionnaires on which menu provides better efficiency (Unit: People)

	Static	Top-3	Top-4	Top-5	Neither	Total
Count(s)	0	4	4	1	1	10

Note. A 4-step WMCP-GA prediction model was integrated into the menu from Top-3 to Top-5. The model was previously trained, and remained unchanged during the task session.

only one participant preferred Top-5. The reason why the number of people who chose Top-3 and Top-4 was the same might be the fact that while Top-4 increased the prediction accuracy when compared to Top-3, it increased the visual search time as well, thus, the total improvement of each was likely the same; Or simply the sample size was too small. Although, Top-5 could offer the highest prediction accuracy, as *n* increased to 5, the highlighting effect might decrease, thus lowering the visual search efficiency.

The objective result is shown in Table 7. The calculation of the average visual search time of static menu interaction utilized all the elements for a particular participant while those of the other 3 menu interactions with prediction only utilized the elements after the fifth. Because we had selected 4-step WMCP-GA in the experiment, the prediction wouldn't show up until the fourth menu had been selected. However, we found out that the fifth visual search time was larger than the following visual search time. We can reason this as once a new change happens suddenly, the user's attention will be disrupted, which will lengthen the time lag. In order to eliminate this effect for the other 3 menu interactions, all the average visual search time was calculated starting from the sixth element.

Figure 6 shows that the average visual search time of a menu item on platforms with prediction is lower than that on the platform without. When compared to the static menu interaction (Bonferroni), Top-3 and Top-4 lowered the visual search time significantly. The visual search time of each had reduced by 20.21% ($p = .024$) and 22.46% ($p = .009$), and Top-5 by 15.12% ($p = .163$). The statistic also suggests that a second search within those predicted items would lengthen the visual search time when *n* exceeds a threshold. For the software we tested, 4 item predictions can reduce the visual search time most.

Table 7. Statistics of the average of visual search times (ms) for the tasks performed of each participant under four different versions of platforms

Participant	Average Visual Search Time			
	Static Menu	Top-3	Top-4	Top-5
1	2846.00	1933.00	1925.00	2021.00
2	3068.00	2923.00	2971.00	3072.00
3	3336.00	2070.00	2100.00	2164.00
4	2690.00	2173.00	2169.00	2235.00
5	3126.00	2482.00	2186.00	2780.00
6	2401.00	1871.00	1669.00	1851.00
7	2894.00	2030.00	1923.00	2069.00
8	2496.00	2634.00	2486.00	2734.00
9	2383.00	1793.00	1778.00	1925.00
10	1961.00	1796.00	1884.00	2242.00
M	2720.10	2170.50	2109.10	2309.30

Figure 6. Average visual search times (ms) for Static menu, and Top-3, Top-4, Top-5 based menu with prediction

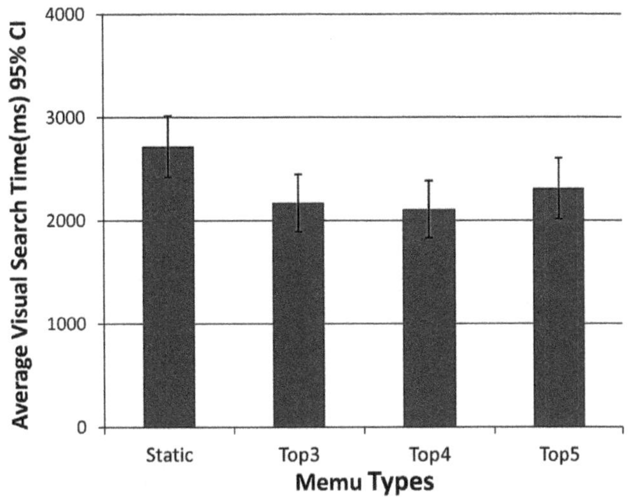

Figure 7. Application scenarios of Markov chain models on desktop (upper) and on mobile devices(down)

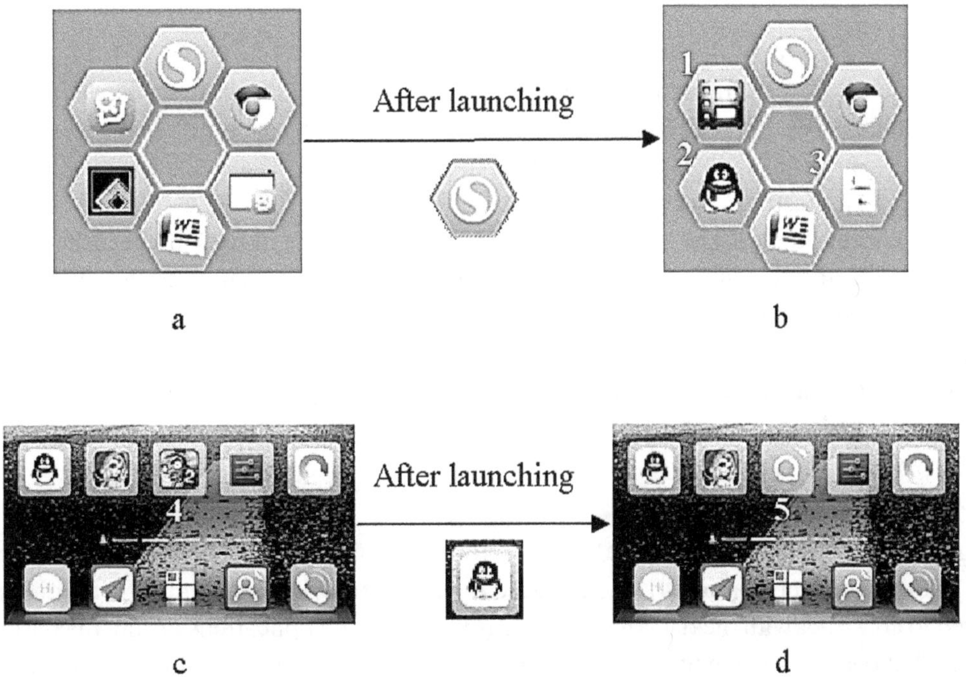

Experiment Summary

From the experiment analyses, these two hypothesizes are fulfilled:

1. The visual search time of menu items with prediction is shorter than that of the static counterpart. Especially, when n equals 4, the visual search time is the lowest (lowered by 22.46%).
2. Users prefer the highlighting menu display to the static menu display. Nine out of ten prefer highlighting.

Therefore, the prediction method can get a satisfied menu prediction result, and the selected method is an effective prediction method for menu item prediction.

Application Scenarios

In order to demonstrate the feasibility of Markov chain based models for menu item prediction in daily usage, we implemented two applications and invited three pilots to use them. However, one pilot refused to participant in this study for security reason. One application, HexaPopApp, is implemented as a plug-in of Rainmeter for desktop and the other, PopApp, is programmed as a widget for android devices (Figure 7). Each application icon denoted a menu item. For simplicity, a top-6 2-step SPMCP model was chosen for HexaPopApp and a top-5 2-step AccessRank - SPMCP model was chosen for PopApp.

HexaPopApp

The usage of HexaPopApp is illustrated in Figure 7(a) and (b). After launching an alternative application of Internet Explorer (IE), three application shortcuts were replaced by a movie application (1), an instant messaging application (2) and the default pdf reader on win8 (3). The selected model was constructed using the data of application launching event from one pilot during the past 46 days.

During this scenario, this pilot planned to have a rest (watching a movie) after sending the essay he wrote via a web mail to his tutor. Since the model predicted and recommended the movie application he usually launched, he could launch app(1) immediately without searching in a bunch of shortcuts on the desktop.

After months of evaluating this application, the pilot reported positive feedback that only a glimpse of HexaPopApp was enough for him to launch the application he tended to use for most of the time.

PopApp

The usage is demonstrated in Figure 7 (c) and (d), after launching an instant messaging application, the shortcut of one game application (4) was replaced by a phone book application (5). This interface was configured to construct with the data of switching front events of applications during the past month. In order to continuously recommend applications in accordance with the taste of users. The model would reconstruct every active hour (times when the android system is in awake mode).

Figure 8 (a) screenshot the monitored performance every five active hours for the pilot in the last five days (M = 81.73%, SD = 9.67%). As the pilot reported that this application saved his time of switching application screens, we believe this application with relatively high accuracy is feasible in daily operation.

Figure 8. The performance every five hours in PopApp (a) and every day in the last month (b)

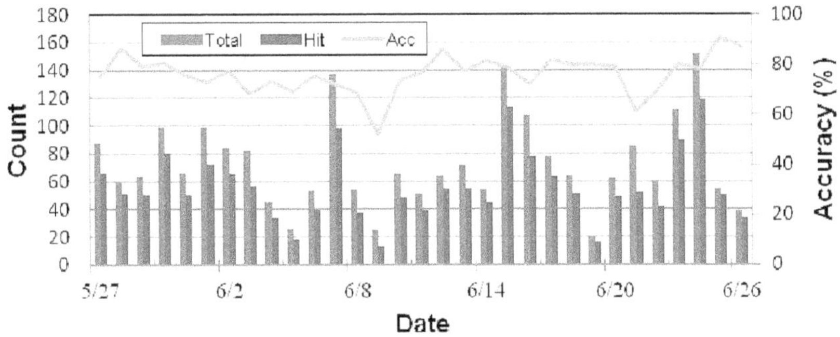

a

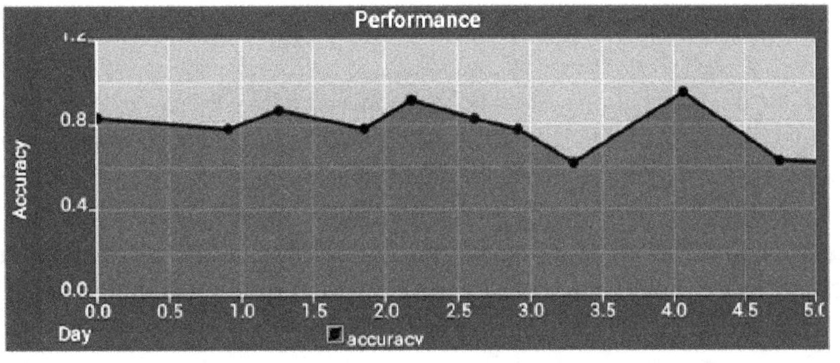

b

Figure 8 (b) illustrated the daily use count of applications and the accuracy ($M = 76.12\%$, $SD = 7.69\%$) of the target application in predicted list in the last month. It performed well except on date 6/9 ($Acc = 52.00\%$) and 6/21 ($Acc = 61.18\%$). The pilot reported that he had installed one new application on 6/9 and had to switch screens to launch this application. And on 6/21 the weekend, he had to use a hardly used navigation application to help a car driving of his relatives. However, the performance ameliorated in the next day, 73.85% on 6/10 and 69.49% on 6/22. This results suggested that this model could perform well for daily operation for routine operation on mobile devices. Nevertheless, once a new application was installed or a sudden event occurred, it failed to adapt immediately but with a delay. However the pilot reported that this delay was acceptable.

FUTURE RESEARCH DIRECTIONS

Technique

There are many technique issues we have not dealt with in this chapter. For example, the construction process of the menu prediction model used in this chapter is offline. When the model construction is completed, it will never update again. Nevertheless, as the well trained prediction model is based on data collected from many users, it can be first deployed as the default configuration of a menu system, and then be updated based on user's selection on the fly. Although this chapter has not shown how to use user item selections or menu clicks to update the prediction model at runtime, we can easily grab those data together to do the same construction process again. This is what we have done in our application scenarios. However, more factors need to be taken into account, such as how the dated and newly data contribute to the update circle, and when the update should be triggered. They all should be proposed based on characteristics of application domains, platforms (e.g., PC or hand-held device) and user abilities. Thus, the prediction model can be gradually directed to users' taste and further reduce navigation time.

Another issue not mentioned here is to reduce navigation time for multi-menu structure, which can be achieved by multiple Markov chain predictions. However, we may try to use HMM to solve this problem as hidden Markov model (HMM) has already been intensively adopted in temporal data mining problems and has good performance.

Recommendation System

From the aspect of a recommendation system (RS), tranditional RSs are based on demographic, content-based, collaborative filtering and/or social information, and have been utilized on varies topics, such as music (Lee, Cho, & Kim, 2010), tourism (Borràs, Moreno, & Valls, 2014), books (Núñez-Valdéz et al., 2012), e-learning (Bobadilla, Serradilla, & Hernando, 2009), markets (Costa-Montenegro, Barragáns-Martínez, & Rey-López, 2012) and web search (McNally, O'Mahony, Coyle, Briggs, & Smyth, 2011). Those information can be implicitly recorded (e.g., movies played) or explicitly generated(e.g., user's rating) by monitoring user behaviors. Large amount of User actions and interactions are stored electronically. This provides researchers an opportunity to study RSs in a great detail.

Recent research (Bobadilla, Ortega, Hernando, & Gutiérrez, 2013) showed that there is a "next generation of RS", with a clear trend of incorporating, local, personal and implicit rather than explicit information to further enhance recommendation. Our chapter belongs to this category. Menu clicks were implicitly and non-intrusively collected from all participants to build a collaborative filtering system. The applications demonstrated in this paper further stressed local and personal information in meeting personal tastes of application menu.

In the era of web 3.0 and internet of things, context-aware recommender system from various devices and sensors (Gavalas & Kenteris, 2011) might help in reducing navigation time and locating items. For example, GPS might help in identifying our time type, whether it's leisure time at home, office time or on the road. Different recommendation sub-system might be of more help. In our application scenarios, if the GPS information has been incorporated, users will be detected on the road, then the navigation application he rarely used might be recommended instead, which could lead to better accuracy and reduced application searching time. Another example is with a portable disease detector. For heart attacks, users might have to navigate through some menus before making emergency phone calls, the time wasted might be vital. While if a detector sensed a possible situation, it could switch the menu item to a one click position as illustrated in the application scenario, then users could get help even when they only have a few seconds to decide.

However, more problems should be considered, such as privacy, cold-start and high levels of sparsity.

Location alike information brought about privacy problems. Privacy preservation is gaining popularity in the RS research field. As noted, one potential participant refused to use our application due to privacy concern. Although no explicit private information is collected, the recording should better leave private information out, especially when social information is used to improve performance.

To solve the problem of high levels of sparsity, dimensionality reduction techniques should be applied. When menu item set is large enough, reduction method based on Matrix Factorization (Koren, Bell, & Volinsky, 2009) would be adequate for processing large matrices.

Cold start problem (Chulyadyo & Leray, 2014) happens when it is impossible to make recommendation because of a lack of information, such as ratings. In our study, the prediction coverage reduced with increasing steps. Therefore, it might be reasonable to use 1-step Markov chain models to handle cold-start problem for each new application session, and then switch to better models after. However, for a new user, when no additional information is available to construct models, recommending process might be delayed. Or it might be possible to use social information to tackle this cold-start problem. menu item RS could be shared in a social network. users could download models in a share center from those who have something in common with him.

Outside of Computer Science

Although most researchers in the field of RS are computer scientists, researchers in other fields, such as physicists and interdisciplinary researchers, are increasingly contributing to RS. For example, psychologists who know more about human behavior scored comparatively high in a recommendation contest of Netflix (Ellenberg, 2008). Thus we believe the accuracy could be further improved utilizing more information or methods from other disciplines, especially from human behaviors, taking human factors into account.

This chapter also has implications for decision making. Selecting menu is analogous to making decisions. The models provided in this chapter can be also used to model characterized decision making behaviors of human. Once decision making model is of high accuracy, we could use the model to simulate decision makings under a given situation, and to help analyze decisive factors of the success and failure of decision makers.

CONCLUSION

In this chapter, we have seen the possibility of using human behavior from the aspect of menu selections to increase the accuracy of menu item prediction algorithms and thus to reduce visual search time or navigation time.

We have examined the efficacy of various prediction algorithms for menu item prediction for a specific painting application, AyuDraw. In the first experiment, we evaluated the prediction ability of three Markov chain based prediction method (ADMCP, SPMCP and WMCP - GA) via SPSS from three aspects: accuracy, coverage and modeling time, and compared the prediction accuracy of ten algorithms and found out that WMCP - GA and AccessRank - WMCP - GA can provide the highest prediction accuracy under different Top-N. Compared to WMCP - GA, AccessRank - WMCP - GA can perform better only when n is small (less than five). In the second experiment, we selected a prediction method based on the results of experiment one, and combined it with the display method of highlighting to verify that the selected prediction method was effective of utilizing clicking behaviors to reduce navigation time.

Two application scenarios of simple models based on Markov chain on mobile device and desktop have been demonstrated. Positive feedbacks about reduced time of locating target menu items indicated the feasibility of models based on Markov chain for menu item prediction.

There is also a need to emphasize on the difference between menu item recommendation and other RSs, thus providing readers a better understanding of the contribution of this work.

First, menu item recommendation is more based on the users' own taste on local item operations, the purpose of which is to reduce navigation time of searching a target item. Users have already formed an intent to open or click that item but they need to find it. Instead, the traditional RS is usually to recommend items users may have interest in, or products and services they may want to buy or use. Usually they do not know the item they exactly want beforehand.

Second, traditional RS needs to consider metrics (del Olmo & Gaudioso, 2008) such as accuracy, coverage, novelty and diversity (Vargas & Castells, 2011) and stability (Adomavicius & Zhang, 2012) in recommendation. However, menu item recommendation does not need to consider novelty and diversity. Because the recommendation is only to help in finding a target item instead of providing users' potential and new items. Thus, we can concentrate more on improving accuracy.

However, The traditional RS can also benefit menu item recommendation. Expert user's launching preferences may help educate novice users. Cold-start users can download those models from a shared configuration center. It is when the traditional RS comes into play, recommending models that benefit users. Social information, such as, sex, age, vocation and country then could be studied to recommend the most appropriate models one user may want to download.

REFERENCES

Adomavicius, G., & Zhang, J. (2012). Stability of Recommendation Algorithms. *ACM Transactions on Information Systems, 30*(4), 23. doi:10.1145/2382438.2382442

Antwarg, L., Lavie, T., Rokach, L., Shapira, B., & Meyer, J. (2013). Highlighting items as means of adaptive assistance. *Behaviour & Information Technology, 32*(8), 761–777. doi:10.1080/0144929X.2011.650710

Bobadilla, J., Ortega, F., Hernando, A., & Gutiérrez, A. (2013). Recommender systems survey. *Knowledge-Based Systems, 46*, 109–132. doi:10.1016/j.knosys.2013.03.012

Bobadilla, J., Serradilla, F., & Hernando, A. (2009). Collaborative filtering adapted to recommender systems of e-learning. *Knowledge-Based Systems, 22*(4), 261–265. doi:10.1016/j.knosys.2009.01.008

Borges, J., & Levene, M. (2000). Data Mining of User Navigation Patterns. In *Web usage analysis and user profiling* (pp. 92–112). Springer Berlin Heidelberg. doi:10.1007/3-540-44934-5_6

Borràs, J., Moreno, A., & Valls, A. (2014). Intelligent tourism recommender systems: A survey. *Expert Systems with Applications, 41*(16), 7370–7389. doi:10.1016/j.eswa.2014.06.007

Chulyadyo, R., & Leray, P. (2014). A Personalized Recommender System from Probabilistic Relational Model and Users' Preferences. *Procedia Computer Science, 35*, 1063–1072. doi:10.1016/j.procs.2014.08.193

Cockburn, A., Gutwin, C., & Greenberg, S. (2007, April). A predictive model of menu performance. In *Proceedings of the SIGCHI conference on Human factors in computing systems*. ACM. doi:10.1145/1240624.1240723

Costa-Montenegro, E., Barragáns-Martínez, A. B., & Rey-López, M. (2012). Which App? A recommender system of applications in markets: Implementation of the service for monitoring users' interaction. *Expert Systems with Applications*, *39*(10), 9367–9375. doi:10.1016/j.eswa.2012.02.131

del Olmo, F. H., & Gaudioso, E. (2008). Evaluation of recommender systems: A new approach. *Expert Systems with Applications*, *35*(3), 790–804. doi:10.1016/j.eswa.2007.07.047

Ellenberg, J. (2008). This psychologist might outsmart the math brains competing for the netflix prize. *Wired Magazine*, 114-122.

Findlater, L., & McGrenere, J. (2004, April). A comparison of static, adaptive, and adaptable menus. In *Proceedings of the SIGCHI conference on Human factors in computing systems* (pp. 89-96). ACM. doi:10.1145/985692.985704

Findlater, L., & McGrenere, J. (2008, April). Impact of screen size on performance, awareness, and user satisfaction with adaptive graphical user interfaces. In *Proceedings of the SIGCHI Conference on Human Factors in Computing Systems*. Florence (pp. 1247-1256). ACM. doi:10.1145/1357054.1357249

Findlater, L., Moffatt, K., McGrenere, J., & Dawson, J. (2009, April). Ephemeral adaptation: The use of gradual onset to improve menu selection performance. In *Proceedings of the SIGCHI Conference on Human Factors in Computing Systems* (pp. 1655-1664). ACM. doi:10.1145/1518701.1518956

Fitchett, S., & Cockburn, A. (2012, May). Accessrank: predicting what users will do next. In *Proceedings of the SIGCHI Conference on Human Factors in Computing Systems* (pp. 2239-2242). ACM. doi:10.1145/2207676.2208380

Gabriel, K. R., & Neumann, J. (1962). A Markov chain model for daily rainfall occurrence at Tel Aviv. *Quarterly Journal of the Royal Meteorological Society*, *88*(375), 90–95. doi:10.1002/qj.49708837511

Gajos, K., Christianson, D., Hoffmann, R., Shaked, T., Henning, K., Long, J. J., & Weld, D. S. (2005). Fast and robust interface generation for ubiquitous applications. In UbiComp 2005: Ubiquitous Computing (pp. 37-55). Springer Berlin Heidelberg. doi:10.1007/11551201_3

Gajos, K. Z., Czerwinski, M., Tan, D. S., & Weld, D. S. (2006, May). Exploring the design space for adaptive graphical user interfaces. In *Proceedings of the working conference on Advanced visual interfaces* (pp. 201-208). ACM. doi:10.1145/1133265.1133306

Gavalas, D., & Kenteris, M. (2011). A web-based pervasive recommendation system for mobile tourist guides. *Personal and Ubiquitous Computing*, *15*(7), 759–770. doi:10.1007/s00779-011-0389-x

Herlocker, J. L., Konstan, J. A., Terveen, L. G., & Riedl, J. T. (2004). Evaluating collaborative filtering recommender systems. *ACM Transactions on Information Systems*, *22*(1), 5–53. doi:10.1145/963770.963772

Kamisaka, D., Muramatsu, S., Yokoyama, H., & Iwamoto, T. (2009, July). Operation prediction for context-aware user interfaces of mobile phones. In *Proceedings of the 2009 Ninth Annual International Symposium on Applications and the Internet* (pp.16-22). IEEE. doi:10.1109/SAINT.2009.12

Kohavi, R. (1995). A study of cross-validation and bootstrap for accuracy estimation and model selection. In Ijcai (Vol. 14, No. 2, pp. 1137-1145).

Koren, Y., Bell, R., & Volinsky, C. (2009). Matrix Factorization Techniques for Recommender Systems. *Computer*, *42*(8), 30–37. doi:10.1109/MC.2009.263

Kurihara, S., Moriyama, K., & Numao, M. (2013, November). Context-aware application prediction and recommendation in mobile devices. In *Proceedings of the 2013 IEEE/WIC/ACM International Joint Conferences on Web Intelligence (WI) and Intelligent Agent Technologies (IAT)-Volume 01*(pp. 494-500). IEEE Computer Society. doi:10.1109/WI-IAT.2013.69

Le-tian, X., & Yuan-shen, Z. (2010). Study on Statistical Experiment of Markov Chain Prediction Methods. *Chinese Journal of Engineering Mathematics*, *27*(2), 313–320.

Lee, D., Choi, J., Kim, J. H., Noh, S. H., Min, S. L., Cho, Y., & Kim, C. S. (1999, May). On the existence of a spectrum of policies that subsumes the least recently used (LRU) and least frequently used (LFU) policies. *Performance Evaluation Review*, *27*(1), 134–143. doi:10.1145/301464.301487

Lee, E., & MacGregor, J. (1985). Minimizing user search time in menu retrieval systems. *Human Factors: The Journal of the Human Factors and Ergonomics Society*, *27*(2), 157–162.

Lee, S. K., Cho, Y. H., & Kim, S. H. (2010). Collaborative filtering with ordinal scale-based implicit ratings for mobile music recommendations. *Information Sciences*, *180*(11), 2142–2155. doi:10.1016/j.ins.2010.02.004

McNally, K., O'Mahony, M. P., Coyle, M., Briggs, P., & Smyth, B. (2011). A case study of collaboration and reputation in social web search. *ACM Transactions on Intelligent Systems and Technology*, *3*(1), 4. doi:10.1145/2036264.2036268

Mitchell, J., & Shneiderman, B. (1989). Dynamic versus static menus: An exploratory comparison. *ACM SIGCHI Bulletin*, *20*(4), 33–37. doi:10.1145/67243.67247

Norman, K. L. (1991). *The psychology of menu selection: Designing cognitive control at the human/computer interface*. Intellect Books.

Núñez-Valdéz, E. R., Lovelle, J. M. C., Martínez, O. S., García-Díaz, V., de Pablos, P. O., & Marín, C. E. M. (2012). Implicit feedback techniques on recommender systems applied to electronic books. *Computers in Human Behavior*, *28*(4), 1186–1193. doi:10.1016/j.chb.2012.02.001

Parkinson, S. R., Sisson, N., & Snowberry, K. (1985). Organization of broad computer menu displays. *International Journal of Man-Machine Studies*, *23*(6), 689–697. doi:10.1016/S0020-7373(85)80063-8

Sears, A., & Shneiderman, B. (1994). Split menus: Effectively using selection frequency to organize menus. *ACM Transactions on Computer-Human Interaction*, *1*(1), 27–51. doi:10.1145/174630.174632

Shih, H. M., & Goonetilleke, R. S. (1998). Effectiveness of menu orientation in Chinese. *Human Factors: The Journal of the Human Factors and Ergonomics Society*, *40*(4), 569–576. doi:10.1518/001872098779649373

Tsandilas, T. (2005, April). An empirical assessment of adaptation techniques. In CHI'05 Extended Abstracts on Human Factors in Computing Systems (pp. 2009-2012). ACM. doi:10.1145/1056808.1057079

Tsandilas, T. (2007, April). Bubbling menus: a selective mechanism for accessing hierarchical drop-down menus. In *Proceedings of the SIGCHI conference on Human factors in computing systems* (pp. 1195-1204). ACM. doi:10.1145/1240624.1240806

Vargas, S., & Castells, P. (2011, October). Rank and relevance in novelty and diversity metrics for recommender systems. In *Proceedings of the fifth ACM conference on Recommender systems* (pp. 109-116). ACM. doi:10.1145/2043932.2043955

KEY TERMS AND DEFINITIONS

Adaptive Menu: A menu in which menu items are generated by computer program based on users' data in order to reduce visual search time or the navigation time of the target menu item. It often needs a display technique.

Genetic Algorithm: An algorithm that mimics natural selection to solve optimization problems. It could be utilized to find feasible weights or select features.

Highlighting Menu: A display technique used to attract users' attention including highlighting the foreground or background of important menu items with colors or shapes, thus lowering search time.

Menu Item Prediction Model: An algorithm or process aimed to generate adaptive menus.

Menu Item: An abstract hyperlink, which could be a text or icon shortcut to an application or an action.

Navigation Time: The time period from forming a search target to actually find the target visually. The search process contains visual search process and any additional operations including switching screens. Visual search time is the same to navigation time without any additional operations.

Visual Search Time: The time period from starting searching on the target scene to actually find the target visually without non-visual operation.

Section 4
Relationship between Information Systems and Networks

Chapter 9
Developing a Successful Facebook Fan Page Based on Costumers' Needs

Aikaterini Manthiou
NEOMA Business School, France

Lanlung Luke Chiang
Yuan Ze University, Taiwan

Liang Rebecca Tang
Iowa State University, USA

ABSTRACT

In this chapter, we determine possible needs that customers are fulfilling by participating in the Facebook fan pages. Understanding the needs that customers aim to pursue is the priority of marketing communications. Therefore, the chapter analyzes a theoretical approach to identify and respond to the needs that consumers seek to satisfy on fan pages. The customer needs are evaluated from functional, hedonic, social, and psychological perspectives and all four of the need components make significant contributions to the overall need on fan pages. Businesses in the food service industry are advised to build an emotional bond with consumers, enhance two-way communication, provide useful information and encourage entertainment on their fan pages. Managers of brands that operate fan pages can be guided by our research with regards to deciding which characteristics or content consumers find important in the these pages.

INTRODUCTION

Over the past years, social media have been considered as the next great wave in technology, business, and social life. Social media is defined as "a second generation of Web development and design, that aims to facilitate communication, secures information sharing, interoperability, and collaboration on the World Wide Web" (Paris, Lee, & Seery, 2010, p. 531). Moreover, social media refers to "a group of

DOI: 10.4018/978-1-4666-8450-8.ch009

Internet-based applications that build on the ideological and technological foundations of Web 2.0, and that allow the creation and exchange of user-generated content." (Kaplan & Haenlein, 2010, p. 61). Social media exists in a variety of forms and serves diverse purposes. Among them, Facebook is currently the world's most successful social networking website (Cortese, 2011). Facebook counts 1.23 billion monthly active users, 945 million mobile users, and 757 million daily users (Protalinski, 2014). Facebook was originally created for individuals' social and entertainment purposes, and is now also widely used as a marketing platform for businesses (Lin & Lu, 2011). Companies are increasingly investing in social media, indicated by worldwide marketing spending on social networking sites of about $4.3 billion (De Vries, Gensler & Leeflang, 2012). The business can send hundreds of thousands of visits, build passionate followings around a brand, and bring in sales and consumers on its Facebook fan page (Weber, 2009). Facebook fan pages have a significant impact on brand-related communications and serves as an effective tool for customer relationship management (Manthiou, Tang, & Bosselman, 2014). A Facebook fan page includes the following attributes: brand/company's overview, contact information, press releases, RSS feeds, Twitter updates, brand news and status updates and customer comments/interaction (Techopedia, 2014a). Moreover, the emergence of brand communities has coincided with the growth in consumer empowerment. They are venues where intense brand loyalty is expressed and fostered, and emotional connection with the brand forged in customers (Hudson & Thal, 2013).

Many businesses in the hospitality industry have realized the importance of Facebook as a marketing tool and seek to take advantage of this opportunity (Lin & Lu, 2011). Recently, restaurant Facebook fan pages were cited as having the highest rates of member participation in the United States when compared to other sectors in the hospitality industry (Top Dining Brands, 2014). Many businesses use these sites to gain consumers' feedback about their products and services, explore new market segments, and connect both existing and prospective consumers, and inquiry about their competitors (Shih, 2009). Facebook is essentially a platform for sharing, and sharing experiences is innate to human nature (Ulusu, 2010). The communication that occurs as a consequence of natural engagement is pertinent, meaningful, and permanent (Cortese, 2011). Many restaurant businesses have realized the importance of fan pages and are ready to take advantage of this opportunity. Although many information sources provide the latest techniques and enterprise-building tactics so that businesses can immediately benefit from the fan pages, industry practitioners should keep in mind that the use of techniques or skills on the fan pages is an aid to customer communications (Shih, 2009). Understanding customer needs or expectations is always the priority of marketing activities (Muñiz & Schau, 2005). Especially in light of the innumerable competitors of fan pages and the minimal switching costs for users, it is imperative to understand customer needs so that restaurants can create effective fan pages that spark customers' interests, meet their expectations, and inspire loyalty to both the fan page and the brand. However, to the authors' knowledge, no previous studies have investigated customer needs on a SNS with fan pages. This study aimed to fill this research gap.

Some previous studies have investigated the application of SNSs in the hospitality and tourism industry. Most of these studies focused on the users' participation and involvement in online communities (e.g. Chung & Buhalis, 2008; Lee, 2007; Qu & Lee, 2011). Wang and Fesenmaier (2004) investigated users' participation in online travel communities and suggested that four fundamental needs drive the extent to which users participate in these virtual communities: (1) functional need; (2) social need; (3) psychological need; (4) hedonic need. Since Facebook can be viewed as an online community, the desires customers seek to satisfy on fan pages can also be explained by these four needs. However, the study of Wang and Fesenmaier (2004) was conducted especially for online travel communities and done before the launch of Facebook (Hills & Argyle, 2002). Therefore some unique features of fan pages were not incorporated into the measures and discussions of the need constructs in Wang and Fesenmaier (2004)'s study. The authors of the present study re-interpreted the four needs as laid out in Wang and Fesenmaier's work and tested the overall need structure with these four needs on a SNS with fan pages.

The purpose of the study is to develop a theoretical foundation to identify and respond to customer needs on fan pages. The results of the present study are of importance for both academics and industry practitioners. The paper contributes to the evolving literature of social media marketing, and provides a theoretical framework for future research investigating customer involvement on Facebook fan pages. From a practical perspective, this study helps restaurant industry understand and predict customer behavior on fan pages, and provides fan page designers guidance to improve communication effectiveness. The specific objectives of this research are to: (1) re-interpret the need paradigm proposed by Wang and Fesenmaier (2004) on fan pages, and (2) assess the second-order model of overall need on fan pages represented by functional, hedonic, social, and psychological aspects. Therefore, the research questions that this research poses are:

1. Is Wang and Fesenmaier's (2004) framework applicable in the context of Facebook fan pages?
2. What needs do consumers fulfill by participating in Facebook fan pages, including functional need, social need, psychological need, and hedonic need?

LITERATURE REVIEW

Social Network Sites (SNSs) and Facebook

In recent years, social networking sites have led us towards a new era of global interconnectedness (Manthiou et al., 2014). Social networking sites can be described as networks of friends for social or professional interactions (Trusov, Bucklin, & Pauwels 2009). Many previous studies have investigated their use, with research questions ranging from their role in identity construction and expression (e.g., Cortese, 2011) to the building and maintenance of social relationship (e.g., Lin and Lu, 2011) and concerns about privacy (e.g., Muñiz & Schau, 2005). Members of social networking sites have the chance to become friends with other members, but they can also become fans of brands on dedicated brand fan pages ((De Vries, Gensler, & Leeflang, 2012). Moreover, social networks (e.g., Friendster, LinkedIn, MySpace, and Facebook) provide many functions, such as allow people to present themselves, share spaces for collaboration, create social connections, and aggregates information exchanges in a web-based environment (Weber, 2009). Some of the SNSs are oriented towards work-related contexts (e.g., LinkedIn.com), connecting those with shared interests such as music or sports (e.g., MySpace.com), romantic relationship initiation (e.g., Friendster.com), or friendship connections (e.g., Facebook.com). A significant number of people connect to SNSs several times a day either to check their profiles or to participate in different online activities (Shin, 2009; Cortese, 2011). It suggests most of users use SNSs to gather, socialize with their peers, and make sense of and help build the culture around them (Lin & Lu, 2011). On the other hand, more and more businesses and institutes use SNSs as a platform to create a continuous dialogue with target consumer groups for marketing their products and/or services (Shin, 2009). In some cases, businesses are finding that their own customers are taking the initiative and starting Facebook fan page (Park, Kee, & Valenzuela, 2009; Manthiou et al., 2014).

Facebook has been viewed as one of the most popular SNSs since its emergence in 2004. It has been viewed as a channel for personal expression and friendship connections. During past years, Facebook has put significant efforts towards creating a marketing communication platform for businesses and organizations (Shih, 2009). Facebook can be a customized two-way communication channel with customers, featuring the general information of the business and its ongoing activities (e.g., announce promos, competitions or company news) (Roy & Shekhar, 2010). Moreover, the Facebook fan pages make the brands' posts more popular contributing to their awareness and image (De Vries, Gensler, & Leeflang, 2012). Using a Facebook fan page can help a brand successfully establish and maintain positive consumer–brand relationships by offering special benefits to members on Facebook (Kang, Tang, & Fiore, 2014).

Customer Needs and Theoretical Foundation of the Study

Needs are the abstract categories that humans require in order to survive, function and thrive (Feenburg & Bakardjieva, 2004). Humans are born with the same needs and learn what will satisfy those needs through the experience of being reared within a specific society, time and place (Goldsmith, 2006). Customer needs have been widely discussed in previous marketing studies (e.g. Andersson & Mossberg, 2004; Arnett & Badrinarayanan, 2005; Herrmann et al., 2000). Marketers have realized that satisfying customer needs is the priority of marketing (Muñiz & Schau, 2005). Ellison *et al.* (2007) showed that the consumption of a given product can simultaneously satisfy more than one need. For example, wearing clothes protects the wearer from cold (physiological), attracts the opposite sex (social), is comfortable to the skin (hedonic), and represents the self-concept and values of the wearer (psychological) (Goldsmith, 2006). From this perspective, a SNS with fan pages can gratify multiple consumer needs.

Several previous studies have investigated the needs or reasons for participation in online communities. Preece (2000) identified five core need attributes customers pursue in online communities, including belongings, active participation and interaction, access to public resources, reciprocity of information, support, and services among users, and share of social conventions, languages and protocols. Hagel and Armstrong (1997) claimed that online communities provide four values to the users, including transaction, interest, fantasy, and relationship.

Wang *et al.* (2002) found that users participate in travel communities for several primary reasons: access to information, voice their own opinions, communicate with others, have fun, meet new people, develop a sense of belonging, and establish and maintain long-term relationships. Wang and Fesenmaier (2004) summarized previous literatures and identified four primary need components users pursue in online travel communities, that is, functional, psychological, social, and hedonic needs. Tuten and Solomon (2014) revealed four significant zones contributing to the social media mix: (1) zone 1 includes the element of social community which describes all the channels of social media focusing on relationships and the common activities users participate in with others who share the same interest or identification; (2) zone 2 refers to the social publishing element which aids in the dissemination of content to audiences; (3) zone 3 underlines the importance of social entertainment which encompasses channels and vehicles that offer opportunities for play and enjoyment; and last (4) zone 4 depicts the social commerce element used to assist users in the process of online buying and selling of products and services. Furthermore, Kang, Tang, and Fiore (2014) determined the correlation between four different benefit dimensions (functional benefits, social–psychological benefits, hedonic benefits, and monetary benefits) obtained by fan page members, members' community participation, brand trust, and brand commitment. The

Figure 1. Theoretical framework of measuring customer needs on fan pages

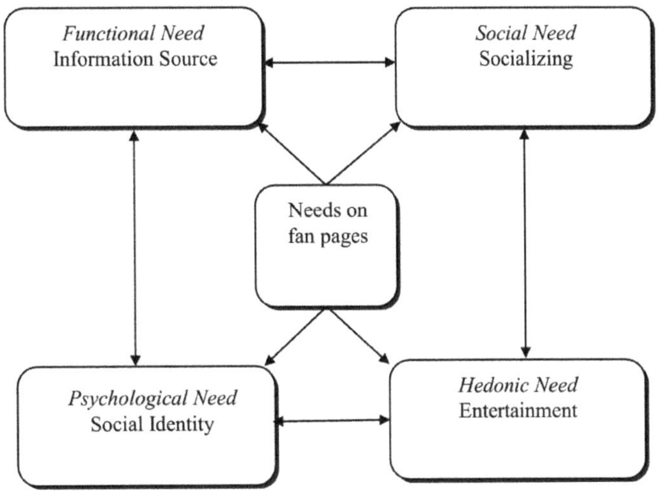

results of their analysis disclosed that fan page members are more likely to visit restaurant Facebook fan pages when they obtain social–psychological and hedonic benefits from their interactions on the pages. Moreover, Manthiou, Tang and Bosselman (2014) indicated that the cognitive and affective components influence consumer attitude towards participation in Facebook fan pages. In particular, social interaction ties impact the attitudes toward fan pages the most, followed by information source, design characteristics, and entertainment.

SNSs with fan pages for businesses are viewed as a type of online community (Stankov et al., 2010), thus, the authors argued that the need paradigm of online travel communities can be adapted to investigate the needs fulfilled on fan pages (Figure 1). Moreover, Ho (2014) revealed that consumers' participation on Facebook has positive and significant impacts on brand trust and community identification. According to Tuten and Solomon (2014), social media venues such as Facebook offer activities that make consumers engaged which results in high levels of brand identification. The four need components are discussed below.

The Functional Need

The term "functional need" refers to needs that are motivated efforts that direct or contribute to a purpose (Chung & Buhalis, 2008). Cothrel (2000) defined functional uses in online communities as contractual needs which are minimum intrinsic informational requirements that users expect to fulfill. The functional needs in online communities include sharing and disseminating information, seeking for learning, and facilitating decision making from other users' support without time and geographical limits (Wang a& Fesenmaier, 2004). Specifically, from the utilitarian perspective, a SNS with fan pages serve as an information source by providing multiple functions (Quan-Haase & Young, 2010; Park *et al.*, 2009). For example, "fan profiles" provide customers' geographic, demographic, and psychological information. "Facebook live" is a live streaming video channel that keep users updated with current happenings. "Notifications" reports important events to users (Smith, 2011). Moreover, in order to increase user engagement, brands should keep updating new wall posts so as to encourage users to like or share this information with their network. Facebook fans are interested in commenting wall posts which are fresh and updated (Leung, Schuckert, & Yeung, 2013).

The Hedonic Need

Hedonic consumption is a form of emotive stimulation, which refers to escapism, diversion, aesthetic enjoyment, or emotional release (Negash et al., 2003). Hedonic pleasure can motivate customers' participation, promote their excitement and concentration, and help them enjoy the experience (Liu & Arnett, 2000). The hedonic need in online communities is mainly accomplished by entertainment and enjoyment (Wang & Fesenmaier, 2004). From a hedonic perspective, community users are viewed as pleasure seekers, who attach importance to the experiential feature of consumptions (Vogt & Fesenmaier, 1998). SNSs with fan pages provide diverse entertainment tools to satisfy the hedonic needs of customers. "Listen with friends" allows users to listen to music at the same time as their friends. Other examples of entertainment can be games such as word-association games, movie quote game, and contests (Smith, 2011). Kang, Tang and Fiore (2014) revealed that fan page members are likely to visit Facebook fan pages when they obtain hedonic benefits from their interactions on these kinds of pages.

The Social Need

Online communities are socially structured, convey social meaning, and meet social needs (Wang *et al.*, 2002). Social needs in online communities describe the development of relationships with other users and the brand through communications and interactions (Angehrn, 1997; Preece, 2000). Social needs in online communities are satisfied by involving into many socializing activities, such as providing help and support, discussing and exchanging ideas, forming relationships, and getting involved with other users (Preece, 2000). Manthiou, Tang and Bosselman (2014) identified that social interaction ties were one of the most influential components of fans' attitude towards a Facebook fan page. The businesses can find different features for socializing on Facebook fan pages (Park *et al.*, 2009). A good example is the "wall". The "wall" which is visible to anyone allows users post and respond messages. The businesses can use the "wall" as an interactive platform to communicate with customers (Weiser, 2001). For instance, budget airlines use their Facebook fan pages to deliver promotional materials, connect with new costumers, invite user engaged into online social activities and conduct information distribution. All these happen through their wall posts (Leung, Schuckert, & Yeung, 2013).

The Psychological Need

The term "psychological need" refers to the emotional aspects of relationships in online communities (Wang *et al.*, 2002). Users in an online community fulfill several primary psychological needs, including a sense of belonging to the community (Dholakia *et al.*, 2009), identity expression through the community (Walther, 1996; Casaló *et al.*, 2010), and a sense of affiliation with other users in the community (Wang *et al.*, 2002). These psychological components described above can be grouped under a general term of social identity. Social identity on fan pages is created by several functions, such as "deals", "networks", and "groups". Businesses can offer online coupons and discounts to customers through the "deals" function (Schmidt *et al.*, 2008). "Networks" and "groups" call customers with similar interests or purposes together to share information and discuss brand-related topics (Stankov *et al.*, 2010).

Social identity describes an integral part of group membership and confers a collective sense of who one is (Hogg & Abrams, 1988). Social identity is a psychological state consisting of the cognitive, affective, and evaluative perspectives (Dholakia *et al.*, 2009). The cognitive component of social identity refers to the perceptions of self-image and personal identity with other members of the community (Bagozzi & Dholakia, 2002). The affective perspective of social identity involves emotional attachment and strong feelings of belonging to the community or affective commitment towards the community (Bagozzi & Dholakia, 2002; Bergami & Bagozzi, 2000). The evaluative component describes an individual's assessment of the value of being a member of the community (Bergami & Bagozzi, 2000; Dholakia *et al.*, 2004). In this study the authors comprehensively investigated social identity from the cognitive, affective, and evaluative perspectives and their satisfaction of the psychological need on fan pages.

Based on the above premises, the study adapted the theoretical framework of user needs in online travel communities proposed by Wang and Fesenmaier (2004) to the context of Facebook fan pages. The new paradigm is illustrated in Figure 1, which achieved the first objective of the study. The functional, hedonic, social, and psychological need components are assessed by the functions of information source, entertainment, socializing, and social identity (respectively). The study further examined how each of the four need categories contributes to the overall need on this SNS with fan pages with a second-order factor model, which adds a critical component to the original need concepts derived by Wang and Fesenmaier (2004) (Figure 2).

Figure 2. Conceptual Model

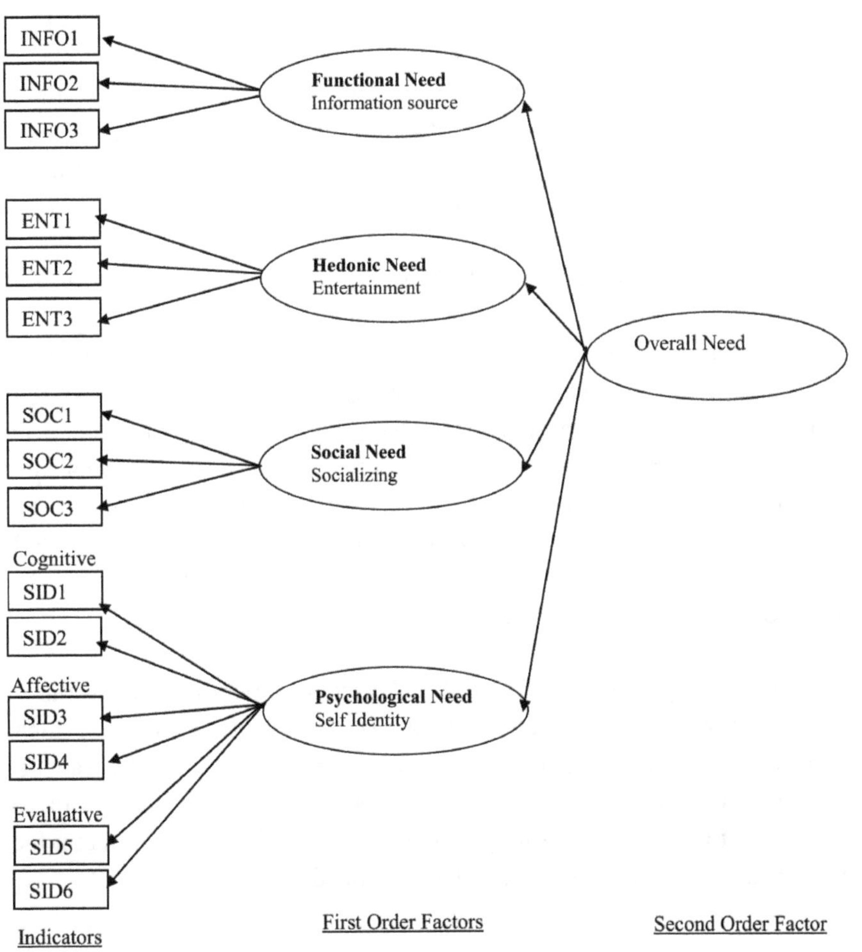

METHODOLOGY

Sample of Facebook Fan Pages

The authors used Facebook to test consumer needs on SNSs. Facebook fan page is an official social network community of a business, the gateway for customers to get close to the brand and its products (Shih, 2009). Facebook has 1.23 billion monthly active users, 945 million mobile users, and 757 million daily users (Protalinski, 2014). Facebook was reported to generate 1.6 billion page views each day, therefore, it could be an effective platform of establishing and maintaining brand-consumer relationships (Needham & Company, 2007). Numerous restaurant Facebook fan pages populate the internet. In order to make the results of the study generalizable, the study investigated the top six of the most popular restaurant brand Facebook fan pages list for 2011, including McDonald's, Subway, Taco Bell, Buffalo Wild Wings, Pizza Hut, and Burger King. In addition, a respondent had the opportunity to indicate any other restaurant brand Facebook page on which he/she was a fan.

Instrument

The final questionnaire included two parts. In the first part, respondents were asked to assess four need components on the specific fan pages, including information source, entertainment, socializing, and social identity. The measurement scales for information source, entertainment, and socializing were adapted from Park *et al.* (2009). The scale of social identity originated from Dholakia *et al.* (2004). Social identity was evaluated from the cognitive, affective, and evaluative perspectives. All the items in part one were measured with a 7-point Likert scale (1=extremely disagree to 7=extremely agree) (Table 1). The second part of the questionnaire was comprised of demographic and behavior information, such as age, gender, and time spent per week on the Facebook page.

A pilot test was conducted to ensure the reliability of the constructs. The survey was administered in a convenience sample of 100 college students in the department of hospitality management at a U.S. Midwestern university. The reliability of the measurements was all above the cutoff of 0.70, which suggested acceptable internal consistency (Nunnally, 1978). Wording was further modified based on the feedback of the respondents. A web-based survey was distributed among 3,000 undergraduate and graduate students at the above-mentioned university. The survey was conducted between April 4 and April 18, 2011. A total of 211 responses were received. The response rate was 7.03%. Of the 211 responses, 175 were returned in a usable format. The data were analyzed by using two statistical software; SPSS and AMOS.

Statistical Methods and Procedures

The study followed a two-step approach to examine the measurement model and the structural model (Anderson & Gerbing, 1988). The first step involved a first-order factor analysis which was employed to test the hypothesized relationships between the measurement items and the four need components (Hu & Bentler, 1999). In the second step, a second-order factor analysis was performed to assess the structural model, with a focus on the relationship between the four first-order latent factors (four need components) and one second-order latent factor (overall need).

The second-order factor models are useful when the first-order factors are distinct and share a significant variance (Bagozzi, 1992). The goal of the higher-order confirmatory factor analysis is to provide a more parsimonious account for the correlations among the lower-order factors (Brown, 2006). The four need components in the study are seemingly different but correlated with each other because they are integral parts of the overall need on the fan pages. Therefore, the two-step approach of the confirmatory factor analysis described above was used to achieve the second research objective of the study.

Table 1. Scale/item measurement properties

Construct and Scale Items	Label	Standardized Factor Loadings	Composite Reliabilities	Cronbach's alphas
Information Source			0.701	0.775
I think of "X" fan page as an important information source.	INFO1	0.794***		
I use the "X" fan page to get useful information about its products/services.	INFO2	0.834***		
I use the "X" fan page to learn about their events.	INFO3	0.597***		
Entertainment			0.785	0.805
It's entertaining to browse the "X" fan page.	ENT1	0.675***		
The "X" fan page is funny.	ENT2	0.821***		
The "X" fan page is exciting.	ENT3	0.813***		
Socializing			0.789	0.840
I use the "X" fan page to talk about its products/services.	SOC1	0.655***		
I feel connected to the brand by being a follower on the "X" fan page.	SOC2	0.949***		
I use the fan pages to get peer support from others who are also interested in the restaurant's products/services.	SOC3	0.810***		
Social Identity			0.720	0.903
Cognitive				
My self-image is consistent with the identity of fans on the "X" fan page.	SID1	0.731***		
My personal identity is consistent with the identity of the other fans on the "X" fan page.	SID2	0.766***		
Affective				
I am attached to the "X" fan page.	SID3	0.723***		
I have strong feelings of belonging towards the "X" fan page.	SID4	0.847***		
Evaluative				
I am a valuable member of the "X" fan page.	SID5	0.767***		
I am an important member of the "X" fan page.	SID6	0.723***		

Note: *** Significant at the .001 level.

Data Collection and Analysis

Table 2. ANOVA test of responses among fan pages

	McDonald's	Subway	Taco Bell	Buffalo Wild Wings	Pizza Hut	Burger King	Other	*F*	*P*
Information Source	3.62	3.9	3.76	3.47	3.56	3.67	4.53	1.715	0.12
Entertainment	3.87	3.52	3.2	4.00	3.45	3.83	3.87	1.366	0.231
Socializing	3.66	3.09	3.25	2.38	3.37	3.82	3.25	0.818	0.558
Social Identity	2.81	2.56	2.69	2.22	2.46	3.43	2.65	0.573	0.752

RESULTS

Respondent Profiles

Among the 175 respondents, 59% were female and 41% were male. Although the age range of respondents varied from 18 to 34, the majority of the respondents was between 20-24 (86.3%). Sixty one percent of the respondents were caucasian. 36.6% spent 31-60 minutes per week on Facebook and 17.7% spent more than 3 hours per week. As indicated in the methdology section, respondents were asked to choose any of the six restaurant fan pages (Mcdonald's, Subway, Taco Bell, Buffalo Wild Wings, Pizza Hut, Burger King) provided on the survey or indicate any other restaurant brand fan page on which they were fans. The response numbers of the several options above were 15(8.6%), 55(31.4%), 45(25.7%), 5(2.9%), 6 (3.4%), 6 (3.4%), and 43 (24.6%), respectively. An ANOVA test was conducted to assess the differences between the responses to the four need constructs. All four constructs included multiple measurement scales. The mean of the measurement scales for each of these four constructs was calculated and used in the ANOVA test. Table 2 shows the result of the ANOVA test, which indicates no significant differences in the responses among the seven options of the restaurant fan pages.

Reliability and Validity

Before analyzing and further evaluating the estimation results, three types of reliability (inter-item reliability, composite reliability of the overall scale, and average variance extracted [AVE]), and two types of validity (discriminant and convergent) were checked (Table 1). Cronbach's alpha was used to test inter-item reliability. In this study, Cronbach's alpha ranged from 0.775 to 0.903. Composite reliability is the reliability of a summated scale, and AVE is the variance in the indicators explained by the common factor. Composite reliability is advised to be greater than 0.7 (Hair *et al.*, 1998) and AVE should be greater than 0.5 (Bagozzi, 1992). Composite reliability of the need constructs in the study ranged from 0.701 to 0.789 and AVE from 0.871 to 0.894. Convergent validity is the degree to which an operation converges with other operations that it theoretically should resemble. Convergent validity is evaluated by confirmatory factor loadings (Anderson & Gerbing, 1988). All confirmatory factor loadings in the study were significant at the 0.001 level, and thus the convergent validity of the measures was satisfactory. Discriminant validity describes the degree to which operationalization diverges from other operationalizations that it theoretically should not resemble (Hair *et al.*, 1998). The squared root of correlations between any pair of constructs in the study were less than AVE, thus supporting discriminant validity (Table 3).

Table 3. Latent variable squared root of correlation matrix

	1	2	3	4
1. Information Source	**0.859**			
2. Entertainment	0.73	**0.876**		
3. Socializing	0.57	0.69	**0.894**	
4. Social Identity	0.71	0.78	0.84	**0.871**

Note: Entries under the diagonals are the squared root of latent construct correlations. Entries on the diagonal are AVEs.
First Order CFA

In the measurement model, the hypothesized relationships between the items and the four first-order factors were examined to determine how well the model fit the data. The results of the initial estimation of the first-order factor model did not provide a satisfactory result: χ^2 (84) = 273.882, p=0.000, χ^2/df =3.260, GFI=0.832, NFI=0 .834, TLI=0.846, CFI=0.877, RMSEA=0.114. Social identity was measured from three perspectives: (1) cognitive (SID1-My self-image is consistent with the identity of fans on "X" fan pages; SID2-My personal identity is consistent with the identity of the other fans on "X" fan pages), (2)

Figure 3. First-order CFA

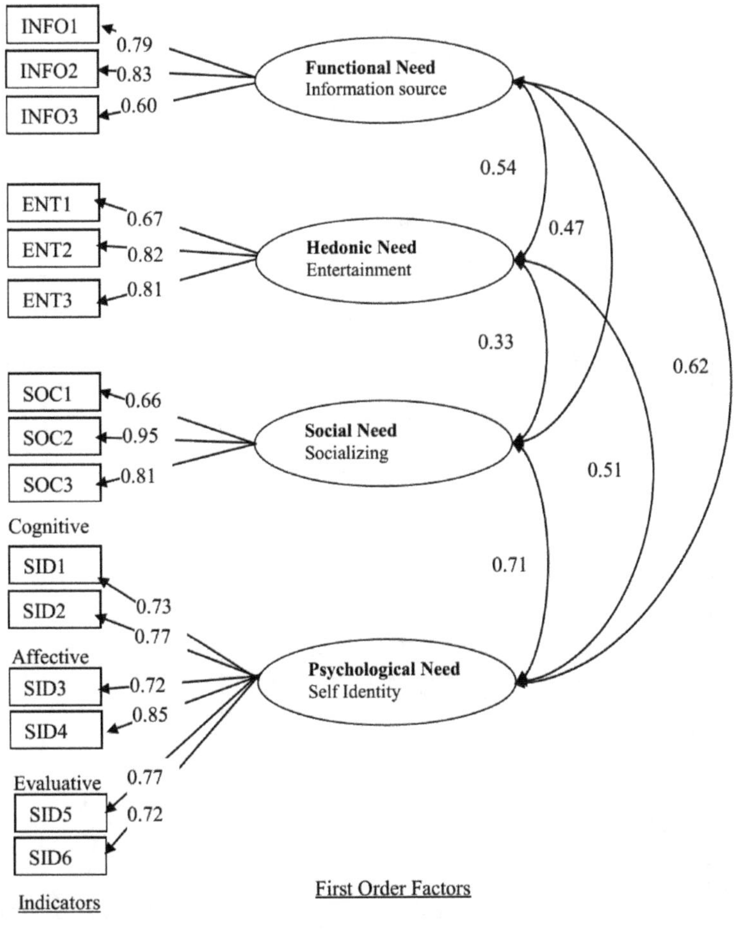

affective (SID3-I am attached to "X" fan pages; SID4-I have strong feelings of belongingness toward "X" fan pages), and (3) evaluative (SID5-I am a valuable member of "X" fan pages; SID6-I am an important member of "X" fan pages). The modification index indicated that the two items of each perspective had high covariances. Hence, the errors of the two items of each perspective were correlated. As a result, the model of fit improved significantly (χ^2 (81) = 143.346, p=0.000, χ^2/df =1.770, GFI=0.903, NFI= 0.913, TLI=0.948, CFI=0.960, RMSEA= 0.067) and indicated an acceptable fit between the model and the data. All the path estimates were highly significant (Figure 3).

Second-Order Factor Analysis

The second-order CFA evaluated the relationship between the four first-order factors (information source, entertainment, socializing, and social identity) and a second-order factor (overall need). The model of fit for the second-order factor model was satisfactory χ^2 (83) = 152.854, p=0.000, χ^2/df =1.842, GFI=0.900, NFI= 0.907, TLI=0.948, CFI=0.943, RMSEA=0.070) (Figure 4). All path estimates were highly significant (Table 4).

Figure 4. Second-Order CFA

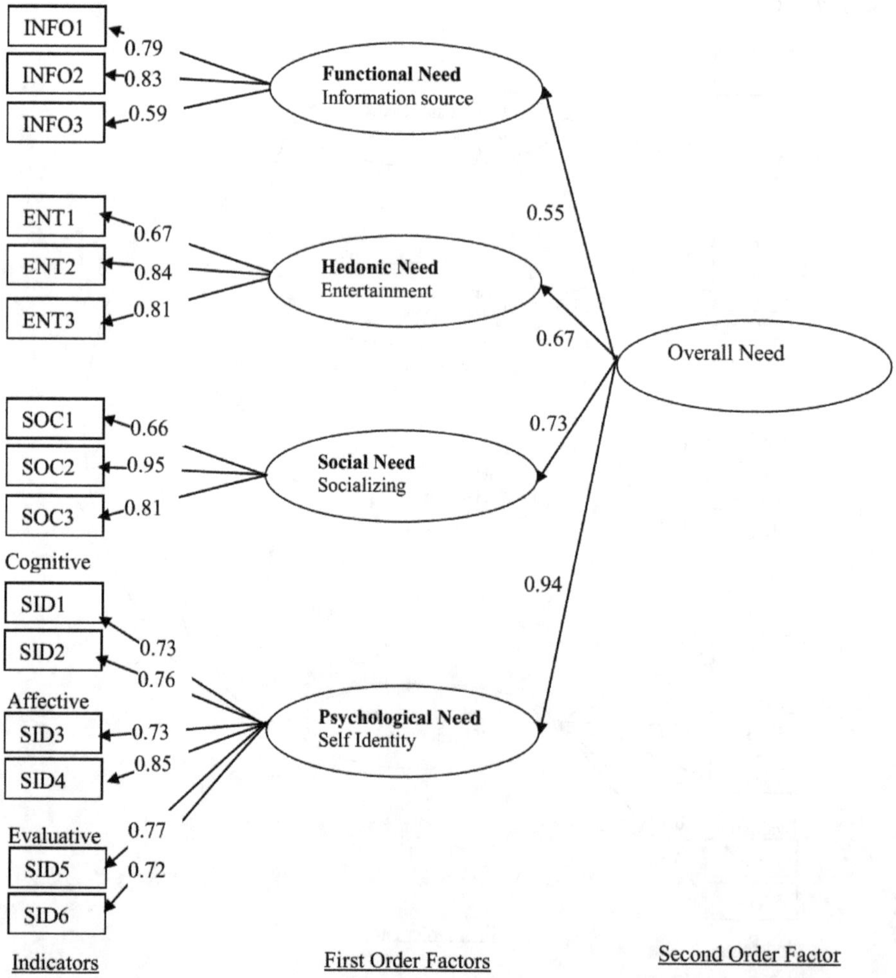

Table 4. Need components and their marketing benefits

Need	Marketing Benefit
Psychological	1. Fostering a sense of belonging among consumers 2. Strengthening consumers' social identity 3. Showing respect, care, and share with customers 4. Helping build long lasting relationships with your customers and creating an interactive community with them.
Social	1. Communicating with both existing and potential customers 2. Experiencing open interactions with customers 3. Responding to negative comments in an effective approach 4. Allowing the brand to engage with the community easily and for free 5. Connecting the company and the staff directly to the customers and fans 6. Driving a lot of traffic to your website through your Facebook fan page.
Hedonic	1. Providing attractive webpage features to spark customers' interest and increase their excitation. 2. Entertaining consumers 3. Offering games for consumers 4. Becoming a sponsor of online virtual games to reach existing or potential customers.
Functional	1. Providing relevant and timely information 2. Sharing updates with consumers 3. Using the fan page as platform to launch new product 4. Alerting your fans of special offers, events and discounts instantly. 4. Gathering customers' comments about the products 5. ROI-related metrics are readily available 6. Listening and observing has the potential to make improvements

DISCUSSION AND IMPLICATIONS

In terms of managerial implications, our findings provide important recommendations to social media marketers. The psychological aspect contributed the most to the model with a regression weight of 0.94 to the overall need, followed by the social aspect (0.73), hedonic aspect (0.67) and utilitarian aspect (0.55). The effective application Facebook fan pages as a marketing tool can assist restaurant brands in creating favorable brand image, encourage consumer loyalty, and build strong customer relationships.

Psychological need was represented by social identity. Social identity is the perception of belonging to the SNSs community where people choose SNSs to facilitate intragroup–intergroup relationships in addition to interpersonal ones (Valkenburg & Peter, 2007). This suggests that SNSs offer people interface to identify within group members who look and act similarly to each other as well as to compare themselves to outgroup members. In order to engage with the any perspectives and increase the explanation power, social identity was measured from cognitive, affective and evaluative perspectives in the study. Cognitive social identity involves the pursuit of consistent personal identity or self-image with others in the community (Bagozzi & Dholakia, 2002). The restaurant brand could consider creating groups on its fan page for customers who show similar characteristics or common interests. The brand can provide the groups with useful and valuable information and lead discussions on topics related to its brand or its products of interest to customers. Active participation in the sharing and dissemination of information fosters a sense of belonging. The formation of groups on restaurant fan pages brings customers closer to the brand and the products (Weber, 2009). Given that customers provide demographic and socio-economic information in the section of "fan profiles" on Facebook, grouping customers with similar characteristics is not a challenge for businesses (Smith, 2011).

Affective social identity is related to customers' attachment and strong feelings of belonging toward the fan pages. Restaurant companies are advised to provide Facebook fans special benefits to strengthen affective social identity. For instance, the Palms Hotel and Casino offered a free night or a room upgrade to customers who checked in on its Facebook page, which increased customers' attachment to its fan page (Smith, 2011). Evaluative social identity increases when customers feel themselves valuable and important to the community. Restaurant brands are advised to respect, care, and share with customers in the marketing activities on their fan pages. For instance, a business can encourage customers to express their opinions and respond in a timely fashion to comments on the "wall" of its Facebook fan page. Consequently, customers would feel that their ideas, opinions and suggestions are considered and appreciated, and that they are an important part of the respective fan page community. When the brand experiences some success with its marketing activities on the fan page, such as an increase in the number of fans, the business is advised to share this success and show gratitudes to their customers for their contribution and support in this process (O'Keefe, 2008; De Vries, Gensler, & Leeflang, 2012). Customers would consider themselves valuable to the success of the fan page. Therefore, successful Facebook fan pages can help members make group identifications and establish affective bonds.

Social needs are fulfilled with socializing activities, the second most important component contributing to overall need. Restaurant businesses can benefit from the interactive culture of the fan pages. Facebook provides multiple tools to facilitate two-way communication between businesses and their customers. One of these tools is the "wall". A business can communicate with any fan page surfers, because it is open to both fans and non-fans (Roy & Shekhar, 2010). The business can post content, answer questions, and get involved in the discussions with customers (Slater & Narver, 2000). Businesses can experience open interactions with customers because in the content of such interactions is not as controllable as that on other marketing channels. Negative comments can be treated as an opportunity to turn a negative to a positive (Smith, 2011). Businesses are advised to respond to negative comments because they are beneficial for the brands to confess their mistakes and tender a solution to the problem instead of concealing and deleting negative posts (Negash *et al.*, 2003). Responding makes customers recognize that the brand is actually concerned about its customers and contributes to brand credibility in the long term (Weiser, 2001).

The third important aspect of overall need on fan pages is the hedonic perspective, which is measured by entertainment. Customers pursue hedonic pleasures by visiting the fan pages (Vogt & Fesenmaier, 1998). Marketers are advised to design attractive website features to spark customers' interest and increase their excitement. Moreover, diverse games attract fun-seekers and motivate them to revisit, such as word association games and movie quote games (O'Keefe, 2008). The success of the games in terms of customer retention can be assessed depending on how active the fans are on these pages. Contests and giveaways can also create a temporary infusion of excitement. Businesses can offer entries in a raffle and award prizes within a limited time period. Consumers respond to the urgency and enjoy a chance of earning a big prize (Smith, 2011).

The functional aspect contributes the least to the overall need, which addresses the Facebook fan page as an information source. Marketers are advised to provide relevant and timely information on restaurant fan pages. Businesses need to refresh announcements, articles, photos, videos, surveys at least once a week. Marketers also need to update promotional information, such as discount codes, today only offers, incentives such as bonuses, or sneak peeks at upcoming events (Smith, 2011). Facebook sends invitations to fans encouraging them to revisit the fan page when the business has any status updates (Stankov *et al.*, 2010). Customers can share updates with others and subscribe to the business's updates by clicking the "Likes" button. From this perspective, the fan pages are an effective tool to maintain customer relationships over the long term (Roy & Shekhar, 2010). According to De Vries, Gensler and Leeflang (2012) the brand post popularity in terms of number of likes and the number of comments depends on vividness, interactivity, the content of the brand post, the top position of a brand post, and the valence of comments.

Moreover, fan pages can also be a platform to launch new products. The updates on fan pages can spark customers' interests and build their anticipation for the new products. Customers comment on the posts about the new products and spread the word to their network of friends on Facebook. For example, Wendy's sponsored a branded virtual gift to spread the word about its new "flavor dipped" sandwiches (Shih, 2009). The event was a fun, visual, and memorable form of word-of-mouth awareness for the new product.

Fifty nine percent of the respondents were female and 41% were male. This finding can be explained by previous studies. Quan-Haase and Young (2010) and Roy and Shekhar (2010) suggested that females use Facebook for finding their old friends and maintaining existing relationship, whereas males were more likely to engage in other web searches (e.g., sports and gaming). The results also indicated that the majority of the respondents was between 20-24 (86.3%) since we collected data from college students. Shin (2009) suggested that young people are the major users of Facebook. Facebook gives users an opportunity to communicate with their friends via messages or chat and also, to track their friends' status messages, walls and other profile changes (Weber, 2009), therefore young people tend to benefit from these functions of Facebook.

For global companies and marketers who want to develop international networks, ethnicities are a significant impact factor on users' behaviors. In this study, 61% were Caucasian among the 175 respondents. The data was collected in a Midwestern university. The Caucasians accounted for the largest proportion of the student population. Facebook is now mainly adopted by the users in the U.S. Facebook is having a hard time in many countries (e.g., Asian countries, Brazil) and it has been blocked or perceived as not reliable for bandwidth (Kim, Jeong, & Lee, 2010). Therefore, users from these areas are inclined to use other SNSs to communicate with their friends and build strong relationship (e.g., Weixin, Orkut, and Myspace) (Lewis & George, 2008). The study also found that the 54.3% of respondents only spent less than 61 minutes per day on Facebook. The result does not suggest college students' addictions to social networking, which is consistent with previous studies such as Meerkerk (2009) and Caplan (2010).

In terms of theoretical implications, customer need has been well-documented in marketing literature. The application in the new context of restaurant fan pages contributes to the body of knowledge on customer need theory. The study also makes substantial contributions to the theoretical development and examination of customer involvement in fan pages in the hospitality industry. This study is the first that assessed customer needs on restaurant fan pages, which provides a theoretical foundation for future research investigating customer attitude and behavior in this online community.

CONCLUSION

The study examined the fundamental framework of the customer need domains on restaurant fan pages. The results of the study indicated that the fan pages can satisfy multiple categories of needs, including functional, hedonic, social, and psychological. By investigating the four need components, marketers can understand the activities customers involve in, predict customer trends, and provide customers with satisfactory experiences on restaurant fan pages. This research provides a theoretical foundation for future research investigating customer attitude and behavior in other social communities.

The authors present the major marketing benefits of four need components in Table 4.

LIMITATIONS AND FUTURE RESEARCH

This study has several limitations that the authors would like to address. First, the sample investigated in the study involved college students at a Midwestern university in the U.S. This group represents only a specific population of consumers. Fan pages cater to diverse audiences with varying expectations and needs. To validate the findings in this study, future research should be directed toward larger and more generalized samples. Geographic, demographic, and socio-economic variables may be included in the analysis because these may influence customer participation on these fan pages. Second, the sample size was small. The main limit of the sample size may pertain to the ability of the sample to represent the population. This small sample size may also have a negative impact on the data analysis, for example, decreased reliability and validity of the estimated coefficient and dependence of goodness-of-fit indexes. Third, this study did not incorporate behavioral intentions or actual behaviors in the model. Future research is advised to consider these behavioral constructs. Fourth, this study focused on Facebook and ignored the importance of other SNSs. The investigation of other SNSs in future studies could build a comprehensive profile of online brand communities. Fifth, future studies should avoid self-selection biases of data collection. Sixth, this study did not include trust as the consequential construct of needs in the conceptual model. Thus, trust is advised to be added in future research. Last, but not least important, the sample of the study focused mainly on restaurant chains. Future studies need to focus attention on smaller, locally-owned restaurants.

REFERENCES

Anderson, J. C., & Gerbing, D. W. (1988). Structural equation modeling in practice: A review and recommended two-step approach. *Psychological Bulletin, 103*(3), 411–423. doi:10.1037/0033-2909.103.3.411

Andersson, T. D., & Mossberg, L. (2004). The dining experience: Do restaurants satisfy customer needs? *Food Service Technology, 4*(4), 171–177. doi:10.1111/j.1471-5740.2004.00105.x

Angehrn, A. (1997). Designing mature Internet business strategies: The ICDT model. *European Management Journal, 15*(4), 361–369. doi:10.1016/S0263-2373(97)00016-9

Arnett, D. B., & Badrinarayanan, V. (2005). Enhancing customer-needs driven CRM strategies: Core selling teams, knowledge management competence, and relationship marketing competence. *Journal of Personal Selling & Sales Management, 25*(4), 329–343.

Bagozzi, R., & Dholakia, U. (2006). Antecedents and purchase consequences of customer participation in small group brand communities. *International Journal of Research in Marketing, 23*(1), 45–61. doi:10.1016/j.ijresmar.2006.01.005

Bagozzi, R. P. (1992). The self-regulation of attitudes, intentions, and behavior. *Social Psychology Quarterly*, *55*(1), 178–204. doi:10.2307/2786945

Bagozzi, R. P., & Dholakia, U. M. (2002). Intentional social action in virtual communities. *Journal of Interactive Marketing*, *16*(2), 2–21. doi:10.1002/dir.10006

Bergami, M., & Bagozzi, R. P. (2000). Self-categorization, affective commitment, and group self-esteem as distinct aspects of social identity in an organization. *The British Journal of Social Psychology*, *39*(4), 555–577. doi:10.1348/014466600164633 PMID:11190685

Brown, T. A. (2006). *Confirmatory factor analysis for applied research*. New York, NY: Guilford.

Caplan, S. E. (2010). Theory and measurement of generalized problematic internet use: A two-step approach. *Computers in Human Behavior*, *26*(5), 1089–1097. doi:10.1016/j.chb.2010.03.012

Casaló, L. V., Flavián, C., & Guinalíu, M. (2010). Determinants of the intention to participate in firm-hosted online travel communities and effects on consumer behavioral intentions. *Tourism Management*, *31*(6), 898–911. doi:10.1016/j.tourman.2010.04.007

Chung, J. Y., & Buhalis, D. (2008). Information needs in online social networks. *Information Technology & Tourism*, *10*(4), 267–281. doi:10.3727/109830508788403123

Cortese, T. (2011). Why niche social networks encourage more meaningful interaction: Opinion. Retrieved May 8, 2011, from http://mashable.com/2011/09/06/social-network-niche-interaction/

Cothrel, J. (2000). Measuring the success of an online community. *Strategy and Leadership*, *28*(2), 17–21. doi:10.1108/10878570010341609

De Vries, L., Gensler, S., & Leeflang, P. (2012). Popularity of Brand Posts on Brand Fan Pages: An Investigation of the Effects of Social Media Marketing. *Journal of Interactive Marketing*, *26*(2), 83–91. doi:10.1016/j.intmar.2012.01.003

Dholakia, U. M., Bagozzi, R. P., & Pearo, L. K. (2004). A social influence model of consumer participation in network- and small-group based virtual communities. *International Journal of Research in Marketing*, *21*(3), 241–263. doi:10.1016/j.ijresmar.2003.12.004

Dholakia, U. M., Blazevic, V., Wiertz, C., & Algesheimer, R. (2009). Communal service delivery: How customers benefit from participation in firm-hosted virtual P3 communities. *Journal of Service Research*, *12*(2), 208–226. doi:10.1177/1094670509338618

Ellison, N. B., Steinfield, C., & Lampe, C. (2007). The benefits of Facebook friends: Social capital and college students' use of online social network sites. *Journal of Computer-Mediated Communication*, *12*(4), 1143–1168. doi:10.1111/j.1083-6101.2007.00367.x

Feenburg, A., & Bakardjieva, M. (2004). Virtual community: No killer implication. *New Media & Society*, *6*(1), 37–43. doi:10.1177/1461444804039904

Fulgoni, G. (2011). *You've got to See it! What social media marketers can learn from broadway success*. Retrieved July 8, 2011, from http://blog.comscore.com/2011/09/what_social_marketers_can_learn_from_broadway.html

Goldsmith, R. E. (2006). Online consumer behavior. In S. Dasgupta (Ed.), *Encyclopedia of Virtual Communities and Technologies* (pp. 338–342). Chicago, IL: Rand McNally. doi:10.4018/978-1-59140-563-4.ch063

Hagel, J., & Armstrong, A. (1997). *Net gain: Expanding markets through virtual communities*. Boston, MA: Harvard Business School Press.

Hair, J. F., Anderson, R. E., Tatham, R. L., & Black, W. C. (1998). *Multivariate Data Analysis with Readings*. Englewood Cliffs, NJ: Prentice Hall.

Hanson, W. (2000). *Principles of internet marketing*. Cincinnati, OH: South-Western College Publishing.

Herrmann, A., Huber, F., & Braunstein, C. (2000). Market-driven product and service design: Bridging the gap between customer needs, quality management, and customer satisfaction. *International Journal of Production Economics*, *66*(1), 77–96. doi:10.1016/S0925-5273(99)00114-0

Hills, P., & Argyle, M. (2002). The Oxford happiness questionnaire: A compact scale for the measurement of psychological well-being. *Personality and Individual Differences*, *33*(7), 1073–1082. doi:10.1016/S0191-8869(01)00213-6

Ho, C.-W. (2014). Consumer behavior on Facebook: Does consumer participation bring positive consumer evaluation of the brand? *EuroMed Journal of Business*, *9*(3), 252–267. doi:10.1108/EMJB-12-2013-0057

Hogg, M. A., & Abrams, D. (1988). *Social identifications: A social psychology of intergroup relations and group processes*. London, UK: Routledge.

Hu, L. T., & Bentler, P. M. (1999). Cutoff criteria for fit indexes in covariance structure analysis: Conventional criteria versus new alternatives. *Structural Equation Modeling*, *6*(1), 21–55. doi:10.1080/10705519909540118

Hudson, S., & Thal, K. (2013). The Impact of social media on the consumer decision process: Implications for tourism marketing. *Journal of Travel & Tourism Marketing*, *30*(1-2), 156–160. doi:10.1080/10548408.2013.751276

Kang, K., Tang, L., & Fiore, A. M. (2014). Enhancing consumer–brand relationships on restaurant Facebook fan pages: Maximizing consumer benefits and increasing active participation. *International Journal of Hospitality Management*, *36*, 145–155. doi:10.1016/j.ijhm.2013.08.015

Kaplan, A. M., & Haenlein, M. (2010). Users of the world, unite! The challenges and opportunities of Social Media. *Business Horizons*, *53*(1), 59–68. doi:10.1016/j.bushor.2009.09.003

Katz, E. (1959). Mass communication research and the study of popular culture: An editorial note on a possible future for this journal. *Studies in Public Communication, 2*(1), 1–6.

Kim, W., Jeong, O.-R., & Lee, S.-W. (2010). On social Web sites. *Information Systems*, *35*(2), 215–236. doi:10.1016/j.is.2009.08.003

Lee, H. (2007). An empirical investigation of travel members' participation behaviors in the online travel community. *Journal of Tourism and Leisure Research*, *19*(2), 355–371.

Leung, R., Schuckert, M., & Yeung, E. (2013). Attracting user social media engagement: A Study of Three Budget Airlines Facebook Pages. In L. Cantoni & Z. Xiang (Eds.), *Information and Communication Technologies in Tourism 2013* (pp. 195–207). Wien: Springer.A. doi:10.1007/978-3-642-36309-2_17

Lewis, C. C., & George, J. F. (2008). Cross-cultural deception in social networking sites and face-to-face communication. *Computers in Human Behavior, 24*(6), 2945–2964. doi:10.1016/j.chb.2008.05.002

Lin, K.-Y., & Lu, H.-P. (2011). Intention to continue using Facebook fan pages from the perspective of social capital theory. *Cyberpsychology, Behavior, and Social Networking, 14*(10), 565–570. doi:10.1089/cyber.2010.0472 PMID:21381968

Liu, C., & Arnett, K. P. (2000). Exploring the factors associated with web site success in the context of electronic commerce. *Information & Management, 38*(1), 23–33. doi:10.1016/S0378-7206(00)00049-5

Manthiou, A., Tang, L., & Bosselman, R. (2014). Reason and reaction: The dual route of the decision-making process on Facebook fan pages. *Electronic Markets*, 1–39.

Meerkerk, G.-J., Van Den Eijnden, R. J. J. M., Vermulst, A. A., & Garretsen, H. F. L. (2009). The compulsive internet use scale (CIUS): Some psychometric properties. *Cyberpsychology & Behavior, 12*(1), 1–6. doi:10.1089/cpb.2008.0181 PMID:19072079

Muñiz, A. Jr, & Schau, H. J. (2005). Religiosity in the abandoned Apple Newton brand community. *The Journal of Consumer Research, 31*(5), 737–747. doi:10.1086/426607

Negash, S., Ryan, T., & Igbaria, M. (2003). Quality and effectiveness in Web-based customer support systems. *Information & Management, 40*(4), 757–768. doi:10.1016/S0378-7206(02)00101-5

Nunnally, J. C. (1978). *Psychometric theory*. New York, NY: McGraw- Hill.

O'Keefe, P. (2008). *Managing online forums*. New York, NY: AMACOM.

Paris, C. M., Lee, W., & Seery, P. (2010). The role of social media in promoting special events: Acceptance of Facebook Events. *Information and Communication Technologies in Tourism, 14*, 531–541.

Park, N., Kee, K. F., & Valenzuela, S. (2009). Being immersed in social networking environment: Facebook groups, uses and gratifications, and social outcomes. *Cyberpsychology & Behavior, 12*(6), 729–733. doi:10.1089/cpb.2009.0003 PMID:19619037

Preece, J. (2000). *Online communities: Designing usability, supporting sociability*. New York, NY: John Wiley.

Protalinski, E. (2014). Facebook passes 1.23 billion monthly active users, 945 million mobile users, and 757 million daily users. Retrieved June 26, 2014, from http://thenextweb.com/facebook/2014/01/29/facebook-passes-1-23-billion- monthly-active-users-945-million-mobile-users-757-million-daily-users/

Qu, H., & Lee, H. (2011). Travelers' social identification and membership behaviors in online travel community. *Tourism Management, 32*(6), 1262–1270. doi:10.1016/j.tourman.2010.12.002

Quan-Haase, A., & Young, A. L. (2010). Uses and gratifications of social media: A comparison of Facebook and instant messaging. *Bulletin of Science, Technology & Society, 30*(5), 350–361. doi:10.1177/0270467610380009

Rheingold, H. (1993). *The virtual community: Homesteading on the electronic frontier. NewYork.* Addison Wesley.

Roy, S. K., & Shekhar, V. (2010). Alternative models of trustworthiness of service providers. *Journal of Global Marketing*, *23*(5), 371–386. doi:10.1080/08911762.2010.521111

Schmidt, S., Cantallops, A. S., & Santos, C. P. (2008). The characteristics of hotel websites and their implications for website effectiveness. *International Journal of Hospitality Management*, *27*(3), 504–516. doi:10.1016/j.ijhm.2007.08.002

Shih, C. C. (2009). *The Facebook era: Tapping online social networks to build better products, reach new audiences, and sell more stuff.* Boston, MA: Pearson Education.

Slater, S., & Narver, J. (2000). Intelligence generation and superior customer value. *Journal of the Academy of Marketing Science*, *28*(1), 120–127. doi:10.1177/0092070300281011

Smith, B. E. (2011). *Sams teach yourself Facebook for business in 10 minutes*. San Antonio, TX: Pearson Education press.

Stankov, U., Lazić, L., & Dragićević, V. (2010). The extent of use of basic Facebook user-generated content by the national tourism organizations in Europe. *European Journal of Tourism Research*, *3*(2), 105–113.

Techopedia, (2014a). *Facebook Page*. Retrieved November 15, 2014, from http://www.techopedia.com/definition/15441/facebook-page

Techopedia, (2014b). *Facebook Fan*. Retrieved November 15, 2014, from http://www.techopedia.com/definition/15436/facebook-fan

Top Dining Brands. (2014). *Fan page list*. Retrieved June 26, 2014, from http://fanpagelist.com/category/brands/dining/

Trusov, M., Bucklin, R., & Pauwels, K. (2009). Effects of Word-of-Mouth Versus Traditional Marketing: Findings from an Internet Social Networking Site. *Journal of Marketing*, *73*(5), 90–102. doi:10.1509/jmkg.73.5.90

Tuten, T., & Solomon, M. (2014). Social Media Marketing. 1st edition. Pearson Education limited, Edinburgh.

Ulusu, Y. (2010). Determinant factors of time spent on Facebook: Brand community engagement and usage types. *Journal of Yasar University*, *18*(5), 2949–2957.

Valkenburg, P. M., & Peter, J. (2007). Internet communication and its relationship to wellbeing: Identifying some underlying mechanisms. *Media Psychology*, *9*(1), 43–58. doi:10.1080/15213260709336802

Vogt, C., & Fesenmaier, D. R. (1998). Expanding the functional tourism information search model: Incorporating aesthetic, hedonic, innovation and sign dimensions. *Annals of Tourism Research*, *25*(5), 551–579. doi:10.1016/S0160-7383(98)00010-3

Wang, Y., & Fesenmaier, D. R. (2004). Modeling participation in an online travel community. *Journal of Travel Research*, *42*(3), 261–270. doi:10.1177/0047287503258824

Wang, Y., Yu, Q., & Fesenmaier, D. R. (2002). Defining the virtual tourist community: Implications for tourism marketing. *Tourism Management, 23*(4), 407–417. doi:10.1016/S0261-5177(01)00093-0

Weber, L. (2009). *Marketing to the social Web: How digital customer communities build your business.* Hoboken, NJ: John Wiley & Sons. doi:10.1002/9781118258125

Weiser, J. (2001). The functions of Internet use and their social and psychological consequence. *Cyberpsychology & Behavior, 4*(6), 723–743. doi:10.1089/109493101753376678 PMID:11800180

KEY TERMS AND DEFINITIONS

Brand Community: Brand communities are defined as specialized, nongeographically bound communities, based on a structured set of social relationships among admirers of a brand (Bagozzi & Dholakia, 2006).

Entertainment: The ability of a medium to fulfill audience needs for escapism, diversion, aesthetic enjoyment, or emotional release (Negash, Ryan, & Igbaria, 2003).

Facebook Page: A Facebook page is a public profile created by businesses, organizations, celebrities and anyone seeking to promote themselves publicly through social media. Facebook pages work much like personal profile pages, except that they have "fans" instead of "friends." These pages are publicly visible online and often post status updates, links, events, photos and videos to their fans' news feeds and walls. Facebook pages provide a way for businesses and other organizations to interact with - rather than just advertise to - potential customers (Techopedia, 2014a).

Facebook Fan: A Facebook fan is a user who likes a particular Facebook page. Users who Like a page are able to receive updates from that page's administrator through status updates, posted content and event invitations. A list of pages a fan has liked will appear on his or her profile page. The term the fan was replaced by Like in 2010. Now fans Like a page and this is shown on their profile rather than that they are a fan (Techopedia, 2014b).

Functional Need: Functional uses in online communities are defined as contractual needs which are minimum intrinsic informational requirements that users expect to fulfill (Cothrel, 2000).

Hedonic Need: Hedonic needs in online communities are mainly accomplished by entertainment and enjoyment (Wang & Fesenmaier, 2004).

Online Community: A group of people who share their consumer experiences via social media, including communicating with other members or the company regarding their concerns and opinions and providing critiques of offered products/services (Rheingold, 1993).

Psychological Need: The term "psychological need" refers to the emotional aspects of relationships in online communities (Wang *et al.*, 2002).

Social Media: "A group of Internet-based applications that build on the ideological and technological foundations of Web 2.0, and that allow the creation and exchange of user-generated content." (Kaplan & Haenlein, 2010, p. 61).

Social Needs: Social needs in online communities describe the development of relationships with other users and the brand through communications and interactions (Preece, 2000).

Chapter 10
Using High-Frequency Interaction Events to Automatically Classify Cognitive Load

Tao Lin
Sichuan University, China

Zhiming Wu
Sichuan University, China

Yu Chen
Sichuan University for Nationalities, China

ABSTRACT

There is still a challenge of creating an evaluation method which can not only unobtrusively collect data without supplement equipment but also objectively, quantitatively and real-time evaluate cognitive load of users based the data. The study explores the possibility of using the features extracted from high-frequency interaction (HFI) events to evaluate cognitive load to respond the challenge. Specifically, back-propagation neural networks, along with two feature selection methods (nBset and SFS), were used as the classifier and it was able to use a set of features to differentiate three cognitive load levels with an accuracy of 74.27%. The main contributions of the research are: (1) knowledge about what detailed features may be predictive of cognitive load changes; (2) demonstrating the potential of using the HFI features in discriminating different cognitive load when suitable classifier and features are adopted.

INTRODUCTION

Cognitive load refers to the amount of mental effort required to process a given amount of information and has been associated with the limited capacity of working memory (Jimison, Pavel, McKanna, & Pavel, 2004; Plass, Moreno, & Brünken, 2010). In many complex systems (e.g. air traffic control, crisis

DOI: 10.4018/978-1-4666-8450-8.ch010

management, military command and control, massively multiplayer role playing games), a competition for the users' attention is going on between many different information items, possibly leading to cognitive overload. This overload originates in the limitations of human attention and constitutes a well-known bottleneck in human information processing(Arciszewski, De Greef, & Van Delft, 2009; Tan, Meyers, & Czerwinski, 2004). For example, whilst playing massively multiplayer role playing games (MMORPGs), players tend to experience cognitive overload because they must learn to deal with the social dynamics around the game in addition to having to interact with the virtual space and game objects (Ang, Zaphiris, & Mahmood, 2007). Prolonged periods of low activity (i.e., underload), on the other hand, also lead to performance degradations because users tend to get out of the information processing loop as they become a passive monitor(Arciszewski et al., 2009; Eirinaki & Vazirgiannis, 2007).

Adaptive systems (e.g.(Coleman & Parker, 1996; de Tjerk, Henryk, & Neerincx, 2010; G. F. Wilson & Russell, 2003)) have been proposed as one of effective solutions to keeping the human within a bandwidth of cognitive load for optimum performance (Arciszewski et al., 2009), and the nature of adaptation is to provide the best match between task demands and cognitive resources of the human(Badros, Nichols, & Borning, 2000; Parasuraman, Mouloua, & Molloy, 1996). Many studies have shown that adaptive systems can enhance performance(Badros et al., 2000; Beaudouin-Lafon, 2001), reduce workload (Funke, Neal, & Paul, 1993; Misue, Eades, Lai, & Sugiyama, 1995), and improve situation awareness (Purchase, Hoggan, & Görg, 2007). However, trigger strategy is still one of the challenging factors in developing successful adaptive systems (Tan et al., 2004) (Rowe, Sibert, & Irwin, 1998), since there still lack effective evaluation methods for cognitive load and a clear understanding of cognitive load. The last decade has witnessed a growing interest in evaluating and understanding users' cognitive load (D.R. Hutchings & Stasko, 2002; Dugald Ralph Hutchings & Stasko, 2004; Lin, Imamiya, & Mao, 2008; Sweeney, Maguire, & Shackel, 1993; Tan et al., 2004; G. M. Wilson & Angela Sasse, 2004). For example, Neerincx (D.R. Hutchings & Stasko, 2002) presented a cognitive task load (CTL) model with the aim of exploring load factors. The CTL model is comprised of three load factors (percentage time occupied, the level of information processing, and task-set switching) and the three factors constitute a three-dimensional space in which all human activities can be projected as a combined factor (i.e., it displays the workload due to all activities combined). Recent several studies have also highlighted the necessity of managing and evaluating cognitive load educational game environments (Kalyuga & Plass, 2009), explored different types of cognitive overload in MMORPGs and presented strategies to overcome them (Ang et al., 2007). Several traditional methods have been adopted, with some success for evaluating cognitive load (Kramer, 1991; F. Paas, Tuovinen, Tabbers, & Van Gerven, 2003), including physiological, subjective (self-report) (F. G. W. C. Paas & Van Merriënboer, 1994) and performance measures (Jameson et al., 2009). The three measures have their own advantages and disadvantages (see section 2 for details) and the current challenge in cognitive load evaluation mainly lies in how to develop a method which can not only unobtrusively gather data without supplement equipment but also objectively, quantitatively and real-time evaluate cognitive load based the data. Also, the method should be suitable for deployment in real-life scenarios.

The potential of behavioral features (e.g. speech features(Ruiz, Taib, & Chen, 2011; Bo Yin, Ruiz, Chen, & Khawaja, 2007), linguistic features (Khawaja, Chen, Owen, & Hickey, 2009), and pen stroke features (Ruiz, Feng, Taib, Handke, & Chen, 2010; Ruiz et al., 2011; Ruiz, Taib, Shi, Choi, & Chen, 2007)) to assessment cognitive load has recently been exploited. Using features from high-frequency interaction (HFI) events to evaluate cognitive load is of main interest in our study, since they offer a convenient way to implicitly capture changes in interaction behavior without affecting task completion.

Interaction events are classified as low-, mid-, or high-frequency (Martínez & Yannakakis, 2010; Quinlan, 1993) according to the typical number of events of a certain type per second (e.g. mouse operations and keystrokes are often referred to as HFI events). Although previous studies (Cocea & Weibelzahl, 2009; Hurst, Hudson, & Mankoff, 2007; Vizer, Zhou, & Sears, 2009) have shown that the features from HFI events can quantify interaction behavior well, its efficacy of evaluating cognitive load have not been explored. If effective, the evaluation method derived from the features from HFI events would be desirable because it: (1) unobtrusively gathers data; (2) does not require extra equipment; (3) is objective and quantitative; (4) has high evaluative bandwidth and therefore (5) is conductive to reflecting dynamics of cognitive load.

The goal of our research was to test the efficacy of interaction behavior quantified by the features from HFI events in evaluating cognitive load of users. The research uses a simplified air traffic control (ATC) game as experimental tasks and an automatic classification method for cognitive load was developed and validated based on machine learning and feature selection techniques. Specifically, back-propagation neural networks, along with two feature selection methods (nBset and SFS), were used as the classifier and it was able to use a set of features to differentiate three cognitive load levels with an accuracy of 74.27%. The main contributions of the research are: (1) knowledge about what detailed features may be predictive of cognitive load changes; (2) demonstrating the potential of using the HFI features in discriminating different cognitive load when suitable classifier and features are adopted.

BACKGROUND

Because of the centrality of working memory load to cognitive load theory, measuring this load has been a high priority for researchers. To provide an introduction for readers unfamiliar with cognitive load evaluation, we briefly introduce the current main cognitive load measures. See a comprehensive review (Sweller, Ayres, & Kalyuga, 2011) for details.

Subjective Measures

Various subjective measures assessing mental workload that use unidimensional to multidimensional scales have been widely employed. There are two most commonly used techniques of subjective mental workload: NASA-Task Load Index (NASA-TLX, (Hart & Staveland, 1988)) and Subjective Workload Assessment Technique (SWAT, (Reid & Nygren, 1988)). NASA-TLX consists of six sub-dimensions: mental demand, physical demand, temporal demand, own performance, mental effort, and frustration. Overall NASA-TLX workload ranges from 0 (no workload) to 100 (maximum workload). SWAT describes three dimensions of mental workload: time load, mental effort and psychological stress. The two subjective mental workload techniques have been suggested to be relatively similar(Miyake, 2001) and are widely used in the field of aeronautics with some success, as shown for instance in a study (Collet, Averty, & Dittmar, 2009) that revealed a positive correlation between the number of aircrafts to control and the NASA-TLX score in air traffic controllers. However, it should be noted that subjective techniques is a good approach to understanding the overall attitudes of participants, but they are usually administered post hoc or during break points, which may disrupt the normal flow of interaction(R. L. Mandryk, Atkins, & Inkpen, 2006; Wierwille & Eggemeier, 1993); furthermore, subjective techniques tend to suffer from low evaluative bandwidth and this may wash out the variability of mental workload during interaction.

Performance Measures

With this group of methods, participants' mental workload is inferred from their observable task performance index such as time to complete a task, type and number of errors, and success rate(Galy, Cariou, & Melan, 2012). Dual-task methodology (Wierwille & Eggemeier, 1993) is often used for performance measures. It usually uses a secondary task in combination with a primary task (Wierwille & Eggemeier, 1993) and the secondary task requires participants to engage in an additional cognitive activity which is secondary to the primary task of learning or problem solving. Task performance on the secondary task is supposed to reflect the level of mental workload imposed by a primary task. Generally, the secondary task entails simple activities requiring sustained attention, such as detecting a visual or auditory signal. Although secondary task performance is a highly sensitive and reliable technique, it has rarely been applied in real-word scenarios (F. Paas et al., 2003) due to some important drawbacks(Chen et al., 2013): (1) they may interfere considerably with the primary task, especially if the primary task is complex or if cognitive resources are limited; (2) performance measures may not reflect subtle changes in mental workload; and (3) they are not suitable for real-time, automated deployment since they have the disadvantage of being calculated post-hoc.

Physiological Measures

Changes in various bodily processes and states have been reported with changes in mental workload. Kramer and Weber (2000) reviewed existing physiological measures of mental workload, including heart rate, heart rate variability (HRV), galvanic skin response (GSR), pupil response, blink rate and duration and event-related brain potentials (ERP). The measurement of cardiac activity is the most popular physiological technique employed in the assessment of mental workload; more especially, Backs(1995) demonstrated the systematic and reliable relationships between HRV and mental workload. Some studies (e.g.,(Fournier, Wilson, & Swain, 1999)) also reported sensitivity of brain event-related changes to different levels of mental workload; particularly, cognitive processing would result in attenuation of the alpha brain electrical rhythm. In addition, pupillary responses seem to be a highly sensitive instrument for tracking fluctuating levels of mental workload (Beatty & Lucero-Wagoner, 2000; Partala & Surakka, 2003). Beatty and Lucero-Wagoner (2000) (Beatty & Lucero-Wagoner, 2000) identified three useful task-evoked pupillary responses (TEPRs): mean pupil dilation, peak dilation, and latency to the peak. These TEPRs typically intensify as a function of cognitive load. One major advantage of physiological measures (Kramer, 1991; Regan L. Mandryk & Atkins, 2007) is the continuous availability of physiological response data, allowing mental workload to be measured at a high rate (high bandwidth) and with a high degree of sensitivity, even in situations in which overt behavior is relatively rare (F. G. Paas, 1992). However, physiological measures typically are intrusive and require supplemental equipment and processing of physiological signals tends to be computationally intensive, making it difficult to evaluate mental workload in real-time. In addition, human expertise tends to be required to interpret the resulting patterns from physiological data.

Behavior Measures

The premise of behavior measures is that observations of behavior features are suitable for mental workload assessment because a user experiencing a high cognitive load will show changes in interactive and communicative behavior, whether voluntary or otherwise(Chen et al., 2013). Some behavior measures (e.g., acoustic and prosodic speech features, linguistic features, symbolic pen gesture features and handwriting features) have been investigated (see (Chen et al., 2013) for a detailed review) and show the potential to indicate mental workload changes.

Behaviors measures are usually task-dependent and have not been explored adequately. A few efforts focused on speech-based and linguistic features ((Khawaja et al., 2009; B. Yin, Ruiz, Chen, & Ambikairajah, 2008; Bo Yin et al., 2007)). Using speech-based and linguistic features to assess mental workload is a particularly wise choice in some real-life speech-based applications such as telephone conversation and voice control, since these data can be easily collected in a non-intrusive and inexpensive way. For example, the Stroop test data reveals an accuracy of 78.9% for classification into low, medium and high load in a speaker-independent scenario(Bo Yin, Chen, Ruiz, & Ambikairajah, 2008). Linguistic features (e.g., pause rates, language complexity, word categories and valence, and grammatical features) were also investigated as an individual set of mental workload indices in the domain where speech and/or conversational transcripts were used as the main input modalities. It should be note that mental workload measurements from the proposed linguistic features tend to require state-of-the-art automatic speech recognition technology with highly accurate, automatic Speech-To-Text (STT) functionality to realize its potential in practical applications(Chen et al., 2013).In addition, the potential of writing features in evaluating cognitive load has begun to be exploited. Although some research on the relationship between cognitive load and handwriting have been reported by psychologists (e.g., (Vanderberg & Lee Swanson, 2007)), but only a few efforts (Ruiz et al., 2010; Ruiz et al., 2011; Ruiz et al., 2007; Yu, Epps, & Chen, 2011) were spent on technically validating the possibility of using writing features (e.g. velocity and shape features) to evaluate cognitive. Ruiz et al. (Ruiz et al., 2007) observed degeneration of pen gesture shape from the "standard form" of the intended gestures as a result of increased cognitive load, and they also further examined the relationship between comprehensive writing behaviors and cognitive load for normal writing, and provided hints for live monitoring and adjustment of cognitive load for pen users(Yu et al., 2011). These explorations are mainly based either on empirical observations of user studies or on linear correlations of some simple writing features with reported cognitive load levels derived from self-reported mental effort or predefined task difficulty.

HFI events have also been used to quantify interaction behavior and indicate some user states (e.g. user cost(Lin et al., 2008), skill levels(Ghazarian & Noorhosseini, 2010; Hurst et al., 2007) and stress(Vizer et al., 2009)), but there still lacks an investigation into whether and how they can be used in evaluating cognitive load. Our pilot study(Lin et al., 2008) investigated eye and hand behavior at various levels of user cost and showed that features of eye and hand movements were to some extent sensitive to changes in user cost. Vizer et al. (Vizer et al., 2009) also analyzed keystroke and linguistic features of spontaneously generated text and classified cognitive and physical stress conditions relative to non-stress conditions with high accuracy rates. The proposed method in the study leverages insights from the above studies and explores the use of HFI events in automatically classifying cognitive load levels.

RESEARCH METHODS

Participants

Forty-eight (12 female) university students aged 18 to 30 participated in the experiment. Before the experiment, all participants were required to complete a background questionnaire about their experience with computer use and with air traffic control (ATC) simulation games, along with personal information such as sex and handedness. All participants used computers daily and were adept at operating a mouse. When asked how often they play ATC simulation games, nine of the participants played often, twenty played occasionally and nineteen never played.

Experimental Tasks

A simplified ATC simulation game was used as the experimental tasks. The mission of players was to land aircraft with any collision and as soon as possible. All operations were performed with a mouse and mainly included two serial steps: (1) move the cursor to the plane to be controlled and select it by left-clicking and (2) hold the mouse button to draw the path players would like the plane to follow. Planes can be landed on either runway and helicopters must land at the helicopter landing zone marked with H. Players were allowed to modify the paths before planes landed on the airport to avoid colliding with each other. Figure 1 shows a screen shot of the task.

Figure 1. A screen shot of the task

Table 1. Difficulty configurations for the three game versions

	Speed	Number	Direction
Low	low and constant	3	fixed
Middle	middle and inconstant	4	fixed
High	high and inconstant	5	random

Three (*low, middle and high*) versions of the game with different difficulty configurations were presented to participants to enable them to experience three different levels of cognitive load. Previous studies (e.g. (Mogford, Guttman, Morrow, & Kopardekar, 1995)) have defined airspace complexity as a construct referring to the rate at which workload increases, and also presented twelve complexity factors which could be used to describe an airspace. Three task difficulty factors (i.e., the *speed* of planes, the *number* of planes which need to be simultaneously controlled on the screen and the incoming *directions* of planes) were chosen to manipulate cognitive load levels in the experiment. Table 1 shows the difficulty configurations for the three versions.

Experimental Procedure

The experimental tasks were performed in an HCI laboratory. Participants were required to play the three versions of the game on a 21" standard desktop display as quickly and as well as possible and each version lasts three minutes. The game screens were captured to help us to characterize some features and phenomena related to user behavior at different levels of cognitive load.

The experiment was divided into four phases: a welcome phase, a practice phase, a game phase, and a debriefing phase. During the welcome phase, all participants were required to sign a consent form with a detailed description of the experiment, its duration, and its research purpose. Each participant also filled out a background questionnaire. During the practice phase, instructions were read to each participant describing the game task rules, and each was given a brief tutorial on how to complete the task. Participants were then allowed to practice for about forty-five minutes and the experimenter observed the practice effects to ensure that participants fully mastered the method of playing the game. After the practice phase, participants have half an hour to reset and then completed the game tasks.

An within-subject experimental design was used in the study and a Latin Square design was adopted to balance the order of the presented game versions. After each game session, participants were required to subjectively rate their perceived mental effort and task difficulty. At the end of game play, explanation of their answers was encouraged and participants discussed their impressions of the experiment.

RESULTS

We were interested in whether some features patterns from HFI events indicating cognitive load changes could be detected using machine learning techniques. In the study, the perceived mental effort and task difficulty were first examined to testify whether the designed three game versions induced different levels of cognitive load. Then, back-propagation neural networks, combined with feature selection techniques were developed to automatically classify the three levels of cognitive load.

Table 2. Means for the self-reported mental effort and task difficulty on a nine-point scale. A response of "1" corresponded to low and "9" to high.

	Low		Middle		High		$F_{(2,141)}$	P
	Mean	*St.Dev*	*Mean*	*St.Dev*	*Mean*	*St.dev*		
Mental effort	2.83	0.63	4.25	0.72	6.18	1.09	187.43	0.001
Difficulty	2.49	0.69	4.18	0.77	7.02	1.27	178.71	0.001

Subjective Evaluation for Cognitive Load and Task Difficulty

Subjective ratings of mental effort and task difficulty were first analyzed to validate whether the different task conditions induced the different levels of cognitive load. A 9-point Likert scale were used to subjectively evaluate mental effort and task difficulty. A one-way ANOVA showed that there were significant differences in mean subjective ratings of mental effort and task difficulty across the three task conditions (see Table 2). Post hoc comparisons also showed that the *high* version was the most difficult and induced the greatest cognitive load, followed by the *middle* and *low* versions (see Table 2). At the debriefing phase, almost all participants also reported that they felt it easy to play the *low* version, and challenging or especially challenging for the *high* versions.

Automatic Classification for Cognitive Load Levels

Classifier

Given the high subjectivity of the reported cognitive load and the highly noisy nature of HFI information, artificial neural networks (ANNs) were employed to classify the different levels of cognitive load. Neural network architectures are motivated by models of our own brains and nerve cells. A neural network can be defined as an interconnection of neurons such that neuron outputs are connected, through weights, to all other neurons including themselves with both lag-free and delay connections allowed(Samarasinghe, 2007). This method has high tolerance for noisy data and the good ability to nonlinear modeling. Our primary goal was to confirm whether the selected HFI features and automated models are able to successfully detect cognitive load changes rather than optimize how well they detect cognitive load. Therefore, we adopted the most commonly used parameter settings when training and testing classification models.

A basic framework of neural network can be designed by selecting architecture, activation function and learning rule. We designed three-layer BPNNs (i.e., input, hidden and output layers) to learn the relation between the selected HFI features (BPNN inputs) and the levels of cognitive load (BPNN output) using the combination of sigmoid function as the activation function and back-propagation as the learning rule. The number of nodes in the hidden layer was set to half of the total number of nodes in the input and output layers. During the learning process the weights are adjusted to result in optimal prediction for classification. The degree of adjustment is affected by the learning rate with higher values corresponding to faster learning. Our BPNNs were trained using the value 0.4 of the learning rate. In addition, the initial connection weights take values from –1 to 1 and all input values are normalized into [0, 1] before they are entered into the BPNNs.

Features from HFI events

Drawing a landing path on the screen in the game is a simple making-decision model which includes four phases: *scanning, projecting, plan*, and *acting*. In the scanning phase, user attention is continuously switched between different stimulus sources (planes) on the screen. The purpose of scanning is to update the mental representation of the visual display and attendant communications systems as the basis for detecting potential collisions. As a potential conflict is detected, a user then projects the future positions of objects and assesses the likelihood of a conflict. During planning phase, the user reviews possible solutions to the problem. Finally, during the acting phase, the user selects and implements a solution.

We defined a series of three types of HFI events (*making-decision, selection* and *drawing events*), which could be easily and accurately labeled, to characterize the making-decision model (the process of drawing a landing path). The *making-decision* event covers scanning, projecting and plan phases and had typically no physical interaction with computers. The *selection* event is to choose the target plane to be controlled and the *drawing* event is to draw its landing path on the screen. The *acting* phase was involved with physical interaction and consisted of the selection and drawing events. Each game session consisted of a number of making-decision, selection and drawing events.

We defined different event attributes for each type of HFI events, and features of a game session were represented using the statistics of event attribute values across the game session. The attributes (e.g., velocity) were calculated from the timestamp and position of the mouse, and these features were further extracted from these attributes. That is, features of a game session were extracted from the three types of the HFI events, respectively. Specifically, for the making-decision event, only a time-related attribute (i.e. lapsed time within the event (T_m)) were adopted and its five statics (i.e., avg(), sigma(), max(), min(), diff()) across a game session were used as a part of features of the game session (see Table 3). For the drawing event, nine cursor motion-related attributes (see Table 4) were adopted and their five statistics (i.e., avg(), sigma(), max(), min(), diff()) across a game session were calculated as a part of the features of the game session; forty-five features (multiply 9 attributes by 5 statistics) were obtained from the drawing event. For the selection event, nine cursor motion-related attributes (see Table 5) were adopted and their five statistics (i.e., avg(), sigma(), max(), min(), diff()) across a game session were calculated as a part of features of the game session. There are also 45 features from the drawing events. Total 95 HFI features were extracted for each game session. Some of the features have been adopted in previous studies (Ghazarian & Noorhosseini, 2010; Hurst et al., 2007) to quantify user behavior and others are from our observation of participants as they completed tasks.

Table 3. The features of a game session from the making-decision events

Feature	Description
Avg(T_m)	Average T_m
Sigma(T_m)	Standard deviation of T_m
Max(T_m)	Maximum T_m
Min(T_m)	Minimum T_m
Diff(T_m)	Difference between maximum and minimum T_m

Table 4. Attributes of a drawing event which were used for calculating the features

Attribute	Description
V_{dx} and V_{dy}	Average velocity of the mouse cursor in X and Y directions respectively during a drawing event
A_{dx} and A_{dy}	Average acceleration of the mouse cursor in X and Y directions respectively during a drawing event
D_d	Distance traveled by the mouse cursor during a drawing event
DC_{dx} and DC_{dy}	Count of direction changes in cursor motion relative to its previous X and Y direction respectively during a drawing event
T_d	Duration of a drawing event
P_d	Count of pauses longer than 100ms(sensible pauses) during a drawing event

Table 5. Attributes of a selection event which were used for calculating the features

Attribute	Description
V_{sx} and V_{sy}	Average velocity of the mouse cursor in X and Y directions respectively during a selection event
A_{sx} and A_{sy}	Average acceleration of the mouse cursor in X and Y directions respectively during a selection event
D_s	Distance traveled by the mouse cursor during a selection event
DC_{sx} and DC_{sy}	Count of direction changes in cursor motion relative to its previous X and Y direction respectively
T_s	Duration of a selection event
P_s	Count of pauses longer than 100ms(sensible pauses) during a selection event

Feature Selection Methods

The quality of the classification model constructed by machine learning techniques depends critically on the set of input data features chosen and there has been evidence that cutting out unnecessary features improves learning quality for evolutionary training of neural networks. However, it is not possible to determine a priori the suitability of any given feature for the final model. Therefore, we used automatic feature set selection algorithms to explore the space of possible input feature sets, searching for the feature sets that generate the highest discrimination among different cognitive load levels. In addition, we would like the classification model to be depend on as few features as possible, both to make it easier to analyze, and to make it more useful for incorporation into future adaptive applications for purpose of e.g. intelligent training systems. Using the extracted 95 features, the n best individual feature selection(nBest) and the sequential forward selection(SFS) methods are applied and compared.

The nBest method is the most popular technique for feature selection and it picks the n individually best features (with regards to a performance function) from the feature subset. The SFS method has been successfully applied to a wide variety of feature selection problems, yielding high performance values with minimal feature subsets. SFS is a bottom-up search procedure where one feature is added at a time to the current feature set. The feature to be added is selected from the subset of the remaining features such that the new feature set generates the maximum value of the performance function over

Table 6. The eight highest performing features for BPNNs ranked by cross-validation performance (P) from left to right.

Features	avg(P_d)	avg(T_m)	avg(DC_{dy})	avg(AC_{dy})	avg(P_s)	max(DC_{dx})	avg(T_d)	diff(V_{dy})
P (%)	66.22	64.67	63.34	63.01	61.90	57.41	53.30	52.55

all candidate features for addition. More advanced methods such as sequential floating forward search (SFFS), genetic feature selection, and Fisher project ion (FP) (Picard, Vyzas, & Healey, 2001) could be used in future experiments and results could be compared to the existing studies. Note that nBest and SFS are not guaranteed to find the optimal feature set since neither searches all possible combinations.

To evaluate the performance of each input feature subset, the collected 144 simple data are randomly divided into three parts, with two parts used for training and one part used for validation. The performance of an BPNN model is measured through the average classification accuracy of the model using the 3-fold cross-validation technique. The feature selection procedure is terminated when an added feature yields equal or lower validation performance than the performance obtained without it, since we are interested in the minimal feature subset that yields the highest performance.

Classification Results

The strength of a single feature (BPNN input) in discriminating cognitive load levels was first examined by applying BPNNs on the 144 simple data. Specifically, the data are randomly partitioned into three groups which are used as 2/3 training and 1/3 validation data subsets with the leave-one-out cross-validation technique to obtain the average classification performance of the BPNN classifier. As observed from Table 6, the best feature for BPNN is the avg(P_d) with a best performance of 66.22% . In addition, the four features (avg(DC_{dy}), avg(T_m), avg(AC_{dy}) and avg(P_s)) also show the potential of discriminating cognitive load levels, whose classification accuracy is better than 60%. The single feature experiments suggest that which features are most predictive of cognitive load levels and the non-linear classier can find meaningful features for cognitive load classification.

Two feature selection methods (nBest and SFS) were also compared within the BPNN approach based on the 95 features. The SFS method generates feature subsets that yield higher validation performance than feature subsets generated by nBest, as presented in Table 7. Overall, The best cross validation performance (74. 27%) is achieved through BPNNs with the feature subset (avg(P_d), avg(T_m), avg(DC_{dy}), avg(P_s), avg(DC_{dx})) while adding more features to the subset does not yield significantly higher performance (see the bottom row of Table 4).

DISCUSSION AND CONCLUSION

Evidence from psychology has suggested that there are links between user behavior and user status. However, how to quantify the links is still a great challenge. The study chose a ATC game as experimental tasks, used the features from HFI events quantitatively characterize users' behavior and explored the possibility of employing machine learning techniques on the features form HFI events to classify cognitive load levels.

Table 7. Classification accuracy (%) of nBest and SFS feature selection methods applied on the BPNN classifier. The highest performance obtained through a minimal feature subset appears in bold.

nBest		SFS	
Feature Subset	**Accuracy**	**Feature Subset**	**Accuracy**
avg(P_d)	66.22	avg(P_d)	66.22
avg(P_d), avg(DC_{dy})	68.15	avg(P_d), avg(T_m)	69.16
avg(P_d), avg(DC_{dy}), avg(AC_{dy})	69.35	avg(P_d), avg(T_m), avg(DC_{dy}),	71.68
avg(P_d), avg(DC_{dy}), avg(AC_{dy}), max(DC_{dx}),	71.09	avg(P_d), avg(T_m), avg(DC_{dy}), avg(P_s)	73.17
avg(P_d), avg(DC_{dy}), avg(AC_{dy}), max(DC_{dx}), diff(V_{dy})	73.02	avg(P_d), avg(T_m), avg(DC_{dy}), avg(P_s), avg(DC_{dx})	**74.27**
avg(P_d), avg(DC_{dy}), avg(AC_{dy}), max(DC_{dx}), diff(V_{dy}), sigma(P_d)	72.91	avg(P_d), avg(T_m), avg(DC_{dy}), avg(P_s), avg(DC_{dx}), avg(T_d)	74.01

The results show the possibility of applying machine learning and feature selection techniques on the features from HFI events to classify reported cognitive load levels.

The strength of a single feature in discriminating cognitive load levels for a non-linear classifier (BPNN) are first examined. There are eight HFI features whose classification accuracy are better than 50%. Notly, avg(P_d) serves the classification purpose best(66.22%), followed by avg(T_m) (64.67%) and avg(DC_{dy}) (63.34%). The captured game screens and interviews with participants in part can help us explain the results. The task difficulty was manipulated by the *speed* of planes, the *number* of planes which need to be simultaneously controlled on the screen and the incoming *directions* of plane. When the game versions became more difficult, participants tended to need more time to choose the target plane and this might lead to an increase in T_m . The increasing difficulty also made the situations of airport change more frequently and caused more landing path collisions, and therefore participants had to continuously rethink and change the direction of the drawing path during the process of drawing a path in order to respond the situations, leading more sensible pauses (P_d) and direction changes (DC_y). The results showed that the different feature patterns seemed to be related to cognitive load levels and might be detected by the BPNN model.

Further, two alternative two feature selection methods (nBest and SFS) for BPNNs were applied, demonstrating the benefit of SFS over nBest for obtaining more accurate classifiers. Five features (avg(P_d), avg(T_m), avg(DC_{dy}), avg(P_s) and avg(DC_{dx})) selected by SFS fed BPNNs and obtain the best classification performance with a cross-validation accuracy of 74.27%. While the classification rates for cognitive load seems not to be comparable to rates obtained using physiological measures, the current approach has advantages over physiological measures because it requires no additional hardware, is unobtrusive, and is less computationally intensive. The results show that some HFI features have the potentials in discriminating cognitive load levels when suitable classifiers are adopted. Non-linear classifiers seem to serve the purpose well.

To our knowledge, machine learning techniques have not been used on HFI features for cognitive load evaluation. The exploratory study extends the state of the art by extracting a unique set of features from HFI events as input to BPNNs for classification of cognitive load levels. Further research are needed to refine the set of features and determine the best machine learning methods for the cognitive load classification purpose.

HFI features were seldom studied previously. The study demonstrates an additional application of these data, demonstrating that that HFI features can be useful as an unobtrusive and easy-to-use method of continuously monitoring cognitive load. These results may spur research into additional applications. For example, the method has implications for research on adaptive interfaces. An adaptive interface, which is aware of the users' current cognitive load, can change its response, presentation and interaction flow to improve users' experience and their task performance. Suppose that a mathematic training system notices that the current user is experiencing high cognitive load while solving a particular mathematical problem through HFI event observation (e.g., mouse movements or keyboard operations). The system might then insert more explanations and suggestive examples, or eliminate unnecessary distractions (e.g., background music) to avoid cognitive overload. Similarly, if the user's current cognitive load is lower than an optimal level, the system might increase the difficulty of the presented problems.

In sum, there has been an increasing need for an objective, unobtrusive, cost-effective evaluation method of monitoring changes of user cognitive load, even in daily interaction situations. The exploratory research confirms the potentials of combining machine learning techniques and HFI features in evaluating cognitive load. The primary contributions are to (1) demonstrate the use of combining machine learning techniques and the HFI features in automatically evaluating cognitive load; and (2) highlight the opportunity of monitoring HFI events for understanding user status. The relative high classification accuracy in the exploratory study warrants further research with the aim of establishing an objective, quantitative, cost-effective and real-time evaluation methodology; furthermore, each of these contributions opens a door for new research, e.g., exploring the relationships between cognitive load and interaction behavior, combining machine learning techniques and interaction behavior data to evaluate cognitive load and using interaction behavior data to quantify user state. This work is also a key step towards a larger goal of being able to detect and understand user need so that that user interfaces can be automatically adapted to fit those needs.

FUTURE WORK

This study presents an initial exploration into how to use HFI features to evaluate cognitive load and the opportunities for future work are vast. Our further research will work toward refining the experimental design, investigating additional analysis methods, and exploring practical applications such as adaptive systems.

Cognitive load in the study was induced by ATC tasks in laboratory settings and participants were drawn from the population at a university. Although our experimental results show the potential of using machine learning methods and HFI features to evaluate cognitive load, its effectiveness should further be validated across multiple applications and fine-grained cognitive load levels in real-word situations. In addition, the size of the data set in the study is relatively small for machine learning and a large number of samples could allow for more accurate classification results.

The study employed nBest and SFS techniques to select features as the input to a classifier (BPNN). While our results confirmed that the approach is viable, a variety of feature selection and machine learning techniques should be explored to develop the best classifier. Also, more potential HFI features should be studied. For example, keystroke and menu-selection events will be explored in future work.

ACKNOWLEDGMENT

We would like to express our gratitude to all participants in the experiment and a special gratitude is expressed to anonymous reviewers for your helpful suggestions and encouragement.

REFERENCES

Ang, C. S., Zaphiris, P., & Mahmood, S. (2007). A model of cognitive loads in massively multiplayer online role playing games. *Interacting with Computers*, *19*(2), 167–179. doi:10.1016/j.intcom.2006.08.006

Arciszewski, H. F. R., De Greef, T. E., & Van Delft, J. H. (2009). Adaptive automation in a naval combat management system. *IEEE Transactions on Systems, Man, and Cybernetics. Part A, Systems and Humans*, *39*(6), 1188–1199. doi:10.1109/TSMCA.2009.2026428

Backs, R. W. (1995). Going Beyond Heart Rate: Autonomic Space and Cardiovascular Assessment of Mental Workload. *The International Journal of Aviation Psychology*, *5*(1), 25–48. doi:10.1207/s15327108ijap0501_3 PMID:11541494

Badros, G. J., Nichols, J., & Borning, A. (2000). *SCWM: An intelligent constraint-enabled window manager*. In *Proceedings of the AAAI Spring Symposium on Smart Graphics*.

Beatty, J., & Lucero-Wagoner, B. (2000). The pupillary system. In J. T. Cacioppo, L. G. Tassinary, & G. G. Berntson (Eds.), *Handbook of psychophysiology*. Cambride, MA: Cambirdge University Press.

Beaudouin-Lafon, M. (2001). *Novel interaction techniques for overlapping windows*. In *Proceedings of the 14th annual ACM symposium on User interface software and technology* (pp. 153-154). ACM. doi:10.1145/502348.502371

Chen, F., Ruiz, N., Choi, E., Epps, J., Khawaja, M. A., Taib, R., & Wang, Y. et al. (2013). Multimodal behavior and interaction as indicators of cognitive load. *ACM Transactions on Interactive Intelligent Systems*, *2*(4), 1–36. doi:10.1145/2395123.2395127

Cocea, M., & Weibelzahl, S. (2009). Log file analysis for disengagement detection in e-Learning environments. *User Modeling and User-Adapted Interaction*, *19*(4), 341–385. doi:10.1007/s11257-009-9065-5

Coleman, M. K., & Parker, D. S. (1996). Aesthetics-based Graph Layout for Human Consumption. *Software, Practice & Experience*, *26*(12), 1415–1438. doi:10.1002/(SICI)1097-024X(199612)26:12<1415::AID-SPE69>3.0.CO;2-P

Collet, C., Averty, P., & Dittmar, A. (2009). Autonomic nervous system and subjective ratings of strain in air-traffic control. *Applied Ergonomics*, *40*(1), 23–32. doi:10.1016/j.apergo.2008.01.019 PMID:18353277

de Tjerk, E. G., Henryk, F. R. A., & Neerincx, M. A. (2010). Adaptive automation based on an object-oriented task model: Implementation and evaluation in a realistic C2 environment. *Journal of Cognitive Engineering and Decision Making*, *4*(2), 152–182. doi:10.1518/155534310X522851

Eirinaki, M., & Vazirgiannis, M. (2007). Web site personalization based on link analysis and navigational patterns. *ACM Transactions on Internet Technology*, *7*(4), 21, es. doi:10.1145/1278366.1278370

Fournier, L. R., Wilson, G. F., & Swain, C. R. (1999). Electrophysiological, behavioral, and subjective indexes of workload when performing multiple tasks: Manipulations of task difficulty and training. *International Journal of Psychophysiology*, *31*(2), 129–145. doi:10.1016/S0167-8760(98)00049-X PMID:9987059

Funke, D. J., Neal, J. G., & Paul, R. D. (1993). An approach to intelligent automated window management. *International Journal of Man-Machine Studies*, *38*(6), 949–983. doi:10.1006/imms.1993.1044

Galy, E., Cariou, M., & Melan, C. (2012). What is the relationship between mental workload factors and cognitive load types? *International Journal of Psychophysiology*, *83*(3), 269–275. doi:10.1016/j.ijpsycho.2011.09.023 PMID:22008523

Ghazarian, A., & Noorhosseini, S. M. (2010). Automatic detection of users' skill levels using high-frequency user interface events. *User Modeling and User-Adapted Interaction*, *20*(2), 109–146. doi:10.1007/s11257-010-9073-5

Hart, S. G., & Staveland, L. E. (1988). Development of NASA-TLX (Task Load Index): Results of empirical and theoretical research. In P. A. Hancock & N. Meshkati (Eds.), *Human Mental Workload* (pp. 139–183). Amsterdam: Elsevier Science. doi:10.1016/S0166-4115(08)62386-9

Hurst, A., Hudson, S. E., & Mankoff, J. (2007). *Dynamic detection of novice vs. skilled use without a task model*. Paper presented at the Proceedings of the SIGCHI conference on Human factors in computing systems, San Jose, California, USA. doi:10.1145/1240624.1240669

Hutchings, D. R., & Stasko, J. (2002). QuickSpace: New operations for the desktop metaphor. In CHI'02 extended abstracts on Human factors in computing systems (pp. 802-803). ACM. doi:10.1145/506443.506605

Hutchings, D. R., & Stasko, J. (2004). *Revisiting display space management: understanding current practice to inform next-generation design*. Paper presented at the Proceedings of Graphics Interface 2004, London, Ontario, Canada.

Jameson, A., Kiefer, J., Müller, C., Großmann-Hutter, B., Wittig, F., & Rummer, R. (2009). Assessment of a user's time pressure and cognitive load on the basis of features of speech. *Resource-Adaptive Cognitive Processes*, 171-204.

Jimison, H., Pavel, M., McKanna, J., & Pavel, J. (2004). Unobtrusive monitoring of computer interactions to detect cognitive status in elders. *IEEE Transactions on Information Technology in Biomedicine*, *8*(3), 248–252. doi:10.1109/TITB.2004.835539 PMID:15484429

Kalyuga, S., & Plass, J. (2009)Evaluating and Managing Cognitive Load in Games. Handbook of research on effective electronic gaming in education, 2, 719-737.

Khawaja, M. Chen, Fang, Owen, Christine, & Hickey, Gregory. (2009). Cognitive Load Measurement from User's Linguistic Speech Features for Adaptive Interaction Design. In Human-Computer Interaction–INTERACT 2009 (pp. 485-489). Springer Berlin Heidelberg.

Kramer, A. F. (1991). Physiological metrics of mental workload: A review of recent progress. *Multiple-task performance*, 279-328.

Lin, T., Imamiya, A., & Mao, X. (2008). Using multiple data sources to get closer insights into user cost and task performance. *Interacting with Computers*, *20*(3), 364–374. doi:10.1016/j.intcom.2007.12.002

Mandryk, R. L., & Atkins, M. S. (2007). A fuzzy physiological approach for continuously modeling emotion during interaction with play technologies. *International Journal of Human-Computer Studies*, *65*(4), 329–347. doi:10.1016/j.ijhcs.2006.11.011

Mandryk, R. L., Atkins, M. S., & Inkpen, K. M. (2006). A continuous and objective evaluation of emotional experience with interactive play environments. *Proceedings of the SIGCHI conference on Human factors in computing systems*, 1027-1036. doi:10.1145/1124772.1124926

Martínez, H. P., & Yannakakis, Georgios N. (2010). *Genetic search feature selection for affective modeling: a case study on reported preferences*. Paper presented at the Proceedings of the 3rd international workshop on Affective interaction in natural environments, Firenze, Italy. doi:10.1145/1877826.1877832

Misue, K., Eades, P., Lai, W., & Sugiyama, K. (1995). Layout Adjustment and the Mental Map. *Journal of Visual Languages and Computing*, *6*(2), 183–210. doi:10.1006/jvlc.1995.1010

Miyake, S. (2001). Multivariate workload evaluation combining physiological and subjective measures. *International Journal of Psychophysiology*, *40*(3), 233–238. doi:10.1016/S0167-8760(00)00191-4 PMID:11228350

Mogford, R. X., Guttman, J. A., Morrow, S. L., & Kopardekar, P. (1995). *The Complexity Construct in Air Traffic Control: A Review and Synthesis of the Literature*. Atlantic City: FAA Technical Center.

Paas, F., Tuovinen, J. E., Tabbers, H., & Van Gerven, P. W. M. (2003). Cognitive load measurement as a means to advance cognitive load theory. *Educational Psychologist*, *38*(1), 63–71. doi:10.1207/S15326985EP3801_8

Paas, F. G. (1992). Training strategies for attaining transfer of problem-solving skill in statistics: A cognitive-load approach. *Journal of Educational Psychology*, *84*(4), 429–434. doi:10.1037/0022-0663.84.4.429

Paas, F. G. W. C., & Van Merriënboer, J. J. G. (1994). Instructional control of cognitive load in the training of complex cognitive tasks. *Educational Psychology Review*, *6*(4), 351–371. doi:10.1007/BF02213420

Parasuraman, R., Mouloua, M., & Molloy, R. (1996). Effects of adaptive task allocation on monitoring of automated systems. *Human Factors: The Journal of the Human Factors and Ergonomics Society*, *38*(4), 665–679. doi:10.1518/001872096778827279 PMID:11536753

Partala, T., & Surakka, V. (2003). Pupil size variation as an indication of affective processing. *International Journal of Human-Computer Studies*, *59*(1-2), 185–198. doi:10.1016/S1071-5819(03)00017-X

Picard, R. W., Vyzas, E., & Healey, J. (2001). Toward machine emotional intelligence: Analysis of affectivephysiological state. *IEEE Transactions on Pattern Analysis and Machine Intelligence*, *23*(10), 1175–1191. doi:10.1109/34.954607

Plass, J. L., Moreno, R., & Brünken, R. (2010). *Cognitive load theory*. Cambridge, New York: Cambridge University Press. doi:10.1017/CBO9780511844744

Purchase, H., Hoggan, E., & Görg, C. (2007). How Important Is the "Mental Map"? – An Empirical Investigation of a Dynamic Graph Layout Algorithm. In Graph drawing (pp. 184-195). Springer Berlin /Heidelberg.

Quinlan, J. R. (1993). *C4.5: programs for machine learning*. San Mateo, Calif.: Morgan Kaufmann Publishers.

Reid, G. B., & Nygren, T. E. (1988). The Subjective Workload Assessment Technique: A Scaling Procedure for Measuring Mental Workload. In A. H. Peter & M. Najmedin (Eds.), Advances in Psychology (Vol. Volume 52, pp. 185-218): North-Holland. doi:10.1016/S0166-4115(08)62387-0

Rowe, D. W., Sibert, J., & Irwin, D. (1998). Heart Rate Variability: Indicator of User State as an Aid to Human-Computer Interaction. *Proceedings of the SIGCHI conference on Human factors in computing systems*, 480-487. doi:10.1145/274644.274709

Ruiz, N., Feng, Q. Q., Taib, R., Handke, T., & Chen, F. (2010). *Cognitive skills learning: pen input patterns in computer-based athlete training*. Paper presented at the International Conference on Multimodal Interfaces and the Workshop on Machine Learning for Multimodal Interaction, Beijing, China. doi:10.1145/1891903.1891955

Ruiz, N., Taib, R., & Chen, F. (2011). *Freeform pen-input as evidence of cognitive load and expertise*. Paper presented at the Proceedings of the 13th international conference on multimodal interfaces, Alicante, Spain. doi:10.1145/2070481.2070511

Ruiz, N., Taib, R., Shi, Y., Choi, E., & Chen, F. (2007). *Using pen input features as indices of cognitive load*. Paper presented at the Proceedings of the 9th international conference on Multimodal interfaces, Nagoya, Aichi, Japan. doi:10.1145/1322192.1322246

Samarasinghe, S. (2007). *Neural networks for applied sciences and engineering: from fundamentals to complex pattern recognition*. Boca Raton, FL: Auerbach.

Sweeney, M., Maguire, M., & Shackel, B. (1993). Evaluating user-computer interaction: A framework. *International Journal of Man-Machine Studies*, *38*(4), 689–711. doi:10.1006/imms.1993.1032

Sweller, J., Ayres, P., & Kalyuga, S. (2011). Measuring Cognitive Load. [). New York, NY: Springer New York.]. *Cognitive Load Theory*, *1*, 71–85.

Tan, D. S., Meyers, B., & Czerwinski, M. (2004). WinCuts: manipulating arbitrary window regions for more effective use of screen space. In CHI'04 extended abstracts on Human factors in computing systems (pp. 1525-1528). ACM. doi:10.1145/985921.986106

Vanderberg, R., & Lee Swanson, H. (2007). Which components of working memory are important in the writing process? *Reading and Writing*, *20*(7), 721–752. doi:10.1007/s11145-006-9046-6

Vizer, L. M., Zhou, L., & Sears, A. (2009). Automated stress detection using keystroke and linguistic features: An exploratory study. *International Journal of Human-Computer Studies*, *67*(10), 870–886. doi:10.1016/j.ijhcs.2009.07.005

Wierwille, Walter W, & Eggemeier, F Thomas. (1993). Recommendations for mental workload measurement in a test and evaluation environment. *Human Factors: The Journal of the Human Factors and Ergonomics Society*, *35*(2), 263–281.

Wilson, G. F., & Russell, C. A. (2003). Real-time assessment of mental workload using psychophysiological measures and artificial neural networks. *Human Factors: The Journal of the Human Factors and Ergonomics Society*, *45*(4), 635–644. doi:10.1518/hfes.45.4.635.27088 PMID:15055460

Wilson, G. M., & Sasse, A. (2004). From doing to being: Getting closer to the user experience. *Interacting with Computers*, *16*(4), 697–705. doi:10.1016/j.intcom.2004.06.001

Yin, B., Chen, F., Ruiz, N., & Ambikairajah, E. (2008). *Speech-based cognitive load monitoring system.* Paper presented at the International Conference on Acoustics, Speech and Signal Processing.

Yin, B., Ruiz, N., & Chen, F., & Khawaja, M. Asif. (2007). *Automatic cognitive load detection from speech features.* Paper presented at the Proceedings of the 19th Australasian conference on Computer-Human Interaction: Entertaining User Interfaces, Adelaide, Australia. doi:10.1145/1324892.1324946

Yin, B., Ruiz, N., Chen, F., & Ambikairajah, E. (2008). *Investigating speech features and automatic measurement of cognitive load.* Paper presented at the IEEE 10th Workshop on Multimedia Signal Processing. doi:10.1109/MMSP.2008.4665218

Yu, K., Epps, J., & Chen, F. (2011). *Cognitive load evaluation of handwriting using stroke-level features.* Paper presented at the Proceedings of the 16th international conference on Intelligent user interfaces, Palo Alto, CA, USA. doi:10.1145/1943403.1943481

ADDITIONAL READING

Beiser, F. C. (2002). *German Idealism: The Struggle Against Subjectivism, 1781-1801*. Harvard University Press.

Gamble, J., Strickland, A., & Thompson, A. (2007). *Crafting & Executing Strategy* (15th ed.). New York: McGraw-Hill.

Kalyuga, S. (2008). *Managing Cognitive Load in Adaptive Multimedia Learning*. Information Science Reference.

Kalyuga, S. (2009). *Cognitive Load Factors in Instructional Design for Advanced Learners*. Nova Science Publishers Inc.

McQuaid John. (2014). Effects of Cognitive Load on Participation of Online Learners. Scholars' Press

KEY TERMS AND DEFINITIONS

Adaptive User Interface (also known as AUI): Is a user interface (UI) which adapts, that is changes, its layout and elements to the needs of the user or context and is similarly alterable by each user.

Attention: Is the behavioral and cognitive process of selectively concentrating on one aspect of the environment while ignoring other things. Attention has also been referred to as the allocation of limited processing resources.

Behavioral Features: Include speech features, linguistic feature, pen stroke features and so on. Subjectivity: is the condition of being a subject: i.e., the quality of possessing perspectives, experiences, feelings, beliefs, desires, and power. It is used as an explanation for what influences and informs people's judgments about truth or reality.

Cognitive Load: Refers to the amount of mental effort required to process a given amount of information and has been associated with the limited capacity of working memory.

Cognitive Overload: Is a situation where others gives too much information or too many tasks to learners simultaneously, resulting in the learner being unable to process this information. In this situation, the language processing demands of an activity go beyond the language processing limits of the learner. It produces anxiety and stress, as well as affecting learning.

Event: In computing, an event is an action or occurrence detected by the program that may be handled by the program.

Performance Measures: Are the processes of collecting, analyzing and/or reporting information regarding the performance of an individual, group, organization, system or component. It can involve studying processes/strategies within organizations, or studying engineering processes/parameters/phenomena, to see whether output are in line with what was intended or should have been achieved.

Physiological Measures: Are techniques used to measure bodily variations either directly or indirectly; examples are measurements of heart rate, mean arterial pressure, and total lung capacity.

Chapter 11
How to Become a Cybercriminal?
An Explanation of Cybercrime Diffusion

Jean-Loup Richet
University of Nantes, France

ABSTRACT

The main purpose of this chapter is to illustrate a landscape of current literature in cybercrime taking into consideration diffusion of innovation theories and economic theory of competition. In this chapter, a narrative review of the literature was carried out, facilitators leading to cybercrime were explored and explained the diffusion of Cybercriminals' best practices. Cybercrime is compatible with young adults lifestyle (familiarity) and requires little knowledge. Moreover, barriers to entry related to costs (psychological, financial), risks and investments are low. This review provides a snapshot and reference base for academics and practitioners with an interest in cybercrime while contributing to a cumulative culture which is desired in the field. This chapter provides insights into barriers to entry into cybercrime and the facilitators of cybercrime.

INTRODUCTION

Cyberspace has created a new dimension of social interaction. It has transcended time and space, and, as such, physical context is no longer linked with social situation. A virtual presence need not be true to the actual persona of its creator in the physical world. This simple fact has had an alarming effect on the negative cyber behaviors of today's youth, who have used the anonymity of the web to indulge in cybercrime or hacking. It has become critical to inquire into and understand the growing criminal cyber-behavior of teenagers. This requires a detailed study of the meanings of and differences between hacking and cybercrime and the visualization and use of these terms by the youth alongside their attitudes towards both.

DOI: 10.4018/978-1-4666-8450-8.ch011

Further, a growing number of scholars state that the Internet presents "some unique opportunities for deviant behavior" (Rogers et al., 2006). Technology has given people the unprecedented ability to hide their identities under cover of anonymity, and they can avoid the penalty for embarrassing or illegitimate activity. Whereas few people (of any age) would be able to walk into a room full of complete strangers and share nude photos of themselves, talk about sex, or discuss illegal use of drugs, they can do it online behind the "protection" of the magically anonymous keyboard. This ability profoundly affects the online behavior of teenagers.

Nevertheless, although some researchers have studied this issue, the factors leading young adults to adopt a web-deviant behavior have received less attention. From this background, the present article sets out to explore the facilitators of hacking and cybercrime. This paper will explain the diffusion of web-deviant behavior amongst young people through an analysis of the literature study while taking into consideration the conceptual model of diffusion of innovation by Greenhalgh et al. (2004).

CYBERCRIME VS. HACKING

Cyberspace transforms the scale and scope of offense; has its own limits, interactional forms, roles, and rules; and it has its own forms of criminal endeavor (Capeller, 2001). According to Yar (2005), the "novel socio-interactional features of the cyberspace environment (primarily the collapse of spatial-temporal barriers, many-to-many connectivity, and the anonymity and plasticity of online identity) [...] make possible new forms and patterns of illicit activity." Anyone who is computer literate can become a cybercriminal.

There is still no clear definition of "cybercrime" (Fafinski et al., 2010). In some cases, cybercrime can encompass the use of computers to assist "traditional" offending but it can also be a crime mediated through technology (Wall, 2007) or an exclusive technological crime, such as a denial-of-service attack). Many criminal law scholars focus on the legalistic framework. For instance, Wall (2001) uses the categories of criminal law to create categories of cybercrime. Others categorize cybercrime as an offense "related to computers, related to content or against the confidentiality, integrity and availability of computer data and systems" (Council of Europe Convention on Cybercrime, 2001).

The use of the term "hacker" has changed over the years from a positive and complimentary definition — the enthusiastic computer programmer who is particularly brilliant — to a negative and pejorative one: the cybercriminal. Nowadays, "cybercriminal" is a term synonymous with "hacker." Hacker, as a term, is commonly used by the mass media to refer to an intruder breaking into computer systems to steal or destroy data. Police describe almost any crime committed through, with, by, or against a computer as "hacking." "For many people, the hacker is an ominous figure, a smart-aleck sociopath ready to burst out of his basement wilderness and savage other people's lives for his own anarchical convenience" (Sterling, 1993).

This concept of "hackers" is still the subject of heated controversy. In response to the common demonization of the term hacker, *The New Hacker's Dictionary* (Raymond & Steele, 1991) has coined the term "cracker." Crackers use their computer-security-related skills to author viruses, trojans, etc., and illegally infiltrate secure systems with the intention of doing harm to the system or criminal intent and to differentiate them from the original and non-criminal hacker. This article will use the term hacker in its original positive meaning and the term cracker for those committing cybercrime.

CONCEPTUAL MODEL AND METHODOLOGY

Originally, hacking was seen as innovative behavior; even the original term "hack" was a slang word used by MIT students in the 1950s to refer to an improvement. According to Rogers (1983), an innovation is "an idea, practice, or object that is perceived as new by an individual or other unit of adoption." Further, the central meaning of innovation relates to improvement or renewal, with novelty being a consequence of this improvement. However, eventually hacking led to many teenagers deviating from using their skills to improve or push the limits of their expertise to using those same skills for cybercrimes, such as Internet extortion and fraud. Our hypothesis is that hacking is an innovation compared to cybercrime and is different from either cyber or traditional criminal behavior. Computer hacking can lead to constructive technological developments; for example, former hackers Dennis Ritchie and Ken Thompson went on to create the UNIX operating system in the 1970s, and hacker Shawn Fanning created Napster. An analysis of literature through a model of diffusion of innovation highlights the facilitators of hacking.

We used the conceptual model of diffusion of innovation by Greenhalgh et al. (2004). The authors conducted a systematic review of empirical literature and theory pertaining to the diffusion of innovation. They studied 6,000 papers, books, and abstracts, focusing on 495 sources for the final report. They formulated a conceptual model from the synthesis of these theoretical and empirical findings; this model categorizes the critical success factors for innovation.

The key attributes of successful innovation indicated in their model involve multiple elements:

1. Compatibility — Must be consistent with the adopter's values and perceived needs.
2. Risk — Zero risk is highly unlikely, so we have modified this need to an acceptable level of risk, where the possibility and value of loss is inferior to the expected positive outcome.
3. Relative advantage — Better than the previous model in terms of effectiveness or cost-effectiveness.
4. Ability to solve user issues — Improving users' performance.
5. Observability — Must have visible benefits.
6. Accessibility — Must be easy to use and require only a little knowledge.
7. Adaptability — Must be modifiable to fit users' needs.
8. Support — Should be open to experiment and have a supporting community providing training and customization.

This research used this framework to analyze the present literature on cybercrime and hacking. An extensive narrative literature review (King & He, 2005) was conducted, synthesizing prior research and analyzing it with regard to the hypothesis. Traditionally, literature reviews target prominent journals and conferences. This approach is relevant to mature research topics but is not pertinent for a literature review on cybercrime, as it is a contemporary phenomenon; therefore, focusing on limited outlets cannot be justified. In light of this, this effort mainly focused on online databases and targeted Business Source Complete, ProQuest, ScienceDirect, Scirus, Scopus, and Web of Science. We conducted keyword and abstract searches across these six databases for all available years. Articles were filtered by scanning titles, abstracts, and excluding duplicates and articles only mentioning hacking or cybercrime that were not focused on these topics. After a full text review, irrelevant articles were discarded and we classified the remaining articles under each of the critical success factors described by Greenhalgh et al.

VALIDITY THREATS

This paper has several limitations. First, although the procedure followed includes searching, filtering, and classifying processes, our approach is vulnerable to subjectivity because it is based on a qualitative interpretation of the literature. Second, the sample was largely based on grey and professional literature. Although the search was oriented towards peer-reviewed and scholarly articles, the literature on hackers is recent, and because it is industry-driven — from the information systems security industry to national agencies — very few research articles were found that also embrace this phenomenon, compared to more mature literature like electronic commerce. Even if untrustworthy studies and non-rigorous articles are excluded from this review, the research method — a narrative literature review — might hinder the ability of the present paper to draw a complete picture of the current development in this field. Finally, the search criteria may be incomplete, as some papers that do not have the term cybercrime in the keywords or abstract may not have been included.

FROM YOUNG HACKERS TO CRACKERS

In this first part, we will study those two critical success factors: risk and observability.

Hacking is a normal part of the lives of most youngsters. Multiple surveys have been carried out on teenagers and hacking (Panda Security, 2009; Trend Micro, 2009; Tufin Technologies, 2010). Although there is suspicion about industry bias, these studies generally come to the same conclusion: casual hacking is almost as established a part of teenage life as downloading music to an iPod. The most common reasons cited for hacking were "for fun," (50%) followed by "curiosity" (30%). Over four out of ten teens have hacked into another person's account to read e-mails, looked at another's bank account details, or logged into another person's social networking profile. The majority of these young hackers are pranksters making fun of their friends, but one in three teens have admitted to being tempted to try hacking or spying on the Internet to make money (with eventual transition to cybercrime). The history of hacking is replete with juvenile hackers. Jonathan James was a fifteen-year-old American hacker arrested in 1999 for a series of intrusions into various systems from NASA to the Department of Defense. Michael Calce was a Canadian high school student who shut down Yahoo, Amazon, eBay, and other websites in 2000.

However, many youngsters have also turned into crackers committing serious criminal offenses. In 2005, eighteen-year-old Phillip Shortman from South Wales was sentenced for having fraudulently stolen £102,000 from eBay users (Griffith, 2005). In 2009, young Australian cracker Anthony Scott Harrison infected more than 3,000 computers to steal credit card and other bank account details with his malware in Australia and abroad (Fewster, 2010). In 2010, twenty-one-year-old Alistair Peckover was jailed for scamming, ID theft, and online fraud of around £110,000 (Blincoe, 2010).

In his economic approach to crime, G. Becker (1968) states that people decide whether to commit a crime through a cost-benefit analysis. According to this perspective, people choose illegal over legal alternatives in the same way they make consumption and economic choices in the marketplace. The decision to commit a crime involves calculations based on an assessment of the availability, risks, and costs of the opportunity and one's ability to reach a desired end. Low earnings or the lack of sufficient opportunities to obtain income are factors behind committing crime. Teenagers have lower earnings, fewer opportunities to obtain income, and may easily discount the future in assessing the opportunity costs of crime (Becker, 1976). Hacking is perceived as riskless and highly compatible to the lifestyles of the computer-savvy teenage population.

Cracking, too, is perceived as low risk or riskless — teenagers discount the future — and its benefits are apparent: how else could a teenager earn nearly £100,000 a year? The mass media indirectly promotes crackers' earnings, and even academic studies show that spam is highly profitable and that botmasters (individuals who are responsible for or maintain a herd of compromised computers) can earn roughly $3,500,000 a year (Kanich et al., 2009; TrendMicro, 2011). Although cybercrime is under-recorded and under-reported (Fafinski et al., 2010), studies highlight crackers' "huge profits" (Bizeul, 2007; Shah & Cole, 2005) For instance, Esthost botnet owner generated at least $14 million in fraudulent advertising fees (FBI, 2011). The aim of this paper is not to give statistical evidence but to point out media emphasis on crackers' revenue — Greenhalgh's observability factor, i.e., visible benefits of online crime compared to traditional crime. In fact, the reality of cybercrimes is under documented, and some researchers criticize statistical abuse and misuse in information security (Ryan & Jefferson, 2003; Florencio & Herley, 2011).

LOW BARRIERS TO ENTRY

In economic theories of competition, barriers to entry are obstacles that make it difficult to enter a given market (Fisher, 1984); technical expertise and accessibility could act as barriers to entry. However, in this case, it is observed that the barriers to entry into the "world of hacking" through technical expertise are low; hence, we could say that Greenhalgh's compatibility factor is present. Young people find hacking an easily accessible activity because hacking is compatible with their lifestyle (Richet, 2012). Computers are in everyone's lives, and the "digital natives" were born with Internet and a keyboard (Prensky, 2001). These digital natives make little distinction between the online and offline realms, the virtual and the real (Palfrey & Gasser, 2008). They may not even know that they behaved illegally (Kallman & Grillo, 1996); for example, searching on the web and downloading a movie from a hacking forum board may seem like a casual and insignificant activity. This is an unfamiliar mindset for many "digital immigrants" born before the 1980s who perceive hacking as being foreign and elusive because they are less familiar with contemporary perceptions of cyberspace. "Hackers are generally teenagers and college kids not engaged in earning a living. They often come from fairly well-to-do middle class backgrounds, and are markedly anti-materialistic" (Sterling, 1993).

KNOWLEDGE DIFFUSION AND INNOVATION

Furthermore, hacking is much less complex than it was a few years ago, as hacking communities have greatly diffused their knowledge through the Internet (Kshetri, 2006; Richet, 2013). The web has, therefore, progressively reduced the complexity of becoming a hacker. With a simple search on Google, anybody can find a large amount of documentation, including hacking how-tos, toolkits, and tips (Long, 2005). For instance, a Google search with the keywords "hacker toolkit" could lead one to a website where he or she could find all of the tools needed for hacking. With the query "how to hack," this researcher found easy entry to a website with a vast number of hacking guides. "By using Google, [one] can gain access to information that may otherwise be hidden" (Billig et al., 2008). The main problem is that the majority of tools used by information systems security auditors can be used by hackers, too, whether it be network protocol packet sniffers, network vulnerability scanners, password recovery tools, web server scanners, or packet crafters. Consequently, hackers disguise themselves with a disclaimer such as "this site is devoted to internet security, and we help people in the security field investigate tools they are unfamiliar with." Similarly, numerous books on hacking were published during the 2000s; there was even a *Hacking for Dummies*. Several hacking tools are available online and require very little knowledge (Krebs, 2006).

Blogs and communities have hugely contributed to information sharing: beginners could benefit from older hackers' knowledge and advice (Imperva, 2011). Teenagers could also easily adopt deviant behavior through experimentation. It is simple to try a network sniffer tool on a private network and then just use it on a public network out of sheer curiosity. A great number of hacking tools are very easy to use or completely user friendly (Krebs, 2006). For instance, with port scanners, the user only has to input the IP address of the computer to scan in order to get detailed descriptions for common ports.

Thus, we could say that cybercrime is accessible, i.e., according to Greenhalgh's definition, cybercrime is easy to do and requires little knowledge. Moreover, online communities not only provide support and training, but also offer customizations and opportunities for innovations (Imperva, 2011).

FITTING USERS' NEEDS

The last factor that contributes to lower barriers is that hacking is now cheaper than ever, and that could easily fit criminals' needs. For instance, cloud computing is a way to easily increase capacity or add capabilities without even investing in infrastructure. Hosted services are delivered over the Internet; they are sold on demand, are "elastic," and are fully managed by the provider. One needs nothing but a personal computer and Internet access. Infrastructure as a Service (IaaS) provides a virtual server with the ability to create, access, and configure virtual servers and storage. Cloud computing allows a student to pay for only as much capacity as is needed and bring more functionalities online as soon as they are required. Cloud computing could be helpful for a cracker as a way to leverage his attack (Subashini & Kavitha, 2011).

In other words, a cracker could use the IaaS massive computing power to brute-force a password (Meer et al., 2009). Brute force software can generate millions of passphrases, encrypt them, and see if they allow access to the network. Thus, cracking a wireless network could be done cheaply and easily (Roth, 2011). Because there are no bandwidth agreements — or detection of servers taking malicious actions — in order to set up an account on an IaaS, an amateur hacker can easily conduct a denial-of-service attack, sending a flood of packets toward the target company's network at low cost and on a large scale, using the high capabilities of the service provider (Ristanpart et al., 2009). In addition, scamming and spamming are facilitated. Indeed, before cloud computing, in order to spam one needed a dedicated server, skills in server management, network configuration and maintenance, knowledge of Internet service provider standards, etc. By comparison, a mail software as a service is a scalable, inexpensive, bulk, and transactional e-mail-sending service for marketing purposes and could be easily set up for spam (Krebs, 2008; Armbrust et al., 2010). On-demand e-mail service providers don't deserve to be blamed for the capabilities they are making accessible to potential criminals, but one should be aware that these services could easily be manipulated to send cheap spam and malware.

As mentioned above, teenagers are digital natives and are accustomed to using computers and the Internet. Cybercrime is thus compatible with their environment. Furthermore, hacking/cracking has become easier than ever nowadays through the diffusion of toolkits, advice, and support from the hacking community. Thanks to the ease of cloud computing, attempting to perform hacking is cheaper and requires fewer skills than ever before.

RELATIVE ADVANTAGES OVER TRADITIONAL CRIME

The psychological cost of committing a crime through the web is low for two main reasons. The first one is that victims of cybercrime are frequently difficult to identify (Phukan, 2002). This means that in massive scamming campaigns, for instance, hackers send an overwhelming amount of e-mails and do not face the victims (they effectively just press "send"). Alternatively, when a hacker creates and distributes malware through the web to steal credit card numbers, he catches "someone" (i.e., numbers) but never sees the victims' faces or physically interacts with them. The second reason is that hackers do not see their behavior as unethical. Scholars report that many students do not consider software piracy as unethical behavior (Im & Van Epps, 1991; Reid et al., 1992). On the contrary, some believe that unethical behavior could help them be successful in life (Davis & Vitell, 1991).

The main advantage of cybercrime over traditional crime is that "such crimes are less likely to be caught and prosecuted [...] only about 5% of cyber-criminals are caught" (Kshetri, 2009). Cybercrimes could be large-scale and highly profitable crimes; even a minuscule response rate to spam is enough to reward spammers with a large profit (Kanich et al., 2009). Moreover, cybercrimes are committed from home. For instance, instead of sifting through rubbish for personal information and searching in the target victim's residential trash (dumpster diving), an identity thief could browse social networking websites for personal details published by his or her target or could search online databases (Delaney, 2005).

Teenage crackers are aware of their victims' weak points, the first one being ignorance. In fact, a majority of computer users are still ignorant of computer security. A survey of home computers and their owners revealed a gap between user perception and the increasing prevalence of threats on the Internet (McAfee, 2010). The National Cyber Security Alliance (NCSA) is a non-profit group that seeks to raise public awareness of cyber security issues. This organization pointed out in 2004 that users thought they were safe from online threat despite the fact that a virus or worm infected 20% of home computers and that snooping programs such as spyware and adware were on an overwhelming 80% of systems (following the interview, technical experts examined home computers, looking for viruses and spyware).

The second weakness of most victims is fear. Businesses are frequently the victims of ransomware (Tsiklis, 2010), computer malware that holds a computer system or data hostage and then demands a ransom for its restoration. Ransomware is like a virus, but instead of destroying a file, it encrypts it, leaving the author of the malware as the only person with the knowledge of the "private decryption key" needed to release the system or data. The author of this crypto-viral extortion attack then offers to recover the key for a fee; businesses prefer to pay instead of being publicized or losing key data (Luo & Liao, 2009). Web extortions are increasing. In 2006, two teenagers attempted to extort $150,000 from MySpace (Evers, 2006). Recently, teens are being extorted in online schemes and both federal prosecutors and child safety experts have seen a rise in online sexual extortion, now called "sextortion." In 2010, Anthony Stancl, an eighteen-year-old student from Wisconsin, was sentenced to fifteen years in prison after prosecutors said he posed as a girl on Facebook to trick male high school classmates into sending him nude cell phone photos, which he then used to extort them (Ellis, 2010).

CONCLUSION

It was established that anonymity of cyberspace has several advantages. A hacker or cracker is less likely to be caught, feels no guilt or remorse after cracking or spamming, profits from his or her victims' ignorance, and inspires fear. Further, in *The Rites of Passage* (1960), Van Gennep states that there are rituals marking transitional phases between childhood and the full inclusion into a social group. Theses rites have three stages: separation, transition, and reincorporation.

Hacking is seen as a rite of passage from childhood to adulthood, as the teenager will come to consider hacking as a childish activity — just for pranks — and enter into the legal economy. This phase would be the departure from the status of teenager, during which the young adult is formally accepted into his or her new role. A great number of hackers are self-taught prodigies and some organizations actually employ hackers as part of their information technology staff. These hackers use their skills to find vulnerabilities in the company's information system(s) so that they can be fixed quickly. This type of hacker activity actually prevents serious cybercrimes. On the other hand, it could be seen that hacking has become a rite of aggregation, the detachment from childish cyber activities to the assumption of a new criminal cyber identity. Similarly, cracking is perceived as a rite of passage from the age of pranks, such as logging into a friend's Facebook profile, to the age of crimes – using another person's ID in on-line fraud. By learning the hacking subculture and language (Holt, 2010), hackers integrate themselves into the hacker social subculture, which is not necessarily criminal. By leaving the legal market, they become part of the cybercriminal underground.

The intention of this paper is to illustrate a landscape of current literature in cybercrime, taking into consideration Greenhalgh et al.'s conceptual model of diffusion of innovation. A narrative review of the literature was carried out, facilitators leading to cybercrime were explored and explained, and how hacking expertise is disseminated was discovered. Beginning with innocent hacking, teenagers could be tempted to crack and spy on the Internet to make money. The shift from non-criminal to criminal activity is facilitated by various factors: hacking compatibility with youth culture, relative advantage over traditional crime, ease of use, and support. This review provides a snapshot and reference base for academics and practitioners with an interest in the diffusion of cybercrime.

Webster and Watson (2002) suggested that the lack of review articles has been hindering the progress of the IS field; this article contributes to a cumulative culture that is desired in any research field, and answer a call for a greater focus on social artifact in information security research (Crossler et al., 2013; Willison and Warkentin, 2013). Although this review cannot claim to be exhaustive, it provides some insights into the diffusion of deviant behavior amongst young adults and into current literature on cybercrime.

ACKNOWLEDGMENT

This manuscript is a summation of enhanced papers published in the International Journal of Technology and Human Interaction (IJTHI) in 2013. The original article, *From Young Hackers to Crackers*, was published in volume 9, issue 3 of the IJTHI. Anabela Mesquita & Chia-Wen Tsai, Editors-in-Chief, IJTHI, selected this article for inclusion in this summation as a chapter titled *How to Become a Cybercriminal? An Explanation of Cybercrime Diffusion*. This chapter has been enhanced in order to allow the readers to benefit from the latest findings in the area since the first publication.

REFERENCES

Armbrust, M., Fox, A., Griffith, R., Joseph, A. D., Katz, R., Konwinski, A., & Zaharia, M. (2010). A view of cloud computing. *Communications of the ACM*, *53*(4), 50–58. doi:10.1145/1721654.1721672

Becker, G. (1968). Crime and punishment: An economic approach. *Journal of Political Economy*, *76*(2), 169–217. doi:10.1086/259394

Becker, G. (1976). *The economic approach to human behavior*. Chicago: University of Chicago Press.

Billig, J., Danilchenko, Y., & Frank, C. (2008). Evaluation of Google hacking. In *Proc. of the 5th Conf. on Information Security Curriculum Development*, 27-32. New York: ACM.

Bizeul, D. (2007). *Russian business networks study*. Retrieved from http://www.bizeul.org/files/RBN_study.pdf

Blicoe, R. (2010). High-living hacker swaps Porsche for porridge. *The Register*. Retrieved from http://www.theregister.co.uk/2010/06/18/hacker_jailed/

Capeller, W. (2001). Not such a neat net: Some comments on virtual criminality. *Social & Legal Studies*, *10*, 229–242.

Council of Europe Convention on Cybercrime. (2001). *Treaties*. Retrieved from http://conventions.coe.int/Treaty/EN/Treaties/html/185.htm

Crossler, R. E., Johnston, A. C., Lowry, P. B., Hu, Q., Warkentin, M., & Baskerville, R. (2013). Future directions for behavioral information security research. *Computers & Security*, (32), 90–101.

Davis, D. L., & Vitell, S. J. (1991). The ethical problems, conflicts, and beliefs of small business information personnel. *Journal of Computer Information Systems*, *22*(4), 53–57.

Delaney, K. (2005, March 29). Identity theft made easier. *Wall Street Journal*, p. B1.

Ellis, R. (2010, August 16). Sextortion: Are your teens being "sexploited" online? *Examiner*. Retrieved from http://www.examiner.com/cyber-safety-in-national/sextortion-are-your-teens-being-sexploited-online

Evers, J. (2006, May 2006). Teens arrested in alleged MySpace extortion scam. *Zdnet*. Retrieved from http://www.zdnet.com/news/teens-arrested-in-alleged-myspace-extortion-scam/148219

Fafinski, S., Dutton, W. H., & Margetts, H. Z. (2010). Mapping and measuring cybercrime. *OII Working Paper, 18.*

FBI. (2011). *Manhattan U.S. attorney charges seven individuals for engineering sophisticated internet fraud scheme that infected millions of computers worldwide and manipulated internet advertising business*. [Press release]. Retrieved from http://www.fbi.gov/newyork/press-releases/2011/manhattan-u.s.-attorney-charges-seven-individuals-for-engineering-sophisticated-internet-fraud-scheme-that-infected-millions-of-computers-worldwide-and-manipulated-internet-advertising-business

Fewster, S. (2010). Computer virus created by Anthony Scott Harrison sparked worldwide alert. *News*. Retrieved from http://www.news.com.au/technology/computer-virus-created-by-anthony-scott-harrison-sparked-worldwide-alert/story-e6frfrnr-1225897175334#ixzz1EiBSPNal

Fisher, F. M. (1979). Diagnosing monopoly. *The Quarterly Review of Economics and Business, 19*(2), 7–33.

Florencio, D., & Herley, C. (2011) Sex, lies and cyber-crime surveys. *Proceedings of the Tenth Workshop on the Economics of Information Security (WEIS)*, George Mason University, USA.

Greenhalgh, T., Robert, G., Macfarlane, F., Bate, P., & Kyriakidou, O. (2004). Diffusion of innovations in service organizations: Systematic review and recommendations. *The Milbank Quarterly, 82*(4), 581–629. doi:10.1111/j.0887-378X.2004.00325.x PMID:15595944

Griffith, H. (2005, June 7). eBay fraud boy con 'really easy.' *BBC News*. Retrieved from http://news.bbc.co.uk/1/hi/wales/4615117.stm

Holt, T. (2010). Examining the role of technology in the formation of deviant subcultures. *Social Science Computer Review, 28*(4), 466–481. doi:10.1177/0894439309351344

Im, J. H., & van Epps, P. D. (1991). Software piracy and software security in business schools: An ethical perspective. *Database, 22*(3), 15–21.

Imperva (2011). Monitoring hacker forums. *Research Note, 5*, 1-14.

Kallman, E. A., & Grillo, J. P. (1996). *Ethical decision making and information technology*. New York: McGraw Hill.

Kanich, C., Kreibich, C., Levchenko, K., Enright, B., Voelker, G., Paxson, V., & Savage, S. (2009). Spamalytics: An empirical analysis of spam marketing conversion. *Proceedings of the 15th ACM Conference on Computer and Communications Security*. Alexandria, Virginia: ACM doi:10.1145/1562164.1562190

King, W. R., & He, J. (2005). Understanding the role and methods of meta-analysis in IS research. *Communications of the Association for Information Systems, 16*, 1.

Krebs, B. (2006, March 16). Hacking made easy. *Washington Post*. Retrieved from http://www.washingtonpost.com/wp-dyn/content/article/2006/03/16/AR2006031600916.html

Krebs, B. (2008, July 1). Amazon: Hey spammers, get off my cloud! *Washington Post*. Retrieved fromhttp://voices.washingtonpost.com/securityfix/2008/07/amazon_hey_spammers_get_off_my.html

Kshetri, N. (2006). The simple economics of cybercrimes. *IEEE Security and Privacy, 4*(1), 33–39. doi:10.1109/MSP.2006.27

Kshetri, N. (2009). Positive externality, increasing returns, and the rise in cybercrimes. *Communications of the ACM, 52*(12), 141. doi:10.1145/1610252.1610288

Long, J. (2005). *Google hacking for penetration testers*. Rockland, MA: Syngress Publishing.

Luo, X., & Liao, Q. (2009). Ransomware: A new cyber hijacking threat to enterprises. In J. Gupta & S. Sharma (Eds.), Handbook of Research on Information Security and Assurance (1–6), Hershey, PA: IGI Global. doi:10.4018/978-1-59904-855-0.ch001

McAfee. (2010). *Threats report: Third quarter 2010*. Retrieved from http://www.mcafee.com/us/resources/reports/rp-quarterly-threat-q3-2010.pdf

Meer, H., Arvanitis, N., & Slaviero, M. (2009). *Clobbering the cloud. Black Hat USA 2009*. Las Vegas, NV: Techweb.

Palfrey, J., & Gasser, U. (2008). *Born digital: Understanding the first generation of digital natives*. New York: Basic Books.

Panda Security. (2009). *Studying or hacking? Today's adolescents could be the hackers of the future*. Retrieved from http://www.pitchengine.com/preview-release.php?id=11537

Phukan, S. (2002). *IT ethics in the internet age: New dimensions. Proc. Informing Science & IT Education Conf. (InSITE)*. Santa Rosa, CA: Informing Science Press.

Prensky, M. (2001).Digital natives, digital immigrants. On the horizon. MCB University Press, Vol. 9 No. 5.

Raymond, E. S., & Steele, G. S. (1991). *The new hacker's dictionary*. Cambridge, MA: MIT Press.

Reid, R. A., Thompson, J. K., & Logsdon, J. L. (1992). Knowledge and attitudes of management students toward software piracy. *Journal of Computer Information Systems*, *33*(1), 46–51.

Richet, J. L. (2012). Market Failure Mechanisms in Cybersecurity. *pre-ICIS workshop on Information Security and Privacy (SIGSEC)*. Paper 26. http://aisel.aisnet.org/wisp2012/26

Richet, J. L. (2013). From Young Hackers to Crackers. *International Journal of Technology and Human Interaction*, *9*(3), 53–62.

Richet, J. L. (2013). Laundering Money Online: a review of cybercriminals methods. *arXiv preprint arXiv:1310.2368*.

Ristanpart, T., Tromer, E., Sacham, H., & Savage, S. (2009) Hey, you, get off of my cloud: Exploring information leakage in third-party compute clouds. *Proceedings of the 16th ACM Conference on Computer and Communications Security*. Chicago, IL: ACM. doi:10.1145/1653662.1653687

Rogers, E. M. (1983). *Diffusion of Innovations*. New York: Free Press.

Rogers, M., Smoak, N., & Liu, J. (2006). Self-reported deviant computer behavior. *Deviant Behavior*, *27*(3), 245–268. doi:10.1080/01639620600605333

Roth, T. (2011). Breaking encryptions. *Black hat conference 2011*. Retrieved from https://media.blackhat.com/bh-dc-1/Roth/BlackHat_DC_2011_Roth_Breaking%20encryptions-Slides.pdf

Ryan, J., & Jefferson, T. I. (2003). The use, misuse, and abuse of statistics in information security research. *Proc. 23rd ASEM National Conference*. St. Louis, MO: ASEM.

Shah, S., & Cole, D. (2011). Spyware/adware. *Black hat conference 2005*. Retrieved from http://www.blackhat.com/html/bh-japan-05/bh-jp-05-en-speakers.html

Snyder, F. (2001). Sites of criminality and sites of governance. *Social & Legal Studies*, *10*, 251–256.

Sterling, B. (1993). *The hacker crackdown: Law and disorder on the electronic frontier*. New York: Bantam.

Subashini, S., & Kavitha, V. (2011). A survey on security issues in service delivery models of cloud computing. *Journal of Network and Computer Applications*, *34*(1), 1–11. doi:10.1016/j.jnca.2010.07.006

Trend Micro. (2009). Brits breeding the next-generation of computer hackers. *Global Security Mag*. Retrieved from http://www.globalsecuritymag.com/Trend-Micro-Brits-Breeding-the,20090403,8329

Trend Micro. (2011). *Big botnet busts*. Retrieved from http://blog.trendmicro.com/big-botnet-busts/

Tsiklis, T. (2010). *Avira-Ransomware threatens with official complaint of piracy*. Retrieved from http://www.virus.gr/portal/en/content/avira-ransomware-threatens-official-complaint-piracy

Tufin Technologies. (2010). *Survey of hacking habits in New York*. Retrieved from http://www.tufin.com/news_events_press_releases.php?index=2010-04-14

Van Gennep, A. (1960). *The rites of passage*. Chicago, IL: The University of Chicago Press.

Wall, D. (2001). Cybercrimes and the internet. In D. Wall (Ed.), *Crime and the internet*. London: Routledge. doi:10.4324/9780203164501_chapter_1

Wall, D. S. (2007). *Cybercrime: The transformation of technology in the networked age*. Cambridge: Polity Press.

Webster, J., & Watson, R. T. (2002). Analyzing the past to prepare for the future: Writing a literature review. *Management Information Systems Quarterly*, *26*(2), iii–xiii.

Willison, R., & Warkentin, M. (2013). Beyond deterrence: An expanded view of employee computer abuse. *Management Information Systems Quarterly*, *37*(1).

Yar, M. (2005). The novelty of cybercrime: An assessment in light of routine activity theory. *European Journal of Criminology*, *2*(4), 407–427. doi:10.1177/147737080556056

KEY TERMS AND DEFINITIONS

Botnet: The word botnet was termed by combining the words robot and network. A botnet is a network of infected computers, that will be used to participate in DDoS attacks.

Crackers: Crackers use their computer-security-related skills to author viruses, trojans, etc., and illegally infiltrate secure systems with the intention of doing harm to the system or criminal intent.

DDoS Attacks: Distributed denial-of-service (DDoS) attacks occurs when a target system or resource is overwhelmed with traffic from multiple sources.

Dumpster Diving: Activity realized by identity thieves; it is the act of sifting through rubbish for personal information and searching in the target victim's residential trash. In the cyberspace, this activity is realized through the browsing of social networking websites for personal details published by the cracker's target.

Ransomware: Computer malware that holds a computer system or data hostage and then demands a ransom for its restoration. Ransomware is like a virus, but instead of destroying a file, it encrypts it, leaving the author of the malware as the only person with the knowledge of the "private decryption key" needed to release the system or data. The author of this crypto-viral extortion attack then offers to recover the key for a fee; businesses prefer to pay instead of being publicized or losing key data.

Chapter 12
From Relationship to Information:
A Study of Twitter and Facebook Usage in Terms of Social Network Size among College Students

Chen Yang
Bowling Green State University, USA

Gi Woong Yun
Bowling Green State University, USA

Louisa Ha
Bowling Green State University, USA

Lanming Chen
Bowling Green State University, USA

ABSTRACT

By measuring social media users' online activities in terms of information broadcasting, information seeking and relational maintenance, this chapter aimed at investigating how college students' SNS usage patterns may affect their online social network sizes. Statistics in this chapter suggested that more information seeking leads to more Twitter followings while more information broadcasting results in a larger size of Twitter followers. The frequency of relationship management, however, did not predict students' social network size on either Twitter or Facebook. Implications of the study were discussed.

INTRODUCTION

Facebook and Twitter are two predominant social networking services (SNSs) in North America, especially among college students. The two SNSs differ from each other in many aspects including the way users' profiles are linked to each other. While Facebook users subscribe to other people's updates based on mutual agreement on being each other's friends, Twitter users can establish a one-way linkage to another user's profile without that person's consent or even notice. In other words, Twitter allows its users to have one-directional profile linkage so that individuals can follow others freely without being

DOI: 10.4018/978-1-4666-8450-8.ch012

followed and may also be followed by other Twitter users without having to follow them back. If two users establish a mutual following connection, it is similar to Facebook's "friends". As a result, there may be a wide range of differences between the sizes of a Twitter user's followings and followers.

The study in this chapter aims at exploring how college students' SNS usage patterns may be correlated to the sizes of their Twitter followings and followers and to the number of Facebook friends. This inquiry has both theoretical and pragmatic significance. First, not many studies have focused on the meaning of the exact numbers of individuals' social network sizes in their interpersonal communication and social life. This is partly due to the fact that, in a traditional social environment, relatives, friends and acquaintances people interact with on a daily basis are trivial and not easily measurable. Since SNSs have made the interaction among friends in their social networks visible and comparatively stable by presenting a chart of countable links, daily interaction dynamics can be monitored, recorded, and tabulated. It is important to investigate what these easily available numbers mean and what factors may influence the size of SNS connections. Second, this question will suggest new applications for the Uses and Gratifications theory in the field of social media. Although SNSs are best known as a virtual platform for people to interact with each other, students' usage of SNSs may go beyond the purpose of maintaining interpersonal relationships in an extended digital world. The current trend is more and more young people, college students in particular, are using SNSs for entertainment, self-promotion, job hunting, news feeding, references and others (Dunne, Lawlor, & Rowley, 2010; Sheldon, 2008; Urist, Dong, & Day, 2009). While it is yet to be discovered whether varied needs can lead to varied structures of social networks for users with diverse motivations, the intentional or unintentional strategies people adopt to increase the size of the network investigated here while dealing with relationship will provide some insights into this topic. Third, this chapter can contribute to the previous literature centering on the influences of individual characteristics on SNS usage patterns related to opinion leadership and opinion seeking in SNS (Shoham & Ruvio, 2008). Better conceptualization of the two constructs will help understand the structural differences of relationship among different social media services and uses.

BACKGROUND

Social Network Services

Social Network Services (SNSs) are web-based services that allow individuals to construct a public or semi-public profile with personal stories in texts, pictures, and videos, as well as views and comments on other profiles within a certain online system (Boyd & Ellison, 2007; Cheung, Chiu, & Lee, 2011; Lewis & West, 2009). According to Boyd and Ellison (2007), the emergence of SNSs represents a breakthrough in the organization of online communities as these networks are no longer interest-based, but person-centered. Specifically, SNSs are structured around profiles which represent users' selves and visualize their social networks to others with the intention of contacting or being contacted (Gross, Acquisti, & Heinz, 2005).

Researchers reported that SNSs are particularly prevalent among college students whose main objectives of SNSs usage are social connection and interaction (Gross et al., 2005; Lewis & West, 2009). SNSs allow them to establish and manage their personal accounts, and to share information including status updates, links, and photos with others through the linkage between profiles (Lewis & West, 2009). Ellison, Steinfield, and Lampe (2007) suggested that SNS usage might help users who have low self-esteem and low life satisfaction to overcome barriers and, therefore, be beneficial to their psychological well-beings. Admittedly, the communication ties fostered by SNSs such as Facebook are characterized by broadness, weakness, low-pressure, and low commitment (Lewis & West, 2009).

Among numerous SNSs, Facebook and Twitter are two remarkable ones in terms of the number of their users as well as their undeniably increasing influence on commerce, politics, and society in general. In 2006, Facebook was ranked the seventh most visited website in terms of total page views (Cassidy, 2006). In September 2011, Twitter announced that it has 100 million active users logging in at least once a month and 50 million active users every day (Taylor, 2011). Different from Facebook and many other SNSs, Twitter specializes in instant updates of information in a short text-based form, known as micro-blogging. Furthermore, Twitter is similar to a news wire service in term of distributing news because the co-founder of Twitter, Biz Stone, has been contemplating the concept of a Twitter news network since 2010 (Berkow, 2010). This feature is reinforced by Twitter's strategy that focuses on asking users for status updates "what's happening" instead of "what you are doing." Marwick and Boyd (2011) claimed that Twitter users rely on technology to reach a networked audience by creating and exchanging content in a many-to-many model. Since Facebook and Twitter have their distinct company mottos, network designs and marketing strategies, it is not surprising to assume that Twitter's popularity in America may be a dramatically different success story from that of Facebook.

SNS Uses and Gratifications

The Uses and Gratifications theory (U&G) proposes that individuals seek various mass media outlets according to their particular needs and, therefore, develop different patterns of media use (Rubin, 1994). Contrary to traditional media research, which assumes the audience to be a passive mass, U&G regards the audience as active consumers of media who select and synthesize certain media content to meet their own interests and needs (Baran & Davis, 1995; Katz, Blumler, & Gurevitch, 1974).

The Uses and Gratification theory has grown into a solid theoretical foundation for a large quantity of studies focusing on audience behavior, thanks to numerous scholars' contributions largely in the field of traditional media. McQuail, Blumler, and Brown (1972) put forward four main clusters of gratifications while investigating television audiences: diversion, personal relationships, personal identity, and surveillance. Diversion includes escape from the constraints of routine, the burdens of problems, and from emotional release. Diversion can happen if people watch a TV show with a story unfolding in an attractive fantasy world; it can also happen if a TV show engages audiences so intensely that they temporarily forget about their daily problems or relieve their oppressive feelings. Personal relationships refer to companionship and social utility. Its gratification happens in two situations: a TV program can foster a vicarious relationship between TV viewers and media personality, a tendency called "para-social relationship" (Horton & Wohl, 1956); the TV program may strengthen social interaction in real life by creating a common activity for family and friends to view the show together or by serving as a subject of daily conversation. Personal identity includes personal reference, reality exploration, and value reinforcement. Personal reference occurs when people use the program content as a reference for their own life or situation; reality exploration occurs when viewers use TV content to stimulate ideas about their own issues in real life; personal identity happens if TV viewers use a program to advocate or reinforce certain values. Surveillance, derived from curiosity and a desire for security, reflects people's disposition to get information and opinions about important events happening in the world. Although the four gratifications were targeted at television viewing behavior only at the time of their creation, this categorization has been acknowledged by Katz, Blumler, and Gurevitch (1973) as it "adequately reflects the full range of functions" (p. 513). Katz, Haas, and Gurevitch (1973) later integrated these gratifications into individuals' need to connect, or disconnect, with different kinds of others, e.g., people and nation.

U&G has been applied to a variety of media types and has recently been brought into the context of digital media, in particular, the study of SNSs (Dunne, Lawlor, & Rowley, 2010). As far as Internet users are concerned, several gratifications have been identified and studied by communications scholars. Papacharissi and Rubin (2000) proposed that email and chatrooms could provide a new platform to satisfy individual's interpersonal communication gratifications. Stafford and Gonier (2004) considered web browsing, information acquisition, interpersonal communication, and socialization as the major gratification related to Internet use. SNS users are not dramatically different from the general Internet users in that they log on social media to fulfill a variety of similar needs. Dunne et al. (2010) contended that U&G fits well into the SNSs context because the interactive nature of such sites and the extremely active role SNSs users play nullifies the traditional criticisms towards the validity of the assumption of active audience (e.g. Lometti, Reeves, & Bybee, 1977). The main gratifications associated with SNSs usage include meeting individuals' need for distributing and obtaining information (Ray, 2007; Song, LaRose, Eastin, & Lin, 2004), fostering supportive relationships (Dominick, 1999), and accumulating their social capital through network-building (Ellison et al., 2007). In addition, SNSs can fulfill users' needs for entertainment, surveillance, and diversion (Eastin, 2002; Ray, 2007). Park and Lee (2014) conducted a survey study of college students' usage of Facebook and found out that four motivations were significantly associated with Facebook intensity, i.e., entertainment, relationship maintenance, self-expression, communication, and impression management.

Based on previous literature, we can classify people's needs of using social media into four basic categories according to their different specializations. The first need is information seeking, i.e., the need for acquiring various information about the society, local community, science and technology, health, politics, shopping and etc. This need is satisfied by reading news or shopping information, checking friends' updates, reading other's posts, and reading comments. The second need is information broadcasting, i.e., the need for spreading news, information, jokes, opinions, personal interests, fashion and others. This need is fulfilled by updating one's status, uploading pictures and videos, posting links to websites, and sharing news, hobbies or personal thoughts on social media. The third need is relational maintenance, i.e., the need to interacting with family, relatives, friends, co-workers, acquaintances and even strangers. This need is satisfied by chatting, emailing, replying to posts, tagging friends in photos, liking and commenting others' posts, adding new friends and etc. The fourth need is entertainment seeking, i.e., the need to entertain oneself by playing games, listening to music, reading jokes and fun stories and others.

The four basic needs identified above showed their compatibility with the previous four main clusters of gratifications raised by McQuail and colleagues (1972) even though their typology was put forward 40 years ago with an intention to analyze television audiences only. Surveillance is about monitoring major events and important information around the world; personal identity emphasizes people's desire to use media's information or knowledge to guide personal life. Therefore, both the needs for surveillance and personal identity are related and similar to information seeking. Diversion, as a way of offering audiences emotional escape from harsh reality, can be associated with the need for entertainment seeking. The gratification of personal relationships also matches the need for relational maintenance. The only need that has not been addressed by McQuail and colleagues (1972) is the need for information broadcasting. Since the four gratifications were proposed in an era dominated by traditional media (such as newspaper, TV and radio) with very little interactive feature, it is understandable that this particular need was not addressed before. The inclusion of a new gratification in our analysis indicates that media users' gratifications can evolve because the development of new technologies makes it possible to identify and cultivate new needs among the audiences.

Driven by the four needs, social media users' online behaviors will demonstrate four different emphases, i.e. information broadcasting, information seeking, relationship maintaining and entertainment seeking. Although the four emphases are not mutually exclusive and sometimes may co-exist, they help media scholars conceptually label SNS users one or more of the following: information broadcasters, information seekers, relationship keepers and entertainment seekers. Entertainment seekers are not the major concern of this chapter since entertainment itself is personal and not interactive in nature. However, many applications for pastime on SNSs became interactive in the sense that the activity of using them for entertainment is automatically shared on social media as a status update or an invitation. In other words, although each of the four needs has its associated activities, the four categories are not clear-cut in serving one particular need. For example, liking a person's post is mainly considered as a way to strengthen friendship with another user on Facebook, but can also be a way of promoting the message in that post. Likewise, listening to music, usually an individual behavior of entertainment can be a social activity since software like Spotify can automatically share people's listening history on Facebook. Despite the few activities bordering on serving multiple gratifications, it is possible that, motivated by the needs for information distributing, information seeking, and relational maintenance, Twitter users could customize their SNS platform by intentionally reconstructing their networks structure, resulting in different sizes of followers and followings. As far as Facebook users are concerned, since many use Facebook as a medium to maintain social networks in a virtual world, we expect the composition of Facebook friends may reflect users' relationship management activities in certain ways.

Opinion Leadership

Opinion leaders are individuals who possess superior status, education, and social prestige and are informally recognized as experts in their fields by their social networks (Li et al., 2013; Weimann, Tustin, Vuuren, & Joubert, 2007). They are usually regarded as information sources and are found to play a critical role in facilitating information exchange, influencing the attitudes and behaviors of others (i.e., opinion followers), and improving group and community performance (Cascio & Shurygailo, 2002; Feick & Price, 1987; Li et al., 2013). Previous literature (Baumgarten, 1975; Rogers, 1983; Shoham & Ruvio, 2008; Summers, 1970; Venkatraman, 1989) has identified opinion leaders as technically competent, socially active, self-conscious, self-confident, and self-centered. They tend to be more involved in social activities, have remarkable individual opinions, and can help accelerate the adoption and diffusion of innovations (Gao et al., 2005). Other scholars found that opinion leaders tend to be more innovative (Myers & Robertson, 1972; Summers, 1970), more venturesome (Taylor, 1977) and less dogmatic (Goldsmith & Goldsmith, 1980; Jacoby, 1971).

Started in the field of sociology and social psychology, opinion leadership was applied to the study of political life and marketing in traditional media channels. However, this construct has recently been introduced into the context of SNSs. According to Kim and Tran (2013), marketers have begun to realize that the opinions people share and the recommendations they make on SNSs regarding certain products and services are, to a large extent, trusted by others. As a result of recognizing the power of Word-of-Mouth (WOM) marketing messages, an increasing number of marketing managers are getting involved in social network marketing in recent years (Kim & Tran, 2013). Since opinion leaders are more influential in leading and causing attitudinal and behavioral changes within certain networks, advertisers need to first identify this special SNS population who can be of tremendous help for their marketing endeavor. Thus, we propose that individuals who are more engaged in information broadcasting behaviors on Twitter are likely to be opinion leaders on the site by attracting a larger size of followers. The following hypothesis is proposed:

H1: The size of followers a SNS user has on Twitter is positively correlated with the user's information broadcasting activities. In other words, the more Twitter users engage in information broadcasting behaviors, the more followers they will have.

Opinion Seeking

Another construct co-existing with opinion leadership is opinion seeking. Within the field of marketing, opinion seekers are defined as individuals who actively search out and rely on information or opinions from knowledgeable interpersonal sources, i.e., opinion leaders, in order to understand current affairs and make purchase decisions (Feick, Price, & Higie, 1986; Flynn, Goldsmith, & Eastman, 1996). Assael (1992) found that opinion seekers tend to place a higher credibility on word-of-mouth messages and recommendations of acquaintances than information obtained from commercial sources (cf. in Shoham & Ruvio, 2008). As observed by Shoham and Ruvio (2008), while opinion leaders are viewed as information sources, opinion seekers are seen as information targets; while opinion leaders use a variety of media and non-media sources, opinion seekers are likely to rely on expert opinions to a greater extent. Therefore, this study proposes that opinion seekers, those who are more involved in information seeking behaviors, are likely to follow more users on Twitter in order to meet their gratification for diverse and accurate information. Specifically, it is hypothesized that:

H2: The size of followings a SNS user has on Twitter is positively correlated with the user's information seeking activities. In other words, the more Twitter users engage in information seeking behaviors, the more Twitter users he/she will follow.

Relationship Maintenance

Because of the one-to-many, instead of one-to-one, mode of communication, SNSs naturally serve as an efficient and convenient vehicle for maintaining and strengthening offline friendships within a large and diverse group of acquaintances (Ross et al., 2009). Young (2011) found out that adults use Facebook status updates and photo postings as an alternative and economical way to engage with their friends and strengthen pre-existing relationships when face-to-face or telephone communications are not possible or efficient.

In particular, some research has showed that college students' main motive for using SNSs is interpersonal communication satisfaction, e.g., relationship formation, maintenance, and enforcement, and social approval and support (Sheldon, 2008; Urista et al., 2009). Ellison and colleagues (2007) argued that, via SNSs, college students manage to keep connections with their previous community members, as well as bond and bridge with new college communities. Cummings, Lee, and Kraut (2006) argued that, when going to college, students' high school friendships are at risk of coming to an end because of increased geographic distance, decreased interaction, and declined psychological connection. Fortunately, engaging in computer-mediated communication with previous communities helps slow down the decline in contact and, therefore, maintains friendship over long distances (Cummings et al., 2006). Since interpersonal relationship development is a reciprocal process that calls for the involvement of both parties, it is hypothesized that the frequency of relational maintenance activities will influence the sizes of both Twitter followers and followings. Based on the reciprocal nature of interpersonal communication, Twitter users who are more engaged in relationship management may have relatively similar sizes of followers and followings. Specifically, the following hypothesis is proposed:

H3: A Twitter user who engages more in relational maintenance activities in SNS is likely to have similar sizes of followers and followings. That is to say, the more relational maintenance behaviors a SNS user has, the smaller difference in size between followings and followers he will have on Twitter.

As far as Facebook is concerned, since Facebook friends are based on mutual approval, we would expect the number of Facebook friends increases with the frequency of users' relational management activities. Specifically, the following hypothesis is proposed:

H4: A Facebook user who engages more in relational maintenance activities is likely to have a larger number of Facebook friends. That is to say, the more relational maintenance behaviors SNS users have, the more Facebook friends they will have.

METHOD

Participants

Because college students are the most likely users of Twitter and Facebook, they are chosen as participants of the study. During the 2012-2013 academic year, a sample of college students from a public university in the Midwest United States were recruited. The data for this study were collected using a self-administered online questionnaire from two large introductory lecture classes and 32 small general education classes. Students who participated received extra class credits as their incentive. On average, the respondents were able to complete the survey in 15-20 minutes. The survey covered a variety of questions concerning college students' opinions and usage of social networks sites, with a focus on the two most popular SNSs, Twitter and Facebook. Demographic data such as gender, age, GPA, ethnicity and others, were also collected. A total of 394 responses from college students were received.

Measures

The following variables are measured by the survey.

- ***Twitter Followers***: the number of people a respondent reported to be followed by on Twitter.
- ***Twitter Followings***: the number of people a respondent reported to follow on Twitter.
- ***Facebook Friends:*** the number of people a respondent reported to have as Facebook friends.

- ***Difference between Twitter followings and followers***: the absolute value of the difference between the sizes of Twitter followings and followers.
- ***Information broadcasting activities:*** the frequency of a user's activities that feature the spread of news, knowledge, opinions and personal life. Respondents were asked to rate the frequency of seven activities including "post news content from other news media," "post pictures taken by myself or people I know," "post picture from other sites," "post videos made by myself or people I know," "post video from other sites," "post comments to, or share something a friend has posted," and "post product review/comments." A 5-level Likert scale was applied to measure the frequency of each activity, indicating the levels of "almost daily," "several times a week," "once a week," "once a month or less," and "never." An average score for the information broadcasting activities was calculated for the data analysis.
- ***Information seeking activities:*** the frequency of a user's activities that feature the search and acquisition of news, knowledge, opinions and other information. Respondents were asked to rate the frequency of various activities including "link to other media sites," "read news posted on the site," "read comments or posts by celebrities, politicians or athletes," "find some useful shopping information, such as coupon or deal information," and "learn something useful from others." A 5-level Likert scale, which is the same with the previous one, was applied to measure the frequency of each activity and an average score was calculated for the data analysis.
- ***Relational maintenance activities:*** the frequency of a user's activities that feature the establishment, maintenance and reinforcement of relationships with family, friends, colleagues and other acquaintances. Respondents were asked to rate the frequencies of activities including "post pictures/videos taken by myself or people I know," "stay in touch with family members," "stay in touch with close friends," "connect with old friends that you lost touch with," "make new friends," "connect with other people who share your hobbies or interests," "find potential romantic partners or people you might like to date," "post comments to, or share something a friend has posted," "send instant messages to or chat with a friend through the social networking site," "tag people in posts, photos or videos," and "receive encouraging messages from others." A 5-level Likert scale, which is the same with the previous ones, was applied to measure the frequency of each activity and an average score was calculated for the data analysis.
- ***Demographic variables:*** age (the actual age of the respondents), gender (male = 1, female = 0), income (from 1 to 5, with 1 indicating "under US$ 30,000 per year" and 5 indicating "over 150,000 per year"), GPA (from 1 to 7, with 1 indicating "higher than 3.5" and 7 indicating "lower than 1"), marital status (sing = 0, married = 1) and ethnicity (white = 1, others = 0).

RESULTS

For H1, a regression analysis model was applied to investigate the relationship between respondents' Twitter followers and their information-broadcasting behavior. Before the regression analysis, a few important demographic variables of the respondents were entered into the model to test their significances. Although it is a student sample, the collected data indicated that the ages of the students ranged from 18 to 38 year old. The 20-year age range suggested that age and marital status might be potential variables that would make a difference and thus were included in the model as control variables too. The data analysis (as in Table 1) suggested that there are significant differences in the number of Twitter followers in terms of

Table 1. Hierarchical Regression Model Predicting the Number of Twitter Followers (N = 394)

	Model 1 (R^2=0.10, F=6.87, p<0.01)			Model 2 (R^2=0.12, F=7.21, p<0.01)		
Variable	β	t	p	β	t	p
Income	.13*	2.44*	.02*	.12*	2.28*	.00
Age	-.19*	-3.68*	.00*	-.18*	-3.42*	.02*
GPA	-.00	-.022	.98	-.00	-.06	.00*
Gender	-.06	-1.23	.22	-.06	-1.18	.95
Marital Status	-.00	-.07	.95	-.00	-.03	.24
Ethnicity	-.24*	-4.58*	.00*	-.22*	-4.22*	.97
Information Broadcasting				.15*	2.90*	.00*

the users' income (beta = 0.13, p = 0.02), age (beta = - 0.19, p < 0.01) and ethnicity (beta = - 0.24, p < 0.01). However, respondents' gender, GPA and marital status did not significantly predict the size of Twitter followers. After adding the variable of information-broadcasting behavior, the regression model remained significant (R-square = 0.35, F=7.21, p <0.01). The frequency of information-broadcasting behavior (beta = 0.15, p < 0.01) significantly predicted the number of people that follow respondents on Twitter. Thus H1 was supported.

For H2, another regression analysis model was applied to investigate the relationship between respondents' Twitter followings and their information-seeking behavior. Before the regression analysis, the same basic demographic characteristics of the respondents were entered into the model to test their significances. The data analysis (as in Table 2) suggested that there are significant differences in the number of Twitter followings in terms of the users' age (beta = - 0.13, p = 0.01) and ethnicity (beta = - 0.26, p < 0.01). However, income was not a significant predictor of the size of Twitters followings, which is different from the model testing H1. Respondents' gender, GPA and marital status did not significantly predict the size of Twitter followings. After adding the variable of information-seeking behavior, the regression model remained significant (R-square = 0.14, F= 8.24, p < 0.01), but age (beta = - 0.10, p = 0.06) became insignificant in predicting the number of Twitter followings. The variable of information-seeking behavior (beta = 0.21, p < 0.01) significantly predicted the number of people the respondents follow on Twitter. Thus H2 was supported too.

Table 2. Hierarchical Regression Model Predicting the Number of Twitter followings (N = 394)

	Model 1 (R^2=0.10, F=6.45, p<0.01)			Model 2 (R^2=0.14, F=8.24, p<0.01)		
Variable	β	t	p	β	t	p
Income	.10	1.91	.06	.09	1.77	.08
Age	-.13*	-2.48*	.01*	-.10	-1.90	.06
GPA	.04	.67	.51	.04	.69	.49
Gender	.02	.46	.65	.05	.97	.33
Marital Status	.01	.25	.80	.02	.29	.77
Ethnicity	-.26*	-5.09*	.00*	-.23*	-4.54*	.00*
Information Seeking				.21*	4.15*	.00*

Table 3. Hierarchical Regression Model Predicting the Difference between Twitter followers and followings (N = 394)

	Model 1 (R^2=0.21, F=2.82, p=0.01)			Model 2 (R^2=0.27, F=4.09, p<0.01)		
Variable	β	t	p	β	t	p
Income	-.10	-1.83	.07	-.08	-1.40	.16
Age	.15*	2.82*	.01*	.14*	2.67*	.01*
GPA	-.08	-1.58	.12	-.09	-1.66	.10
Gender	-.02	-.42	.68	-.03	-.51	.61
Marital Status	-.01	-.13	.90	-.01	-.11	.91
Ethnicity	-.11*	-2.07*	.04*	-.09	-1.62	.11
Relation Maintenance				.17*	3.35*	.00*

For H3, a regression model was applied to investigate how the difference between Twitter followings and followers is related to the respondents' relational maintenance behavior. Before the regression analysis, the same basic demographic characteristics of the respondents were entered in the model to test their significances. The data analysis (as in Table 3) suggested that there are significant differences between Twitter followings and followers in terms of the users' income (beta = 0.15, p < 0.01) and ethnicity (beta = - 0.11, p = 0.04). However, age was not a significant predictor, which is different from the testing of H1 and H2. Respondents' gender, GPA and marital status did not significantly predict the difference between Twitter followers and followings. After adding the variable of relational maintenance behavior, the regression model remained significant (R-square = 0.07, F = 4.09, p < 0.01). The variable of relational management behavior (beta = 0.17, p < 0.00) significantly predicted the difference between Twitter followings and followers. However, since the standardized coefficient for relational maintenance activities was larger than zero, the analysis suggested that the difference in size between Twitter followings and followers increases with more frequent relational maintenance activities. Thus H3 was rejected.

For H4, a regression model was applied to investigate how the number of Facebook friends is related to the respondents' relational maintenance behavior. Before the regression analysis, the same basic demographic characteristics of the respondents were entered in the model to test their significances. The

Table 4. Hierarchical Regression Model Predicting the Number of Facebook friends (N = 394)

	Model 1 (R^2=0.23, F=3.27, p<0.01)			Model 2 (R^2=0.23, F=2.80, p<0.01)		
Variable	β	t	p	β	t	p
Income	-.07	-1.36	.17	-.07	-1.35	.18
Age	-.14*	-2.71*	.01*	-.15*	-2.70*	.01*
GPA	-.02	-.34	.74	-.02	-.33	.74
Gender	-.07	-1.28	.201	-.07	-1.28	.20
Marital Status	.02	.40	.69	.02	.41	.69
Ethnicity	.10	1.97	.05	.10	1.93	.05
Relation Maintenance				-.01	-.12	.90

data analysis (as in Table 4) suggested that there are significant differences in the number of Facebook friends in terms of the users' age (beta= - 0.14, $p < 0.01$). However, income, GPA, gender, ethnicity and marital status were not significant in predicting the sizes of Facebook friends. After adding the variable of relational management behavior into the analysis, the regression model's significance actually decreased, with adjusted R-square dropped from 0.04 ($F = 3.27$, $p < 0.01$) to 0.03 ($F = 2.80$, $p < 0.01$). The analysis indicated the frequency of relational management activities (beta = - 0.01, p=0.90) did not predict the number of college students' Facebook friends at all. Thus, H4 was not supported by the data.

DISCUSSIONS

Based on the data analysis, several implications deserve the attention of communication scholars interested in social media.

First, the study reconfirmed that young minority groups are playing an important role in social media. Previous research has discovered that young minorities, especially African Americans, are more active than white Americans in using social media. Pew Research Center's (2013) "Internet and American Life Project" reported that 26% of black internet users surveyed said they use Twitter, compared with 19% of Hispanics and 14% of white users. The predominance of black users in SNSs, Twitter mostly, became a social media phenomenon known as Black Twitter (Brock, 2012). In addition, Pew research also found that 23% of blacks surveyed use Instagram in comparison to18% of Latinos and 11% of whites. In accordance with previous research, results based on this survey substantiated that non-white users have significantly higher frequencies of information broadcasting, information seeking and relational maintenance activities than white users. This finding corresponds to a previous study (Fang & Ha, 2011) arguing that ethnic users were more likely to be heavy users of SNS. Although the current study did not provide any information about whether the followers and followings of minority Twitter users are also ethnic minorities, the large number of ethnic Twitter users with enormous network sizes may indicate there is a strong virtual bonding network among ethnic students on SNS.

Second, the study provided evidence that various types of social media in the market may serve different functions. Specifically, the findings suggested that Facebook and Twitter appeal to different audiences for distinct gratification purposes. The inclusivity of Facebook's social network can be traced back to its initial marketing strategy. Founded in early 2004, Facebook was originally a Harvard-only SNS that required a harvard.edu email address to join (Cassidy, 2006). Even after its expansion to other universities, Facebook still required a university email address. Its history has predefined Facebook's closed networking feature, which encourages "an intimate, private community" (Boyd & Ellison, 2007, p. 218). Although most online social networks, like Facebook, take the form of "Friends," "Contact", and "Fans" that resemble the bi-directional and strong-tie feature of an offline friendship, others obviously do not follow the pattern (Boyd & Ellison, 2007). Twitter's "followings" and "followers" are a typical exception. Brock (2012) argued that, in contrast to Facebook, Twitter highlights the feature of "comment," or Tweet, but not its profiles or networks. This feature may have contributed to the phenomenon of Black Twitter, which Brock (2012) referred to as "a user-generated source of culturally relevant online content, combining social network elements and broadcast principles to share information" (p. 530). Honeycutt and Herring's (2009) study of Twitter's "@" feature argued that the adoption of "@" (i.e. the at sign) facilitated information flow to others by reinforcing addressivity. Brock (2012) suggested that the reason behind the uniqueness of Twitter may be its temporal, electronic, and structural discourse mediation, which encourages weak tie (Granovetter, 1973). The current study showed consistency with the above arguments with the fact that information-driven usage (information seeking and broadcasting) and relationship-driven usage (relational maintenance) demonstrated dramatically different predicting powers for social network sizes on Facebook and Twitter.

Third, the study suggested that, with the increasing activities of information seeking and broadcasting, the social networking feature of Facebook and Twitter may become secondary among the heaviest SNS users enjoying a large friend base. The fact that Facebook user's relationship management activities failed to predict the size of their friend base is intriguingly surprising. A possible explanation is that the social network on Facebook is limited to individual's circle of family, relatives, collogues, and friends in real life and thus will not grow along with increased online relational maintenance activities. This makes sense because many Facebook users treat Facebook as their private space and are careful about expanding their online social network to random Internet acquaintances. As Ellison, Steinfield, and Lampe (2007) argued, a majority of peoples use Facebook to maintain existing offline relationships or strengthen offline connections rather than seek new friendships. On the other hand, the frequency of relational maintenance activities on Twitter is positively correlated with the difference of sizes between followers and followings, which is the opposite to the third hypothesis that smaller differences in size between a Twitter user's followings and followers lead to more activities of relational management. Although the reasons behind are yet to be discovered, it may indicate that some social networking services, despite the main function implied by its name, are transforming from relationship-centered services to information-oriented multiple-purpose websites. In other words, there may be a trend for social media to turn into a major platform catering to the needs of information-driven users such as lobbyists, marketers, online shoppers, and news readers. Boyd and Ellison (2008) have preferred the term "Social Network Sites" to "Social Networking Sites" because they believed "networking" emphasizes relationship initiation but many users are not necessarily networking or looking for new people on social media. With informational functions further weakening social media's networking features, some changes in SNS use patterns are occurring, as demonstrated by news consumption activities. Purcell et al. (2010) discovered that SNSs, along with email, have become a source of news for 75 percent of online news consumers in the United States. A study of the top 25 US news websites also showed that Facebook has become a major source of traffic for the five most visited sites with the help of its social recommendation function (Olmstead et al., 2011). As Dye (2007) pointed out, SNSs have created a new generation of people whose identities are defined by their online networks as well as the user-generated content.

CONCLUSION

The findings in this chapter supported the hypothesis that the sizes of Twitter followers and followings are significantly influenced by SNS users' information-driven needs, indicating college students are either searching for or distributing information (or even doing both at the same time) when using Twitter. In order to cater to that particular need, college students take the initiative to adjust their online behavior and actively seek various channels to achieve these gratifications. This process is reflected by the significant correlation between the sizes of Twitter social networks and different SNS usage patterns related to information. Particularly, information seekers on Twitter actively follow more Twitter accounts, either public or individual ones, to keep themselves updated with the latest news about friends' life as well as major political and social events and etc. On the other hand, it may not be easy for information broadcasters on

Twitter to ask other users to follow them. However, since most information broadcasters are specialized in a specific target audience, these users can always attract more followers by posting relevant information that may intrigue potential followers. For example, an information broadcaster who specializes in American football may increase the size of his/her Twitter followers by intentionally focusing his/her tweets on NFL-related topics. Since many followers in the broadcaster's Twitter network may share the same interest, tweets on American football are likely to get more exposure and may eventually attract more followers via his/her existing followers' activities of commenting and retweeting. This process of getting more followers may seem slow and passive at first sight, but is effective as demonstrated by the cases of viral posts. Consistent with the classical Uses and Gratifications theory, the findings provided evidence that social media users take full advantage of the hashtagged tweets to meet their own gratifications, leading to customized sizes of their Twitter followers and followers.

This chapter also contributed to the scarce literature regarding the differentiating traits between opinion seekers and opinion leaders on SNSs. Although research in traditional media has indicated that opinion leaders are typically younger with a higher level of education, higher income and greater social mobility than regular users and opinion seekers (Gatigon & Robertson, 1985; Midgley & Dowling, 1978; Robertson, Zielinski & Ward, 1984; Rogers, 1983), it is uncertain whether these findings can be replicated in an Internet environment. The current study showed that the varying sizes of followings and followers among Twitter users are significantly correlated with their information-oriented behavior: opinion seekers are likely to follow more people while opinion leaders are likely to attract more followers. Moreover, both opinion seekers and leaders tend to be younger minorities with higher household income. Even though the research used a student sample, age still made a difference in users' social network sizes on both Twitter and Facebook: younger SNS users are not only likely to have a larger size of followers, but also are likely to have a larger number of friends on Facebook. This finding gave mixed support to the literature of opinion leadership since typical opinion leaders has been defined as tech-savvy young people with superior status, education, and social prestige. Although income and age did appear to be important in predicting a typical opinion leader in this study, ethnic minorities seemed more likely to bear the traits of SNS opinion leadership, which somewhat contradicts our traditional perception of opinion leaders as a middle-class Caucasian white collars.

Nevertheless, the study in this chapter still has its limitations. First, the measures of the three SNS user needs may not be specific enough since the questionnaire asked about the frequencies of different SNS usage activities in general but did not differentiate Twitter from Facebook. It is possible that a SNS user's information broadcasting, information seeking and relational maintenance activities differ dramatically between Facebook and Twitter. However, since the series of questions were understood as a general scale for individual SNS usage habits, these measurements are likely to roughly reflect users' daily usage pattern as a whole and thus should not cause a threat to the validity of the measurements. Second, the target population of this study is college students in American universities, which makes the findings a bit difficult to be generalized to a larger population. For example, although education does not prove to be a significant predictor of opinion leadership, its role may have been underplayed due to the use of a student sample. However, as this study focuses on relationships among the variables rather than describing the college population, it is acceptable to have a convenient sample for this study. It is unknown whether the relationship between SNS usage patterns and social network sizes is still applicable among young people other than college students, not to mention the general population. Further research needs to be done regarding the relationship between SNS usage pattern and people's social network structure among the general population.

REFERENCES

Assael, H. (1992). *Consumer Behavior and Marketing Action*. Boston: PWS-Kent Pub.

Baran, S., & Davis, D. (1995). *Mass communication theory*. Belmont, CA: Wadsworth.

Baumgarten, S. A. (1975). The innovative communicator in the diffusion research. *JMR, Journal of Marketing Research*, *12*(1), 12–18. doi:10.2307/3150653

Berkow, J. (2010). *FP Tech Desk: The Coming Twitter News Network*. Retrieved from http://business.financialpost.com/2010/11/23/fp-tech-desk-the-coming-twitter-news-network/?__lsa=93af-dddb

Boyd, D., & Ellison, N. (2007). Social network sites: Definition, history, and scholarship. *Journal of Computer-Mediated Communication*, *13*(1), 210–230. doi:10.1111/j.1083-6101.2007.00393.x

Brock, A. (2012). From the blackhand side: Twitter as a cultural conversation. *Journal of Broadcasting & Electronic Media*, *56*(4), 529–549. doi:10.1080/08838151.2012.732147

Cascio, W. F., & Shurygailo, S. (2002). E-leadership and virtual teams. *Organizational Dynamics*, *31*(2), 362–376.

Cassidy, J. (2006). Me media: How hanging out on the Internet became big business. *The New Yorker*, *82*(13), 50.

Catignon, H., & Thomas, S. R. (1985). A Propositional Inventory for New Diffusion Research. *The Journal of Consumer Research*, *11*(4), 849–867. doi:10.1086/209021

Cheung, C. K., Chiu, P., & Lee, M. O. (2011). Online social networks: Why do students use Facebook? *Computers in Human Behavior*, *27*(4), 1337–1343. doi:10.1016/j.chb.2010.07.028

Cummings, J., Lee, J., & Kraut, R. (2006). Communication technology and friendship during the transition from high school to college. In R. E. Kraut, M. Brynin, & S. Kiesler (Eds.), *Computers, phones, and the Internet: Domesticating information technology* (pp. 265–278). New York, NY: Oxford University Press.

Dominick, J. (1999). Who do you think you are? Personal home pages and self-presentation on the World Wide Web. *Journalism & Mass Communication Quarterly*, *76*(4), 646–658. doi:10.1177/107769909907600403

Dunne, A., Lawlor, M.-A., & Rowley, J. (2010). Young people's use of online social networking sites – a uses and gratifications perspective. *Journal of Research in Interactive Marketing*, *4*(1), 46–58. doi:10.1108/17505931011033551

Dye, J. (2007). Meet generation C: Creatively connecting through content. *EContent*, *30*(4), 38.

Eastin, M. (2002). *From Internet use to the unique Internet user: Assessing information, entertainment and social cognitive models of Internet use*. Paper presented to the National Communication Association Conference, New Orleans, LA.

Ellison, N. B., Steinfield, C., & Lampe, C. (2007). The benefits of Facebook "friends:" Social capital and college students' use of online social network sites. *Journal of Computer-Mediated Communication*, *12*(4), 1143–1168. doi:10.1111/j.1083-6101.2007.00367.x

Fang, L., & Ha, L. (2011). *Who are the heavy users of Social Network Sites among College Students? A Study of Social Network Sites and College Students*. Paper presented at the annual meeting of the Association for Education in Journalism and Mass Communication, St. Louis, MO.

Feick, L. F., & Price, L. L. (1987). The market maven: A diffuser of market information. *Journal of Marketing*, *51*(1), 83–97. doi:10.2307/1251146

Feick, L. F., Price, L. L., & Higie, R. A. (1986). People who use people: The other side of opinion leadership. *Advances in Consumer Research. Association for Consumer Research (U. S.)*, *13*(1), 301–305.

Flynn, L. R., Goldsmith, R. E., & Eastman, J. K. (1996). Opinion leaders and opinion seekers: Two new measurement scales. *Journal of the Academy of Marketing Science*, *24*(2), 137–147. doi:10.1177/0092070396242004

Gao, J., Zhang, M., Jiang, F., & Wang, X. (2005). Analysis of opinion leader in on-line communities. In R. Khosla, R. J. Howlett, & L. C. Jain (Eds.), *Lecture notes in artificial intelligence. Knowledge-based intelligent information and engineering systems, Pt. 1: Proceedings* (pp. 1153–1159). Berlin: Springer.

Goldsmith, E. B., & Ronald, E. G. (1980). Dogmatism and confidence as related factors in evaluation of new products. *Psychological Reports*, *47*(3), 1068–1070. doi:10.2466/pr0.1980.47.3f.1068

Granovetter, M. S. (1973). The strength of weak ties. *American Journal of Sociology*, *78*(6), 1360–1380. doi:10.1086/225469

Gross, R., Acquisti, A., & Heinz, H. J., III. (2005). Information revelation and privacy in online social networks. *Proceedings of the 2005 ACM workshop on Privacy in the Electronic Society*, 71-80. New York. doi:10.1145/1102199.1102214

Honeycutt, C., & Herring, S. C. (2009). Beyond microblogging: Conversation and collaboration via Twitter. *Proceedings of the 42nd Hawaii International Conference on System Sciences*, 1–10.

Horton, D., & Wohl, R. (1956). Mass communication and para-social interaction: Observations on intimacy at a distance. *Psychiatry*, *19*(3), 215–229. PMID:13359569

Jacoby, J. (1971). Personality and innovation proneness. *JMR, Journal of Marketing Research*, *8*(2), 244–247. doi:10.2307/3149771

Katz, E., Blumler, J. G., & Gurevitch, M. (1973). Uses and Gratifications research. *Public Opinion Quarterly*, *37*(4), 509–523. doi:10.1086/268109

Katz, E., Gurevitch, M., & Haas, H. (1973). On the use of the mass media for important things. *American Sociological Review*, *38*(2), 164–181. doi:10.2307/2094393

Kim, Y. S., & Tran, V. L. (2013). Assessing the ripple effects of online opinion leaders with trust and distrust metrics. *Expert Systems with Applications*, *40*(9), 3500–3511. doi:10.1016/j.eswa.2012.12.058

Lewis, J., & West, A. (2009). 'Friending': London-based undergraduates' experience of Facebook. *New Media & Society*, *11*(7), 1209–1229. doi:10.1177/1461444809342058

Li, Y., Ma, S., Zhang, Y., Huang, R., & Kinshuk, . (2013). An improved mix framework for opinion leader identification in online learning communities. *Knowledge-Based Systems*, *43*, 43–51. doi:10.1016/j.knosys.2013.01.005

Lometti, G. E., Reeves, B., & Bybee, C. R. (1977). Investigating the assumptions of uses and gratifications research. *Communication Review*, *4*(3), 321–338.

Marwick, A., & Boyd, D. (2011). I Tweet Honestly, I tweet passionately: Twitter users, context collapse, and the imagined audience. *New Media & Society*, *13*(1), 114–133. doi:10.1177/1461444810365313

McQuail, D., Blumler, J. G., & Brown, J. R. (1972). The television audience: A revised perspective. In D. McQuail (Ed.), *Sociology of mass communications* (pp. 135–164). Harmondsworth, UK: Penguin.

Midgley, D. F., & Dowling, G. R. (1978). Innovativeness: The Concept and Its Measurement. *The Journal of Consumer Research*, *4*(4), 229–242. doi:10.1086/208701

Myers, J. H., & Robertson, T. S. (1972). Dimensions of opinion leadership. *JMR, Journal of Marketing Research*, *9*(1), 41–46. doi:10.2307/3149604

Olmsead, K., Mitchell, A., & Rosenstiel, T. (2011). Navigating news online: Here people go, how they get there and what lures them away. *Pew Research Journalism Project*. Retrieved from http://www.journalism.org/2011/05/09/navigating-news-online/

Papacharissi, Z., & Rubin, A. (2000). Predictors of Internet usage. *Journal of Broadcasting & Electronic Media*, *44*(2), 175–196. doi:10.1207/s15506878jobem4402_2

Park, N., & Lee, S. (2014). College students' motivations for Facebook use and psychological outcomes. *Journal of Broadcasting & Electronic Media*, *58*(4), 601–620. doi:10.1080/08838151.2014.966355

Pew Research Center. (2013). Pew Research points out different (social) strokes for different folks. *Advertising Age*, *84*(7), 5.

Purcell, K., Rainie, L., Mitchell, A., Rosenstiel, T., & Olmstead, K. (2010). Understanding the participatory news consumer. *Pew Internet and American Life Project*. Retrieved from http://www.pewinternet.org/Reports/2010/Online-News.aspx

Ray, M. B. (2007). Needs, motives, and behaviors in Computer-Mediated Communication: An inductive exploration of social networking websites. Paper presented at *the annual meeting of International Communication Association*, San Francisco, CA.

Robertson, T. S., Zielinski, J., & Ward, S. (1984). *Consumer Behavior*. Glenview, IL: Scott, Foresman.

Rogers, E. M. (1983). *Diffusion of Innovations*. New York, NY: The Free Press.

Ross, C., Orr, E. S., Sisic, M., Arseneault, J. M., Simmering, M. G., & Orr, R. R. (2009). Personality and motivations associated with Facebook use. *Computers in Human Behavior*, *25*(2), 578–586. doi:10.1016/j.chb.2008.12.024

Rubin, A. M. (1994). Media uses and effects: A uses-and-gratifications perspective. In J. Bryant & D. Zillmann (Eds.), *Media effects: Advances in theory and research* (pp. 417–436). Hillsdale, NJ: Lawrence Erlbaum Associates, Inc.

Sheldon, P. (2008). Student favorite: Facebook and motives for its use. *Southwestern Mass Communication Journal, 23*(2), 39–53.

Shoham, A., & Ruvio, A. (2008). Opinion leaders and followers: A replication and extension. *Psychology and Marketing, 25*(3), 280–297. doi:10.1002/mar.20209

Song, I., LaRose, R., Eastin, M. S., & Lin, C. A. (2004). Internet gratifications and internet addiction: On the uses and abuses of new media. *Cyberpsychology & Behavior, 7*(4), 384–394. doi:10.1089/cpb.2004.7.384 PMID:15331025

Stafford, T., & Gonier, D. (2004). What Americans like about being online. *Communications of the ACM, 47*(11), 107–112. doi:10.1145/1029496.1029502

Summers, J. O. (1970). The identity of women's clothing fashion opinion leaders. *JMR, Journal of Marketing Research, 7*(2), 178–185. doi:10.2307/3150106

Taylor, C. (2011). *Twitter has 100 million active users.* Retrieved July 28th, 2013 from http://mashable.com/2011/09/08/twitter-has-100-million-active-users/

Taylor, J. W. (1977). A striking characteristic of innovators. *JMR, Journal of Marketing Research, 14*(1), 104–107. doi:10.2307/3151066

Urista, M. A., Dong, Q., & Day, K. N. (2009). Explaining why young adults use MySpace and Facebook through uses and gratifications theory. *Human Communication, 12*(2), 215–229.

Venkatraman, M. P. (1989). Opinion leaders, adopters, and communication adopters: A role analysis. *Psychology and Marketing, 6*(1), 51–68. doi:10.1002/mar.4220060104

Weimann, G., Tustin, D. H., Vuuren, D. V., & Joubert, J. P. R. (2007). Looking for opinion leaders: Traditional vs. modern measures in traditional societies. *International Journal of Public Opinion Research, 19*(2), 173–190. doi:10.1093/ijpor/edm005

Young, K. (2011). Social ties, social networks and the Facebook experience. *International Journal of Emerging Technologies & Society, 9*(1), 20–34.

KEY TERMS AND DEFINITIONS

Facebook Friend: An individual Facebook user whose Facebook profile is mutually connected to another individual Facebook user.

Information Broadcasting: A social media user's activities of spreading knowledge, news and other information among his/her online social network.

Information Seeking: A social media user's activities of searching and acquiring knowledge, news and other information from his/her online social network.

Opinion Leader: An individual whose opinions are trusted and valued by people in his/her social network due to this person's social status, education, authority, expertise or popularity.

Relational Maintenance: A social media user's activities of establishing, keeping and strengthening social relations online.

Social Network Size: The total number of individual profiles connected in a certain way to one social media user.

Social Networking Service: An online platform for people to build and develop social networks and relations via communication in the cyberspace, also known as SNS or social networking site.

Twitter Follower: A Twitter user who subscribes to the Twitter user of interest with an intention of receiving Twitter updates (Tweets) by that particular user.

Twitter Following: A Twitter user whom the Twitter user of interest subscribes to with an intention of receiving Twitter updates (Tweets) by that particular user.

Uses and Gratifications: A communication theory assuming that people are active audiences who use media to meet their specific needs.

Chapter 13
Do College Students Benefit from Their Social Media Experience?
Social Media Involvement and Its Impact on College Students' Self-Efficacy Perception

Ling Fang
Bowling Green State University, USA

Louisa Ha
Bowling Green State University, USA

ABSTRACT

As young people are increasingly dependent on Social Networking sites (SNS) to socialize, seek information, and self-broadcast, their SNS consumption has been found to be associated with social capital and social support in a positive way especially among individuals with low psychological assets. This exploratory study investigated SNS involvement in relation to college students' perceived self-efficacy change afterwards based on the social cognitive theory and literature review on social media effects studies. Undergraduate students (N = 395) in a Midwest U.S. public university participated in a web survey in September 2012. Results indicated a positive potential of involving in SNS activities to strengthen users' self-efficacy. Research data also point to the mediating roles of social support and social learning on SNS involvement and self-efficacy, especially among low self-efficacy college students with homogenous SNS networks. Theoretical and practical implications are discussed.

DOI: 10.4018/978-1-4666-8450-8.ch013

INTRODUCTION

College is a major life transition for young people, who usually face with challenges in different domains of life and need to gain various kinds of support to cope with problems and adjust to transitions. As the foundation of human agency, the self-efficacy belief is predictive of individuals' achievements and coping capability in adversities (Bandura, 1986). Although previous research on social media effects paid more attention to self-esteem, perception of self-efficacy, as a useful behavioral indicator and college educational goal, deserves more attention. In an age of building their self-efficacy, young people may constantly reassess themselves based on information they process from real-life or virtual environment. Applying the social cognitive theory in social media settings, this study examines students' self-reported self-efficacy change after using SNS as a proxy measure to explore the short-term positive effect of SNS involvement.

REVIEW OF LITERATURE

SNS Attributes, Uses and Gratifications

Social networking sites are the latest online communication tools that allow users to create a public or private profile to interact with people in their networks (Boyd & Ellison, 2007). The first social networking site was founded back in the 90s but the success of the sites was obtained with the rise of MySpace, Facebook, and Twitter. The popularity of SNS has now become one of the Internet's most visible and intriguing phenomena. As of December 2012, there are more than a billion monthly active Facebook users and 618 million daily active users. Boyd and Ellison (2007) defined SNS as "web-based services that allow individuals to (1) construct a public or semi-public profile within a bounded system, (2) articulate a list of other users with whom they share a connection, and (3) view and traverse their list of connections and those made by others within the system". Young people have been in the forefront of SNS and spending time on social networking sites appears to be part of most U.S. young adults' daily activities (Ellison, Steinfield, & Lampe, 2007; Pew Research Center, 2010). As of December 2012, 86% of Internet users between 18 and 29 use Facebook (Pew Research Center, 2012). A recent study of 2,359 college students found that they spent almost two hours daily on Facebook (Bosker, 2013). The popularity of SNS can be explained from its unique attributes and capability of satisfying diversified needs. Four major and unique SNS attributes contribute to its popularity among college students are summarized as: richness, sociability, publicness, and participation.

Richness. The rich functionalities of SNS possesses contributes to its capability to satisfy diversified needs and its popularity. As a mediated communication channel with both attributes of mass media and interpersonal communication technologies, SNS not only serves as a multimedia platform providing traditional media content like news, picture, audio, or video, but also offers a combination of computer-mediated communication tools like instant messaging, blog, interest groups, bulletin boards, games, customized applications, and email (Ha & Hu, 2013; Mount & Garcia Martinez, 2014).

Sociability. Besides providing media content like traditional media, SNS also enables a many-to-many interactive communication model among SNS users and serves as a virtual social place for relationship initiation, maintenance, and development (Rafaeli, 1998). The friend list, as a fundamental SNS component, displays users' social connections. Other major SNS features, like searching, personal profile,

Table 1. Motivations, SNS uses, user behaviors, and SNS attributes

Motivations	SNS Uses	User Behaviors	SNS Attributes
Information, entertainment, mood management	Information-seeking	Consuming content: social searching and social browsing	Richness
Connectivity	Socializing	Online interaction and participation: posting and replying on Wall, chatting, involving in SNS group and community building	Sociability
Self-expression and actualization	Self-broadcasting	Producing content: status updating and posting	Publicness and participation

instant messaging, and bulletin boards posting facilitate interpersonal communication to an even greater degree (Ha & Hu, 2013; Tang, 2013). The unique sharing function enables users to send and receive media content or messages within their social networks and promotes the growth of a network of mutual friends that has potential for massive expansion (Ancu & Cozma, 2009; Ha & Hu, 2013).

Publicness. Although users can communicate privately within SNS, they can also communicate in ways that are visible to everyone else within the network (Hall, 2009).

SNS is a virtual individualized public sphere, where people of similar interest get together to communicate, share, and discuss ideas (Raacke, & Bonds-Raacke, 2008; Boyd & Ellison, 2007). Users' personal profiles can be used to manage their impression in public.

Participation. User-generated SNS content depends on users' contribution. Every SNS user can be a content producer and editor other than a content consumer. Content created will appear in public page and be automatically kept in personal page in terms of "Timeline" and won't disappear unless deleted by authors. Research findings have shown that ease of use and user control leave people greater satisfaction of use from their SNS experience (Shao, 2009). The participatory nature of SNS results in diversified social media content.

Based on the theory of the niche (Feaster, 2009), media compete to provide gratifications and obtain limited resources in terms of time, money, and attention from users. Social media with rich and unique attributes are successful in satisfying young people' diversified communication needs and thus dominant their daily life. Early studies on SNS uses and gratifications indicate two major types of SNS uses: socializing and information seeking (Fang & Ha, 2011; Quan-Haase and Young, 2010; Ray, 2007; Urista et al., 2009; Hall, 2009). Recent studies found self-presentation and impression management as another important motivation for college students to use SNS (Gonzales & Hancock, 2011; Mehdizadeh, 2010; Sun, & Wu, 2012). Therefore, three major uses and gratifications most prevalent among college students are: information seeking, socializing, and self-broadcasting (Shao, 2009; Smock et al., 2011; Yartey & Ha, 2013). Shao (2009) illustrated three levels of SNS use in relation to different gratifications in a hierarchical model. At the first level, individuals consume SNS content to fulfill their information, entertainment, and mood management needs. At the next level, they participate through interacting with the content and other users to enhance social connections and build virtual communities. In the top level, users produce their own content for self-expression and actualization. Built on Shao's hierarchical model, three types of prevalent SNS uses and gratifications have been associated with SNS attributes.

Information-seeking. Compared to direct communication and impression management, it has been found that SNS users devoted most of their time to either non-specific passive social browsing like reading newsfeeds and goal-oriented social searching such as viewing friends' profiles (Wise et al., 2010). In the experiment, users spent more time on social searching than on social browsing on SNS. Since SNS combines mass media and interpersonal communication technologies attributes together, informational-seeking on SNS is not limited to social news but also information from users' social circles, like friends' information and updates.

Socializing. Although family members continue to play a crucial role in adolescents' well-being, peers become increasingly important in adolescence (Parker & Asher, 1993). Young people are increasingly dependent on SNS to socialize with friends to satisfy social connection and belongingness needs (Chen, & Marcus, 2012; Sheldon, Abad, & Hinsch, 2011). A vast majority of college students use SNS for making new friends and maintaining existing personal relationships (Chen, & Marcus, 2012; Ellison et al., 2007; Raacke & Bonds-Raacke's, 2008). Sheldon, Abad and Hinsch (2011) found in their correlational data the coexistence of Facebook use in relations to connectedness satisfaction and dissatisfaction. Based on experimental results, they indicated that dissatisfaction with existing connection as a motivation to use Facebook and satisfaction with Facebook connection as a gratification obtained. The fact that individuals with lower initial disconnection are less likely to reduce their Facebook use confirms connectedness as an important motivation for SNS social use. A bunch of mediated communication tools on SNS, like instant messaging, bulletin boards, and SNS group, enable users a greater level of interactivity with their social networks.

Self-broadcasting. Motivated by self-expression and actualization needs, users produce content on SNS to present and broadcast themselves (Shao, 2009). Matsuba (2006) found that college students with less clearly defined self-concepts were more likely to use online media. Researchers paid special attention to students' self-disclosure behavior on SNS (Chen & Marcus, 2012; Mansson, & Myers, 2011). Mansson and Myers (2011) identified 29 types of expressed affection in 214 undergraduate students' interaction with close friends through Facebook, including verbal, nonverbal, and social support messages. The public and participatory attributes of SNS enable users to produce their own content in terms of creating or changing personal profile, updating status, and posting diaries, pictures or videos to present themselves publicly in their online social circles.

Social capital, social support, and subjective well-being are the three common SNS social psychological effect covers outcome variables of SNS use in previous research.

SNS Involvement and Social Capital

Social capital refers to resources a person is allowed to draw from other members in his or her network. A distinction of social capital into bridging and bonding social capital has been widely adopted in current social media research (Putnam, 2000). Bridging social capital are loose connections between individuals and is regarded as a source of informational support but typically not emotional support, while bonding social capital refers to connections between emotionally close relationships (Granovetter, 1973). Network characteristic has been found to be associated with social capital (Bessiere et al, 2008): while weak ties predict bridging social capital, strong tie enhances bonding social capital. Students' SNS use has been found to positively associated with social capital in general, and most obvious for bridging social capital (Ellison, Steinfield, & Lampe, 2007). The positive effect has been explained by low transaction cost interacting with both strong and weak ties via SNS (Ellison et al., 2010).

SNS Involvement and Social Support

Built on studies on SNS social capital, review of emerging research on social support use of SNS suggests that SNS serves as a channel to broadcast support needs and to obtain emotional support (Bessiere et al, 2008; Kim & Lee, 2011; Wright, 2012) as well as informational support (Bessiere et al, 2008; Ellison, 2011; Wright et al., 2013). Self-broadcasting use of SNS via status updating lowers barriers to broadcast support needs, which directly contributes to social support on SNS (Vitak & Ellison, 2013). In addition, SNS social capital in terms of network size, strong/weak tie preference, homophily, network convergence, have been found to be associated with social support received. An inverted U-shape curvilinear relationship between the number of Facebook friends and perceived social support suggested the quantity of friends only partially explains support received (Kim & Lee, 2011). Strong tie-oriented network associated with higher similarity and network convergence is associated with higher emotional support (Bessiere et al, 2008; Wright, 2012), while more heterogeneous and less similar weak tie-oriented network tend to serve as useful sources of information and advice (Bessiere et al, 2008; Ellison, 2011; Wright et al., 2013). However, recent research suggests that the support weak ties provide is more than informational support. Wright (2012) found that online weak ties were regarded as being less risky to disclose information to, less judgmental, and more objective. Similarly, a recent panel survey found that communication with strong ties among Facebook users who lost their jobs actually increased their stress of unemployment, while communication with weak ties didn't (Burke & Kraut, 2013). Therefore, both strong ties and weak ties on SNS could be support providers. Although precious research doesn't reach a conclusion on how different network characteristics are indicative of potential support received from social media, SNS social capital based on various network characteristics can be a predictor of social support available on SNS.

SNS Involvement and Subjective Well-Being

A positive association between SNS involvement and psychological well-being mediated by social support has been suggested in previous research. Honest self-presentation (Kim, & Lee, 2011) and self-disclosure on SNS (Vitak, & Ellison, 2013) have been found to be associated with subjective well-being indirectly through social support. Social support received from SNS use then buffers against stress and depression (Grieve et al., 2013; Wright et al., 2013) and contributes to life satisfaction (Apaolaza et al., 2013; Grieve et al., 2013; Lee, Lee, & Kwon, 2011; Kim, & Lee, 2011). Perceptions of attitude and background similarity and attraction of potential support providers were predictive of increased emotional support from Facebook, and further predictive of lower levels of perceived stress (Wright, 2012). Besides, positive self-presentation (Kim & Lee, 2011) and viewing personal profile (Toma & Hancock, 2013) have direct effects on subjective well-being.

SNS Use and Social Psychological Effects: From Negative to Positive

The literature of the social psychological effects of SNS use on individuals has evolved from purely negative to more positive effects. The negative effects were based on an "everyone gets poorer" hypothesis.

"Everyone gets poorer" hypothesis. Early research findings on social psychological effects of online media use suggest a negative alienation effect for individuals with both high and low social psychological assets. Based on a longitudinal survey of 256 people from 93 different households during first two

years of being online, Kraut and his colleagues (1998) found in their HomeNet study that greater use of the Internet was associated with declines in participants' communication with family members in the household, decline in the size of their social circle, and increases in their depression and loneliness. Hence they called this an Internet paradox because Internet use reduced people's social capital, replacing their face-to-face communication with offline strong ties by the time they spent online communicating with weak ties However, in a later longitudinal US survey, Bessiere and his colleges (2008) found that only those people with original higher or medium social support levels who use the Internet to meet new people and talk in online groups show high depression scores. Internet does not affect the depression levels of those with originally low levels of social support.. In contrast, other scholars conducted research on new media use among users with low social psychological assets and claim for the addictive nature of social media and a potential vicious circle in individuals' social media use and psychological well-being (Karaiskos et al., 2010; Kuss, & Griffiths, 2011). Lonely individuals and adolescents with weak offline social connections were more likely to involve in problematic online media use, which resulted in negative life outcomes, like fewer face-to-face interactions and higher levels of loneliness, supporting the "poor-get-poorer" hypothesis.

"Everyone gets richer" hypothesis. As young people are increasingly dependent on SNS to maintain and develop relationships, research shows a positive trend of social media use in relation to social psychological effects. Using SNS to cope with negative emotions and to obtain social support gains researchers' attention in recent studies. Emotional support has been included as one of six gratifications obtained from SNS in a factor analysis of an online survey among 437 students (Zhang, Tang, & Leung, 2011). Two tendencies of positive SNS social psychological effects can be summarized as: a "rich gets richer" pattern supported by social augmentation hypothesis (Baumeister & Leary, 1995) and a "poor gets richer" pattern explained by the social compensation hypothesis (McKenna & Bargh, 2000). According to the social augmentation hypothesis, online media enable users to connect with similar interests individuals and to build a supportive community (Parks & Floyd, 1996). Enlarged social networks with strengthened social ties contribute to higher levels of social capital among users (Ellison et al., 2007). Compared to socially and psychologically rich people, individuals with lower offline social psychological assets seem to benefit more from their online interaction on SNS. Research has shown that positive associations between SNS use and social capital are more prominent among minority students (Valenzuela, Park, & Kee, 2009) as well as individuals with lower self-esteem (Steinfield, Ellison, & Lampe, 2008; Tazghini & Siedlecki, 2013) and life satisfaction (Ellison et al., 2007). Zywica and Danowski (2008) found support for both social compensation and social enhancement hypotheses in their study and argue that "everyone gets richer" by using Facebook. Among those 614 college students, those more extroverted and with higher self-esteem are more popular both offline and on Facebook, while those more introverted and with lower self-esteem are less popular offline and try to look popular on Facebook.

Social Cognitive Theory and Self-Efficacy

The social cognitive theory (Bandura, 1986) was originally labeled social learning theory (Bandura, 1977). Both of these two social psychological theories were proposed to study human behavior under social environment, or how human behavior is influenced by social interaction: Human behavior is not only intrigued by inner motivations but learned through observation and modeling. Based on social learning theory, "social cognitive theory broadens the scope of modeling influences and the functions it serves" (Bandura, 2012, p.351). Human behaviors are examined in a continuous reciprocal interaction between environmental, cognitive, behavioral determinants. According to social cognitive theory, human behavior is regulated by both external social environmental factors and internal cognitive factors. Take social media users' behavior as an example: it is a product of both users' online social interaction and their motivations to use it. Furthermore, the self-efficacy concept, as a key element of social cognitive theory, serves to connect information processed from environment with consequential behavioral outcomes. Bandura (1986) defined self-efficacy as "people's judgments of their capabilities to organize and execute courses of action required to attain designated types of performances" (p.391) and summarized four sources of self-efficacy: enactive performance attainments, vicarious experiences of observing others' performances, verbal persuasion, and physiological states. Firstly, the performance attainment is the most influential source of efficacy information. Success raises self-efficacy; failure lowers it. In addition, self-efficacy perception can also be derived from observation of similar others' performance. Visualizing peers' successful achievement raises observers' belief in their own ability to master comparable tasks, especially when people have little prior experience on areas they are evaluated (Bandura, 1986). Thirdly, self-efficacy beliefs can be influenced by the verbal messages and social persuasions individuals receive from others. These messages can help one to exert the extra effort and persistence required to succeed if envisioned success is attainable. Lastly, people evaluate their own capability partially based on their physiological status. Positive moods like optimism enhance self-efficacy, whereas negative moods such as depression; diminish it (Bandura, 1982).

As social interaction plays an important role in social media use, social cognitive theory can serve as a useful framework to explain SNS users' cognitive and behavioral changes resulting from online social interaction. A few recent studies have paid attention to self-efficacy as a gratification obtained from SNS use (Ahn, 2011; Mauri, et al., 2011; Shao, 2009; Wohn et al., 2013). Some scholars suggested that SNS peer interaction could promote social learning and enhance users' self-efficacy (Ahn, 2011; Wohn et al., 2013; Yu et al., 2010). For example, Wohn and his colleagues (2013) argued that social media served as a social learning platform for first-generation college students and provide them connections to successfully role models from similar background. Besides, a few health communication scholars examined the association between social media use and coping self-efficacy among individuals facing pressing health concerns like obesity (Ballantine & Stephenson, 2011) and diabetes (Greene et al., 2011). However, in general, whether college students benefit from their online interaction through observational learning hasn't been fully explored.

HYPOTHESES, RESEARCH QUESTIONS, AND THE RESEARCH MODEL

Based on review of literature of SNS effects and Bandura (1986)'s statements about sources of self-efficacy in social cognitive theory, a hypothesized research model (see Figure 1 in Appendices) focusing on SNS social support and social learning has been developed to explore SNS use in relation to self-efficacy among college students. Self-efficacy obtained serves as a proxy measure of short-term SNS effect in this study. Therefore, the first and the major hypothesis has been proposed:

H1: College students' involvement in SNS is positively associated with perceived increase in self-efficacy.

Previous studies suggest that users with more social resources on SNS are more likely to obtained social support and benefit psychologically after using SNS (Bessiere et al, 2008; Ellison, 2011; Wright, 2012; Wright et al., 2013). SNS social capital is defined as social resources obtained from SNS and is

Figure 1. Research Questions, Hypotheses, and Research Model

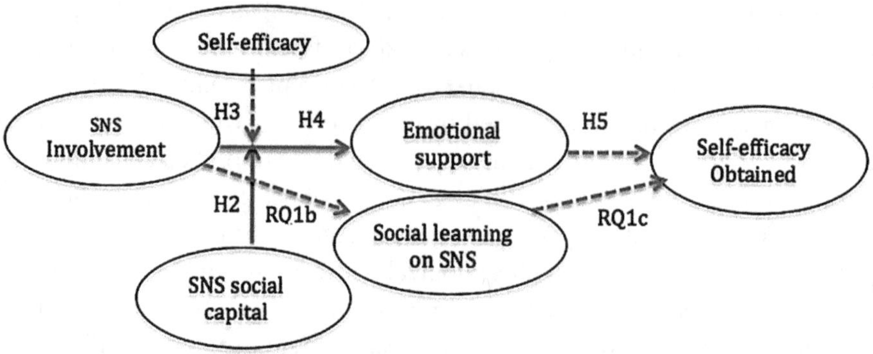

measured in terms of network size and diversity of friend types in this study. High SNS social capital users refer to individuals with larger and heterogeneous SNS network, while low SNS social capital users are those with smaller and homogeneous SNS network. Users with high level of SNS social capital are expected to be more likely to receive social support from SNS and obtained higher self-efficacy afterwards.

H2: The positive association between students' SNS involvement and perceived self-efficacy change after using SNS is stronger among students with high level of SNS social capital than low level of SNS social capital.

According to the social compensation hypothesis, positive associations between social media use and social capital are more prominent among individuals with lower self-esteem and life satisfaction. The emotional stability has also been found a negative predictor of social media use controlling for socio-demographics and life satisfaction (Correa et al., 2010). Online media lower barriers to interaction and encourage more self-disclosure among users, therefore, are especially attractive for lonely, anxious, or depressed people (Tidwell & Walther, 2002). Based on the social compensation hypothesis, the third research question has been proposed to test whether low-efficacy individuals, who more frequently experience negative emotions and emotion instability, are more likely to depend on SNS to cope with negative emotions and gain self-efficacy from their SNS experience than high self-efficacy individuals.

H3: The positive association between students' SNS involvement and perceived self-efficacy change after using SNS is stronger among students with low level of self-efficacy than students with high level of self-efficacy.

Students were categorized into four groups: high SNS social capital and high self-efficacy, high SNS social capital and low self-efficacy, low SNS social capital and high self-efficacy, low SNS social capital and low self-efficacy. If significant association between SNS involvement and self-efficacy is found in any group, the following research questions and hypotheses were further investigated to confirm whether SNS social support and social learning mediate SNS involvement and self-efficacy. The relationship between SNS involvement and social support has been indicated in SNS effect research. Encouragement, as a central component of social support, has been regarded as a contributor of self-efficacy in social cognitive theory. Therefore, H4 and H5 are proposed as follows:

H4: There is a positive association between SNS involvement and SNS social support.
H5: There is a positive association between SNS social support and perceived change in self-efficacy.

Social learning behavior on SNS hasn't been examined empirically, whether social learning is actually existed in social media settings and whether it is related to self-efficacy remains largely unknown. The following exploratory questions are proposed.

RQ1a: Do college students learn something useful from others while using SNS?
RQ1b: Is students' SNS involvement associated with their frequency of learning something useful on SNS?
RQ1c: Is students' frequency of learning something useful on SNS associated with increase in self-efficacy?

METHODS

This study is based on a web survey of college students in a public university in the Midwest from September 6 to 30, 2012. To recruit participants with a variety of majors and class standings, 36 small general education and two large introductory lecture classes were used and students received extra credit for participating in the study. The survey took approximately 15 minutes to complete. A total of 394 students participated in the study. The respondents were asked their opinion about media use, psychological status, and demographic information.

Measurements

SNS involvement: SNS involvement, a concept proposed by Ha and Hu in 2012, is adopted in this study to measure the degree to which student users indulge in social media. Compared to time spent on social media, SNS involvement is regarded as a better measurement of students' social media use because it also takes the activity dimension into consideration and indicates how active users are. It is measured as the product of time spent on SNS, including MySpace, Facebook, YouTube, Twitter, and other social networking sites, and respondents' frequency of updating in a typical week.

Self-efficacy: Self-efficacy refers to perceived change in capability or motivation to complete tasks, socialize, and involve in public issues after using SNS. Respondents are asked in the questionnaire whether or not they feel more confident and capable in completing tasks, socializing, and involving in public issues after using social network sites. Self-efficacy is calculated by summing up three specific self-efficacy obtained after using social media.

SNS social capital: Social capital accumulated from SNS is measured by the product of network size and the diversity of friend types. Specifically, student respondents are asked to check from eleven types of people they have in their Facebook friend list, including 1) parents, 2) spouse or current romantic partner, 3) siblings and relatives,4)coworkers, 5) neighbors, 6) classmates in college, 7) people they went to high school with, 8) members of a group they belong to, 9) people they have never met in person, 10) people they have met in person only one time, and 11) other. The sum of the total number of friend types respondents checked in the category is computed as the *diversity of friend types*. The network size is measured by adding up the number of friends respondents reported on their Facebook, Twitter, and other social networking sites. Based on the median value of a product of network size and friend type diversity), respondents were split into low SNS social capital users and high SNS social capital users.

Self-efficacy status: The self-efficacy scale was adopted from Schwarzer and his colleagues (1997). Respondents are categorized into low self-efficacy and high self-efficacy users based on the median value (Median = 31).

SNS emotional support. SNS emotional support is measured by respondents' reported frequency of receiving encouraging messages from others on SNS.

Social learning on SNS: Social learning behavior on SNS is measured by respondents' reported frequency of learning something useful from others on SNS.

FINDINGS

Among the 394 students with an average age of 19.5, 201 (57.3%) were females, and 150 (42.7%) were males. There are 288 (81.4%) Caucasian students and 66 (18.6%) minority students, including 41 (11.6%) African American students, 7 (2%) Hispanic students, 3 (0.8%) Asian students, 3 (0.8%) Native American students, and 12 (3.4%) other students. With an average self-efficacy obtained score of 1.85 (n = 314, SD = 1.04), 143 students (45.3%) reported an increase of academic self-efficacy; 255 (80.7%) reported an increase in social self-efficacy; and 187 (59.4%) reported an increase of political self-efficacy after their social media use. The average hour per week they spend on SNS is 25.81 hours, including 9.64 hours on Facebook, 8.39 hours on Twitter, 5.42 hours on Youtube, 0.1 hours on Linkedln, 3.1 hours on other SNS. Most students (67.5%) update their SNS once to several times per week (see Table 2 below for detailed descriptive analysis results).

Path analysis was run to test the research model (as shown in Figure 1) step by step. All regression coefficients reported were standardized coefficients. Firstly, a weak relationship was found existed between SNS involvement and perceived self-efficacy change (β = .149, p = .013). Secondly, the moderating effect of SNS social capital and self-efficacy status were tested by creating two sets of groups in AMOS research model respectively. Self-efficacy status was identified as a significant moderator (χ^2 = 3.88, df = 1, p = .049). The positive relationship between SNS involvement and perceived self-efficacy change was stronger among low self-efficacy users (β= .255, p = .002) than high self-efficacy users (β= .122, p = n.s.). In contrast, SNS social capital measured by the product of SNS network size and diversity

Table 2. Descriptive Analysis Result

Variable Type	Variable Names	Mean (n, SD) or Distribution (n)
Dependent Variables	Self-efficacy Obtained Academic Self-efficacy obtained Social Self-efficacy obtained Political Self-efficacy obtained	1.85 (n = 314, SD = 1.04) Self-efficacy Increased: 45.3% (n = 143) Self-efficacy Increased: 80.7% (n = 255) Self-efficacy Increased: 59.4% (n = 187)
Independent Variables	SNS involvement	111.97 (n = 274, SD = 204.3)
	Frequency of learning something useful	2.84 (n = 313, SD = 1.29)
Moderators	Self-efficacy	31.32 (n = 340, SD = 5.36)
	SNS social capital	5928.31 (n = 238, SD = 5539.65)
Mediator	Frequency of receiving encouraging messages	2.97 (n = 311, SD = 1.24)

Note: Missing data were excluded from the analysis

of SNS friend types didn't serve as a significant moderator ($c^2 = 3.05$, df = 1, p = .081). However, the diversity of SNS friend types was found as a significant moderator ($\chi^2 = 4.37$, df = 1, p = .037). The association between SNS involvement and perceived self-efficacy change was more prominent among users with homogenous online social network (β= .276, p = .002) than users with heterogeneous social network (β= .103, p = n. s.). Therefore, the first three hypotheses were overall supported by the data. A positive association between SNS involvement and self-efficacy obtained was found and it was more prominent among low self-efficacy users with homogenous networks. The data set was then separated into four groups: low self-efficacy users with homogeneous social network, low self-efficacy users with heterogeneous social network, high self-efficacy users with homogeneous social network, and high self-efficacy users with heterogeneous social network. The association between SNS involvement and perceived self-efficacy change was found significant among low self-efficacy users with homogeneous social network (β= .447, p < .001) and high self-efficacy users with homogeneous social network (β= .28, p = .031), but not among low self-efficacy users with heterogeneous social network (β = .19, p = n.s.) and high self-efficacy users with heterogeneous social network (β = .159, p = n.s.). Therefore, following research questions and hypotheses related to two potential mediators were only tested among low and high self-efficacy SNS users with homogeneous social network.

Before adding SNS emotional support and social learning as two mediators into the model, RQ1a regarding SNS users' participation in online social learning behavior was addressed. According to the results, among 313 student respondents who answered the question, a large majority (81.5%) reported learning useful from other on SNS. Only 18.5% (n = 58) of the students indicated that they never learn something useful from others on SNS. Among those who reported learning something useful, 23.3% (n = 73) reported learn something useful from others on SNS once a month or less, 25.9% (n = 81) selected "once a week", 19.8% (n = 62) selected "several times a week", and 12.5% (n = 39) chose "almost daily". Being certain about SNS users' participation in social learning behavior online, both SNS social learning and emotional support were then added as two mediators into the model to test their mediating roles between SNS involvement and perceived self-efficacy change in H4, H5, RQ1b, and RQ1c. According to AMOS analysis output, the direct path between SNS involvement and perceived self-efficacy change became weaker among low self-efficacy users with homogeneous social network (β= .313, p = .036, n = 52) and high self-efficacy users with homogeneous social network (β= .142, p = n.s., n = 60). Specifically, among low self-efficacy users with homogeneous social network, their SNS involvement was positively associated with SNS emotional support (β= .425, p = .001) and SNS social learning (β = .447, p < .001), which were then positively related to perceived self-efficacy change (β= .129, p = n.s.; β = .197, p = n.s.). Due to the non-significance of the second half of the indirect path, the mediating roles of SNS social learning and SNS emotional support were not effective among low self-efficacy users with homogeneous social network. SNS involvement among high self-efficacy users with homogeneous social network was positively associated with SNS social learning (β = .159, p = n.s.) and SNS emotional support (β = .269, p = .038), which were then positively related to perceived self-efficacy change (β = .424, p < .001; β = .313, p = .004). Thus, SNS emotional support, but SNS social learning, was an effective mediator between SNS involvement and perceived self-efficacy change among high self-efficacy users with homogeneous social network. Therefore, H4 hypothesizing a significant positive association between SNS involvement and SNS emotional support was supported for both low and high self-efficacy SNS users with homogeneous social network, while H5 hypothesizing a significant positive association between SNS emotional support and perceived self-efficacy change was only supported among high self-efficacy users with homogeneous social network. Additionally, answers to RQ1b regarding the association between students' SNS involvement and their social learning was addressed with a significant positive relationship among low self-efficacy SNS users with homogeneous social network and a non-significant positive one among high self-efficacy SNS users with homogeneous social network. Regarding RQ1c, the association between SNS social learning and perceived self-efficacy change was found significant among high self-efficacy users with homogeneous social network but not among low self-efficacy users with homogeneous social network.

DISCUSSION

There are a few major findings in this study. First, the weak but positively significant association between SNS involvement and self-efficacy obtained suggests a positive potential of using SNS to boost self-efficacy, especially among homogeneous social network users. Future studies may further investigate specific SNS activities that positively relate to users' self-efficacy enhancement. For instance, previous research indicated that self-disclosure and participating in SNS groups were positively related to social capital and social support (Gil de Zúñiga, 2012; Vitak, & Ellison, 2013). In addition, the research data points to the potential mediating roles of social support and social learning on SNS involvement and self-efficacy, suggesting two possible approaches of using SNS to enhance users' self-efficacy. The fact that frequency of receiving encouraging messages on SNS among high self-efficacy users with homogeneous social network was found as an effective mediator provides empirical support for the social support approach.

Besides above major findings, this study provides some theoretical implications. Firstly, the empirical data supports the hypothesis that self-efficacy change is a gratification obtained from SNS involvement especially among low self-efficacy users. Although the self-efficacy concept has been suggested as a gratification obtained from SNS use in previous studies (Ahn, 2011; Mauri, et al., 2011; Shao, 2009; Wohn et al., 2013), this is the first study that takes it as the dependent variable and provides empirical support. That low self-efficacy users are more likely to gain social support and obtain self-efficacy via their SNS involvement supports the uses and gratifications theory. Low self-efficacy users, who usually have a higher need to increase their perceived self-efficacy, can benefit more on SNS to boost their self-efficacy than high self-efficacy users by learning or receiving encouragement from close SNS friends. In addition, the obvious positive associations between SNS emotional support and social learning and perceived increase in self-efficacy among high self-efficacy users with homogeneous network supports Bandura (1982)'s social cognitive theory on sources of self-efficacy perception.

This study also provided a few practical implications regarding social media involvement as a potential coping strategy among low self-efficacy users. Low self-efficacy users were found more likely to perceive self-efficacy increase after using SNS. One possible explanation is that low self-efficacy users were more likely to selectively involve in SNS activities that enhance their coping with stress or adversity, such as spending time connecting with close friends who emotionally support them. Besides, findings of this study suggest a positive potential of using SNS as a social learning channel, it has been argued that diversified modeling is superior to exposure to the same performance by a single role model (Bandura, 1986, p.404). It will be interesting to explore in future studies SNS use for diversified modeling and how it relates to perceived self-efficacy change after using, especially among users who have difficulties in real-life interaction.

However, as an exploratory study, measurements for quite a few important concepts were underdeveloped, such as perceived self-efficacy change or the self-efficacy obtained, SNS social capital, and SNS social support. Future studies may include more networking characteristics to measure SNS social capital, such as network homophily, centrality, density, and reciprocity (Mesch & Talmud, 2010). Social support measure may include both emotional support and informational support. Lastly, as this study were based on a cross-sectioned survey and a convenient student sample, findings should not be over generalized.

REFERENCES

Ahn, J. (2011). The effect of social network sites on adolescents' social and academic development: Current theories and controversies. *Journal of the American Society for Information Science and Technology*, *62*(8), 1435–1445. doi:10.1002/asi.21540

Apaolaza, V., Hartmann, P., Medina, E., Barrutia, J., & Echebarria, C. (2013). The relationship between socializing on the Spanish online networking site Tuenti and teenagers' subjective wellbeing: The roles of self-esteem and loneliness. *Computers in Human Behavior*, *29*(4), 1282–1289. doi:10.1016/j.chb.2013.01.002

Ballantine, P. W., & Stephenson, R. J. (2011). Help me, I'm fat! Social support in online weight loss networks. *Journal of Consumer Behaviour*, *10*(6), 332–337. doi:10.1002/cb.374

Bandura, A. (1977). *Social learning theory*. Englewood Cliffs, NJ: Prentice Hall.

Bandura, A. (1982). Self-efficacy mechanism in human agency. *The American Psychologist*, *37*(2), 122–147. doi:10.1037/0003-066X.37.2.122

Bandura, A. (1986). *Social foundations of thought and action: A social cognitive theory*. Bnglewood Cliffs, N. J.: Prentice-Hall.

Bandura, A. (2012). Social cognitive theory. In P. M. Van Lange, A. W. Kruglanski, & E. Higgins (Eds.), *Handbook of theories of social psychology* (Vol. 1, pp. 349–373). Thousand Oaks, CA: Sage Publications Ltd. doi:10.4135/9781446249215.n18

Baumeister, R. F., & Leary, M. R. (1995). The need to belong: Desire for interpersonal attachments as a fundamental human motivation. *Psychological Bulletin*, *117*(3), 497–529. doi:10.1037/0033-2909.117.3.497 PMID:7777651

Bessiere, K., Kiesler, S., Kraut, R., & Boneva, B. S. (2008). Effect of Internet use and social resources on changes in depression. *Information Communication and Society*, *11*(1), 47–70. doi:10.1080/13691180701858851

Bosker, B. (2013). *Wealthier College Students Share, Connect More On Facebook: Study*. Retrieved from http://www.huffingtonpost.com/2013/06/25/college-students- facebookstudy_n_3497733.html

Boyd, D. M., & Ellison, N. B. (2007). Social network sites: Definition, history, and scholarship. *Journal of Computer-Mediated Communication*, *13*(1), 210–230. doi:10.1111/j.1083-6101.2007.00393.x

Burke, M., & Kraut, R. (2013). *Using Facebook after losing a job: Different benefits of strong and weak ties.* Paper presented in Computer Supported Cooperative Work Conference, San Antonio, TX.

Chen, B., & Marcus, J. (2012). Students' self-presentation on Facebook: An examination of personality and self-construal factors. *Computers in Human Behavior, 28*(6), 2091–2099. doi:10.1016/j.chb.2012.06.013

Correa, T., Hinsley, A., & Gil De Zuniga, H. (2010). Who interacts on the Web? The intersection of users' personality and social media use (English). *Computers in Human Behavior, 26*(2), 247–253. doi:10.1016/j.chb.2009.09.003

Ellison, N., Lampe, C., & Steinfield, C. (2010). With a little help from my friends: social network sites and social capital. In Z. Papacharissi (Ed.), *The Networked Self: Identity, Community and Culture on Social Network Sites* (pp. 124–145). New York: Routledge.

Ellison, N. B., Steinfield, C., & Lampe, C. (2007). The Benefits of Facebook "Friends:" Social Capital and College Students' Use of Online Social Network Sites. *Journal of Computer-Mediated Communication, 12*(4), 1143–1168. doi:10.1111/j.1083-6101.2007.00367.x

Ellison, N. B., Steinfield, C., & Lampe, C. (2011). Connection strategies: Social capital implications of Facebook-enabled communication practices. *New Media & Society, 13*(6), 873–892. doi:10.1177/1461444810385389

Fang, L., & Ha, L. (2011). *Who are the heavy users of Social Network Sites among College Students? A Study of Social Network Sites and College Students.* Paper presented to the Communication Technology Division, Association for Education in Journalism and Mass Communication Association Annual Convention, St. Louis, MO.

Feaster, J. C. (2009). The repertoire niches of interpersonal media: Competition and coexistence at the level of the individual. *New Media & Society, 11*(6), 965–984. doi:10.1177/1461444809336549

Gil de Zúñiga, H., Jung, N., & Valenzuela, S. (2012). Social Media Use for News and Individuals' Social Capital, Civic Engagement and Political Participation. *Journal of Computer-Mediated Communication, 17*(3), 319–336. doi:10.1111/j.1083-6101.2012.01574.x

Gonzales, A. L., & Hancock, J. T. (2011). Mirror, Mirror on my Facebook Wall: Effects of Exposure to Facebook on Self-Esteem. *Cyberpsychology, Behavior, and Social Networking, 14*(1/2), 79–83. doi:10.1089/cyber.2009.0411 PMID:21329447

Granovetter, M. (1973). The Strength of Weak Ties. *American Journal of Sociology, 78*(6), 1360–1380. doi:10.1086/225469

Greene, J. A., Choudhry, N. K., Kilabuk, E., & Shrank, W. H. (2011). Online Social Networking by Patients with Diabetes: A Qualitative Evaluation of Communication with Facebook. *Journal of General Internal Medicine, 26*(3), 287–292. doi:10.1007/s11606-010-1526-3 PMID:20945113

Grieve, R., Indian, M., Witteveen, K., Anne Tolan, G., & Marrington, J. (2013). Face-to-face or Facebook: Can social connectedness be derived online? *Computers in Human Behavior, 29*(3), 604–609. doi:10.1016/j.chb.2012.11.017

Ha, L., & Hu, X. (2013). Social Media Involvement among College Students and General Population: Implications to Media Management. In Handbook of Social Media Management: Value Chain and Business Models in Changing Media Markets. Springer. doi:10.1007/978-3-642-28897-5_44

Hall, A. (2009). *College students' motives for using social networking sites and their relationships to personality traits*. Conference paper – International Communication Association, 1 – 38.

Jackson, L. A., Von Eye, A., Witt, E. A., Zhao, Y., & Fitzgerald, H. E. (2011). A longitudinal study of the effects of Internet use and videogame playing on academic performance and the roles of gender, race and income in these relationships. Computers in Human Behavior, 27(1), 228-239.

Karaiskos, D., Tzavellas, E., Balta, G., & Paparrigopoulos, T. (2010). Social network addiction: A new clinical disorder? *European Psychiatry*, *25*, 855. doi:10.1016/S0924-9338(10)70846-4

Kim, J., & Lee, J. (2011). The Facebook paths to happiness: effects of the number of Facebook friends and self-presentation on subjective well-being. *Cyberpsychology, Behavior And Social Networking*, *14*(6), 359-364.

Kraut, R., Patterson, M., Lundmark, V., Kiesler, S., Mukhopadhyay, T., & Scherlis, W. (1998). Internet paradox: A social technology that reduces social involvement and psychological well-being? *The American Psychologist*, *53*(9), 1017–1031. doi:10.1037/0003-066X.53.9.1017 PMID:9841579

Kuss, D. J., & Griffiths, M. D. (2011). Online social networking and addiction--a review of the psychological literature. *International Journal of Environmental Research and Public Health*, *8*(9), 3528–3552. doi:10.3390/ijerph8093528 PMID:22016701

Lee, G., Lee, J., & Kwon, S. (2011). Use of social-networking sites and subjective well-being: A study in South Korea. *Cyberpsychology, Behavior, and Social Networking*, *14*(3), 151–155. doi:10.1089/cyber.2009.0382 PMID:20649450

Mansson, D. H., & Myers, S. A. (2011). An initial examination of college students' expressions of affection through Facebook. *The Southern Communication Journal*, *76*(2), 155–168. doi:10.1080/10417940903317710

Matsuba, M. (2006). Searching for self and relationships online. *Cyberpsychology & Behavior: The Impact Of The Internet, Multimedia And Virtual Reality On Behavior. Society*, *9*(3), 275–284.

Mauri, M., Cipresso, P., Balgera, A., Villamira, M., & Riva, G. (2011). Why is Facebook so successful? Psychophysiological measures describe a core flow state while using Facebook. *Cyberpsychology, Behavior, and Social Networking*, *14*(12), 723–731. doi:10.1089/cyber.2010.0377 PMID:21879884

McKena, K. Y. A., & Bargh, J. (2000). Plan 9 from cyberspace: The implications of the Internet for personality and social psychology. *Personality and Social Psychology Review*, *4*(1), 57–75. doi:10.1207/S15327957PSPR0401_6

Mehdizadeh, S. (2010). Self-Presentation 2.0: Narcissism and Self-Esteem on Facebook. *Cyberpsychology, Behavior, and Social Networking*, *13*(4), 357–364. doi:10.1089/cyber.2009.0257 PMID:20712493

Mesch, G. S., & Talmud, I. (2010). *Wired Youth: The social world of adolescence in the information age*. London, New York: Routledge.

Mount, M., & Garcia Martinez, M. (2014). Social media: A tool for open innovation. *California Management Review*, *56*(4), 124–143. doi:10.1525/cmr.2014.56.4.124

Parker, J. G., & Asher, S. R. (1993). Friendship and friendship quality in middle childhood: Links with peer group acceptance and feelings of loneliness and social dissatisfaction. *Developmental Psychology*, *29*(4), 611–621. doi:10.1037/0012-1649.29.4.611

Parks, M. R., & Floyd, K. (1996). Making friends in cyberspace. *Journal of Communication*, *46*(1), 80–97. doi:10.1111/j.1460-2466.1996.tb01462.x

Pew Research Center. (2010). *Social media & mobile Internet use among teens and young adults*. Retrieved March 2, 2010, from http://www.pewinternet.org/Reports/2010/Social- Media-and-Young-Adults.aspx

Pew Research Center. (2012). *The Demographics of Social Media Users*. Retrieved from http://www.pewinternet.org/Reports/2013/Social-media-users.aspx

Putnam, R. (2000). *Bowling alone: Collapse and revival of American community*. New York: Simon & Schuster. doi:10.1145/358916.361990

Quan-Haase, A., & Young, A. L. (2010). Uses and gratifications of social media: A comparison of Facebook and Instant Messaging. *Technology & Society*, *30*(5), 350–361.

Raacke, J., & Bonds-Raacke, J. (2008). MySpace and Facebook: Applying the uses and gratifications theory to exploring friend-networking sites. *Cyberpsychology & Behavior*, *11*(2), 169–174. doi:10.1089/cpb.2007.0056 PMID:18422409

Rafaeli, S. (1998). Interactivity. In R. P. Hawkins, J. M. Wiemann, & S. Pingree (Eds.), *Advancing Communication Science: Merging Mass and Interpersonal Processes* (pp. 110–134). Beverly Hills, California: Sage.

Ray, M. B. (2007). *Needs, Motives, and Behaviors in computer-mediated communication: An inductive exploration of social networking websites*. Conference paper—International Communication Association, 2007 Annual Conventioin.

Schwarzer, R., Bäßler, J., Kwiatek, P., Schröder, K., & Zhang, J. (1997). The assessment of optimistic self-beliefs: Comparison of the German, Spanish, and Chinese versions of the General Self-efficacy Scale. *Applied Psychology*, *46*(1), 69–88. doi:10.1111/j.1464-0597.1997.tb01096.x

Shao, G. (2009). Understanding the appeal of user-generated media: A uses and gratification perspective. *Internet Research*, *19*(1), 7–25. doi:10.1108/10662240910927795

Sheldon, K. M., Abad, N., & Hinsch, C. (2011). A two-process view of Facebook use and relatedness need-satisfaction: Disconnection drives use, and connection rewards it. *Journal of Personality and Social Psychology*, *100*(4), 766–775. doi:10.1037/a0022407 PMID:21280967

Steinfield, C., Ellison, N. B., & Lampe, C. (2008). Social capital, self-esteem, and use of online social network sites: A longitudinal analysis. *Journal of Applied Developmental Psychology*, *29*(6), 434–445. doi:10.1016/j.appdev.2008.07.002

Sun, T., & Wu, G. (2012). Traits, Predictors, and Consequences of Facebook Self- Presentation. *Social Science Computer Review, 30*(4), 419–433. doi:10.1177/0894439311425978

Tang, T. (2013). *The role of structure in social media use: Locating synthesis in measuring social media structures*. Paper presented in 2013 Eastern Communication Association, Theory and Methodology session.

Tazghini, S., & Siedlecki, K. L. (2013). A mixed method approach to examining Facebook use and its relationship to self-esteem. *Computers in Human Behavior, 29*(3), 827–832. doi:10.1016/j.chb.2012.11.010

Tidwell, L. C., & Walther, J. B. (2002). Computer-mediated communication effects on disclosure, impressions, and interpersonal evaluations: Getting to know one another a bit at a time. *Computers in Human Behavior, 28*, 317–348.

Toma, C. L., & Hancock, J. T. (2013). Self-Affirmation Underlies Facebook Use. *Personality and Social Psychology Bulletin, 39*(3), 321–331. doi:10.1177/0146167212474694 PMID:23359086

Urista, M. A., Dong, Q., & Day, K. D. (2009). Explaining why young adults use MySpace and Facebook through uses and gratifications theory. *Human Communication, 12*(2), 215–229.

Valenzuela, S., Park, N., & Kee, K. F. (2009). Is there social capital in a social network site?: Facebook use and college students' life satisfaction, Trust, and Participation. *Journal of Computer-Mediated Communication, 14*(4), 875–901. doi:10.1111/j.1083-6101.2009.01474.x

Vitak, J., & Ellison, N. B. (2013). 'There's a network out there you might as well tap': Exploring the benefits of and barriers to exchanging informational and support-based resources on Facebook. *New Media & Society, 15*(2), 243–259. doi:10.1177/1461444812451566

Wise, K., Alhabash, S., & Park, H. (2010). Emotional responses during social information seeking on Facebook. *Cyberpsychology, Behavior, and Social Networking, 13*(5), 555–562. doi:10.1089/cyber.2009.0365 PMID:20950180

Wohn, D., Ellison, N., Khan, M., Fewins-Bliss, R., & Gray, R. (2013). The role of social media in shaping first-generation high school students' college aspirations: A social capital lens. *Computers & Education, 63*, 424–436. doi:10.1016/j.compedu.2013.01.004

Wright, K. (2012). Similarity, network convergence, and availability of emotional support as predictors of strong-tie/weak-tie support network preference on Facebook. *The Southern Communication Journal, 77*(5), 389–402. doi:10.1080/1041794X.2012.681003

Wright, K., Rosenberg, J., Egbert, N., Ploeger, N., Bernard, D., & King, S. (2013). Communication competence, social support, and depression among college students: A model of Facebook and face-to-face support network influence. *Journal of Health Communication, 18*(1), 41–57. doi:10.1080/10810730.2012.688250 PMID:23030518

Yu, A. Y., Tian, S. W., Vogel, D., & Chi-Wai Kwok, R. (2010). Can learning be virtually boosted? An investigation of online social networking impacts. *Computers & Education, 55*(4), 1494–1503. doi:10.1016/j.compedu.2010.06.015

Zhang, Y., Tang, L., & Leung, L. (2011). Gratifications, collective self-esteem, online emotional openness, and traitlike communication apprehension as predictors of Facebook uses. *Cyberpsychology, Behavior, and Social Networking*, *14*(12), 733–739. doi:10.1089/cyber.2010.0042 PMID:21651417

Zywica, J., & Danowski, J. (2008). The Faces of Facebookers: Investigating Social Enhancement and Social Compensation Hypotheses; Predicting Facebook™ and Offline Popularity from Sociability and Self-Esteem, and Mapping the Meanings of Popularity with Semantic Networks. *Journal of Computer-Mediated Communication*, *14*(1), 1–34. doi:10.1111/j.1083-6101.2008.01429.x

KEY TERMS AND DEFINITIONS

Self-Efficacy: A key element of social cognitive theory that serves to connect information processed from one's environment with consequential behavioral outcomes.

Social Cognitive Theory: States that human behavior is regulated by both external social environmental factors and internal cognitive factors.

Social Networking Sites: The latest online communication tools—including MySpace, Facebook, YouTube, Twitter, and others—that allow users to create a public or private profile to interact with people in their networks.

APPENDIX

Survey Questions (Only Including Related Questions)

1. How many hours do you spend on social network sites in a typical week (including reading, posting, forwarding messages, etc.)? Write 0 if you don't use that site.
   a. Facebook ____
   b. Linkedln ____
   c. Twitter ____
   d. Youtube ____
   e. Other social network sites (Please specify) ____
2. How frequently do you update your social network page?
   a. Several times an hour
   b. Every several hours
   c. Everyday
   d. Once to several times a week
   e. Between once a week and once a month
   f. Less than once a month
   g. Hardly ever update
3. Do you feel more confident and capable in the following areas after using social networking cites?
   a. Completing tasks Yes__ No__
   b. Socializing Yes__ No__
   c. Involving in public issues Yes__ No__
4. Approximately how many "Friends" or "Followers" do you have on the main social networking sites that you use? Write 0 if you don't use that social networking site.
   a. Facebook ______
   b. Twitter ______
   c. Other social networking sites ______
5. Please check the type of people you have on your Facebook friend list in the following categories (Check all that apply).
   __ a. Parents
   __ b. Spouse or current romantic partner
   __ c. Siblings and relatives (Other than parents or spouse)
   __ d. Co-workers from your current or previous job
   __ e. Neighbors
   __ f. Classmates in college
   __ g. People you went to high school with
   __ h. Members of a group you belong to
   __ i. People you have never met in person
   __ j. People you have met in person only one time
   __ k. Other

7. Please check the categories that best represent yourself for each statement below.
    a. I can always manage to solve difficult problems if I try hard enough
    Not at all true Barely true Moderately true Exactly true
    b. If someone opposes me, I can find means and ways to get what I want
    Not at all true Barely true Moderately true Exactly true
    c. It is easy for me to stick to my aims and accomplish my goals
    Not at all true Barely true Moderately true Exactly true
    d. I am confident that I could deal efficiently with unexpected events
    Not at all true Barely true Moderately true Exactly true
    e. Thanks to my resourcefulness, I know how to handle unforeseen situations
    Not at all true Barely true Moderately true Exactly true
    f. I can solve most problems if I invest that necessary effort
    Not at all true Barely true Moderately true Exactly true
    g. I can remain calm when facing difficulties because I can rely on my coping abilities
    Not at all true Barely true Moderately true Exactly true
    h. When I am confronted with a problem, I can usually find several solutions
    Not at all true Barely true Moderately true Exactly true
    i. If I am in a bind, I can usually think of something to do
    Not at all true Barely true Moderately true Exactly true
    j. No matter what comes my way, I'm usually able to handle it
    Not at all true Barely true Moderately true Exactly true
8. How frequent do you do anything below on your social networking sites?
    a. Learn something useful from others
    Almost daily Several times a week Once a week Once a month or less Never
    b. Receive encouraging messages from others
    Almost daily Several times a week Once a week Once a month or less Never
9. What is your age? ____
10. What's your gender? Male ____ Female ____
11. What's your ethnicity?
    a. Caucasian
    b. African-American
    c. Hispanic
    d. Asian
    e. Native American
    f. Other

Chapter 14
What Motivates Online Shoppers to "Like" Brands' Facebook Fan Pages?

Mohammad Hatim Abuljadail
Bowling Green State University, USA

Louisa Ha
Bowling Green State University, USA

Fang Wang
Bowling Green State University, USA

Liu Yang
Bowling Green State University, USA

ABSTRACT

Brands' Facebook fan pages have been frequently used as a marketing tool to reach to more individuals; however, Facebook users' motivations to participate in those fan pages are still unclear. This paper investigates the motives that stimulate online shoppers to "like" brands' Facebook fan pages. This study is interested in knowing if online shoppers' proclivity to "like" brands' Facebook fan pages is based on their hedonic or utilitarian motivations. The authors propose a model based on hedonic and utilitarian motivations and uses and gratifications theoretical framework. An online survey was conducted among college students who shopped online in Northwest Ohio (N=198). The findings show that utilitarian motivations have positive significant relationships with "liking" brands' Facebook fan pages.

INTRODUCTION

The emergence of new media and the growing popularity of social network sites (SNS) such as Facebook, LinkedIn, Twitter, and Google have a great impact on people's lives (Xu, Ryan, Prybutok & Wen, 2012). There were 864 million daily active Facebook users in September, 2014 (Facebook, 2014). Organizations set up fan pages to reach Facebook users and cultivate consumer loyalty. In 2012, there were more than 1.5 million organization fan pages on Facebook and 20 million users "like" those Facebook fan pages (JeanJean, 2012).

The strength of the online social network sites as a marketing tool is that they are structured around users (Mislovem Marcon, Gummadi, Druschel & Bhattacharjee, 2007). This structure helps to utilize the interconnectedness of the users to reach to more customers with relatively less effort and low costs.

DOI: 10.4018/978-1-4666-8450-8.ch014

Therefore, marketers increasingly show interests in using Facebook for their marketing.In 2013, 92% of marketers reported that they utilize Facebook as a marketing tool. (SocialMediaExaminer, 2013). Despite the potential and contribution of social networking sites to brands, few studies have investigated specifically the use of fan pages on Facebook (Wallace, Buil, de Chernatony & Hogan, 2014). As so many organizations utilize fan pages on Facebook, a theoretical framework is needed to help organizations to understand underlying reasons behind users' motivations to participate in brands' Facebook pages.

Many previous studies have investigated users' participation motives in traditional brand communities. Muniz & O'guinn (2000) study traditional brand communities and they define it as specialized, non-geographically bound communities, based on a structured set of social relationships among customers of a brand. They state that brand community is marked by a shared consciousness, rituals and traditions and a sense of moral responsibility. However new online communities, including social networking sites, are more powerful. Through social networks sites users are capable of interacting with other users and with the organization itself, with relatively little effort. Consequently, users' participation in brands' social media pages can be classified into two groups in regards to their devotion to the brand: devoted and casual users. Devoted users are those who are more "committed" to the brand and visit the brand communities frequently. Furthermore, previous studies have explored consumption behavior from the utilitarian and hedonic dimensions. Utilitarian motivation is defined as goal-oriented and rational motivation, which is concerned with instrumental value and effectiveness of the consumption, while hedonic refers to seeking fun, play, enjoyment and experiences of the consumption (Babin, Darden & Griffin,1994; Batra and Ahtola 1990; Voss, Spangenberg & Grohmann, 2003; Wertenbroch and Dhar 2000). The hedonic and utilitarian model has been used to explain consumption or usage behaviors, such as using the internet for information seeking, entertainment, socializing, and status seeking behavior (Cotte, Chowdhury, Ratneshwar & Ricci, 2006; Hartman, Shim, Barber & O'Brien, 2006; Maenpaa, Kanto, Kuusela & Paul, 2006). In the context of internet usage behavior, companies' online communities represent an important research area, as they have been shown to have a positive effect on brand loyalty (Jang et al., 2008). Previous studies have emphasized the importance of the utilitarian value of online communities for users (Bateman, Gray & Butler, 2011; Casaló, Flavian, & Guinaliu, 2010; Sanchez-Franco & Rondan-Cataluña, 2010).

Facebook offers the tools to help organizations to create online communities that foster the relationship between the brand and individuals. Facebook fan pages are similar to users' personal profiles, but are designed for organizations, and they are open to the public. Users can become fans or follow the brands' fan pages on Facebook by "liking" the page. Due to the community attributes of Facebook fan pages, members are able to find other members who share their interest (Koh & Kim, 2003; Muñiz & O'Guinn, 2001) and are able to interact with them through "liking", posting, commenting and sharing content. Many brands have been heavily using Facebook fan pages as a marketing communication tool; it is important to recruit fans for their Facebook page because fans are those that receive updates automatically from the page and traceable for the organizations. However, many marketers are still trying to search for the magic key to attract fans for their Facebook pages. This paper aims to investigate these motives of "liking" brands' Facebook fan pages by developing a model based on hedonic and utilitarian motivations approach and uses and gratifications perspective to help organizations better understand the reasons behind their customers' "liking" their brands' Facebook fan pages and develop their Facebook page content strategies accordingly. The research questions are as follows:

RQ1: What is the relationship between utilitarian motivation of social media use and "liking" brands' Facebook fan pages?

RQ2: What is the relationship between hedonic motivation of social media use and "liking" brands' Facebook fan pages?

LITERATURE REVIEW

Online Brand Communities

One of the online communities' important attributes is its ability to function without the restrictions of geographic and times limits, differentiating it from traditional (offline) communities (Bagozzi & Dholakia, 2002; Kozinets 2002). Online communities usually depend on users' shared interests (Bagozzi & Dholakia, 2002). Carlson, Suter, and Brown (2008) indicate that "social brand communities" consist of members who may never interact face-to-face, but can interact in that virtual community with other members, who share the same interests. According to Woisetschlager, D.M, Hartleb and Blut (2008), individuals join and participate in a community in order to discuss their shared experiences with other people who are interested in the same brand. Similarly, Alagoz, S, Ekici and Islek (2011) emphasize the importance of brand community and its effects on bonding customers' relationships. Thompson and Sinha (2008) indicate that the higher participation and longer term membership in a brand community increases the likelihood of product adoption from that particular brand. For example, members of Intel online forums are more likely to use new Intel products, and they are less willing to use products by AMD (the rival brand) unless they could not find Intel products. (Thompson & Sinha, 2008).

Brodie, Ilic, Juric and Hollebeek (2011) discuss the complexity of the consumer engagement process which is determined by the level of consumer participation over time in that particular online brand community. Sicilia and Palazon (2008) suggest that online communities can promote the development of loyal consumers around a particular brand when certain values are provided in the online community, such as functional, social and experimental values.

Brands' Facebook fan pages have attributes that can be classified as online communities. Although a brand representative is usually assigned to manage the page and to produce the content, Facebook fan pages also enable users to produce content on the page. In addition, another attribute that classifies brands' Facebook fan pages as communities is that its members usually share a common interest-- that shared common interest in the brand is perhaps one of the main reasons why they "like" the page, and why they keep coming back to interact with those of similar interests.

Facebook Use

Individuals use Facebook for different reasons; such as maintaining relationships (Bosch, 2009; Lampe, Ellison, & Steinfield, 2006; Lewis & West, 2009), making new relationships (Lampe et al., 2006; Sheldon, 2008), increasing user's popularity (Urista, Qingwen & Day, (2009), expressing oneself (Joinson, 2008)., having fun (Lewis & West, 2009; Sheldon, 2008), passing time (Joinson, 2008; Sheldon, 2008)., learning (Bosch, 2009; Pempek et al., 2009), and completing tasks (Young and Quan-Haase, 2009). Users who participate in Facebook for maintaining relationships are more likely to read friends' news feeds, send messages to their friends or give comments to their existing friends. People who use Facebook for developing new relationships are more likely to search for other people's information on Facebook. Users could listen to music, watch videos, read news shared by others, or play games on Facebook for fun or passing time. People who perceive Facebook as a self-broadcasting platform may often update their status or post pictures to express themselves or make themselves more popular.

Facebook fans and non-fans have different patterns in using social media. Hu et al. (2013) have found that Facebook brand page fans are more active than non-fans on social network sites. Fans are more likely to stay in touch with their friends, find romantic partners, post news and pictures, links to other websites, write comments, send instant messages, and tag people than non-fans. In regards to entertainment related activities, fans are more likely to read posts of celebrities, politicians or athletes, and play games than non-fans. In terms of brand and shopping information, fans are more likely to search for shopping information than non-fans (Hu et al., 2013). Therefore, building upon existing knowledge of Facebook fans' behavior, the purpose of this study is to investigate specifically the relationship between hedonic and utilitarian motivations and the "liking" brands' Facebook fan pages.

Facebook Fan Pages Use

Facebook's popularity has attracted companies to use Facebook as a new brand promotion channel. The strategies to attract more people to "like" the retailers' Facebook fan page play an important role in many companies' marketing plans. Companies could develop new customers, make existing customers more loyal, and interact with their fans through Facebook fan pages. O'Connor's (2013) study shows that there is a statistically significant correlation between the number of fans in brands' Facebook fan pages of the most popular brands and their company stock prices. O'Connor (2013) suggests that the social media sites' popularity could serve as a behavioral indicator of brand affinity, customer loyalty, or brand performance.

Brands' Facebook fan pages can provide texts, sale announcements, discounts, advertisements, videos or links about their products or services to their fans. Subsequently, their fans can interact with other fans or with the organization itself. Fans' participation in a brand's fan pages automatically appears in fans' news feed and in their friends' news feed as well, which becomes free word-of-mouth for retailers or brands (Hu et al., 2013). Facebook fan pages provide various communication tools such as liking, sharing, and posting comments. Individuals could use these features to interact with other fans who share the same interests, which might lead to sharing, liking, or reposting messages that promote the brand. This interaction with other fans who share the same interest could encourage fans to participate more in the fan page. Previous studies indicate that there is a positive relationship between SNS users' common and shared values and the likelihood of joining Facebook fan pages (Lin & Lu, 2011; Li & Bernoff, 2008; Powell, 2009; Qualman, 2009). Lin and Lu's (2011) findings indicate that social interaction ties, shared values and trust in the information offered by the Facebook fan page encourage users to keep using Facebook fan pages.

Chiu, Hsu and Wang (2006) explore the importance of the consistency and trustworthiness of the information available on social network sites; their findings indicate that the consistency of information is related to the likelihood of individuals to use the site more. Similarly, previous studies show that participation in online communities is positively related to the benefits perceived from that community (Morgan & Hunt, 1994). In addition, several studies have explored the benefits fans get from Facebook fan pages. Kanga, Tang and Fiorec (2013) show that there are several benefits users gain from participating in Facebook fan pages: functional benefits, social psychological benefits, hedonic benefits and monetary benefits. Functional benefits refer to goal oriented functions, such as information exchange. Social benefits refer to the support and help fans get from Facebook fan communities (Wang & Fesenmaier, 2004). Psychological benefits refer to the feelings of belonging to the fan communities (Wang & Fesenmaier, 2004). Hedonic benefits are classified into two: self-hedonic benefit and community-hedonic benefit (Akkinen & Tuunainen, 2004). Self-hedonic benefit refers to having fun, relaxation and killing time. Community-hedonic benefits refer to playing games or interacting with other fans. Monetary benefits refer to how fans could get discounts or coupons through Facebook fan pages. All these benefits stimulate fans to actively participate in the fan communities. The more benefits fans get from Facebook fan pages, the more likely they will be to continue to participate in the fan pages.

THEORETICAL FRAMEWORK

Uses and Gratifications Theory (U&G)

Uses and gratifications is defined as the social and psychological roots of needs and wants, that create expectations of the media and other communication sources used, which leads to different kinds of exposure to the media resulting in needs and wants gratifications and other consequences (Katz, Blumler & Gurevitch, 1974). The uses and gratifications theory emphasizes the audience's use of media and other forms of communication such as interpersonal communication to satisfy their desires and needs (Rubin, 2002). Uses and gratifications theory is a psychological perspective that differs from the traditional media consumption approaches that assume media consumers as passive. This theory proposes that media use is determined by individuals' needs and motives to communicate, the psychological and social environment, the media use, media use alternative, communication behavior, and consequences of such behavior (Rubin, 1994).

Previous U&G studies have examined how the use of media fulfills cognitive and emotional needs, including personal and entertainment needs (Rubin, 2002). SNS have bridged mass media and interpersonal communication (Papacharissi, 2010), which attracted many scholars to investigate the uses and gratifications of those new media. Stafford and Gonier (2004) claim that there are gratifications, caused by the internet, that trigger behaviors, such as web searching, information seeking, and interpersonal and social engagement.

Internet use serves cognitive (utilitarian) and affective (hedonic) needs and wants on the mass media and interpersonal levels. Interpersonal utility functions refer to the cognitive and rational interpersonal needs (Leung, 2007). In addition, internet use can also be triggered by hedonic motivations such as relaxation, entertainment, encouragement, and status (Urista et al., 2009). Individuals use the internet for different reasons such as entertainment, information seeking, and socializing (Mutsuba, 2006).

The uses and gratifications theory suggests that people use media to satisfy their various needs and wants (Blumler & Katz, 1974), and media users are active consumers of media (Katz et al., 1974). Facebook and other SNS platforms provide the capacity for users to fulfill those various needs and wants within one massive network (Urista et al., 2009).

Hedonic and Utilitarian Motivations

Individuals' goals and activities are often categorized into hedonic or utilitarian (Khan, Dhar & Wertenbroch, 2005). Hedonic experiences can be fun, affective, and impulsive (Holbrook & Hirschman, 1982). On the other hand, utilitarian activities are functional, sensible, and useful (Botti & McGill, 2011). Utilitarian actions are easier to justify compared to hedonic activities (Wertenbroch & Dhar, 2000). Nonetheless, hedonic experiences are affectively richer than utilitarian experience; and attitudes and behaviors are more likely to be shaped by affective factors (Botti & McGill, 2011). Inclinations to hedonic activities are emotionally driven, while those toward utilitarian activities are usually cognitively driven (Holbrook & Havlena, 1986; Shiv & Fedorikhin, 1999).

Motivational triggers determine if an experience is hedonic or utilitarian in nature (Khan et al., 2005). Individuals consume based on two motives: hedonic reasons that are emotionally driven, and instrumental (utilitarian) motives that are rational and concerned with expectations of consequences (Batra & Ahtola, 1990). Therefore, hedonic activities are internally and innately motivated and inherently rewarding and that's why it is sought. On the other hand, utilitarian activities are externally motivated and they are considered instrumental to achieving the goals that are not inherently rewarding, but are usually focused on long-term goals or externally rewarding goals (Holbrook & Hirschman, 1982; Wertenbroch & Dhar, 2000). The difference between internal and external triggers is based mainly on the goal of pursuing the activity; the goal orientation might or might not be self-motivated (Moller, Arlen, Ryan & Deci, 2006).

Previous studies have focused on motivations of online shopping behaviors (Bridges & Florsheim, 2008), design of online shopping space in relation to shopping motivations (Kourouthanassis, Giaglis & Vrechopoulos, 2007), and concluded that utilitarian motivations aspects are the main aspects to motivate people to make purchasing decisions online (Mikalef, Giannakos & Pateli, 2013). Hyllegard, Ogle, Yan and Reitz (2011) findings show that utilitarian motives positively predict product and brand consumption.

On the other hand, Kim and Shim (2002) argue that individuals browse the internet not only for information seeking about particular products, but also emotional fulfillment. Both hedonic and utilitarian motivations have been used in previous studies in different contexts, where either utilitarian or hedonic motivations were the main forces of particular behaviors. However, few studies have investigated the motivations of participating in SNS through the utilitarian and hedonic motivations perspective and the particular motives that could predict particular participation behaviors on SNS especially in brand communities such as Facebook fan pages.

This study focuses on "liking" brands' Facebook fan pages and the triggers that motivate this particular behavior. For the purpose of this study, the authors will use the following definitions: utilitarian motivations are the social media use that are goal oriented and rational behavior, while hedonic motivations are the usage that are concerned with aspects of fun, playfulness and enjoyment of social media (Babin, Darden & Griffin; 1994; Batra & Ahtola 1990). This study proposes a model that examines the relationship between online shoppers' motivations (hedonic and utilitarian) and "liking" behavior. Utilitarian and hedonic motivations may both predict the liking of Facebook fan pages. Based on the literature review and the theoretical framework, we pose the following research questions on examine the specific relationship between liking of Facebook fan pages and type of motivations of social media use:

RQ1: What is the relationship between utilitarian motivation of social media use and "liking" brands' Facebook fan pages?

RQ2: What is the relationship between hedonic motivation of social media use and "liking" brands' Facebook fan pages?

RESEARCH METHOD

Data and Sampling

The data for this study was acquired via a self-administered web survey of college students from September 6th to 30th, 2012. Thirty-six small general education classes and two large introductory lecture courses, with a variety of majors and class standings from a public university in the Midwest United States, were used to recruit respondents. Students received extra credit for participating in the study and 394 responses were received. Only those who shopped online were asked if they "like" Facebook fan pages to examine the fan page use's relationship with online shopping. In this study, 196 respondents did not shop online, so the final sample size was 198. The questionnaire consisted of items measuring respondents' social media use motivations, and "Liking" levels of brands' Facebook fan pages. The authors inquired the respondents' frequency of "liking" brands' Facebook fan pages.

Measurements

"Liking": To measure "liking" brands' Facebook fan pages, respondents were asked to indicate their frequency of "liking" the brands' official fan pages on Facebook. The question is: Did you "Like" (become a fan of) official pages of any online retailers on Facebook? The levels of frequency are: 1) *Yes, all the time;* 2) *Yes, some of them; 3) Yes, only a few of them; and 4) No, I have never been a fan of these official pages on Facebook.* The authors coded reversely to be consistent with the positive indication. Among the 198 respondents, the average score of "liking" brands' Facebook fan pages is 2.9 with standard deviation of 0.83 (M=2.9, SD=.83).

Utilitarian motivations: To measure utilitarian motivations, the respondents were asked to indicate their frequency of (1) *reading news posted on the social media site*, (2) *finding some useful shopping information, such as coupons or deal information* in a 5-point Likert Scale. In order to be consistent with the positive way of indication in the study, the authors recoded the frequency with 5 indicating the highest level (almost daily) and 1 indicating the lowest level (never). An index created by averaging two items indicates the levels of utilitarian motivations of social media use (M=3.50, SD=0.97).

Hedonic motivations: To measure hedonic motivations, the respondents were asked to indicate their frequency of using different social media activities in the 5-point scale ranging from *"almost daily"* to *"never."* In order to be consistent with the positive way of indication in the study, the authors recoded the frequency with 5 indicating the highest level (almost daily) and 1 indicating the lowest level (never). The items include: (1) *stay in touch with family members, including all communications with them on SNS; (2) stay in touch with close friends, including all communications with them on SNS; (3) connect with old friends that you lost touch with; (4) make new friends;(5) connect with other people who share your hobbies or interests; (6) find potential romantic partners or people you might like to date; (7) send instant messages to or chat with a friend.* The index of averaging the above items indicates the levels of hedonic motivation of social media use (M=3.21, SD=0.82).

Online shopping: To measure online shopping, the respondents were asked to indicate their frequency of online shopping (number of times) in a typical month. Among 198 respondents, the average score of online shopping is 3.64 with standard deviation of 6.96 (M=3.64, SD=6.96).

Number of years of using Facebook member: Respondents were asked to give a specific number to the question: "How many years have you been a member of Facebook"? Among 198 respondents, the average experience of being a member in Facebook was 5.49 years with 1.84 standard deviation (M=5.49, SD=1.84).

Number of friends on Facebook: Respondents were asked to give an approximate number of the question: "Approximately how many "friends" do you have on Facebook"? Among 198 respondents, the average score of Facebook network size was 63 (M= 63, SD= 43).

Number of hours per week of Facebook use: Respondents were asked to report the number of hours they spend on Facebook in a typical week. The average score of Facebook use was 9.67 hours (M= 9.67, SD= 15.81).

Facebook update: A 7-point Likert Scale was designed to measure Facebook update frequency. The respondents were asked to report their frequency of Facebook updates by asking the question: "*How frequently do you update your Facebook page*"? In the scale, 7 indicates "*Several times an hour*", 6 indicates "*Every several hours*", 5 is "*Everyday*", 4 is "*Once to several times a week*", 3 is "*Between once a week and once a month,*" 2 is "*Less than once a month,*" and 1 is "*Hardly every update*" . The average score of Facebook update frequency was 4.12 (M= 4.12, SD=1.57).

RESULTS

In this study, 198 college students were included in the final analysis. A multiple hierarchical regression is used to test if online shoppers' utilitarian motivation and hedonic motivation can predict their "Liking" behavior of brands' Facebook fan pages, controlling the demographic variables including age, personal disposable income, and students' GPA, as well as participants' Facebook use including the number of years of using Facebook, the number of friends on Facebook, the number of hours spent on Facebook per week, and Facebook update frequency. Five outlier cases were removed from the analysis after Mahalanobis test. Participants ranged in age from 18 to 38 with average of 20 and standard deviation of 2.4 (*M=20, SD=2.4*), with 81 (40.9%) males and 117 (59.1%) females. Participants' average GPA was 2.28 with the standard deviation .96(M=2.28, SD=0.96). Four levels of disposable personal income ranged from 1 to 4 with an average of 1.44 with the standard deviation of 0.73 (M=1.44, SD=0.73).

As listed in table 1, multiple regression test shows that the overall model is significant in predicting liking behavior with an adjusted R square of .124 ($R^2_{adjusted}$=.124, $F_{(10,182)}$ =3.73, p<.001), see table 1. Among the controlling variables, the results show that number of friends on Facebook (*Beta*= 0.15, t=2.07, p <.05) and Facebook update frequency (*Beta*= 0.21, t=2.68, p <.01) can significantly predict liking behavior among college students. On average, those who have more friends on Facebook are more likely to "like" brand's Facebook fan pages. Those who have a higher frequency in updating their information on Facebook are more likely to click "like" on brand's Facebook fan pages. Approximately one unit increase of the number of friends on Facebook will result in 0.15 increase of "liking". One unit increase of Facebook update frequency will contribute to 0.21 increase of liking brand's Facebook fan pages.

Multiple regression also revealed that utilitarian motivation has a positive significant relationship with the "liking" behavior of brands' Facebook fan pages, when variables such as demographics and Facebook use are controlled; it predicts the "liking" behavior of brands' Facebook fan pages (*Beta*= 0.18, t=2.17,

Table 1. Predictors of "Liking" Brands' Facebook Fan Pages

Variables	*β*	*t*
Age	-.048	-.66
Personal Income	-.062	-.87
GPA	-.077	-1.12
Online Shopping	.12	1.64
Number of Years of Using Facebook	-.023	-.32
Number of Friends on Facebook	.15*	2.07
Facebook Use Hours/Week	.05	.65
Facebook Update Frequency	.21**	2.68
Utilitarian Motivation	.18*	2.17
Hedonic Motivation	.03	.34
Adjusted R^2	.124	

Note. N=193. * p<.05. **p<.01.

p <.05). On average, one unit increase of utilitarian motivation of social media use will lead to 0.18 increase of the "Liking" behavior of brands' Facebook fan pages. However, there was not a significant relationship between hedonic motivation of social media use and "liking" brands' Facebook fan pages (*Beta*= 0.29, t=0.34, p =0.73), when controlling demographic variables and Facebook use variables.

DISCUSSION

The findings of this study provide useful insights into understanding "what motivates online shoppers to "like" brands' Facebook fan pages". The authors investigate the relationship between online shoppers motivations (hedonic and utilitarian) of social media use to "liking" brands' Facebook fan pages. The results show a statistically significant relationship between utilitarian motivations of social media use and "liking" brands' Facebook fan pages, but not between hedonic motivations and liking of Facebook brand fan pages (see Table 1).

The findings of the study show a difference between utilitarian and hedonic motivations. Hedonic motivations of social media use appear to have no statistically significant relationship with "liking" brands' Facebook fan pages. The results are consistent with previous studies that claim that the utilitarian motivations aspects are more significant than hedonic motivations to stimulate people's participation behavior in brands' communities (Bridges & Florsheim, 2008; Hyllegard et al., 2011; Kourouthanassis et al., 2007; Mikalef et al., 2013). Individuals' behaviors that are based on utilitarian motivations will be more likely to seek valuable information from their participation through Facebook fan page (Kim et al. 2002). The results of this study emphasize the importance of providing useful information to consumers in increasing the liking of brands' Facebook fan pages.

IMPLICATIONS TO MARKETERS

The results show a difference between utilitarian and hedonic motivations in relation to "liking" brands' Facebook fan pages. Utilitarian motivations predict the "liking" behavior on brands' Facebook fan pages while hedonic motivations did not have a statistical significant relationship with "liking" brands' Facebook fan pages. One challenge the results of this study pose to marketers on Facebook is how to capitalize on the utilitarian motivation of online shoppers. Marketers should evaluate what kinds of product information (including distribution/retailers, location, product updates, and sales announcement or even customer experience testimonials) are considered important to consumers, post the type that they deemed important and encourage users to "like" the fan page to receive automatic updates from the marketer. Marketers should utilize their Facebook pages to provide information useful to consumers to entice them "like" their pages such as discounts, special sales announcements, product features, and production demonstration videos and pictures. In addition, the results of the study shows that active Facebook users who have larger network size and updating their Facebook page more frequently are more likely to be liking the brand Facebook page. These types of fan attributes is conducive to spreading the brand's information for the market through these fans' networks of friends.

LIMITATIONS AND SUGGESTIONS FOR FUTURE RESEARCH

There are some limitations in this study; the study limits to online shoppers' Facebook brands' fan pages. The fan pages are limited to commercial brand Facebook pages, not non-profit organization Facebook fan pages. Future research can compare the liking of commercial and non-profit organizations' Facebook fan pages.

In addition, non-online shoppers Facebook users were excluded from the study so the participants' probability of liking fan pages is probably higher in this study. Many specific uses of fan pages were not examined and future studies should include specific features of fan pages and incentives such as discount coupons to sign up more fans. A content analysis of the features used by Facebook pages with many likes will be a good direction of future research to establish best practices of creating "likes" from brand pages. Furthermore, in regards to the sample, this study focused on a younger population. While the authors think that the younger population is important due to their heavy use of social media, it is still important to capture different generations, such as older social media users. Thus, future studies need to explore the behavior of other age groups.

Despite these limitations, this paper provides empirical evidence on the typology of motives that influence Facebook users the most to "like" brands' Facebook fan pages. This findings have implications to both marketing practitioners and researchers. Marketing practitioners must know the motives of their brand's fans participation in order to better serve their consumers which also reflects on the brand's success and loyalty.

REFERENCES

Akkinen, M., & Tuunainen, V. (2004). The sources of value creation in consumer–oriented online communities. In conference "Frontiers of e-Business Research", Tampere, FI.

Alagoz, S., Ekici, N., & Islek, M. (2011). Brand communities in the axis of scialaization customers: Sample of Volkswagen Beetle owners, Turkey. (English). *Ege Academic Review,* 465-477.

Babin, B. J., Darden, W. R., & Griffin, M. (1994). Work and/or fun: Measuring hedonic and utilitarian shopping value. *The Journal of Consumer Research, 20*(4), 644–656. doi:10.1086/209376

Bagozzi, R. P., & Dholakia, U. M. (2002). Intentional social action in virtual communities. *Journal of Interactive Marketing, 16*(2), 2–21. doi:10.1002/dir.10006

Bateman, P. J., Gray, P. H., & Butler, B. S. (2011). Research Note—The Impact of Community Commitment on Participation in Online Communities. *Information Systems Research, 22*(4), 841–854. doi:10.1287/isre.1090.0265

Batra, R., & Ahtola, O. T. (1990). Measuring the Hedonic and Utilitarian Sources of Consumer Attitudes. *Marketing Letters, 2*(April), 159–170.

Blumler, J. G., & Katz, E. (1974). *The uses of mass communication.* Newbury Park, CA: Sage.

Bosch, T. E. (2009). Using online social networking for teaching and learning: Facebook use at the university of cape town. *Communication: South African Journal for Communication Theory and Research, 35*(2), 185–200.

Botti, S., & McGill, A. L. (2011). The locus of choice: Personal causality and satisfaction with hedonic and utilitarian decisions. *The Journal of Consumer Research, 37*(6), 1065–1078. doi:10.1086/656570

Bridges, E., & Florsheim, R. (2008). Hedonic and utilitarian shopping goals: The online experience. *Journal of Business Research, 61*(4), 309–314. doi:10.1016/j.jbusres.2007.06.017

Brodie, R. J., Ilic, A., Juric, B., & Hollebeek, L. (2011). Consumer engagement in a virtual brand community: An exploratory analysis. *Journal of Business Research,* 1–10.

Carlson, B. D., Suter, T. A., & Brown, T. J. (2008). Social versus psychological brand community: The role of psychological sense of brand community. *Journal of Business Research, 61*(4), 284–291. doi:10.1016/j.jbusres.2007.06.022

Casaló, L. V., Flavián, C., & Guinalíu, M. (2010). Antecedents and consequences of consumer participation in on-line communities: The case of the travel sector. *International Journal of Electronic Commerce, 15*(2), 137–167. doi:10.2753/JEC1086-4415150205

Chiu, C. M., Hsu, M. H., & Wang, E. T. G. (2006). Understanding knowledge sharing in virtual communities: An integration of social capital and social cognitive theories. *Decision Support Systems, 42*(3), 1872–1888. doi:10.1016/j.dss.2006.04.001

Cotte, J., Chowdhury, T. G., Ratneshwar, S., & Ricci, L. M. (2006). Pleasure or utility? Time planning style and web usage behaviors. *Journal of Interactive Marketing, 20*(1), 45–57. doi:10.1002/dir.20055

Facebook (2014). Facebook newsroom. http://newsroom.fb.com/company-info/

Facebook fanpages: Maximizing consumer benefits and increasing active participation. (2014). *International Journal of Hospitality Management*, *36*, 145–155. doi:10.1016/j.ijhm.2013.08.015

Hartman, J. B., Shim, S., Barber, B., & O'Brien, M. (2006). Adolescents' utilitarian and hedonic Web consumption behavior: Hierarchical influence of personal values and innovativeness. *Psychology and Marketing*, *23*(10), 813–839. doi:10.1002/mar.20135

Holbrook, M. B., & Havlena, W. J. (1986). The varieties of consumption experience: Comparing two typologies of emotion in consumer behavior. *The Journal of Consumer Research*, 394–404.

Holbrook, M. B., & Hirschman, E. C. (1982). The experiential aspects of consumption: Consumer fantasies, feelings, and fun. *The Journal of Consumer Research*, *9*(2), 132–140. doi:10.1086/208906

Hu, X., Ha, L., Mo, S. M., & Xu, Y. (2013). Who are the fans of Facebook fan pages? An electronic word-of-mouth communication perspective. A paper presented to the International Conference on Internet Studies, September 7-8, Hong Kong China.

Hyllegard, K. H., Ogle, J., Yan, R., & Reitz, A. R. (2011). An exploratory study of college students' fanning behavior on Facebook. *College Student Journal*, *45*(3), 601–616.

Jang, H., Olfman, L., Ko, I., Koh, J., & Kim, K. (2008). The influence of on-line brand community characteristics on community commitment and brand loyalty. *International Journal of Electronic Commerce*, *12*(3), 57–80. doi:10.2753/JEC1086-4415120304

Jeanjean, L. (2012). 5 tips to increase your facebook fan base. *Journal of Financial Planning*, *1*(1), 20–21.

Joinson, A. N. (2008). 'Looking at', 'Looking up' or 'Keeping up with' people? Motives and uses of Facebook. In Proceedings of the 26th annual SIGCHI conference on human factors in computing systems (pp. 1027–1036). New York: ACM.

Kanga, J., Tang, L., & Fiorec, A.M. (2013). *Enhancing consumer–brand relationships on restaurant*. Academic Press.

Katz, E., Blumler, J. G., & Gurevitch, M. (1974). Uses and gratifications research. *Public Opinion Quarterly*, *37*(4), 509–524. doi:10.1086/268109

Khan, U., Dhar, R., & Wertenbroch, K. (2005). A behavioral decision theory perspective on hedonic and utilitarian choice. *Inside consumption: Frontiers of research on consumer motives, goals, and desires*, 144-165.

Kim, Y. M., & Shim, K. Y. (2002). The influence of Internet shopping mall characteristics and user traits on purchase intent. *Irish Marketing Review*, *15*, 25–34.

Koh, J., Kim, Y. G., & Kim, Y. G. (2003). Sense of virtual community: A conceptual framework and empirical validation. *International Journal of Electronic Commerce*, *8*(2), 75–94.

Kourouthanassis, P. E., Giaglis, G. M., & Vrechopoulos, A. P. (2007). Enhancing user experience through pervasive information systems: The case of pervasive retailing. *International Journal of Information Management*, *27*(5), 319–335. doi:10.1016/j.ijinfomgt.2007.04.005

Kozinets, R. V. (2002). The field behind the screen: Using netnography for marketing research in online communities. *JMR, Journal of Marketing Research, 39*(1), 61–72. doi:10.1509/jmkr.39.1.61.18935

Lampe, C., Ellison, N., & Steinfield, C. (2006). A Facebook in the crowd: Social searching vs. social browsing. In *Proceedings of the 2006 20th anniversary conference on computer supported cooperative work* (pp. 167–170). New York: ACM. doi:10.1145/1180875.1180901

Leung, L. (2007). Unwillingness-to-communicate and college students' motives in SMS mobile messaging. *Telematics and Informatics, 24*(2), 115–129. doi:10.1016/j.tele.2006.01.002

Lewis, J., & West, A. (2009). 'Friending': London-based undergraduates' experience of Facebook. *New Media & Society, 11*(7), 1209–1229. doi:10.1177/1461444809342058

Li, C., & Bernoff, J. (2008). *Groundswell: winning in a world transformed by social technologies*. Boston, MA: Harvard Business School Press.

Lin, K.-Y., & Lu, H.-P. (2011). Intention to continue using Facebook fan pages from the perspective of social capital theory. *Cyberpsychology, Behavior, and Social Networking, 14*(10), 565–570. doi:10.1089/cyber.2010.0472 PMID:21381968

Maenpaa, K., Kanto, A., Kuusela, H., & Paul, P. (2006). More hedonic versus less hedonic consumption behaviour in advanced internet bank services. *Journal of Financial Services Marketing, 11*(1), 4–16. doi:10.1057/palgrave.fsm.4760004

Matsuba, K. (2006). Searching for self and relationships online. *Cyberpsychology & Behavior, 9*(3), 275–284. doi:10.1089/cpb.2006.9.275 PMID:16780395

Mikalef, P., Giannakos, M., & Pateli, A. (2013). Shopping and word-of-mouth intentions on social media. *Journal of Theoretical and Applied Electronic Commerce Research, 8*(1), 17–34. doi:10.4067/S0718-18762013000100003

Mislove, A., Marcon, M., Gummadi, K. P., Druschel, P., & Bhattacharjee, B. (2007). Measurement and analysis of online social networks. In *Proceedings of the 7th ACM SIGCOMM conference on Internet measurement* (pp. 29-42). ACM. doi:10.1145/1298306.1298311

Moller, A. C., Ryan, R. M., & Deci, E. L. (2006). Self-Determination Theory and Public Policy: Improving the Quality of Consumer Decisions without Using Coercion. *Journal of Public Policy & Marketing, 25*(1), 104–116. doi:10.1509/jppm.25.1.104

Morgan, R. M., & Hunt, S. D. (1994). The commitment–trust theory of relationship marketing. *Journal of Marketing, 58*(3), 20–38. doi:10.2307/1252308

Muniz, A. M. Jr, & O'guinn, T. C. (2001). Brand community. *The Journal of Consumer Research, 27*(4), 412–432. doi:10.1086/319618

O'Connor, A. J. (2013). The Power of Popularity An Empirical Study of the Relationship Between Social Media Fan Counts and Brand Company Stock Prices. *Social Science Computer Review, 31*(2), 229–235. doi:10.1177/0894439312448037

Papacharissi, Z. (2010). Uses and gratifications. In *D. Stacks & SalwenAn Integrated Approach to Communication Theory and Research. D. Stacks & Salwen* (2nd ed.). New York: Routledge.

Pempek, T. A., Yermolayeva, Y. A., & Calvert, S. (2009). College students' social networking experiences on Facebook. *Journal of Applied Developmental Psychology*, *30*(3), 227–238. doi:10.1016/j.appdev.2008.12.010

Powell, J. (2009). *33 million people in the room: how to create, influence, and run a successful business with social networking*. Upper Saddle River, NJ: FT Press.

Qualman, E. (2009). *Socialnomics: how social media transforms the way we live and do business*. Hoboken, NJ: John Wiley & Sons, Inc.

Rubin, A. (1994). Media uses and effects: A uses-and-gratifications perspective. In J. Bryant & D. ZiIlmann (Eds.), Media effects: Advances in theory and research (pp. 417-436). Hillsdale, N J: Lawrence Erlbaum Associates, Inc.

Rubin, A. (2002). The uses-and-gratifications perspective of media effects. In J. Bryant & D. Zillmann (Eds.), *Media effects: Advances in theory and research* (2nd ed., pp. 525–548). Hillsdale, NJ: Lawrence Erlbaum Associates, Inc.

Sanchez-Franco, M. J., & Rondan-Cataluña, F. J. (2010). Virtual travel communities and customer loyalty: Customer purchase involvement and web site design. *Electronic Commerce Research and Applications*, *9*(2), 171–182. doi:10.1016/j.elerap.2009.05.004

Sheldon, P. (2008a). Student favourite: Facebook and motives for its use. *Southwestern Mass Communication Journal*, *23*(2), 39–53.

Shiv, B., & Fedorikhin, A. (1999). Heart and mind in conflict: The interplay of affect and cognition in consumer decision making. *The Journal of Consumer Research*, *26*(3), 278–292. doi:10.1086/209563

Sicilia, M., & Palazón, M. (2008). Brand communities on the internet: A case study of Coca-Cola's Spanish virtual community. Corporate Communications. *International Journal (Toronto, Ont.)*, *13*(3), 255–270.

Socialmediaexaminer. Social media marketing industry report 2013, Retrieved December 1st 2013, from http://www.socialmediaexaminer.com/SocialMediaMarketingIndustryReport2013.pdf

Stafford, T., & Gonier, D. (2004). What Americans like about being online. *Communications of the ACM*, *47*(11), 107–112. doi:10.1145/1029496.1029502

Thompson, S. A., & Sinha, R. K. (2008). Brand communities and new product adoption: The influence and limits of oppositional loyalty. *Journal of Marketing*, *72*(6), 65–80. doi:10.1509/jmkg.72.6.65

Urista, M. A., Qingwen, D., & Day, K. D. (2009). Explaining Why Young Adults Use MySpaceand Facebook Through Uses and Gratifications Theory. *Human Communication*, *12*(2), 215–229.

Voss, K. E., Spangenberg, E. R., & Grohmann, B. (2003). Measuring the hedonic and utilitarian dimensions of consumer attitude. *JMR, Journal of Marketing Research*, *40*(3), 310–320. doi:10.1509/jmkr.40.3.310.19238

Wallace, E., Buil, I., de Chernatony, L., & Hogan, M. (2014). Who "Likes" You … and Why? A Typology of Facebook Fans. *Journal of Advertising Research*, *54*(1), 92–109. doi:10.2501/JAR-54-1-092-109

Wang, Y. C., & Fesenmaier, D. (2004a). Towards understanding members' general participation in and active contribution to an online travel community. *Tourism Management*, *25*(6), 709–722. doi:10.1016/j.tourman.2003.09.011

Wang, Y. C., & Fesenmaier, D. (2004b). Modeling participation in an online travel community. *Journal of Travel Research*, *42*(3), 261–270. doi:10.1177/0047287503258824

Wertenbroch, K., & Dhar, R. (2000). Consumer choice between hedonic and utilitarian goods. *JMR, Journal of Marketing Research*, 60–71.

Woisetschläger, D. M., Hartleb, V., & Blut, M. (2008). How to make brand communities work: Antecedents and consequences of consumer participation. *Journal of Relationship Marketing*, *7*(3), 237–256. doi:10.1080/15332660802409605

Xu, C., Ryan, S., Prybutok, V., & Wen, C. (2012). It is not for fun: An examination of social network site usage. *Information & Management*, *49*(5), 210–217. doi:10.1016/j.im.2012.05.001

Young, A. L., & Quan-Haase, A. (2009). Information revelation and internet privacy concerns on social network sites: A case study of Facebook. In *Proceedings of the Fourth international conference on communities and technologies* (pp. 265–274). doi:10.1145/1556460.1556499

KEY TERMS AND DEFINITIONS

Hedonic Motivations: Hedonic motivations refers to the motives that are driven by emotional satisfaction such as happiness, enjoyment and fantasy when "liking" a brand's Facebook fan page.

Like: The act of clicking the "like" button on the brand's Facebook fan page.

Official Brand's Facebook Fan Page: A public page that is created and managed by brands to establish public presence on Facebook. These official pages enable any interested Facebook member to become a fan, which allow members to interact with the brand and receive the brand's updates.

Online Brand Communities: Specialized, non-geographically bound communities, based on a structured set of social relationships among customers of a brand.

Social Network Site: Social Network Sites is a platform that is designed to build social networks among people who have mutual interests, friends, activities or backgrounds.

Uses and Gratifications: Uses and gratifications is defined as the social and psychological roots of needs and wants, that creates expectations of the media and other communication sources used, which leads to different kinds of exposure to the media resulting in needs and wants gratifications and other consequences.

Utilitarian Motivations: Utilitarian motivations refers to the motives that are rational and goal oriented when "liking" a brand's Facebook fan page.

Related References

To continue our tradition of advancing information science and technology research, we have compiled a list of recommended IGI Global readings. These references will provide additional information and guidance to further enrich your knowledge and assist you with your own research and future publications.

Aayeshah, W., & Bebawi, S. (2014). The Use of Facebook as a Pedagogical Platform for Developing Investigative Journalism Skills. In G. Mallia (Ed.), *The Social Classroom: Integrating Social Network Use in Education* (pp. 83–99). Hershey, PA: Information Science Reference; doi:10.4018/978-1-4666-4904-0.ch005

Adi, A., & Scotte, C. G. (2013). Barriers to Emerging Technology and Social Media Integration in Higher Education: Three Case Studies. In M. Pătruţ & B. Pătruţ (Eds.), *Social Media in Higher Education: Teaching in Web 2.0* (pp. 334–354). Hershey, PA: Information Science Reference; doi:10.4018/978-1-4666-2970-7.ch017

Agazzi, E. (2012). How Can the Problems of An Ethical Judgment on Science and Technology Be Correctly Approached? In R. Luppicini (Ed.), *Ethical Impact of Technological Advancements and Applications in Society* (pp. 30–38). Hershey, PA: Information Science Reference; doi:10.4018/978-1-4666-1773-5.ch003

Agina, A. M., Tennyson, R. D., & Kommers, P. (2013). Understanding Children's Private Speech and Self-Regulation Learning in Web 2.0: Updates of Vygotsky through Piaget and Future Recommendations. In P. Ordóñez de Pablos, H. Nigro, R. Tennyson, S. Gonzalez Cisaro, & W. Karwowski (Eds.), *Advancing Information Management through Semantic Web Concepts and Ontologies* (pp. 1–53). Hershey, PA: Information Science Reference; doi:10.4018/978-1-4666-2494-8.ch001

Ahrens, A., Bassus, O., & Zaščerinska, J. (2014). Enterprise 2.0 in Engineering Curriculum. In M. Cruz-Cunha, F. Moreira, & J. Varajão (Eds.), *Handbook of Research on Enterprise 2.0: Technological, Social, and Organizational Dimensions* (pp. 599–617). Hershey, PA: Business Science Reference; doi:10.4018/978-1-4666-4373-4.ch031

Akputu, O. K., Seng, K. P., & Lee, Y. L. (2014). Affect Recognition for Web 2.0 Intelligent E-Tutoring Systems: Exploration of Students' Emotional Feedback. In J. Pelet (Ed.), *E-Learning 2.0 Technologies and Web Applications in Higher Education* (pp. 188–215). Hershey, PA: Information Science Reference; doi:10.4018/978-1-4666-4876-0.ch010

Al-Hajri, S., & Tatnall, A. (2013). A Socio-Technical Study of the Adoption of Internet Technology in Banking, Re-Interpreted as an Innovation Using Innovation Translation. In A. Tatnall (Ed.), *Social and Professional Applications of Actor-Network Theory for Technology Development* (pp. 207–220). Hershey, PA: Information Science Reference; doi:10.4018/978-1-4666-2166-4.ch016

Al Hujran, O., Aloudat, A., & Altarawneh, I. (2013). Factors Influencing Citizen Adoption of E-Government in Developing Countries: The Case of Jordan. [IJTHI]. *International Journal of Technology and Human Interaction*, *9*(2), 1–19. doi:10.4018/jthi.2013040101

Alavi, R., Islam, S., Jahankhani, H., & Al-Nemrat, A. (2013). Analyzing Human Factors for an Effective Information Security Management System. [IJSSE]. *International Journal of Secure Software Engineering*, *4*(1), 50–74. doi:10.4018/jsse.2013010104

Altun, N. E., & Yildiz, S. (2013). Effects of Different Types of Tasks on Junior ELT Students' Use of Communication Strategies in Computer-Mediated Communication. [IJCALLT]. *International Journal of Computer-Assisted Language Learning and Teaching*, *3*(2), 17–40. doi:10.4018/ijcallt.2013040102

Amaldi, P., & Smoker, A. (2013). An Organizational Study into the Concept of "Automation Policy" in a Safety Critical Socio-Technical System. [IJSKD]. *International Journal of Sociotechnology and Knowledge Development*, *5*(2), 1–17. doi:10.4018/jskd.2013040101

An, I. S. (2013). Integrating Technology-Enhanced Student Self-Regulated Tasks into University Chinese Language Course. [IJCALLT]. *International Journal of Computer-Assisted Language Learning and Teaching*, *3*(1), 1–15. doi:10.4018/ijcallt.2013010101

Andacht, F. (2013). The Tangible Lure of the Technoself in the Age of Reality Television. In R. Luppicini (Ed.), *Handbook of Research on Technoself: Identity in a Technological Society* (pp. 360–381). Hershey, PA: Information Science Reference; doi:10.4018/978-1-4666-2211-1.ch020

Anderson, A., & Petersen, A. (2012). Shaping the Ethics of an Emergent Field: Scientists' and Policymakers' Representations of Nanotechnologies. In R. Luppicini (Ed.), *Ethical Impact of Technological Advancements and Applications in Society* (pp. 219–231). Hershey, PA: Information Science Reference; doi:10.4018/978-1-4666-1773-5.ch017

Anderson, J. L. (2014). Games and the Development of Students' Civic Engagement and Ecological Stewardship. In J. Bishop (Ed.), *Gamification for Human Factors Integration: Social, Education, and Psychological Issues* (pp. 199–215). Hershey, PA: Information Science Reference; doi:10.4018/978-1-4666-5071-8.ch012

Ann, O. C., Lu, M. V., & Theng, L. B. (2014). A Face Based Real Time Communication for Physically and Speech Disabled People. In Assistive Technologies: Concepts, Methodologies, Tools, and Applications (pp. 1434-1460). Hershey, PA: Information Science Reference. doi:10.4018/978-1-4666-4422-9.ch075

Aricak, O. T., Tanrikulu, T., Siyahhan, S., & Kinay, H. (2013). Cyberbullying: The Bad and the Ugly Side of Information Age. In M. Pătruţ & B. Pătruţ (Eds.), *Social Media in Higher Education: Teaching in Web 2.0* (pp. 318–333). Hershey, PA: Information Science Reference; doi:10.4018/978-1-4666-2970-7.ch016

Ariely, G. (2011). Boundaries of Socio-Technical Systems and IT for Knowledge Development in Military Environments. [IJSKD]. *International Journal of Sociotechnology and Knowledge Development*, *3*(3), 1–14. doi:10.4018/jskd.2011070101

Ariely, G. (2013). Boundaries of Socio-Technical Systems and IT for Knowledge Development in Military Environments. In J. Abdelnour-Nocera (Ed.), *Knowledge and Technological Development Effects on Organizational and Social Structures* (pp. 224–238). Hershey, PA: Information Science Reference; doi:10.4018/978-1-4666-2151-0.ch014

Arjunan, S., Kumar, D. K., Weghorn, H., & Naik, G. (2014). Facial Muscle Activity Patterns for Recognition of Utterances in Native and Foreign Language: Testing for its Reliability and Flexibility. In Assistive Technologies: Concepts, Methodologies, Tools, and Applications (pp. 1462-1480). Hershey, PA: Information Science Reference. doi:10.4018/978-1-4666-4422-9.ch076

Arling, P. A., Miech, E. J., & Arling, G. W. (2013). Comparing Electronic and Face-to-Face Communication in the Success of a Long-Term Care Quality Improvement Collaborative. [IJRQEH]. *International Journal of Reliable and Quality E-Healthcare*, *2*(1), 1–10. doi:10.4018/ijrqeh.2013010101

Asghari-Oskoei, M., & Hu, H. (2014). Using Myoelectric Signals to Manipulate Assisting Robots and Rehabilitation Devices. In Assistive Technologies: Concepts, Methodologies, Tools, and Applications (pp. 970-990). Hershey, PA: Information Science Reference. doi:10.4018/978-1-4666-4422-9.ch049

Aspradaki, A. A. (2013). Deliberative Democracy and Nanotechnologies in Health. [IJT]. *International Journal of Technoethics*, *4*(2), 1–14. doi:10.4018/jte.2013070101

Asselin, S. B. (2014). Assistive Technology in Higher Education. In Assistive Technologies: Concepts, Methodologies, Tools, and Applications (pp. 1196-1208). Hershey, PA: Information Science Reference. doi:10.4018/978-1-4666-4422-9.ch062

Auld, G., & Henderson, M. (2014). The Ethical Dilemmas of Social Networking Sites in Classroom Contexts. In G. Mallia (Ed.), *The Social Classroom: Integrating Social Network Use in Education* (pp. 192–207). Hershey, PA: Information Science Reference; doi:10.4018/978-1-4666-4904-0.ch010

Awwal, M. A. (2012). Influence of Age and Genders on the Relationship between Computer Self-Efficacy and Information Privacy Concerns. [IJTHI]. *International Journal of Technology and Human Interaction*, *8*(1), 14–37. doi:10.4018/jthi.2012010102

Ballesté, F., & Torras, C. (2013). Effects of Human-Machine Integration on the Construction of Identity. In R. Luppicini (Ed.), *Handbook of Research on Technoself: Identity in a Technological Society* (pp. 574–591). Hershey, PA: Information Science Reference; doi:10.4018/978-1-4666-2211-1.ch030

Baporikar, N. (2014). Effective E-Learning Strategies for a Borderless World. In J. Pelet (Ed.), *E-Learning 2.0 Technologies and Web Applications in Higher Education* (pp. 22–44). Hershey, PA: Information Science Reference; doi:10.4018/978-1-4666-4876-0.ch002

Bardone, E. (2011). Unintended Affordances as Violent Mediators: Maladaptive Effects of Technologically Enriched Human Niches. [IJT]. *International Journal of Technoethics*, *2*(4), 37–52. doi:10.4018/jte.2011100103

Basham, R. (2014). Surveilling the Elderly: Emerging Demographic Needs and Social Implications of RFID Chip Technology Use. In M. Michael & K. Michael (Eds.), *Uberveillance and the Social Implications of Microchip Implants: Emerging Technologies* (pp. 169–185). Hershey, PA: Information Science Reference; doi:10.4018/978-1-4666-4582-0.ch007

Bates, M. (2013). The Ur-Real Sonorous Envelope: Bridge between the Corporeal and the Online Technoself. In R. Luppicini (Ed.), *Handbook of Research on Technoself: Identity in a Technological Society* (pp. 272–292). Hershey, PA: Information Science Reference; doi:10.4018/978-1-4666-2211-1.ch015

Bauer, K. A. (2012). Transhumanism and Its Critics: Five Arguments against a Posthuman Future. In R. Luppicini (Ed.), *Ethical Impact of Technological Advancements and Applications in Society* (pp. 232–242). Hershey, PA: Information Science Reference; doi:10.4018/978-1-4666-1773-5.ch018

Bax, S. (2011). Normalisation Revisited: The Effective Use of Technology in Language Education. [IJCALLT]. *International Journal of Computer-Assisted Language Learning and Teaching*, *1*(2), 1–15. doi:10.4018/ijcallt.2011040101

Baya'a, N., & Daher, W. (2014). Facebook as an Educational Environment for Mathematics Learning. In G. Mallia (Ed.), *The Social Classroom: Integrating Social Network Use in Education* (pp. 171–190). Hershey, PA: Information Science Reference; doi:10.4018/978-1-4666-4904-0.ch009

Bayerl, P. S., & Janneck, M. (2013). Professional Online Profiles: The Impact of Personalization and Visual Gender Cues on Online Impression Formation. [IJSKD]. *International Journal of Sociotechnology and Knowledge Development*, *5*(3), 1–16. doi:10.4018/ijskd.2013070101

Bell, D., & Shirzad, S. R. (2013). Social Media Business Intelligence: A Pharmaceutical Domain Analysis Study. [IJSKD]. *International Journal of Sociotechnology and Knowledge Development*, *5*(3), 51–73. doi:10.4018/ijskd.2013070104

Bergmann, N. W. (2014). Ubiquitous Computing for Independent Living. In Assistive Technologies: Concepts, Methodologies, Tools, and Applications (pp. 679-692). Hershey, PA: Information Science Reference. doi:10.4018/978-1-4666-4422-9.ch033

Bertolotti, T. (2011). Facebook Has It: The Irresistible Violence of Social Cognition in the Age of Social Networking. [IJT]. *International Journal of Technoethics*, *2*(4), 71–83. doi:10.4018/jte.2011100105

Berzsenyi, C. (2014). Writing to Meet Your Match: Rhetoric and Self-Presentation for Four Online Daters. In H. Lim & F. Sudweeks (Eds.), *Innovative Methods and Technologies for Electronic Discourse Analysis* (pp. 210–234). Hershey, PA: Information Science Reference; doi:10.4018/978-1-4666-4426-7.ch010

Best, L. A., Buhay, D. N., McGuire, K., Gurholt, S., & Foley, S. (2014). The Use of Web 2.0 Technologies in Formal and Informal Learning Settings. In G. Mallia (Ed.), *The Social Classroom: Integrating Social Network Use in Education* (pp. 1–22). Hershey, PA: Information Science Reference; doi:10.4018/978-1-4666-4904-0.ch001

Bhattacharya, S. (2014). Model-Based Approaches for Scanning Keyboard Design: Present State and Future Directions. In Assistive Technologies: Concepts, Methodologies, Tools, and Applications (pp. 1497-1515). Hershey, PA: Information Science Reference. doi:10.4018/978-1-4666-4422-9.ch078

Bibby, S. (2011). Do Students Wish to 'Go Mobile'?: An Investigation into Student Use of PCs and Cell Phones. [IJCALLT]. *International Journal of Computer-Assisted Language Learning and Teaching*, *1*(2), 43–54. doi:10.4018/ijcallt.2011040104

Bishop, J. (2014). The Psychology of Trolling and Lurking: The Role of Defriending and Gamification for Increasing Participation in Online Communities Using Seductive Narratives. In J. Bishop (Ed.), *Gamification for Human Factors Integration: Social, Education, and Psychological Issues* (pp. 162–179). Hershey, PA: Information Science Reference; doi:10.4018/978-1-4666-5071-8.ch010

Bishop, J., & Goode, M. M. (2014). Towards a Subjectively Devised Parametric User Model for Analysing and Influencing Behaviour Online Using Neuroeconomics. In J. Bishop (Ed.), *Gamification for Human Factors Integration: Social, Education, and Psychological Issues* (pp. 80–95). Hershey, PA: Information Science Reference; doi:10.4018/978-1-4666-5071-8.ch005

Biswas, P. (2014). A Brief Survey on User Modelling in Human Computer Interaction. In Assistive Technologies: Concepts, Methodologies, Tools, and Applications (pp. 102-119). Hershey, PA: Information Science Reference. doi:10.4018/978-1-4666-4422-9.ch006

Black, D. (2013). The Digital Soul. In R. Luppicini (Ed.), *Handbook of Research on Technoself: Identity in a Technological Society* (pp. 157–174). Hershey, PA: Information Science Reference; doi:10.4018/978-1-4666-2211-1.ch009

Blake, S., Winsor, D. L., Burkett, C., & Allen, L. (2014). iPods, Internet and Apps, Oh My: Age Appropriate Technology in Early Childhood Educational Environments. In K-12 Education: Concepts, Methodologies, Tools, and Applications (pp. 1650-1668). Hershey, PA: Information Science Reference. doi:10.4018/978-1-4666-4502-8.ch095

Boghian, I. (2013). Using Facebook in Teaching. In M. Pătruţ & B. Pătruţ (Eds.), *Social Media in Higher Education: Teaching in Web 2.0* (pp. 86–103). Hershey, PA: Information Science Reference; doi:10.4018/978-1-4666-2970-7.ch005

Boling, E. C., & Beatty, J. (2014). Overcoming the Tensions and Challenges of Technology Integration: How Can We Best Support our Teachers? In K-12 Education: Concepts, Methodologies, Tools, and Applications (pp. 1504-1524). Hershey, PA: Information Science Reference. doi:10.4018/978-1-4666-4502-8.ch087

Bonanno, P. (2014). Designing Learning in Social Online Learning Environments: A Process-Oriented Approach. In G. Mallia (Ed.), *The Social Classroom: Integrating Social Network Use in Education* (pp. 40–61). Hershey, PA: Information Science Reference; doi:10.4018/978-1-4666-4904-0.ch003

Bongers, B., & Smith, S. (2014). Interactivating Rehabilitation through Active Multimodal Feedback and Guidance. In Assistive Technologies: Concepts, Methodologies, Tools, and Applications (pp. 1650-1674). Hershey, PA: Information Science Reference. doi:10.4018/978-1-4666-4422-9.ch087

Bottino, R. M., Ott, M., & Tavella, M. (2014). Serious Gaming at School: Reflections on Students' Performance, Engagement and Motivation. [IJGBL]. *International Journal of Game-Based Learning, 4*(1), 21–36. doi:10.4018/IJGBL.2014010102

Brad, S. (2014). Design for Quality of ICT-Aided Engineering Course Units. [IJQAETE]. *International Journal of Quality Assurance in Engineering and Technology Education, 3*(1), 52–80. doi:10.4018/ijqaete.2014010103

Braman, J., Thomas, U., Vincenti, G., Dudley, A., & Rodgers, K. (2014). Preparing Your Digital Legacy: Assessing Awareness of Digital Natives. In G. Mallia (Ed.), *The Social Classroom: Integrating Social Network Use in Education* (pp. 208–223). Hershey, PA: Information Science Reference; doi:10.4018/978-1-4666-4904-0.ch011

Bratitsis, T., & Demetriadis, S. (2013). Research Approaches in Computer-Supported Collaborative Learning. [IJeC]. *International Journal of e-Collaboration, 9*(1), 1–8. doi:10.4018/jec.2013010101

Brick, B. (2012). The Role of Social Networking Sites for Language Learning in UK Higher Education: The Views of Learners and Practitioners. [IJCALLT]. *International Journal of Computer-Assisted Language Learning and Teaching, 2*(3), 35–53. doi:10.4018/ijcallt.2012070103

Burke, M. E., & Speed, C. (2014). Knowledge Recovery: Applications of Technology and Memory. In M. Michael & K. Michael (Eds.), *Uberveillance and the Social Implications of Microchip Implants: Emerging Technologies* (pp. 133–142). Hershey, PA: Information Science Reference; doi:10.4018/978-1-4666-4582-0.ch005

Burton, A. M., Liu, H., Battersby, S., Brown, D., Sherkat, N., Standen, P., & Walker, M. (2014). The Use of Motion Tracking Technologies in Serious Games to Enhance Rehabilitation in Stroke Patients. In J. Bishop (Ed.), *Gamification for Human Factors Integration: Social, Education, and Psychological Issues* (pp. 148–161). Hershey, PA: Information Science Reference; doi:10.4018/978-1-4666-5071-8.ch009

Burusic, J., & Karabegovic, M. (2014). The Role of Students' Personality Traits in the Effective Use of Social Networking Sites in the Educational Context. In G. Mallia (Ed.), *The Social Classroom: Integrating Social Network Use in Education* (pp. 224–243). Hershey, PA: Information Science Reference; doi:10.4018/978-1-4666-4904-0.ch012

Busch, C. D., Lorenzo, A. M., Sánchez, I. M., González, B. G., García, T. P., Riveiro, L. N., & Loureiro, J. P. (2014). In-TIC for Mobile Devices: Support System for Communication with Mobile Devices for the Disabled. In Assistive Technologies: Concepts, Methodologies, Tools, and Applications (pp. 345-356). Hershey, PA: Information Science Reference. doi:10.4018/978-1-4666-4422-9.ch017

Bute, S. J. (2013). Integrating Social Media and Traditional Media within the Academic Environment. In M. Pătruţ & B. Pătruţ (Eds.), *Social Media in Higher Education: Teaching in Web 2.0* (pp. 75–85). Hershey, PA: Information Science Reference; doi:10.4018/978-1-4666-2970-7.ch004

Butler-Pascoe, M. E. (2011). The History of CALL: The Intertwining Paths of Technology and Second/Foreign Language Teaching. [IJCALLT]. *International Journal of Computer-Assisted Language Learning and Teaching, 1*(1), 16–32. doi:10.4018/ijcallt.2011010102

Cabrera, L. (2012). Human Implants: A Suggested Framework to Set Priorities. In R. Luppicini (Ed.), *Ethical Impact of Technological Advancements and Applications in Society* (pp. 243–253). Hershey, PA: Information Science Reference; doi:10.4018/978-1-4666-1773-5.ch019

Cacho-Elizondo, S., Shahidi, N., & Tossan, V. (2013). Intention to Adopt a Text Message-based Mobile Coaching Service to Help Stop Smoking: Which Explanatory Variables? [IJTHI]. *International Journal of Technology and Human Interaction, 9*(4), 1–19. doi:10.4018/ijthi.2013100101

Caldelli, R., Becarelli, R., Filippini, F., Picchioni, F., & Giorgetti, R. (2014). Electronic Voting by Means of Digital Terrestrial Television: The Infrastructure, Security Issues and a Real Test-Bed. In Assistive Technologies: Concepts, Methodologies, Tools, and Applications (pp. 905-915). Hershey, PA: Information Science Reference. doi:10.4018/978-1-4666-4422-9.ch045

Camacho, M. (2013). Making the Most of Informal and Situated Learning Opportunities through Mobile Learning. In M. Pătruţ & B. Pătruţ (Eds.), *Social Media in Higher Education: Teaching in Web 2.0* (pp. 355–370). Hershey, PA: Information Science Reference; doi:10.4018/978-1-4666-2970-7.ch018

Camilleri, V., Busuttil, L., & Montebello, M. (2014). MOOCs: Exploiting Networks for the Education of the Masses or Just a Trend? In G. Mallia (Ed.), *The Social Classroom: Integrating Social Network Use in Education* (pp. 348–366). Hershey, PA: Information Science Reference; doi:10.4018/978-1-4666-4904-0.ch018

Campos, P., Noronha, H., & Lopes, A. (2013). Work Analysis Methods in Practice: The Context of Collaborative Review of CAD Models. [IJSKD]. *International Journal of Sociotechnology and Knowledge Development*, *5*(2), 34–44. doi:10.4018/jskd.2013040103

Cao, G. (2013). A Paradox Between Technological Autonomy and Ethical Heteronomy of Philosophy of Technology: Social Control System. [IJT]. *International Journal of Technoethics*, *4*(1), 52–66. doi:10.4018/jte.2013010105

Carofiglio, V., & Abbattista, F. (2013). BCI-Based User-Centered Design for Emotionally-Driven User Experience. In M. Garcia-Ruiz (Ed.), *Cases on Usability Engineering: Design and Development of Digital Products* (pp. 299–320). Hershey, PA: Information Science Reference; doi:10.4018/978-1-4666-4046-7.ch013

Carpenter, J. (2013). Just Doesn't Look Right: Exploring the Impact of Humanoid Robot Integration into Explosive Ordnance Disposal Teams. In R. Luppicini (Ed.), *Handbook of Research on Technoself: Identity in a Technological Society* (pp. 609–636). Hershey, PA: Information Science Reference; doi:10.4018/978-1-4666-2211-1.ch032

Carroll, J. L. (2014). Wheelchairs as Assistive Technology: What a Special Educator Should Know. In Assistive Technologies: Concepts, Methodologies, Tools, and Applications (pp. 623-633). Hershey, PA: Information Science Reference. doi:10.4018/978-1-4666-4422-9.ch030

Casey, L. B., & Williamson, R. L. (2014). A Parent's Guide to Support Technologies for Preschool Students with Disabilities. In Assistive Technologies: Concepts, Methodologies, Tools, and Applications (pp. 1340-1356). Hershey, PA: Information Science Reference. doi:10.4018/978-1-4666-4422-9.ch071

Caviglione, L., Coccoli, M., & Merlo, A. (2013). On Social Network Engineering for Secure Web Data and Services. In L. Caviglione, M. Coccoli, & A. Merlo (Eds.), *Social Network Engineering for Secure Web Data and Services* (pp. 1–4). Hershey, PA: Information Science Reference; doi:10.4018/978-1-4666-3926-3.ch001

Chadwick, D. D., Fullwood, C., & Wesson, C. J. (2014). Intellectual Disability, Identity, and the Internet. In Assistive Technologies: Concepts, Methodologies, Tools, and Applications (pp. 198-223). Hershey, PA: Information Science Reference. doi:10.4018/978-1-4666-4422-9.ch011

Chao, L., Wen, Y., Chen, P., Lin, C., Lin, S., Guo, C., & Wang, W. (2012). The Development and Learning Effectiveness of a Teaching Module for the Algal Fuel Cell: A Renewable and Sustainable Battery. [IJTHI]. *International Journal of Technology and Human Interaction*, *8*(4), 1–15. doi:10.4018/jthi.2012100101

Charnkit, P., & Tatnall, A. (2013). Knowledge Conversion Processes in Thai Public Organisations Seen as an Innovation: The Re-Analysis of a TAM Study Using Innovation Translation. In A. Tatnall (Ed.), *Social and Professional Applications of Actor-Network Theory for Technology Development* (pp. 88–102). Hershey, PA: Information Science Reference; doi:10.4018/978-1-4666-2166-4.ch008

Chen, E. T. (2014). Challenge and Complexity of Virtual Team Management. In E. Nikoi & K. Boateng (Eds.), *Collaborative Communication Processes and Decision Making in Organizations* (pp. 109–120). Hershey, PA: Business Science Reference; doi:10.4018/978-1-4666-4478-6.ch006

Chen, R., Xie, T., Lin, T., & Chen, Y. (2013). Adaptive Windows Layout Based on Evolutionary Multi-Objective Optimization. [IJTHI]. *International Journal of Technology and Human Interaction, 9*(3), 63–72. doi:10.4018/jthi.2013070105

Chen, W., Juang, Y., Chang, S., & Wang, P. (2012). Informal Education of Energy Conservation: Theory, Promotion, and Policy Implication. [IJTHI]. *International Journal of Technology and Human Interaction, 8*(4), 16–44. doi:10.4018/jthi.2012100102

Chino, T., Torii, K., Uchihira, N., & Hirabayashi, Y. (2013). Speech Interaction Analysis on Collaborative Work at an Elderly Care Facility. [IJSKD]. *International Journal of Sociotechnology and Knowledge Development, 5*(2), 18–33. doi:10.4018/jskd.2013040102

Chiu, M. (2013). Gaps Between Valuing and Purchasing Green-Technology Products: Product and Gender Differences. [IJTHI]. *International Journal of Technology and Human Interaction, 8*(3), 54–68. doi:10.4018/jthi.2012070106

Chivukula, V., & Shur, M. (2014). Web-Based Experimentation for Students with Learning Disabilities. In Assistive Technologies: Concepts, Methodologies, Tools, and Applications (pp. 1156-1172). Hershey, PA: Information Science Reference. doi:10.4018/978-1-4666-4422-9.ch060

Coakes, E., Bryant, A., Land, F., & Phippen, A. (2011). The Dark Side of Technology: Some Sociotechnical Reflections. [IJSKD]. *International Journal of Sociotechnology and Knowledge Development, 3*(4), 40–51. doi:10.4018/IJSKD.2011100104

Cole, I. J. (2013). Usability of Online Virtual Learning Environments: Key Issues for Instructors and Learners. In C. Gonzalez (Ed.), *Student Usability in Educational Software and Games: Improving Experiences* (pp. 41–58). Hershey, PA: Information Science Reference; doi:10.4018/978-1-4666-1987-6.ch002

Colombo, B., Antonietti, A., Sala, R., & Caravita, S. C. (2013). Blog Content and Structure, Cognitive Style and Metacognition. [IJTHI]. *International Journal of Technology and Human Interaction, 9*(3), 1–17. doi:10.4018/jthi.2013070101

Constantinides, M. (2011). Integrating Technology on Initial Training Courses: A Survey Amongst CELTA Tutors. [IJCALLT]. *International Journal of Computer-Assisted Language Learning and Teaching, 1*(2), 55–71. doi:10.4018/ijcallt.2011040105

Cook, R. G., & Crawford, C. M. (2013). Addressing Online Student Learning Environments and Socialization Through Developmental Research. In M. Khosrow-Pour (Ed.), *Cases on Assessment and Evaluation in Education* (pp. 504–536). Hershey, PA: Information Science Reference; doi:10.4018/978-1-4666-2621-8.ch021

Corritore, C. L., Wiedenbeck, S., Kracher, B., & Marble, R. P. (2012). Online Trust and Health Information Websites. [IJTHI]. *International Journal of Technology and Human Interaction, 8*(4), 92–115. doi:10.4018/jthi.2012100106

Covarrubias, M., Bordegoni, M., Cugini, U., & Gatti, E. (2014). Supporting Unskilled People in Manual Tasks through Haptic-Based Guidance. In Assistive Technologies: Concepts, Methodologies, Tools, and Applications (pp. 947-969). Hershey, PA: Information Science Reference. doi:10.4018/978-1-4666-4422-9.ch048

Coverdale, T. S., & Wilbon, A. D. (2013). The Impact of In-Group Membership on e-Loyalty of Women Online Shoppers: An Application of the Social Identity Approach to Website Design. [IJEA]. *International Journal of E-Adoption, 5*(1), 17–36. doi:10.4018/jea.2013010102

Crabb, P. B., & Stern, S. E. (2012). Technology Traps: Who Is Responsible? In R. Luppicini (Ed.), *Ethical Impact of Technological Advancements and Applications in Society* (pp. 39–46). Hershey, PA: Information Science Reference; doi:10.4018/978-1-4666-1773-5.ch004

Crespo, R. G., Martíne, O. S., Lovelle, J. M., García-Bustelo, B. C., Díaz, V. G., & Ordoñez de Pablos, P. (2014). Improving Cognitive Load on Students with Disabilities through Software Aids. In Assistive Technologies: Concepts, Methodologies, Tools, and Applications (pp. 1255-1268). Hershey, PA: Information Science Reference. doi:10.4018/978-1-4666-4422-9.ch066

Croasdaile, S., Jones, S., Ligon, K., Oggel, L., & Pruett, M. (2014). Supports for and Barriers to Implementing Assistive Technology in Schools. In Assistive Technologies: Concepts, Methodologies, Tools, and Applications (pp. 1118-1130). Hershey, PA: Information Science Reference. doi:10.4018/978-1-4666-4422-9.ch058

Cucchiarini, C., & Strik, H. (2014). Second Language Learners' Spoken Discourse: Practice and Corrective Feedback through Automatic Speech Recognition. In H. Lim & F. Sudweeks (Eds.), *Innovative Methods and Technologies for Electronic Discourse Analysis* (pp. 169–189). Hershey, PA: Information Science Reference; doi:10.4018/978-1-4666-4426-7.ch008

Dafoulas, G. A., & Saleeb, N. (2014). 3D Assistive Technologies and Advantageous Themes for Collaboration and Blended Learning of Users with Disabilities. In Assistive Technologies: Concepts, Methodologies, Tools, and Applications (pp. 421-453). Hershey, PA: Information Science Reference. doi:10.4018/978-1-4666-4422-9.ch021

Dai, Z., & Paasch, K. (2013). A Web-Based Interactive Questionnaire for PV Application. [IJSKD]. *International Journal of Sociotechnology and Knowledge Development, 5*(2), 82–93. doi:10.4018/jskd.2013040106

Daradoumis, T., & Lafuente, M. M. (2014). Studying the Suitability of Discourse Analysis Methods for Emotion Detection and Interpretation in Computer-Mediated Educational Discourse. In H. Lim & F. Sudweeks (Eds.), *Innovative Methods and Technologies for Electronic Discourse Analysis* (pp. 119–143). Hershey, PA: Information Science Reference; doi:10.4018/978-1-4666-4426-7.ch006

Davis, B., & Mason, P. (2014). Positioning Goes to Work: Computer-Aided Identification of Stance Shifts and Semantic Themes in Electronic Discourse Analysis. In H. Lim & F. Sudweeks (Eds.), *Innovative Methods and Technologies for Electronic Discourse Analysis* (pp. 394–413). Hershey, PA: Information Science Reference; doi:10.4018/978-1-4666-4426-7.ch018

Dogoriti, E., & Pange, J. (2014). Considerations for Online English Language Learning: The Use of Facebook in Formal and Informal Settings in Higher Education. In G. Mallia (Ed.), *The Social Classroom: Integrating Social Network Use in Education* (pp. 147–170). Hershey, PA: Information Science Reference; doi:10.4018/978-1-4666-4904-0.ch008

Donegan, M. (2014). Features of Gaze Control Systems. In Assistive Technologies: Concepts, Methodologies, Tools, and Applications (pp. 1055-1061). Hershey, PA: Information Science Reference. doi:10.4018/978-1-4666-4422-9.ch054

Douglas, G., Morton, H., & Jack, M. (2012). Remote Channel Customer Contact Strategies for Complaint Update Messages. [IJTHI]. *International Journal of Technology and Human Interaction, 8*(2), 43–55. doi:10.4018/jthi.2012040103

Drake, J. R., & Byrd, T. A. (2013). Searching for Alternatives: Does Your Disposition Matter? [IJTHI]. *International Journal of Technology and Human Interaction, 9*(1), 18–36. doi:10.4018/jthi.2013010102

Driouchi, A. (2013). ICTs and Socioeconomic Performance with Focus on ICTs and Health. In ICTs for Health, Education, and Socioeconomic Policies: Regional Cases (pp. 104-125). Hershey, PA: Information Science Reference. doi:10.4018/978-1-4666-3643-9.ch005

Driouchi, A. (2013). Social Deficits, Social Cohesion, and Prospects from ICTs. In ICTs for Health, Education, and Socioeconomic Policies: Regional Cases (pp. 230-251). Hershey, PA: Information Science Reference. doi:10.4018/978-1-4666-3643-9.ch011

Driouchi, A. (2013). Socioeconomic Reforms, Human Development, and the Millennium Development Goals with ICTs for Coordination. In ICTs for Health, Education, and Socioeconomic Policies: Regional Cases (pp. 211-229). Hershey, PA: Information Science Reference. doi:10.4018/978-1-4666-3643-9.ch010

Drula, G. (2013). Media and Communication Research Facing Social Media. In M. Pătruţ & B. Pătruţ (Eds.), *Social Media in Higher Education: Teaching in Web 2.0* (pp. 371–392). Hershey, PA: Information Science Reference; doi:10.4018/978-1-4666-2970-7.ch019

Druzhinina, O., Hvannberg, E. T., & Halldorsdottir, G. (2013). Feedback Fidelities in Three Different Types of Crisis Management Training Environments. [IJSKD]. *International Journal of Sociotechnology and Knowledge Development, 5*(2), 45–62. doi:10.4018/jskd.2013040104

Eason, K., Waterson, P., & Davda, P. (2013). The Sociotechnical Challenge of Integrating Telehealth and Telecare into Health and Social Care for the Elderly. [IJSKD]. *International Journal of Sociotechnology and Knowledge Development, 5*(4), 14–26. doi:10.4018/ijskd.2013100102

Edenius, M., & Rämö, H. (2011). An Office on the Go: Professional Workers, Smartphones and the Return of Place. [IJTHI]. *International Journal of Technology and Human Interaction, 7*(1), 37–55. doi:10.4018/jthi.2011010103

Eke, D. O. (2011). ICT Integration in Nigeria: The Socio-Cultural Constraints. [IJTHI]. *International Journal of Technology and Human Interaction, 7*(2), 21–27. doi:10.4018/jthi.2011040103

Evett, L., Ridley, A., Keating, L., Merritt, P., Shopland, N., & Brown, D. (2014). Designing Serious Games for People with Disabilities: Game, Set, and Match to the Wii. In J. Bishop (Ed.), *Gamification for Human Factors Integration: Social, Education, and Psychological Issues* (pp. 97–105). Hershey, PA: Information Science Reference; doi:10.4018/978-1-4666-5071-8.ch006

Evmenova, A. S., & Behrmann, M. M. (2014). Communication Technology Integration in the Content Areas for Students with High-Incidence Disabilities: A Case Study of One School System. In Assistive Technologies: Concepts, Methodologies, Tools, and Applications (pp. 26-53). Hershey, PA: Information Science Reference. doi:10.4018/978-1-4666-4422-9.ch003

Evmenova, A. S., & King-Sears, M. E. (2014). Technology and Literacy for Students with Disabilities. In Assistive Technologies: Concepts, Methodologies, Tools, and Applications (pp. 1269-1291). Hershey, PA: Information Science Reference. doi:10.4018/978-1-4666-4422-9.ch067

Ewais, A., & De Troyer, O. (2013). Usability Evaluation of an Adaptive 3D Virtual Learning Environment. [IJVPLE]. *International Journal of Virtual and Personal Learning Environments*, *4*(1), 16–31. doi:10.4018/jvple.2013010102

Farrell, H. J. (2014). The Student with Complex Education Needs: Assistive and Augmentative Information and Communication Technology in a Ten-Week Music Program. In K-12 Education: Concepts, Methodologies, Tools, and Applications (pp. 1436-1472). Hershey, PA: Information Science Reference. doi:10.4018/978-1-4666-4502-8.ch084

Fathulla, K. (2012). Rethinking Human and Society's Relationship with Technology. [IJSKD]. *International Journal of Sociotechnology and Knowledge Development*, *4*(2), 21–28. doi:10.4018/jskd.2012040103

Fidler, C. S., Kanaan, R. K., & Rogerson, S. (2011). Barriers to e-Government Implementation in Jordan: The Role of Wasta. [IJTHI]. *International Journal of Technology and Human Interaction*, *7*(2), 9–20. doi:10.4018/jthi.2011040102

Fischer, G., & Herrmann, T. (2013). Socio-Technical Systems: A Meta-Design Perspective. In J. Abdelnour-Nocera (Ed.), *Knowledge and Technological Development Effects on Organizational and Social Structures* (pp. 1–36). Hershey, PA: Information Science Reference; doi:10.4018/978-1-4666-2151-0.ch001

Foreman, J., & Borkman, T. (2014). Learning Sociology in a Massively Multi-Student Online Learning Environment. In J. Bishop (Ed.), *Gamification for Human Factors Integration: Social, Education, and Psychological Issues* (pp. 216–224). Hershey, PA: Information Science Reference; doi:10.4018/978-1-4666-5071-8.ch013

Fornaciari, F. (2013). The Language of Technoself: Storytelling, Symbolic Interactionism, and Online Identity. In R. Luppicini (Ed.), *Handbook of Research on Technoself: Identity in a Technological Society* (pp. 64–83). Hershey, PA: Information Science Reference; doi:10.4018/978-1-4666-2211-1.ch004

Fox, J., & Ahn, S. J. (2013). Avatars: Portraying, Exploring, and Changing Online and Offline Identities. In R. Luppicini (Ed.), *Handbook of Research on Technoself: Identity in a Technological Society* (pp. 255–271). Hershey, PA: Information Science Reference; doi:10.4018/978-1-4666-2211-1.ch014

Fox, W. P., Binstock, J., & Minutas, M. (2013). Modeling and Methodology for Incorporating Existing Technologies to Produce Higher Probabilities of Detecting Suicide Bombers. [IJORIS]. *International Journal of Operations Research and Information Systems*, *4*(3), 1–18. doi:10.4018/joris.2013070101

Franchi, E., & Tomaiuolo, M. (2013). Distributed Social Platforms for Confidentiality and Resilience. In L. Caviglione, M. Coccoli, & A. Merlo (Eds.), *Social Network Engineering for Secure Web Data and Services* (pp. 114–136). Hershey, PA: Information Science Reference; doi:10.4018/978-1-4666-3926-3.ch006

Frigo, C. A., & Pavan, E. E. (2014). Prosthetic and Orthotic Devices. In Assistive Technologies: Concepts, Methodologies, Tools, and Applications (pp. 549-613). Hershey, PA: Information Science Reference. doi:10.4018/978-1-4666-4422-9.ch028

Fuhrer, C., & Cucchi, A. (2012). Relations Between Social Capital and Use of ICT: A Social Network Analysis Approach. [IJTHI]. *International Journal of Technology and Human Interaction*, *8*(2), 15–42. doi:10.4018/jthi.2012040102

Galinski, C., & Beckmann, H. (2014). Concepts for Enhancing Content Quality and eAccessibility: In General and in the Field of eProcurement. In Assistive Technologies: Concepts, Methodologies, Tools, and Applications (pp. 180-197). Hershey, PA: Information Science Reference. doi:10.4018/978-1-4666-4422-9.ch010

Galván, J. M., & Luppicini, R. (2012). The Humanity of the Human Body: Is Homo Cybersapien a New Species? [IJT]. *International Journal of Technoethics*, *3*(2), 1–8. doi:10.4018/jte.2012040101

García-Gómez, A. (2013). Technoself-Presentation on Social Networks: A Gender-Based Approach. In R. Luppicini (Ed.), *Handbook of Research on Technoself: Identity in a Technological Society* (pp. 382–398). Hershey, PA: Information Science Reference; doi:10.4018/978-1-4666-2211-1.ch021

Gill, L., Hathway, E. A., Lange, E., Morgan, E., & Romano, D. (2013). Coupling Real-Time 3D Landscape Models with Microclimate Simulations. [IJEPR]. *International Journal of E-Planning Research*, *2*(1), 1–19. doi:10.4018/ijepr.2013010101

Godé, C., & Lebraty, J. (2013). Improving Decision Making in Extreme Situations: The Case of a Military Decision Support System. [IJTHI]. *International Journal of Technology and Human Interaction*, *9*(1), 1–17. doi:10.4018/jthi.2013010101

Griol, D., Callejas, Z., & López-Cózar, R. (2014). Conversational Metabots for Educational Applications in Virtual Worlds. In Assistive Technologies: Concepts, Methodologies, Tools, and Applications (pp. 1405-1433). Hershey, PA: Information Science Reference. doi:10.4018/978-1-4666-4422-9.ch074

Griol Barres, D., Callejas Carrión, Z., Molina López, J. M., & Sanchis de Miguel, A. (2014). Towards the Use of Dialog Systems to Facilitate Inclusive Education. In Assistive Technologies: Concepts, Methodologies, Tools, and Applications (pp. 1292-1312). Hershey, PA: Information Science Reference. doi:10.4018/978-1-4666-4422-9.ch068

Groba, B., Pousada, T., & Nieto, L. (2014). Assistive Technologies, Tools and Resources for the Access and Use of Information and Communication Technologies by People with Disabilities. In Assistive Technologies: Concepts, Methodologies, Tools, and Applications (pp. 246-260). Hershey, PA: Information Science Reference. doi:10.4018/978-1-4666-4422-9.ch013

Groß, M. (2013). Personal Knowledge Management and Social Media: What Students Need to Learn for Business Life. In M. Pătruţ & B. Pătruţ (Eds.), *Social Media in Higher Education: Teaching in Web 2.0* (pp. 124–143). Hershey, PA: Information Science Reference; doi:10.4018/978-1-4666-2970-7.ch007

Gu, L., Aiken, M., Wang, J., & Wibowo, K. (2011). The Influence of Information Control upon Online Shopping Behavior. [IJTHI]. *International Journal of Technology and Human Interaction, 7*(1), 56–66. doi:10.4018/jthi.2011010104

Hainz, T. (2012). Value Lexicality and Human Enhancement. [IJT]. *International Journal of Technoethics, 3*(4), 54–65. doi:10.4018/jte.2012100105

Harnesk, D., & Lindström, J. (2014). Exploring Socio-Technical Design of Crisis Management Information Systems. In Crisis Management: Concepts, Methodologies, Tools and Applications (pp. 514-530). Hershey, PA: Information Science Reference. doi:10.4018/978-1-4666-4707-7.ch023

Hicks, D. (2014). Ethics in the Age of Technological Change and its Impact on the Professional Identity of Librarians. In *Technology and Professional Identity of Librarians: The Making of the Cybrarian* (pp. 168–187). Hershey, PA: Information Science Reference; doi:10.4018/978-1-4666-4735-0.ch009

Hicks, D. (2014). Technology, Profession, Identity. In *Technology and Professional Identity of Librarians: The Making of the Cybrarian* (pp. 1–20). Hershey, PA: Information Science Reference; doi:10.4018/978-1-4666-4735-0.ch001

Hirata, M., Yanagisawa, T., Matsushita, K., Sugata, H., Kamitani, Y., Suzuki, T., . . . Yoshimine, T. (2014). Brain-Machine Interface Using Brain Surface Electrodes: Real-Time Robotic Control and a Fully Implantable Wireless System. In Assistive Technologies: Concepts, Methodologies, Tools, and Applications (pp. 1535-1548). Hershey, PA: Information Science Reference. doi:10.4018/978-1-4666-4422-9.ch080

Hodge, B. (2014). Critical Electronic Discourse Analysis: Social and Cultural Research in the Electronic Age. In H. Lim & F. Sudweeks (Eds.), *Innovative Methods and Technologies for Electronic Discourse Analysis* (pp. 191–209). Hershey, PA: Information Science Reference; doi:10.4018/978-1-4666-4426-7.ch009

Hoey, J., Poupart, P., Boutilier, C., & Mihailidis, A. (2014). POMDP Models for Assistive Technology. In Assistive Technologies: Concepts, Methodologies, Tools, and Applications (pp. 120-140). Hershey, PA: Information Science Reference. doi:10.4018/978-1-4666-4422-9.ch007

Hogg, S. (2014). An Informal Use of Facebook to Encourage Student Collaboration and Motivation for Off Campus Activities. In G. Mallia (Ed.), *The Social Classroom: Integrating Social Network Use in Education* (pp. 23–39). Hershey, PA: Information Science Reference; doi:10.4018/978-1-4666-4904-0.ch002

Holmqvist, E., & Buchholz, M. (2014). A Model for Gaze Control Assessments and Evaluation. In Assistive Technologies: Concepts, Methodologies, Tools, and Applications (pp. 332-343). Hershey, PA: Information Science Reference. doi:10.4018/978-1-4666-4422-9.ch016

Hsiao, S., Chen, D., Yang, C., Huang, H., Lu, Y., Huang, H., & Lin, Y. et al. (2013). Chemical-Free and Reusable Cellular Analysis: Electrochemical Impedance Spectroscopy with a Transparent ITO Culture Chip. [IJTHI]. *International Journal of Technology and Human Interaction, 8*(3), 1–9. doi:10.4018/jthi.2012070101

Hsu, M., Yang, C., Wang, C., & Lin, Y. (2013). Simulation-Aided Optimal Microfluidic Sorting for Monodispersed Microparticles. [IJTHI]. *International Journal of Technology and Human Interaction, 8*(3), 10–18. doi:10.4018/jthi.2012070102

Huang, W. D., & Tettegah, S. Y. (2014). Cognitive Load and Empathy in Serious Games: A Conceptual Framework. In J. Bishop (Ed.), *Gamification for Human Factors Integration: Social, Education, and Psychological Issues* (pp. 17–30). Hershey, PA: Information Science Reference; doi:10.4018/978-1-4666-5071-8.ch002

Huseyinov, I. N. (2014). Fuzzy Linguistic Modelling in Multi Modal Human Computer Interaction: Adaptation to Cognitive Styles using Multi Level Fuzzy Granulation Method. In Assistive Technologies: Concepts, Methodologies, Tools, and Applications (pp. 1481-1496). Hershey, PA: Information Science Reference. doi:10.4018/978-1-4666-4422-9.ch077

Hwa, S. P., Weei, P. S., & Len, L. H. (2012). The Effects of Blended Learning Approach through an Interactive Multimedia E-Book on Students' Achievement in Learning Chinese as a Second Language at Tertiary Level. [IJCALLT]. *International Journal of Computer-Assisted Language Learning and Teaching*, *2*(1), 35–50. doi:10.4018/ijcallt.2012010104

Iglesias, A., Ruiz-Mezcua, B., López, J. F., & Figueroa, D. C. (2014). New Communication Technologies for Inclusive Education in and outside the Classroom. In Assistive Technologies: Concepts, Methodologies, Tools, and Applications (pp. 1675-1689). Hershey, PA: Information Science Reference. doi:10.4018/978-1-4666-4422-9.ch088

Inghilterra, X., & Ravatua-Smith, W. S. (2014). Online Learning Communities: Use of Micro Blogging for Knowledge Construction. In J. Pelet (Ed.), *E-Learning 2.0 Technologies and Web Applications in Higher Education* (pp. 107–128). Hershey, PA: Information Science Reference; doi:10.4018/978-1-4666-4876-0.ch006

Ionescu, A. (2013). Cyber Identity: Our Alter-Ego? In R. Luppicini (Ed.), *Handbook of Research on Technoself: Identity in a Technological Society* (pp. 189–203). Hershey, PA: Information Science Reference; doi:10.4018/978-1-4666-2211-1.ch011

Jan, Y., Lin, M., Shiao, K., Wei, C., Huang, L., & Sung, Q. (2013). Development of an Evaluation Instrument for Green Building Literacy among College Students in Taiwan. [IJTHI]. *International Journal of Technology and Human Interaction*, *8*(3), 31–45. doi:10.4018/jthi.2012070104

Jawadi, N. (2013). E-Leadership and Trust Management: Exploring the Moderating Effects of Team Virtuality. [IJTHI]. *International Journal of Technology and Human Interaction*, *9*(3), 18–35. doi:10.4018/jthi.2013070102

Jiménez-Castillo, D., & Fernández, R. S. (2014). The Impact of Combining Video Podcasting and Lectures on Students' Assimilation of Additional Knowledge: An Empirical Examination. In J. Pelet (Ed.), *E-Learning 2.0 Technologies and Web Applications in Higher Education* (pp. 65–87). Hershey, PA: Information Science Reference; doi:10.4018/978-1-4666-4876-0.ch004

Jin, L. (2013). A New Trend in Education: Technoself Enhanced Social Learning. In R. Luppicini (Ed.), *Handbook of Research on Technoself: Identity in a Technological Society* (pp. 456–473). Hershey, PA: Information Science Reference; doi:10.4018/978-1-4666-2211-1.ch025

Johansson, L. (2012). The Functional Morality of Robots. In R. Luppicini (Ed.), *Ethical Impact of Technological Advancements and Applications in Society* (pp. 254–262). Hershey, PA: Information Science Reference; doi:10.4018/978-1-4666-1773-5.ch020

Johansson, L. (2013). Robots and the Ethics of Care. [IJT]. *International Journal of Technoethics*, *4*(1), 67–82. doi:10.4018/jte.2013010106

Johri, A., Dufour, M., Lo, J., & Shanahan, D. (2013). Adwiki: Socio-Technical Design for Mananging Advising Knowledge in a Higher Education Context. [IJSKD]. *International Journal of Sociotechnology and Knowledge Development*, *5*(1), 37–59. doi:10.4018/jskd.2013010104

Jones, M. G., Schwilk, C. L., & Bateman, D. F. (2014). Reading by Listening: Access to Books in Audio Format for College Students with Print Disabilities. In Assistive Technologies: Concepts, Methodologies, Tools, and Applications (pp. 454-477). Hershey, PA: Information Science Reference. doi:10.4018/978-1-4666-4422-9.ch022

Kaba, B., & Osei-Bryson, K. (2012). An Empirical Investigation of External Factors Influencing Mobile Technology Use in Canada: A Preliminary Study. [IJTHI]. *International Journal of Technology and Human Interaction*, *8*(2), 1–14. doi:10.4018/jthi.2012040101

Kampf, C. E. (2012). Revealing the Socio-Technical Design of Global E-Businesses: A Case of Digital Artists Engaging in Radical Transparency. [IJSKD]. *International Journal of Sociotechnology and Knowledge Development*, *4*(4), 18–31. doi:10.4018/jskd.2012100102

Kandroudi, M., & Bratitsis, T. (2014). Classifying Facebook Usage in the Classroom or Around It. In G. Mallia (Ed.), *The Social Classroom: Integrating Social Network Use in Education* (pp. 62–81). Hershey, PA: Information Science Reference; doi:10.4018/978-1-4666-4904-0.ch004

Kidd, P. T. (2014). Social Networking Technologies as a Strategic Tool for the Development of Sustainable Production and Consumption: Applications to Foster the Agility Needed to Adapt Business Models in Response to the Challenges Posed by Climate Change. In Sustainable Practices: Concepts, Methodologies, Tools and Applications (pp. 974-987). Hershey, PA: Information Science Reference. doi:10.4018/978-1-4666-4852-4.ch054

Kirby, S. D., & Sellers, D. M. (2014). The Live-Ability House: A Collaborative Adventure in Discovery Learning. In Assistive Technologies: Concepts, Methodologies, Tools, and Applications (pp. 1626-1649). Hershey, PA: Information Science Reference. doi:10.4018/978-1-4666-4422-9.ch086

Kitchenham, A., & Bowes, D. (2014). Voice/Speech Recognition Software: A Discussion of the Promise for Success and Practical Suggestions for Implementation. In Assistive Technologies: Concepts, Methodologies, Tools, and Applications (pp. 1005-1011). Hershey, PA: Information Science Reference. doi:10.4018/978-1-4666-4422-9.ch051

Konrath, S. (2013). The Empathy Paradox: Increasing Disconnection in the Age of Increasing Connection. In R. Luppicini (Ed.), *Handbook of Research on Technoself: Identity in a Technological Society* (pp. 204–228). Hershey, PA: Information Science Reference; doi:10.4018/978-1-4666-2211-1.ch012

Koutsabasis, P., & Istikopoulou, T. G. (2013). Perceived Website Aesthetics by Users and Designers: Implications for Evaluation Practice. [IJTHI]. *International Journal of Technology and Human Interaction*, *9*(2), 39–52. doi:10.4018/jthi.2013040103

Kraft, E., & Wang, J. (2012). An Exploratory Study of the Cyberbullying and Cyberstalking Experiences and Factors Related to Victimization of Students at a Public Liberal Arts College. In R. Luppicini (Ed.), *Ethical Impact of Technological Advancements and Applications in Society* (pp. 113–131). Hershey, PA: Information Science Reference; doi:10.4018/978-1-4666-1773-5.ch009

Kulman, R., Stoner, G., Ruffolo, L., Marshall, S., Slater, J., Dyl, A., & Cheng, A. (2014). Teaching Executive Functions, Self-Management, and Ethical Decision-Making through Popular Videogame Play. In Assistive Technologies: Concepts, Methodologies, Tools, and Applications (pp. 771-785). Hershey, PA: Information Science Reference. doi:10.4018/978-1-4666-4422-9.ch039

Kunc, L., Míkovec, Z., & Slavík, P. (2013). Avatar and Dialog Turn-Yielding Phenomena. [IJTHI]. *International Journal of Technology and Human Interaction*, *9*(2), 66–88. doi:10.4018/jthi.2013040105

Kuo, N., & Dai, Y. (2012). Applying the Theory of Planned Behavior to Predict Low-Carbon Tourism Behavior: A Modified Model from Taiwan. [IJTHI]. *International Journal of Technology and Human Interaction*, *8*(4), 45–62. doi:10.4018/jthi.2012100103

Kurt, S. (2014). Accessibility Issues of Educational Web Sites. In Assistive Technologies: Concepts, Methodologies, Tools, and Applications (pp. 54-62). Hershey, PA: Information Science Reference. doi:10.4018/978-1-4666-4422-9.ch004

Kuzma, J. (2013). Empirical Study of Cyber Harassment among Social Networks. [IJTHI]. *International Journal of Technology and Human Interaction*, *9*(2), 53–65. doi:10.4018/jthi.2013040104

Kyriakaki, G., & Matsatsinis, N. (2014). Pedagogical Evaluation of E-Learning Websites with Cognitive Objectives. In D. Yannacopoulos, P. Manolitzas, N. Matsatsinis, & E. Grigoroudis (Eds.), *Evaluating Websites and Web Services: Interdisciplinary Perspectives on User Satisfaction* (pp. 224–240). Hershey, PA: Information Science Reference; doi:10.4018/978-1-4666-5129-6.ch013

Lee, H., & Baek, E. (2012). Facilitating Deep Learning in a Learning Community. [IJTHI]. *International Journal of Technology and Human Interaction*, *8*(1), 1–13. doi:10.4018/jthi.2012010101

Lee, W., Wu, T., Cheng, Y., Chuang, Y., & Sheu, S. (2013). Using the Kalman Filter for Auto Bit-rate H.264 Streaming Based on Human Interaction. [IJTHI]. *International Journal of Technology and Human Interaction*, *9*(4), 58–74. doi:10.4018/ijthi.2013100104

Li, Y., Guo, N. Y., & Ranieri, M. (2014). Designing an Online Interactive Learning Program to Improve Chinese Migrant Children's Internet Skills: A Case Study at Hangzhou Minzhu Experimental School. In Z. Yang, H. Yang, D. Wu, & S. Liu (Eds.), *Transforming K-12 Classrooms with Digital Technology* (pp. 249–265). Hershey, PA: Information Science Reference; doi:10.4018/978-1-4666-4538-7.ch013

Lin, C., Chu, L., & Hsu, H. (2013). Study on the Performance and Exhaust Emissions of Motorcycle Engine Fuelled with Hydrogen-Gasoline Compound Fuel. [IJTHI]. *International Journal of Technology and Human Interaction*, *8*(3), 69–81. doi:10.4018/jthi.2012070107

Lin, L. (2013). Multiple Dimensions of Multitasking Phenomenon. [IJTHI]. *International Journal of Technology and Human Interaction*, *9*(1), 37–49. doi:10.4018/jthi.2013010103

Lin, T., Li, X., Wu, Z., & Tang, N. (2013). Automatic Cognitive Load Classification Using High-Frequency Interaction Events: An Exploratory Study. [IJTHI]. *International Journal of Technology and Human Interaction*, *9*(3), 73–88. doi:10.4018/jthi.2013070106

Lin, T., Wu, Z., Tang, N., & Wu, S. (2013). Exploring the Effects of Display Characteristics on Presence and Emotional Responses of Game Players. [IJTHI]. *International Journal of Technology and Human Interaction*, *9*(1), 50–63. doi:10.4018/jthi.2013010104

Lin, T., Xie, T., Mou, Y., & Tang, N. (2013). Markov Chain Models for Menu Item Prediction. [IJTHI]. *International Journal of Technology and Human Interaction*, *9*(4), 75–94. doi:10.4018/ijthi.2013100105

Lin, X., & Luppicini, R. (2011). Socio-Technical Influences of Cyber Espionage: A Case Study of the GhostNet System. [IJT]. *International Journal of Technoethics*, *2*(2), 65–77. doi:10.4018/jte.2011040105

Linek, S. B., Marte, B., & Albert, D. (2014). Background Music in Educational Games: Motivational Appeal and Cognitive Impact. In J. Bishop (Ed.), *Gamification for Human Factors Integration: Social, Education, and Psychological Issues* (pp. 259–271). Hershey, PA: Information Science Reference; doi:10.4018/978-1-4666-5071-8.ch016

Lipschutz, R. D., & Hester, R. J. (2014). We Are the Borg! Human Assimilation into Cellular Society. In M. Michael & K. Michael (Eds.), *Uberveillance and the Social Implications of Microchip Implants: Emerging Technologies* (pp. 366–407). Hershey, PA: Information Science Reference; doi:10.4018/978-1-4666-4582-0.ch016

Liu, C., Zhong, Y., Ozercan, S., & Zhu, Q. (2013). Facilitating 3D Virtual World Learning Environments Creation by Non-Technical End Users through Template-Based Virtual World Instantiation. [IJVPLE]. *International Journal of Virtual and Personal Learning Environments*, *4*(1), 32–48. doi:10.4018/jvple.2013010103

Liu, F., Lo, H., Su, C., Lou, D., & Lee, W. (2013). High Performance Reversible Data Hiding for Mobile Applications and Human Interaction. [IJTHI]. *International Journal of Technology and Human Interaction*, *9*(4), 41–57. doi:10.4018/ijthi.2013100103

Liu, H. (2012). From Cold War Island to Low Carbon Island: A Study of Kinmen Island. [IJTHI]. *International Journal of Technology and Human Interaction*, *8*(4), 63–74. doi:10.4018/jthi.2012100104

Lixun, Z., Dapeng, B., & Lei, Y. (2014). Design of and Experimentation with a Walking Assistance Robot. In Assistive Technologies: Concepts, Methodologies, Tools, and Applications (pp. 1600-1605). Hershey, PA: Information Science Reference. doi:10.4018/978-1-4666-4422-9.ch084

Low, R., Jin, P., & Sweller, J. (2014). Instructional Design in Digital Environments and Availability of Mental Resources for the Aged Subpopulation. In Assistive Technologies: Concepts, Methodologies, Tools, and Applications (pp. 1131-1154). Hershey, PA: Information Science Reference. doi:10.4018/978-1-4666-4422-9.ch059

Luczak, H., Schlick, C. M., Jochems, N., Vetter, S., & Kausch, B. (2014). Touch Screens for the Elderly: Some Models and Methods, Prototypical Development and Experimental Evaluation of Human-Computer Interaction Concepts for the Elderly. In Assistive Technologies: Concepts, Methodologies, Tools, and Applications (pp. 377-396). Hershey, PA: Information Science Reference. doi:10.4018/978-1-4666-4422-9.ch019

Luor, T., Lu, H., Johanson, R. E., & Yu, H. (2012). Minding the Gap Between First and Continued Usage of a Corporate E-Learning English-language Program. [IJTHI]. *International Journal of Technology and Human Interaction*, *8*(1), 55–74. doi:10.4018/jthi.2012010104

Luppicini, R. (2013). The Emerging Field of Technoself Studies (TSS). In R. Luppicini (Ed.), *Handbook of Research on Technoself: Identity in a Technological Society* (pp. 1–25). Hershey, PA: Information Science Reference; doi:10.4018/978-1-4666-2211-1.ch001

Magnani, L. (2012). Material Cultures and Moral Mediators in Human Hybridization. In R. Luppicini (Ed.), *Ethical Impact of Technological Advancements and Applications in Society* (pp. 1–20). Hershey, PA: Information Science Reference; doi:10.4018/978-1-4666-1773-5.ch001

Maher, D. (2014). Learning in the Primary School Classroom using the Interactive Whiteboard. In K-12 Education: Concepts, Methodologies, Tools, and Applications (pp. 526-538). Hershey, PA: Information Science Reference. doi:10.4018/978-1-4666-4502-8.ch031

Manolache, M., & Patrut, M. (2013). The Use of New Web-Based Technologies in Strategies of Teaching Gender Studies. In M. Pătruţ & B. Pătruţ (Eds.), *Social Media in Higher Education: Teaching in Web 2.0* (pp. 45–74). Hershey, PA: Information Science Reference; doi:10.4018/978-1-4666-2970-7.ch003

Manthiou, A., & Chiang, L.Liang (Rebecca) Tang. (2013). Identifying and Responding to Customer Needs on Facebook Fan Pages. [IJTHI]. *International Journal of Technology and Human Interaction, 9*(3), 36–52. doi:10.4018/jthi.2013070103

Marengo, A., Pagano, A., & Barbone, A. (2013). An Assessment of Customer's Preferences and Improve Brand Awareness Implementation of Social CRM in an Automotive Company. [IJTD]. *International Journal of Technology Diffusion, 4*(1), 1–15. doi:10.4018/jtd.2013010101

Martin, I., Kear, K., Simpkins, N., & Busvine, J. (2013). Social Negotiations in Web Usability Engineering. In M. Garcia-Ruiz (Ed.), *Cases on Usability Engineering: Design and Development of Digital Products* (pp. 26–56). Hershey, PA: Information Science Reference; doi:10.4018/978-1-4666-4046-7.ch002

Martins, T., Carvalho, V., & Soares, F. (2014). An Overview on the Use of Serious Games in Physical Therapy and Rehabilitation. In Assistive Technologies: Concepts, Methodologies, Tools, and Applications (pp. 758-770). Hershey, PA: Information Science Reference. doi:10.4018/978-1-4666-4422-9.ch038

Mathew, D. (2013). Online Anxiety: Implications for Educational Design in a Web 2.0 World. In M. Pătruţ & B. Pătruţ (Eds.), *Social Media in Higher Education: Teaching in Web 2.0* (pp. 305–317). Hershey, PA: Information Science Reference; doi:10.4018/978-1-4666-2970-7.ch015

Mazzanti, I., Maolo, A., & Antonicelli, R. (2014). E-Health and Telemedicine in the Elderly: State of the Art. In Assistive Technologies: Concepts, Methodologies, Tools, and Applications (pp. 693-704). Hershey, PA: Information Science Reference. doi:10.4018/978-1-4666-4422-9.ch034

Mazzara, M., Biselli, L., Greco, P. P., Dragoni, N., Marraffa, A., Qamar, N., & de Nicola, S. (2013). Social Networks and Collective Intelligence: A Return to the Agora. In L. Caviglione, M. Coccoli, & A. Merlo (Eds.), *Social Network Engineering for Secure Web Data and Services* (pp. 88–113). Hershey, PA: Information Science Reference; doi:10.4018/978-1-4666-3926-3.ch005

McColl, D., & Nejat, G. (2013). A Human Affect Recognition System for Socially Interactive Robots. In R. Luppicini (Ed.), *Handbook of Research on Technoself: Identity in a Technological Society* (pp. 554–573). Hershey, PA: Information Science Reference; doi:10.4018/978-1-4666-2211-1.ch029

McDonald, A., & Helmer, S. (2011). A Comparative Case Study of Indonesian and UK Organisational Culture Differences in IS Project Management. [IJTHI]. *International Journal of Technology and Human Interaction, 7*(2), 28–37. doi:10.4018/jthi.2011040104

McGee, E. M. (2014). Neuroethics and Implanted Brain Machine Interfaces. In M. Michael & K. Michael (Eds.), *Uberveillance and the Social Implications of Microchip Implants: Emerging Technologies* (pp. 351–365). Hershey, PA: Information Science Reference; doi:10.4018/978-1-4666-4582-0.ch015

McGrath, E., Lowes, S., McKay, M., Sayres, J., & Lin, P. (2014). Robots Underwater! Learning Science, Engineering and 21st Century Skills: The Evolution of Curricula, Professional Development and Research in Formal and Informal Contexts. In K-12 Education: Concepts, Methodologies, Tools, and Applications (pp. 1041-1067). Hershey, PA: Information Science Reference. doi:10.4018/978-1-4666-4502-8.ch062

Meissonierm, R., Bourdon, I., Amabile, S., & Boudrandi, S. (2012). Toward an Enacted Approach to Understanding OSS Developer's Motivations. [IJTHI]. *International Journal of Technology and Human Interaction, 8*(1), 38–54. doi:10.4018/jthi.2012010103

Melius, J. (2014). The Role of Social Constructivist Instructional Approaches in Facilitating Cross-Cultural Online Learning in Higher Education. In J. Keengwe, G. Schnellert, & K. Kungu (Eds.), *Cross-Cultural Online Learning in Higher Education and Corporate Training* (pp. 253–270). Hershey, PA: Information Science Reference; doi:10.4018/978-1-4666-5023-7.ch015

Melson, G. F. (2013). Building a Technoself: Children's Ideas about and Behavior toward Robotic Pets. In R. Luppicini (Ed.), *Handbook of Research on Technoself: Identity in a Technological Society* (pp. 592–608). Hershey, PA: Information Science Reference; doi:10.4018/978-1-4666-2211-1.ch031

Mena, R. J. (2014). The Quest for a Massively Multiplayer Online Game that Teaches Physics. In T. Connolly, T. Hainey, E. Boyle, G. Baxter, & P. Moreno-Ger (Eds.), *Psychology, Pedagogy, and Assessment in Serious Games* (pp. 292–316). Hershey, PA: Information Science Reference; doi:10.4018/978-1-4666-4773-2.ch014

Meredith, J., & Potter, J. (2014). Conversation Analysis and Electronic Interactions: Methodological, Analytic and Technical Considerations. In H. Lim & F. Sudweeks (Eds.), *Innovative Methods and Technologies for Electronic Discourse Analysis* (pp. 370–393). Hershey, PA: Information Science Reference; doi:10.4018/978-1-4666-4426-7.ch017

Millán-Calenti, J. C., & Maseda, A. (2014). Telegerontology®: A New Technological Resource for Elderly Support. In Assistive Technologies: Concepts, Methodologies, Tools, and Applications (pp. 705-719). Hershey, PA: Information Science Reference. doi:10.4018/978-1-4666-4422-9.ch035

Miscione, G. (2011). Telemedicine and Development: Situating Information Technologies in the Amazon. [IJSKD]. *International Journal of Sociotechnology and Knowledge Development, 3*(4), 15–26. doi:10.4018/jskd.2011100102

Miwa, N., & Wang, Y. (2011). Online Interaction Between On-Campus and Distance Students: Learners' Perspectives. [IJCALLT]. *International Journal of Computer-Assisted Language Learning and Teaching, 1*(3), 54–69. doi:10.4018/ijcallt.2011070104

Moore, M. J., Nakano, T., Suda, T., & Enomoto, A. (2013). Social Interactions and Automated Detection Tools in Cyberbullying. In L. Caviglione, M. Coccoli, & A. Merlo (Eds.), *Social Network Engineering for Secure Web Data and Services* (pp. 67–87). Hershey, PA: Information Science Reference; doi:10.4018/978-1-4666-3926-3.ch004

Morueta, R. T., Gómez, J. I., & Gómez, Á. H. (2012). B-Learning at Universities in Andalusia (Spain): From Traditional to Student-Centred Learning. [IJTHI]. *International Journal of Technology and Human Interaction*, *8*(2), 56–76. doi:10.4018/jthi.2012040104

Mosindi, O., & Sice, P. (2011). An Exploratory Theoretical Framework for Understanding Information Behaviour. [IJTHI]. *International Journal of Technology and Human Interaction*, *7*(2), 1–8. doi:10.4018/jthi.2011040101

Mott, M. S., & Williams-Black, T. H. (2014). Media-Enhanced Writing Instruction and Assessment. In J. Keengwe, G. Onchwari, & D. Hucks (Eds.), *Literacy Enrichment and Technology Integration in Pre-Service Teacher Education* (pp. 1–16). Hershey, PA: Information Science Reference; doi:10.4018/978-1-4666-4924-8.ch001

Mulvey, F., & Heubner, M. (2014). Eye Movements and Attention. In Assistive Technologies: Concepts, Methodologies, Tools, and Applications (pp. 1030-1054). Hershey, PA: Information Science Reference. doi:10.4018/978-1-4666-4422-9.ch053

Muro, B. F., & Delgado, E. C. (2014). RACEM Game for PC for Use as Rehabilitation Therapy for Children with Psychomotor Disability and Results of its Application. In Assistive Technologies: Concepts, Methodologies, Tools, and Applications (pp. 740-757). Hershey, PA: Information Science Reference. doi:10.4018/978-1-4666-4422-9.ch037

Muwanguzi, S., & Lin, L. (2014). Coping with Accessibility and Usability Challenges of Online Technologies by Blind Students in Higher Education. In Assistive Technologies: Concepts, Methodologies, Tools, and Applications (pp. 1227-1244). Hershey, PA: Information Science Reference. doi:10.4018/978-1-4666-4422-9.ch064

Najjar, M., Courtemanche, F., Hamam, H., Dion, A., Bauchet, J., & Mayers, A. (2014). DeepKøver: An Adaptive Intelligent Assistance System for Monitoring Impaired People in Smart Homes. In Assistive Technologies: Concepts, Methodologies, Tools, and Applications (pp. 634-661). Hershey, PA: Information Science Reference. doi:10.4018/978-1-4666-4422-9.ch031

Nap, H. H., & Diaz-Orueta, U. (2014). Rehabilitation Gaming. In J. Bishop (Ed.), *Gamification for Human Factors Integration: Social, Education, and Psychological Issues* (pp. 122–147). Hershey, PA: Information Science Reference; doi:10.4018/978-1-4666-5071-8.ch008

Neves, J., & Pinheiro, L. D. (2012). Cyberbullying: A Sociological Approach. In R. Luppicini (Ed.), *Ethical Impact of Technological Advancements and Applications in Society* (pp. 132–142). Hershey, PA: Information Science Reference; doi:10.4018/978-1-4666-1773-5.ch010

Nguyen, P. T. (2012). Peer Feedback on Second Language Writing through Blogs: The Case of a Vietnamese EFL Classroom. [IJCALLT]. *International Journal of Computer-Assisted Language Learning and Teaching*, *2*(1), 13–23. doi:10.4018/ijcallt.2012010102

Ninaus, M., Witte, M., Kober, S. E., Friedrich, E. V., Kurzmann, J., Hartsuiker, E., & Wood, G. et al. (2014). Neurofeedback and Serious Games. In T. Connolly, T. Hainey, E. Boyle, G. Baxter, & P. Moreno-Ger (Eds.), *Psychology, Pedagogy, and Assessment in Serious Games* (pp. 82–110). Hershey, PA: Information Science Reference; doi:10.4018/978-1-4666-4773-2.ch005

Olla, V. (2014). An Enquiry into the use of Technology and Student Voice in Citizenship Education in the K-12 Classroom. In K-12 Education: Concepts, Methodologies, Tools, and Applications (pp. 892-913). Hershey, PA: Information Science Reference. doi:10.4018/978-1-4666-4502-8.ch053

Orange, E. (2013). Understanding the Human-Machine Interface in a Time of Change. In R. Luppicini (Ed.), *Handbook of Research on Technoself: Identity in a Technological Society* (pp. 703–719). Hershey, PA: Information Science Reference; doi:10.4018/978-1-4666-2211-1.ch036

Palmer, D., Warren, I., & Miller, P. (2014). ID Scanners and Überveillance in the Night-Time Economy: Crime Prevention or Invasion of Privacy? In M. Michael & K. Michael (Eds.), *Uberveillance and the Social Implications of Microchip Implants: Emerging Technologies* (pp. 208–225). Hershey, PA: Information Science Reference; doi:10.4018/978-1-4666-4582-0.ch009

Papadopoulos, F., Dautenhahn, K., & Ho, W. C. (2013). Behavioral Analysis of Human-Human Remote Social Interaction Mediated by an Interactive Robot in a Cooperative Game Scenario. In R. Luppicini (Ed.), *Handbook of Research on Technoself: Identity in a Technological Society* (pp. 637–665). Hershey, PA: Information Science Reference; doi:10.4018/978-1-4666-2211-1.ch033

Patel, K. K., & Vij, S. K. (2014). Unconstrained Walking Plane to Virtual Environment for Non-Visual Spatial Learning. In Assistive Technologies: Concepts, Methodologies, Tools, and Applications (pp. 1580-1599). Hershey, PA: Information Science Reference. doi:10.4018/978-1-4666-4422-9.ch083

Patrone, T. (2013). In Defense of the 'Human Prejudice'. [IJT]. *International Journal of Technoethics*, *4*(1), 26–38. doi:10.4018/jte.2013010103

Peevers, G., Williams, R., Douglas, G., & Jack, M. A. (2013). Usability Study of Fingerprint and Palmvein Biometric Technologies at the ATM. [IJTHI]. *International Journal of Technology and Human Interaction*, *9*(1), 78–95. doi:10.4018/jthi.2013010106

Pellas, N. (2014). Theoretical Foundations of a CSCL Script in Persistent Virtual Worlds According to the Contemporary Learning Theories and Models. In E. Nikoi & K. Boateng (Eds.), *Collaborative Communication Processes and Decision Making in Organizations* (pp. 72–107). Hershey, PA: Business Science Reference; doi:10.4018/978-1-4666-4478-6.ch005

Perakslis, C. (2014). Willingness to Adopt RFID Implants: Do Personality Factors Play a Role in the Acceptance of Uberveillance? In M. Michael & K. Michael (Eds.), *Uberveillance and the Social Implications of Microchip Implants: Emerging Technologies* (pp. 144–168). Hershey, PA: Information Science Reference; doi:10.4018/978-1-4666-4582-0.ch006

Pereira, G., Brisson, A., Dias, J., Carvalho, A., Dimas, J., Mascarenhas, S., & Paiva, A. et al. (2014). Non-Player Characters and Artificial Intelligence. In T. Connolly, T. Hainey, E. Boyle, G. Baxter, & P. Moreno-Ger (Eds.), *Psychology, Pedagogy, and Assessment in Serious Games* (pp. 127–152). Hershey, PA: Information Science Reference; doi:10.4018/978-1-4666-4773-2.ch007

Pérez Pérez, A., Callejas Carrión, Z., López-Cózar Delgado, R., & Griol Barres, D. (2014). On the Use of Speech Technologies to Achieve Inclusive Education for People with Intellectual Disabilities. In Assistive Technologies: Concepts, Methodologies, Tools, and Applications (pp. 1106-1117). Hershey, PA: Information Science Reference. doi:10.4018/978-1-4666-4422-9.ch057

Peschl, M. F., & Fundneider, T. (2014). Theory U and Emergent Innovation: Presencing as a Method of Bringing Forth Profoundly New Knowledge and Realities. In O. Gunnlaugson, C. Baron, & M. Cayer (Eds.), *Perspectives on Theory U: Insights from the Field* (pp. 207–233). Hershey, PA: Business Science Reference; doi:10.4018/978-1-4666-4793-0.ch014

Petrovic, N., Jeremic, V., Petrovic, D., & Cirovic, M. (2014). Modeling the Use of Facebook in Environmental Higher Education. In G. Mallia (Ed.), *The Social Classroom: Integrating Social Network Use in Education* (pp. 100–119). Hershey, PA: Information Science Reference; doi:10.4018/978-1-4666-4904-0.ch006

Phua, C., Roy, P. C., Aloulou, H., Biswas, J., Tolstikov, A., Foo, V. S., . . . Xu, D. (2014). State-of-the-Art Assistive Technology for People with Dementia. In Assistive Technologies: Concepts, Methodologies, Tools, and Applications (pp. 1606-1625). Hershey, PA: Information Science Reference. doi:10.4018/978-1-4666-4422-9.ch085

Potts, L. (2011). Balancing McLuhan With Williams: A Sociotechnical View of Technological Determinism. [IJSKD]. *International Journal of Sociotechnology and Knowledge Development*, *3*(2), 53–57. doi:10.4018/jskd.2011040105

Potts, L. (2013). Balancing McLuhan With Williams: A Sociotechnical View of Technological Determinism. In J. Abdelnour-Nocera (Ed.), *Knowledge and Technological Development Effects on Organizational and Social Structures* (pp. 109–114). Hershey, PA: Information Science Reference; doi:10.4018/978-1-4666-2151-0.ch007

Potts, L. (2014). Sociotechnical Uses of Social Web Tools during Disasters. In Crisis Management: Concepts, Methodologies, Tools and Applications (pp. 531-541). Hershey, PA: Information Science Reference. doi:10.4018/978-1-4666-4707-7.ch024

Proença, R., Guerra, A., & Campos, P. (2013). A Gestural Recognition Interface for Intelligent Wheelchair Users. [IJSKD]. *International Journal of Sociotechnology and Knowledge Development*, *5*(2), 63–81. doi:10.4018/jskd.2013040105

Quilici-Gonzalez, J. A., Kobayashi, G., Broens, M. C., & Gonzalez, M. E. (2012). Ubiquitous Computing: Any Ethical Implications? In R. Luppicini (Ed.), *Ethical Impact of Technological Advancements and Applications in Society* (pp. 47–59). Hershey, PA: Information Science Reference; doi:10.4018/978-1-4666-1773-5.ch005

Rambaree, K. (2014). Computer-Aided Deductive Critical Discourse Analysis of a Case Study from Mauritius with ATLAS-ti 6.2. In H. Lim & F. Sudweeks (Eds.), *Innovative Methods and Technologies for Electronic Discourse Analysis* (pp. 346–368). Hershey, PA: Information Science Reference; doi:10.4018/978-1-4666-4426-7.ch016

Ratan, R. (2013). Self-Presence, Explicated: Body, Emotion, and Identity Extension into the Virtual Self. In R. Luppicini (Ed.), *Handbook of Research on Technoself: Identity in a Technological Society* (pp. 322–336). Hershey, PA: Information Science Reference; doi:10.4018/978-1-4666-2211-1.ch018

Rechy-Ramirez, E. J., & Hu, H. (2014). A Flexible Bio-Signal Based HMI for Hands-Free Control of an Electric Powered Wheelchair. [IJALR]. *International Journal of Artificial Life Research*, *4*(1), 59–76. doi:10.4018/ijalr.2014010105

Reiners, T., Wood, L. C., & Dron, J. (2014). From Chaos Towards Sense: A Learner-Centric Narrative Virtual Learning Space. In J. Bishop (Ed.), *Gamification for Human Factors Integration: Social, Education, and Psychological Issues* (pp. 242–258). Hershey, PA: Information Science Reference; doi:10.4018/978-1-4666-5071-8.ch015

Reinhardt, J., & Ryu, J. (2013). Using Social Network-Mediated Bridging Activities to Develop Socio-Pragmatic Awareness in Elementary Korean. [IJCALLT]. *International Journal of Computer-Assisted Language Learning and Teaching*, *3*(3), 18–33. doi:10.4018/ijcallt.2013070102

Revuelta, P., Jiménez, J., Sánchez, J. M., & Ruiz, B. (2014). Automatic Speech Recognition to Enhance Learning for Disabled Students. In Assistive Technologies: Concepts, Methodologies, Tools, and Applications (pp. 478-493). Hershey, PA: Information Science Reference. doi:10.4018/978-1-4666-4422-9.ch023

Ribeiro, J. C., & Silva, T. (2013). Self, Self-Presentation, and the Use of Social Applications in Digital Environments. In R. Luppicini (Ed.), *Handbook of Research on Technoself: Identity in a Technological Society* (pp. 439–455). Hershey, PA: Information Science Reference; doi:10.4018/978-1-4666-2211-1.ch024

Richet, J. (2013). From Young Hackers to Crackers. [IJTHI]. *International Journal of Technology and Human Interaction*, *9*(3), 53–62. doi:10.4018/jthi.2013070104

Rigas, D., & Almutairi, B. (2013). An Empirical Investigation into the Role of Avatars in Multimodal E-government Interfaces. [IJSKD]. *International Journal of Sociotechnology and Knowledge Development*, *5*(1), 14–22. doi:10.4018/jskd.2013010102

Rodríguez, W. R., Saz, O., & Lleida, E. (2014). Experiences Using a Free Tool for Voice Therapy based on Speech Technologies. In Assistive Technologies: Concepts, Methodologies, Tools, and Applications (pp. 508-523). Hershey, PA: Information Science Reference. doi:10.4018/978-1-4666-4422-9.ch025

Rothblatt, M. (2013). Mindclone Technoselves: Multi-Substrate Legal Identities, Cyber-Psychology, and Biocyberethics. In R. Luppicini (Ed.), *Handbook of Research on Technoself: Identity in a Technological Society* (pp. 105–122). Hershey, PA: Information Science Reference; doi:10.4018/978-1-4666-2211-1.ch006

Rowe, N. C. (2012). The Ethics of Cyberweapons in Warfare. In R. Luppicini (Ed.), *Ethical Impact of Technological Advancements and Applications in Society* (pp. 195–207). Hershey, PA: Information Science Reference; doi:10.4018/978-1-4666-1773-5.ch015

Russo, M. R. (2014). Emergency Management Professional Development: Linking Information Communication Technology and Social Communication Skills to Enhance a Sense of Community and Social Justice in the 21st Century. In Crisis Management: Concepts, Methodologies, Tools and Applications (pp. 651-665). Hershey, PA: Information Science Reference. doi:10.4018/978-1-4666-4707-7.ch031

Sajeva, S. (2011). Towards a Conceptual Knowledge Management System Based on Systems Thinking and Sociotechnical Thinking. [IJSKD]. *International Journal of Sociotechnology and Knowledge Development*, *3*(3), 40–55. doi:10.4018/jskd.2011070103

Sajeva, S. (2013). Towards a Conceptual Knowledge Management System Based on Systems Thinking and Sociotechnical Thinking. In J. Abdelnour-Nocera (Ed.), *Knowledge and Technological Development Effects on Organizational and Social Structures* (pp. 115–130). Hershey, PA: Information Science Reference; doi:10.4018/978-1-4666-2151-0.ch008

Saleeb, N., & Dafoulas, G. A. (2014). Assistive Technologies and Environmental Design Concepts for Blended Learning and Teaching for Disabilities within 3D Virtual Worlds and Learning Environments. In Assistive Technologies: Concepts, Methodologies, Tools, and Applications (pp. 1382-1404). Hershey, PA: Information Science Reference. doi:10.4018/978-1-4666-4422-9.ch073

Salvini, P. (2012). Presence, Reciprocity and Robotic Mediations: The Case of Autonomous Social Robots. [IJT]. *International Journal of Technoethics*, *3*(2), 9–16. doi:10.4018/jte.2012040102

Samanta, I. (2013). The Impact of Virtual Community (Web 2.0) in the Economic, Social, and Political Environment of Traditional Society. In S. Saeed, M. Khan, & R. Ahmad (Eds.), *Business Strategies and Approaches for Effective Engineering Management* (pp. 262–274). Hershey, PA: Business Science Reference; doi:10.4018/978-1-4666-3658-3.ch016

Samanta, S. K., Woods, J., & Ghanbari, M. (2011). Automatic Language Translation: An Enhancement to the Mobile Messaging Services. [IJTHI]. *International Journal of Technology and Human Interaction*, *7*(1), 1–18. doi:10.4018/jthi.2011010101

Sarkar, N. I., Kuang, A. X., Nisar, K., & Amphawan, A. (2014). Hospital Environment Scenarios using WLAN over OPNET Simulation Tool. [IJICTHD]. *International Journal of Information Communication Technologies and Human Development*, *6*(1), 69–90. doi:10.4018/ijicthd.2014010104

Sarré, C. (2013). Technology-Mediated Tasks in English for Specific Purposes (ESP): Design, Implementation and Learner Perception. [IJCALLT]. *International Journal of Computer-Assisted Language Learning and Teaching*, *3*(2), 1–16. doi:10.4018/ijcallt.2013040101

Saykili, A., & Kumtepe, E. G. (2014). Facebook's Hidden Potential: Facebook as an Educational Support Tool in Foreign Language Education. In G. Mallia (Ed.), *The Social Classroom: Integrating Social Network Use in Education* (pp. 120–146). Hershey, PA: Information Science Reference; doi:10.4018/978-1-4666-4904-0.ch007

Sayoud, H. (2011). Biometrics: An Overview on New Technologies and Ethic Problems. [IJT]. *International Journal of Technoethics*, *2*(1), 19–34. doi:10.4018/jte.2011010102

Scott, C. R., & Timmerman, C. E. (2014). Communicative Changes Associated with Repeated Use of Electronic Meeting Systems for Decision-Making Tasks. In E. Nikoi & K. Boateng (Eds.), *Collaborative Communication Processes and Decision Making in Organizations* (pp. 1–24). Hershey, PA: Business Science Reference; doi:10.4018/978-1-4666-4478-6.ch001

Scott, K. (2013). The Human-Robot Continuum of Self: Where the Other Ends and Another Begins. In R. Luppicini (Ed.), *Handbook of Research on Technoself: Identity in a Technological Society* (pp. 666–679). Hershey, PA: Information Science Reference; doi:10.4018/978-1-4666-2211-1.ch034

Shasek, J. (2014). ExerLearning®: Movement, Fitness, Technology, and Learning. In J. Bishop (Ed.), *Gamification for Human Factors Integration: Social, Education, and Psychological Issues* (pp. 106–121). Hershey, PA: Information Science Reference; doi:10.4018/978-1-4666-5071-8.ch007

Shen, J., & Eder, L. B. (2011). An Examination of Factors Associated with User Acceptance of Social Shopping Websites. [IJTHI]. *International Journal of Technology and Human Interaction*, *7*(1), 19–36. doi:10.4018/jthi.2011010102

Shrestha, P. (2012). Teacher Professional Development Using Mobile Technologies in a Large-Scale Project: Lessons Learned from Bangladesh. [IJCALLT]. *International Journal of Computer-Assisted Language Learning and Teaching*, *2*(4), 34–49. doi:10.4018/ijcallt.2012100103

Silvana de Rosa, A., Fino, E., & Bocci, E. (2014). Addressing Healthcare On-Line Demand and Supply Relating to Mental Illness: Knowledge Sharing About Psychiatry and Psychoanalysis Through Social Networks in Italy and France. In A. Kapoor & C. Kulshrestha (Eds.), *Dynamics of Competitive Advantage and Consumer Perception in Social Marketing* (pp. 16–55). Hershey, PA: Business Science Reference; doi:10.4018/978-1-4666-4430-4.ch002

Smith, M., & Murray, J. (2014). Augmentative and Alternative Communication Devices: The Voices of Adult Users. In Assistive Technologies: Concepts, Methodologies, Tools, and Applications (pp. 991-1004). Hershey, PA: Information Science Reference. doi:10.4018/978-1-4666-4422-9.ch050

Smith, P. A. (2013). Strengthening and Enriching Audit Practice: The Socio-Technical Relevance of "Decision Leaders". In J. Abdelnour-Nocera (Ed.), *Knowledge and Technological Development Effects on Organizational and Social Structures* (pp. 97–108). Hershey, PA: Information Science Reference; doi:10.4018/978-1-4666-2151-0.ch006

So, J. C., & Lam, S. Y. (2014). Using Social Networks Communication Platform for Promoting Student-Initiated Holistic Development Among Students. [IJISSS]. *International Journal of Information Systems in the Service Sector*, *6*(1), 1–23. doi:10.4018/ijisss.2014010101

Söderström, S. (2014). Assistive ICT and Young Disabled Persons: Opportunities and Obstacles in Identity Negotiations. In Assistive Technologies: Concepts, Methodologies, Tools, and Applications (pp. 1084-1105). Hershey, PA: Information Science Reference. doi:10.4018/978-1-4666-4422-9.ch056

Son, J., & Rossade, K. (2013). Finding Gems in Computer-Assisted Language Learning: Clues from GLoCALL 2011 and 2012 Papers. [IJCALLT]. *International Journal of Computer-Assisted Language Learning and Teaching*, *3*(4), 1–8. doi:10.4018/ijcallt.2013100101

Sone, Y. (2013). Robot Double: Hiroshi Ishiguro's Reflexive Machines. In R. Luppicini (Ed.), *Handbook of Research on Technoself: Identity in a Technological Society* (pp. 680–702). Hershey, PA: Information Science Reference; doi:10.4018/978-1-4666-2211-1.ch035

Spillane, M. (2014). Assistive Technology: A Tool for Inclusion. In Assistive Technologies: Concepts, Methodologies, Tools, and Applications (pp. 1-11). Hershey, PA: Information Science Reference. doi:10.4018/978-1-4666-4422-9.ch001

Stahl, B. C., Heersmink, R., Goujon, P., Flick, C., van den Hoven, J., Wakunuma, K., & Rader, M. et al. (2012). Identifying the Ethics of Emerging Information and Communication Technologies: An Essay on Issues, Concepts and Method. In R. Luppicini (Ed.), *Ethical Impact of Technological Advancements and Applications in Society* (pp. 61–79). Hershey, PA: Information Science Reference; doi:10.4018/978-1-4666-1773-5.ch006

Stern, S. E., & Grounds, B. E. (2011). Cellular Telephones and Social Interactions: Evidence of Interpersonal Surveillance. [IJT]. *International Journal of Technoethics*, *2*(1), 43–49. doi:10.4018/jte.2011010104

Stinson, J., & Gill, N. (2014). Internet-Based Chronic Disease Self-Management for Youth. In Assistive Technologies: Concepts, Methodologies, Tools, and Applications (pp. 224-245). Hershey, PA: Information Science Reference. doi:10.4018/978-1-4666-4422-9.ch012

Stockwell, G. (2011). Online Approaches to Learning Vocabulary: Teacher-Centred or Learner-Centred? [IJCALLT]. *International Journal of Computer-Assisted Language Learning and Teaching, 1*(1), 33–44. doi:10.4018/ijcallt.2011010103

Stradella, E. (2012). Personal Liability and Human Free Will in the Background of Emerging Neuroethical Issues: Some Remarks Arising From Recent Case Law. [IJT]. *International Journal of Technoethics, 3*(2), 30–41. doi:10.4018/jte.2012040104

Stubbs, K., Casper, J., & Yanco, H. A. (2014). Designing Evaluations for K-12 Robotics Education Programs. In K-12 Education: Concepts, Methodologies, Tools, and Applications (pp. 1342-1364). Hershey, PA: Information Science Reference. doi:10.4018/978-1-4666-4502-8.ch078

Suki, N. M., Ramayah, T., Ming, M. K., & Suki, N. M. (2011). Factors Enhancing Employed Job Seekers Intentions to Use Social Networking Sites as a Job Search Tool. [IJTHI]. *International Journal of Technology and Human Interaction, 7*(2), 38–54. doi:10.4018/jthi.2011040105

Sweeney, P., & Moore, C. (2012). Mobile Apps for Learning Vocabulary: Categories, Evaluation and Design Criteria for Teachers and Developers. [IJCALLT]. *International Journal of Computer-Assisted Language Learning and Teaching, 2*(4), 1–16. doi:10.4018/ijcallt.2012100101

Szeto, A. Y. (2014). Assistive Technology and Rehabilitation Engineering. In Assistive Technologies: Concepts, Methodologies, Tools, and Applications (pp. 277-331). Hershey, PA: Information Science Reference. doi:10.4018/978-1-4666-4422-9.ch015

Tamim, R. (2014). Technology Integration in UAE Schools: Current Status and Way Forward. In K-12 Education: Concepts, Methodologies, Tools, and Applications (pp. 41-57). Hershey, PA: Information Science Reference. doi:10.4018/978-1-4666-4502-8.ch004

Tan, R., Wang, S., Jiang, Y., Ishida, K., & Fujie, M. G. (2014). Motion Control of an Omni-Directional Walker for Walking Support. In Assistive Technologies: Concepts, Methodologies, Tools, and Applications (pp. 614-622). Hershey, PA: Information Science Reference. doi:10.4018/978-1-4666-4422-9.ch029

Tankari, M. (2014). Cultural Orientation Differences and their Implications for Online Learning Satisfaction. In J. Keengwe, G. Schnellert, & K. Kungu (Eds.), *Cross-Cultural Online Learning in Higher Education and Corporate Training* (pp. 20–61). Hershey, PA: Information Science Reference; doi:10.4018/978-1-4666-5023-7.ch002

Tchangani, A. P. (2014). Bipolarity in Decision Analysis: A Way to Cope with Human Judgment. In A. Masegosa, P. Villacorta, C. Cruz-Corona, M. García-Cascales, M. Lamata, & J. Verdegay (Eds.), *Exploring Innovative and Successful Applications of Soft Computing* (pp. 216–244). Hershey, PA: Information Science Reference; doi:10.4018/978-1-4666-4785-5.ch012

Tennyson, R. D. (2014). Computer Interventions for Children with Disabilities: Review of Research and Practice. In Assistive Technologies: Concepts, Methodologies, Tools, and Applications (pp. 841-864). Hershey, PA: Information Science Reference. doi:10.4018/978-1-4666-4422-9.ch042

Terrell, S. S. (2011). Integrating Online Tools to Motivate Young English Language Learners to Practice English Outside the Classroom. [IJCALLT]. *International Journal of Computer-Assisted Language Learning and Teaching, 1*(2), 16–24. doi:10.4018/ijcallt.2011040102

Tiwary, U. S., & Siddiqui, T. J. (2014). Working Together with Computers: Towards a General Framework for Collaborative Human Computer Interaction. In Assistive Technologies: Concepts, Methodologies, Tools, and Applications (pp. 141-162). Hershey, PA: Information Science Reference. doi:10.4018/978-1-4666-4422-9.ch008

Tomas, J., Lloret, J., Bri, D., & Sendra, S. (2014). Sensors and their Application for Disabled and Elderly People. In Assistive Technologies: Concepts, Methodologies, Tools, and Applications (pp. 357-376). Hershey, PA: Information Science Reference. doi:10.4018/978-1-4666-4422-9.ch018

Tomasi, A. (2013). A Run for your [Techno]Self. In R. Luppicini (Ed.), Handbook of Research on Technoself: Identity in a Technological Society (pp. 123-136). Hershey, PA: Information Science Reference. doi:10.4018/978-1-4666-2211-1.ch007

Tootell, H., & Freeman, A. (2014). The Applicability of Gaming Elements to Early Childhood Education. In J. Bishop (Ed.), *Gamification for Human Factors Integration: Social, Education, and Psychological Issues* (pp. 225–241). Hershey, PA: Information Science Reference; doi:10.4018/978-1-4666-5071-8.ch014

Tsai, C. (2011). How Much Can Computers and Internet Help?: A Long-Term Study of Web-Mediated Problem-Based Learning and Self-Regulated Learning. [IJTHI]. *International Journal of Technology and Human Interaction*, *7*(1), 67–81. doi:10.4018/jthi.2011010105

Tsai, W. (2013). An Investigation on Undergraduate's Bio-Energy Engineering Education Program at the Taiwan Technical University. [IJTHI]. *International Journal of Technology and Human Interaction*, *8*(3), 46–53. doi:10.4018/jthi.2012070105

Tsiakis, T. (2013). Using Social Media as a Concept and Tool for Teaching Marketing Information Systems. In M. Pătruţ & B. Pătruţ (Eds.), *Social Media in Higher Education: Teaching in Web 2.0* (pp. 24–44). Hershey, PA: Information Science Reference; doi:10.4018/978-1-4666-2970-7.ch002

Tu, C., McIsaac, M. S., Sujo-Montes, L. E., & Armfield, S. (2014). Building Mobile Social Presence for U-Learning. In F. Neto (Ed.), *Technology Platform Innovations and Forthcoming Trends in Ubiquitous Learning* (pp. 77–93). Hershey, PA: Information Science Reference; doi:10.4018/978-1-4666-4542-4.ch005

Valeria, N., Lu, M. V., & Theng, L. B. (2014). Collaborative Virtual Learning for Assisting Children with Cerebral Palsy. In Assistive Technologies: Concepts, Methodologies, Tools, and Applications (pp. 786-810). Hershey, PA: Information Science Reference. doi:10.4018/978-1-4666-4422-9.ch040

Van Leuven, N., Newton, D., Leuenberger, D. Z., & Esteves, T. (2014). Reaching Citizen 2.0: How Government Uses Social Media to Send Public Messages during Times of Calm and Times of Crisis. In Crisis Management: Concepts, Methodologies, Tools and Applications (pp. 839-857). Hershey, PA: Information Science Reference. doi:10.4018/978-1-4666-4707-7.ch041

Vargas-Hernández, J. G. (2013). International Student Collaboration and Experiential Exercise Projects as a Professional, Inter-Personal and Inter-Institutional Networking Platform. [IJTEM]. *International Journal of Technology and Educational Marketing*, *3*(1), 28–47. doi:10.4018/ijtem.2013010103

Velicu, A., & Marinescu, V. (2013). Usage of Social Media by Children and Teenagers: Results of EU KIDS Online II. In M. Pătruţ & B. Pătruţ (Eds.), *Social Media in Higher Education: Teaching in Web 2.0* (pp. 144–178). Hershey, PA: Information Science Reference; doi:10.4018/978-1-4666-2970-7.ch008

Vidaurre, C., Kübler, A., Tangermann, M., Müller, K., & Millán, J. D. (2014). Brain-Computer Interfaces and Visual Activity. In Assistive Technologies: Concepts, Methodologies, Tools, and Applications (pp. 1549-1570). Hershey, PA: Information Science Reference. doi:10.4018/978-1-4666-4422-9.ch081

Viswanathan, R. (2012). Augmenting the Use of Mobile Devices in Language Classrooms. [IJCALLT]. *International Journal of Computer-Assisted Language Learning and Teaching*, *2*(2), 45–60. doi:10.4018/ijcallt.2012040104

Wallgren, L. G., & Hanse, J. J. (2012). A Two-Wave Study of the Impact of Job Characteristics and Motivators on Perceived Stress among Information Technology (IT) Consultants. [IJTHI]. *International Journal of Technology and Human Interaction*, *8*(4), 75–91. doi:10.4018/jthi.2012100105

Wang, H. (2014). A Guide to Assistive Technology for Teachers in Special Education. In Assistive Technologies: Concepts, Methodologies, Tools, and Applications (pp. 12-25). Hershey, PA: Information Science Reference. doi:10.4018/978-1-4666-4422-9.ch002

Wang, S., Ku, C., & Chu, C. (2013). Sustainable Campus Project: Potential for Energy Conservation and Carbon Reduction Education in Taiwan. [IJTHI]. *International Journal of Technology and Human Interaction*, *8*(3), 19–30. doi:10.4018/jthi.2012070103

Wang, Y., & Tian, J. (2013). Negotiation of Meaning in Multimodal Tandem Learning via Desktop Videoconferencing. [IJCALLT]. *International Journal of Computer-Assisted Language Learning and Teaching*, *3*(2), 41–55. doi:10.4018/ijcallt.2013040103

Wareham, C. (2011). On the Moral Equality of Artificial Agents. [IJT]. *International Journal of Technoethics*, *2*(1), 35–42. doi:10.4018/jte.2011010103

Warwick, K., & Gasson, M. N. (2014). Practical Experimentation with Human Implants. In M. Michael & K. Michael (Eds.), *Uberveillance and the Social Implications of Microchip Implants: Emerging Technologies* (pp. 64–132). Hershey, PA: Information Science Reference; doi:10.4018/978-1-4666-4582-0.ch004

Welch, K. C., Lahiri, U., Sarkar, N., Warren, Z., Stone, W., & Liu, C. (2014). Affect-Sensitive Computing and Autism. In Assistive Technologies: Concepts, Methodologies, Tools, and Applications (pp. 865-883). Hershey, PA: Information Science Reference. doi:10.4018/978-1-4666-4422-9.ch043

Wessels, B., Dittrich, Y., Ekelin, A., & Eriksén, S. (2014). Creating Synergies between Participatory Design of E-Services and Collaborative Planning. In Assistive Technologies: Concepts, Methodologies, Tools, and Applications (pp. 163-179). Hershey, PA: Information Science Reference. doi:10.4018/978-1-4666-4422-9.ch009

White, E. L. (2014). Technology-Based Literacy Approach for English Language Learners. In K-12 Education: Concepts, Methodologies, Tools, and Applications (pp. 723-740). Hershey, PA: Information Science Reference. doi:10.4018/978-1-4666-4502-8.ch042

Whyte, K. P., List, M., Stone, J. V., Grooms, D., Gasteyer, S., Thompson, P. B., & Bouri, H. et al. (2014). Uberveillance, Standards, and Anticipation: A Case Study on Nanobiosensors in U.S. Cattle. In M. Michael & K. Michael (Eds.), *Uberveillance and the Social Implications of Microchip Implants: Emerging Technologies* (pp. 260–279). Hershey, PA: Information Science Reference; doi:10.4018/978-1-4666-4582-0.ch012

Wilson, S., & Haslam, N. (2013). Reasoning about Human Enhancement: Towards a Folk Psychological Model of Human Nature and Human Identity. In R. Luppicini (Ed.), *Handbook of Research on Technoself: Identity in a Technological Society* (pp. 175–188). Hershey, PA: Information Science Reference; doi:10.4018/978-1-4666-2211-1.ch010

Woodhead, R. (2012). What is Technology? [IJSKD]. *International Journal of Sociotechnology and Knowledge Development*, *4*(2), 1–13. doi:10.4018/jskd.2012040101

Woodley, C., & Dorrington, P. (2014). Facebook and the Societal Aspects of Formal Learning: Optional, Peripheral, or Essential. In G. Mallia (Ed.), *The Social Classroom: Integrating Social Network Use in Education* (pp. 269–291). Hershey, PA: Information Science Reference; doi:10.4018/978-1-4666-4904-0.ch014

Yamazaki, T. (2014). Assistive Technologies in Smart Homes. In Assistive Technologies: Concepts, Methodologies, Tools, and Applications (pp. 663-678). Hershey, PA: Information Science Reference. doi:10.4018/978-1-4666-4422-9.ch032

Yan, Z., Chen, Q., & Yu, C. (2013). The Science of Cell Phone Use: Its Past, Present, and Future. [IJCBPL]. *International Journal of Cyber Behavior, Psychology and Learning*, *3*(1), 7–18. doi:10.4018/ijcbpl.2013010102

Yang, Y., Wang, X., & Li, L. (2013). Use Mobile Devices to Wirelessly Operate Computers. [IJTHI]. *International Journal of Technology and Human Interaction*, *9*(1), 64–77. doi:10.4018/jthi.2013010105

Yartey, F. N., & Ha, L. (2013). Like, Share, Recommend: Smartphones as a Self-Broadcast and Self-Promotion Medium of College Students. [IJTHI]. *International Journal of Technology and Human Interaction*, *9*(4), 20–40. doi:10.4018/ijthi.2013100102

Yaseen, S. G., & Al Omoush, K. S. (2013). Investigating the Engage in Electronic Societies via Facebook in the Arab World. [IJTHI]. *International Journal of Technology and Human Interaction*, *9*(2), 20–38. doi:10.4018/jthi.2013040102

Yeo, B. (2012). Sustainable Economic Development and the Influence of Information Technologies: Dynamics of Knowledge Society Transformation. [IJSKD]. *International Journal of Sociotechnology and Knowledge Development*, *4*(3), 54–55. doi:10.4018/jskd.2012070105

Yu, L., & Ureña, C. (2014). A Review of Current Approaches of Brain Computer Interfaces. In Assistive Technologies: Concepts, Methodologies, Tools, and Applications (pp. 1516-1534). Hershey, PA: Information Science Reference. doi:10.4018/978-1-4666-4422-9.ch079

Zelenkauskaite, A. (2014). Analyzing Blending Social and Mass Media Audiences through the Lens of Computer-Mediated Discourse. In H. Lim & F. Sudweeks (Eds.), *Innovative Methods and Technologies for Electronic Discourse Analysis* (pp. 304–326). Hershey, PA: Information Science Reference; doi:10.4018/978-1-4666-4426-7.ch014

Compilation of References

Adams, P. (2011). ICT and pedagogy: Opportunities missed? *Education*, *39*(1), 21–33.

Adomavicius, G., & Zhang, J. (2012). Stability of Recommendation Algorithms. *ACM Transactions on Information Systems*, *30*(4), 23. doi:10.1145/2382438.2382442

Ahn, J. (2011). The effect of social network sites on adolescents' social and academic development: Current theories and controversies. *Journal of the American Society for Information Science and Technology*, *62*(8), 1435–1445. doi:10.1002/asi.21540

Ajzen, I. (1991). The Theory of Planned Behavior. *Organizational Behavior and Human Decision Processes*, *50*(2), 179–212. doi:10.1016/0749-5978(91)90020-T

Ajzen, I., & Fishbein, M. (1980). *Understanding Attitudes and Predicting Social*. Englewood Cliffs, NJ: Prentice-Hall.

Akkinen, M., & Tuunainen, V. (2004). The sources of value creation in consumer–oriented online communities. In conference "Frontiers of e-Business Research", Tampere, FI.

Alagoz, S., Ekici, N., & Islek, M. (2011). Brand communities in the axis of scialaization customers: Sample of Volkswagen Beetle owners, Turkey. (English). *Ege Academic Review*, 465-477.

Albaili, M. A. (1993). Inferred hemispheric thinking style, gender, and academic major among United Arab Emirates college students. *Perceptual and Motor Skills*, *76*(3), 971–977. doi:10.2466/pms.1993.76.3.971 PMID:8321616

Alter, S. (1977). A Taxonomy of Decision Support Systems. *Sloan Management Review*, *19*(1), 39–57.

Alter, S. (2004). A work system view of DSS in its fourth decade. *Decision Support Systems*, *38*(3), 319–327. doi:10.1016/j.dss.2003.04.001

Anderson, J. C., & Gerbing, D. W. (1988). Structural Equation Modeling in Practice: A Review and Recommended Two-step Approach. *Psychological Bulletin*, *103*(3), 411–423. doi:10.1037/0033-2909.103.3.411

Andersson, T. D., & Mossberg, L. (2004). The dining experience: Do restaurants satisfy customer needs? *Food Service Technology*, *4*(4), 171–177. doi:10.1111/j.1471-5740.2004.00105.x

Andrade, J. (2010). What does doodling do? *Applied Cognitive Psychology*, *24*(1), 100–106. doi:10.1002/acp.1561

Andrews, C., & North, C. (2013). The Impact of Physical Navigation on Spatial Organization for Sensemaking. *Visualization and Computer Graphics. IEEE Transactions on*, *19*(12), 2207–2216.

Andrews, L., Cacho-Elizondo, S., Drenna, J., & Tossan, V. (2013). Consumer Acceptance of an SMS-assisted Smoking Cessation Intervention: A Multi Country Study. *Journal of Health Marketing Quarterly*, *30*(1), 47–62. doi:10.1080/07359683.2013.758015 PMID:23458481

Ang, C. S., Zaphiris, P., & Mahmood, S. (2007). A model of cognitive loads in massively multiplayer online role playing games. *Interacting with Computers*, *19*(2), 167–179. doi:10.1016/j.intcom.2006.08.006

Angehrn, A. (1997). Designing mature Internet business strategies: The ICDT model. *European Management Journal*, *15*(4), 361–369. doi:10.1016/S0263-2373(97)00016-9

Angelini, P., Cortese, P. F., Di Battista, G., & Patrignani, M. (2013). Topological morphing of planar graphs. *Theoretical Computer Science*, *514*, 2–20. doi:10.1016/j.tcs.2013.08.018

Anthony, R. N. (1965). *Planning and Control Systems: A framework for analysis*. Cambridge, MA: Harvard University Graduate School of Business Administration.

Antonietti, A. (2003). Cognitive styles assessment. In R. Fernandez-Ballesteros (Ed.), *Encyclopaedia of psychological assessment* (Vol. 1, pp. 248–253). London: Sage. doi:10.4135/9780857025753.n54

Antonietti, A., Calcaterra, A., Colombo, B., & Giorgetti, M. (2003). *Attorno al computer. Strumenti e ricerche su nuove tecnologie e apprendimento* [Around the computer. Instruments and studies about the relation between new technologies and learning]. Rome: Carocci.

Antonietti, A., & Colombo, B. (2008). Computer-supported learning tools: A bi-circular bi-directional framework. *New Ideas in Psychology*, *26*(1), 120–142. doi:10.1016/j.newideapsych.2007.08.001

Antonietti, A., Confalonieri, E., & Marchetti, A. (Eds.). (2014). *Metarepresentation and narrative in educational settings: A cultural framework for reflective thinking*. New York: Cambridge University Press.

Antonietti, A., Fabio, R. A., Boari, G., & Bonanomi, A. (2005). Il questionario "Style of Learning and Thinking" (SOLAT): Dati psicometrici per una validazione e standardizzazione della versione italiana[The questionnaire "Style of Learning and Thinking" (SOLAT): Psychometric measures to validate and standardise the Italian version]. *TPM. Testing Psicometria Metodologia*, *12*, 299–316.

Antwarg, L., Lavie, T., Rokach, L., Shapira, B., & Meyer, J. (2013). Highlighting items as means of adaptive assistance. *Behaviour & Information Technology*, *32*(8), 761–777. doi:10.1080/0144929X.2011.650710

Apaolaza, V., Hartmann, P., Medina, E., Barrutia, J., & Echebarria, C. (2013). The relationship between socializing on the Spanish online networking site Tuenti and teenagers' subjective wellbeing: The roles of self-esteem and loneliness. *Computers in Human Behavior*, *29*(4), 1282–1289. doi:10.1016/j.chb.2013.01.002

Archambault, D., & Purchase, H. C. (2013). Mental map preservation helps user orientation in dynamic graphs. In W. Didimo & M. Patrignani (Eds.), Lecture Notes in Computer Science: Vol. 7704. *Graph Drawing* (pp. 475–486). Berlin, Germany: Springer Berlin Heidelberg. doi:10.1007/978-3-642-36763-2_42

Arciszewski, H. F. R., De Greef, T. E., & Van Delft, J. H. (2009). Adaptive automation in a naval combat management system. *IEEE Transactions on Systems, Man, and Cybernetics. Part A, Systems and Humans*, *39*(6), 1188–1199. doi:10.1109/TSMCA.2009.2026428

Armbrust, M., Fox, A., Griffith, R., Joseph, A. D., Katz, R., Konwinski, A., & Zaharia, M. (2010). A view of cloud computing. *Communications of the ACM*, *53*(4), 50–58. doi:10.1145/1721654.1721672

Arnett, D. B., & Badrinarayanan, V. (2005). Enhancing customer-needs driven CRM strategies: Core selling teams, knowledge management competence, and relationship marketing competence. *Journal of Personal Selling & Sales Management*, *25*(4), 329–343.

Arnott, D., & Pervan, G. (2005). A critical analysis of decision support systems research. *Journal of Information Technology*, *20*(2), 67–68. doi:10.1057/palgrave.jit.2000035

Arnott, D., & Pervan, G. (2008). Eight key issues for the decision support systems discipline. *Decision Support Systems*, *44*(3), 657–672. doi:10.1016/j.dss.2007.09.003

Ascari, A., Bakshi, A. & Grijpink, F. (2010). mHealth: A New Vision for Healthcare: McKinsey & Company, Inc., and GSMA.

Assael, H. (1992). *Consumer Behavior and Marketing Action*. Boston: PWS-Kent Pub.

Aubry, M., Lièvre, P., & Hobbs, B. (2010). Project management in extreme environments. *Project Management Journal*, *41*(3), 2–4. doi:10.1002/pmj.20185

Babin, B. J., Darden, W. R., & Griffin, M. (1994). Work and/or fun: Measuring hedonic and utilitarian shopping value. *The Journal of Consumer Research*, *20*(4), 644–656. doi:10.1086/209376

Backs, R. W. (1995). Going Beyond Heart Rate: Autonomic Space and Cardiovascular Assessment of Mental Workload. *The International Journal of Aviation Psychology*, *5*(1), 25–48. doi:10.1207/s15327108ijap0501_3 PMID:11541494

Baddeley, A. D. (1996). Exploring the central executive. *Quarterly Journal of Experimental Psychology*, *49A*(1), 5–28. doi:10.1080/713755608

Badros, G. J., Nichols, J., & Borning, A. (2000). SCWM: an intelligent constraint-enabled window manager. In Proceedings of the AAAI Spring Symposium on Smart Graphics (pp. 76–83). Palo Alto, CA: AAAI Press; Retrieved from http://scwm.mit.edu

Badros, G. J., Nichols, J., & Borning, A. (2000). *SCWM: An intelligent constraint-enabled window manager*. In *Proceedings of the AAAI Spring Symposium on Smart Graphics*.

Bagozzi, R. P. (1992). The self-regulation of attitudes, intentions, and behavior. *Social Psychology Quarterly*, *55*(1), 178–204. doi:10.2307/2786945

Bagozzi, R. P., & Dholakia, U. M. (2002). Intentional social action in virtual communities. *Journal of Interactive Marketing*, *16*(2), 2–21. doi:10.1002/dir.10006

Bagozzi, R. P., & Yi, Y. (1991). Multitrait-Multimethod Matrices in Consumer Research. *JMR, Journal of Marketing Research*, *17*, 426–439.

Bagozzi, R., & Dholakia, U. (2006). Antecedents and purchase consequences of customer participation in small group brand communities. *International Journal of Research in Marketing*, *23*(1), 45–61. doi:10.1016/j.ijresmar.2006.01.005

Baker, J., Jones, D., & Burkman, J. (2009). Using Visual Representations of Data to Enhance Sensemaking in Data Exploration Tasks. *Journal of the Association for Information Systems*, *10*(7), 533–559.

Balkis, M., & Isiker, G. B. (2005). The relationship between thinking styles and personality types. *Social Behavior and Personality*, *33*(3), 283–294. doi:10.2224/sbp.2005.33.3.283

Ballantine, P. W., & Stephenson, R. J. (2011). Help me, I'm fat! Social support in online weight loss networks. *Journal of Consumer Behaviour*, *10*(6), 332–337. doi:10.1002/cb.374

Bandura, A. (1977). *Social learning theory*. Englewood Cliffs, NJ: Prentice Hall.

Bandura, A. (1982). Self-efficacy mechanism in human agency. *The American Psychologist*, *37*(2), 122–147. doi:10.1037/0003-066X.37.2.122

Bandura, A. (1986). *Social foundations of thought and action: A social cognitive theory*. Bnglewood Cliffs, N. J.: Prentice-Hall.

Bandura, A. (2012). Social cognitive theory. In P. M. Van Lange, A. W. Kruglanski, & E. Higgins (Eds.), *Handbook of theories of social psychology* (Vol. 1, pp. 349–373). Thousand Oaks, CA: Sage Publications Ltd. doi:10.4135/9781446249215.n18

Baran, S., & Davis, D. (1995). *Mass communication theory*. Belmont, CA: Wadsworth.

Barker, V. (2009). Older adolescents' motivations for social network site use: The influence of gender, group identity, and collective self-esteem. *Cyberpsychology & Behavior*, *12*(2), 209–213. doi:10.1089/cpb.2008.0228 PMID:19250021

Barley, R. A. (1997). Behavior inhibition, sustained attention, and executive functions: Constructing a unifying theory of ADHD. *Psychological Bulletin*, *121*(1), 65–94. doi:10.1037/0033-2909.121.1.65 PMID:9000892

Bateman, P. J., Gray, P. H., & Butler, B. S. (2011). Research Note—The Impact of Community Commitment on Participation in Online Communities. *Information Systems Research*, *22*(4), 841–854. doi:10.1287/isre.1090.0265

Batra, R., & Ahtola, O. T. (1990). Measuring the Hedonic and Utilitarian Sources of Consumer Attitudes. *Marketing Letters*, *2*(April), 159–170.

Baumeister, R. F., & Leary, M. R. (1995). The need to belong: Desire for interpersonal attachments as a fundamental human motivation. *Psychological Bulletin*, *117*(3), 497–529. doi:10.1037/0033-2909.117.3.497 PMID:7777651

Baumgarten, S. A. (1975). The innovative communicator in the diffusion research. *JMR, Journal of Marketing Research, 12*(1), 12–18. doi:10.2307/3150653

Baym, K. N., Zhang, B. Y., & Lin, M. C. (2004). Social interactions across media. *New Media & Society, 6*(3), 299–318. doi:10.1177/1461444804041438

Beatty, J., & Lucero-Wagoner, B. (2000). The pupillary system. In J. T. Cacioppo, L. G. Tassinary, & G. G. Berntson (Eds.), *Handbook of psychophysiology*. Cambride, MA: Cambirdge University Press.

Beaudouin-Lafon, M. (2001). Novel interaction techniques for overlapping windows. In *Proceedings of the 14th Annual ACM Symposium on User Interface Software and Technology* (pp. 153–154). New York, NY: ACM. doi:10.1145/502348.502371

Becker, G. (1968). Crime and punishment: An economic approach. *Journal of Political Economy, 76*(2), 169–217. doi:10.1086/259394

Becker, G. (1976). *The economic approach to human behavior*. Chicago: University of Chicago Press.

Beilock, S. L., & Goldin-Meadow, S. (2010). Gesture changes thought by grounding it in action. *Psychological Science, 21*(11), 1605–1610. doi:10.1177/0956797610385353 PMID:20889932

Bell, B. A., & Feiner, S. K. (2000). Dynamic space management for user interfaces. In AckermanM.EdwardsK. (Eds.), *Proceedings of the 13th annual ACM Symposium on User Interface Software and Technology* (pp. 239–248). New York, NY: ACM. doi:10.1145/354401.354790

Bergami, M., & Bagozzi, R. P. (2000). Self-categorization, affective commitment, and group self-esteem as distinct aspects of social identity in an organization. *The British Journal of Social Psychology, 39*(4), 555–577. doi:10.1348/014466600164633 PMID:11190685

Bergman, D. S. (2010). *Mental map preservation principles and related measurements for quantitative evaluation of force-directed graph layout algorithm behaviour* (Master's thesis). School of Interactive Arts and Technology, Simon Fraser University. Retrieved from http://summit.sfu.ca/item/10032

Berkow, J. (2010). *FP Tech Desk: The Coming Twitter News Network*. Retrieved from http://business.financialpost.com/2010/11/23/fp-tech-desk-the-coming-twitter-news-network/?__lsa=93af-dddb

Bert, F., Giacometti, M., Gualano, M. R., & Siliquini, R. (2014). Smartphones and health promotion: A review of the evidence. *Journal of Medical Systems, 38*(1), 1–11. doi:10.1007/s10916-013-9995-7 PMID:24346929

Bessiere, K., Kiesler, S., Kraut, R., & Boneva, B. S. (2008). Effect of Internet use and social resources on changes in depression. *Information Communication and Society, 11*(1), 47–70. doi:10.1080/13691180701858851

Bi, X., Bae, S.-H., & Balakrishnan, R. (2010). Effects of interior bezels of tiled-monitor large displays on visual search, tunnel steering, and target selection. Paper presented at the Computer Human Interaction. doi:10.1145/1753326.1753337

Billig, J., Danilchenko, Y., & Frank, C. (2008). Evaluation of Google hacking. In *Proc. of the 5th Conf. on Information Security Curriculum Development*, 27-32. New York: ACM.

Bizeul, D. (2007). *Russian business networks study*. Retrieved from http://www.bizeul.org/files/RBN_study.pdf

Blicoe, R. (2010). High-living hacker swaps Porsche for porridge. *The Register*. Retrieved from http://www.theregister.co.uk/2010/06/18/hacker_jailed/

Bloch, P. H. (1982). Involvement beyond the Purchase Process: Conceptual Issues and Empirical Investigation. *Advances in Consumer Research. Association for Consumer Research (U. S.), 9*(1), 413–417.

Blood, R. (2000). *Weblogs: A history and perspective*. Retrieved from http://rebeccablood.net/essays/weblog_hystory.html

Bluedorn, A. C. (2001). Time and organizational culture. In N. M. Ashkanasy, C. P. M. Wilderom, & M. F. Peterson (Eds.), *Handbook of organizational culture and climate*. Thousand Oaks, CA.

Bluedorn, A. C. (2002). *The Human Organization of Time: Temporal Realities and Experience*. Stanford, CA: Stanford University Press.

Blumler, J. G., & Katz, E. (1974). *The uses of mass communication*. Newbury Park, CA: Sage.

Bly, S. A., & Rosenberg, J. K. (1986). A comparison of tiled and overlapping windows. In *Proceedings of the SIGCHI Conference on Human Factors in Computing Systems* (pp. 101–106). New York, NY: ACM. doi:10.1145/22627.22356

Bobadilla, J., Ortega, F., Hernando, A., & Gutiérrez, A. (2013). Recommender systems survey. *Knowledge-Based Systems*, *46*, 109–132. doi:10.1016/j.knosys.2013.03.012

Bobadilla, J., Serradilla, F., & Hernando, A. (2009). Collaborative filtering adapted to recommender systems of e-learning. *Knowledge-Based Systems*, *22*(4), 261–265. doi:10.1016/j.knosys.2009.01.008

Borges, J., & Levene, M. (2000). Data Mining of User Navigation Patterns. In *Web usage analysis and user profiling* (pp. 92–112). Springer Berlin Heidelberg. doi:10.1007/3-540-44934-5_6

Borràs, J., Moreno, A., & Valls, A. (2014). Intelligent tourism recommender systems: A survey. *Expert Systems with Applications*, *41*(16), 7370–7389. doi:10.1016/j.eswa.2014.06.007

Bosch, T. E. (2009). Using online social networking for teaching and learning: Facebook use at the university of cape town. *Communication: South African Journal for Communication Theory and Research*, *35*(2), 185–200.

Bosker, B. (2013). *Wealthier College Students Share, Connect More On Facebook: Study*. Retrieved from http://www.huffingtonpost.com/2013/06/25/college-students-facebookstudy_n_3497733.html

Botti, S., & McGill, A. L. (2011). The locus of choice: Personal causality and satisfaction with hedonic and utilitarian decisions. *The Journal of Consumer Research*, *37*(6), 1065–1078. doi:10.1086/656570

Boyd, D. (2008). Why youth love social networking sites: The role of networked publics in teenage social life. In D. Buckingham (Ed.), *Youth, identity, and digital media* (pp. 119–142). Cambridge, MA: MIT Press.

Boyd, D. (2011). Social network sites as networked publics: Affordances, dynamics, and implications. In Z. Papacharissi (Ed.), *A networked self: Identity, community, and culture on social network sites* (pp. 39–58). New York: Routledge.

Boyd, D., & Ellison, N. (2007). Social network sites: Definition, history, and scholarship. *Journal of Computer-Mediated Communication*, *13*(1), 210–230. doi:10.1111/j.1083-6101.2007.00393.x

Bracken, C. C., & Skalski, P. (2009). Telepresence and Video Games: The Impact of Image Quality. *PsychNology Journal*, *7*(1), 101–112.

Bransford, J., Pellegrino, J., & Donovan, M. (1999). *How people learn: Bridging research and practice*. Washington, DC: National Research Council, National Academy Press.

Bridges, E., & Florsheim, R. (2008). Hedonic and utilitarian shopping goals: The online experience. *Journal of Business Research*, *61*(4), 309–314. doi:10.1016/j.jbusres.2007.06.017

Briones-Vozmediano, E., Vives-Cases, C., & Peiró-Pérez, R. (2012). Gender Sensitivity in National Health Plans in Latin America and the European Union. *Health Policy (Amsterdam)*, *106*(1), 88–96. doi:10.1016/j.healthpol.2012.03.001 PMID:22465154

Broadbent, D. (1958). *Perception and communication*. Oxford: Pergamon. doi:10.1037/10037-000

Brock, A. (2012). From the blackhand side: Twitter as a cultural conversation. *Journal of Broadcasting & Electronic Media*, *56*(4), 529–549. doi:10.1080/08838151.2012.732147

Brodie, R. J., Ilic, A., Juric, B., & Hollebeek, L. (2011). Consumer engagement in a virtual brand community: An exploratory analysis. *Journal of Business Research*, 1–10.

Brown, T. A. (2006). *Confirmatory factor analysis for applied research*. New York, NY: Guilford.

Bruner, J. S. (1998). Narrative and metanarrative in the construction of Self. In M. Ferrari & R. J. Sternberg (Eds.), *Self awareness: Its nature and development* (pp. 308–331). New York: Guilford Press.

Bryant, S. (2013). *More than half of the time spent on Internet is via mobile*. Retrieved June 19, 2014 from http://www.ad-ology.com/half-time-spent-internet-via-mobile/

Burke, M., & Kraut, R. (2013). *Using Facebook after losing a job: Different benefits of strong and weak ties*. Paper presented in Computer Supported Cooperative Work Conference, San Antonio, TX.

Bury, K. F., Davies, S. E., & Darnell, M. J. (1985). *Window management: a review of issues and some results from user testing* (Technical Report No. HFC-53).

Cacho-Elizondo, S., Shahidi, N., & Tossan, V. (2013). Intention to adopt a text message-based mobile coaching service to help stop smoking: Which explanatory variables? *International Journal of Technology and Human Interaction*, *9*(4), 1–19. doi:10.4018/ijthi.2013100101

Cakir, H. (2013). Use of blogs in pre-service teacher education to improve student engagement. *Computers & Education*, *68*, 244–252. doi:10.1016/j.compedu.2013.05.013

Calcaterra, A., Antonietti, A., & Underwood, J. (2005). Cognitive style, hypermedia navigation and learning. *Computers & Education*, *44*(4), 441–457. doi:10.1016/j.compedu.2004.04.007

Campbell, S. W. (2008). Mobile technology and the body: Apparategeist, fashion, & function. In J. Katz (Ed.), *Handbook of mobile communication studies* (pp. 153–164). Cambridge, MA: MIT Press. doi:10.7551/mitpress/9780262113120.003.0012

Campbell, S. W., & Russo, T. C. (2003). The social construction of mobile telephony: An application of the social influence model to perceptions and uses of mobile phones within personal communication networks. *Communication Monographs*, *70*(4), 317–334. doi:10.1080/0363775032000179124

Capeller, W. (2001). Not such a neat net: Some comments on virtual criminality. *Social & Legal Studies*, *10*, 229–242.

Caplan, S. E. (2010). Theory and measurement of generalized problematic internet use: A two-step approach. *Computers in Human Behavior*, *26*(5), 1089–1097. doi:10.1016/j.chb.2010.03.012

Carlson, B. D., Suter, T. A., & Brown, T. J. (2008). Social versus psychological brand community: The role of psychological sense of brand community. *Journal of Business Research*, *61*(4), 284–291. doi:10.1016/j.jbusres.2007.06.022

Carneiro Moro, S. M., Ribeiro Cortez, P. A., & Ferreira Rita, P. M. R. (2014). Business intelligence in banking: A literature analysis from 2002 to 2013 using text mining and latent Dirichlet allocation. *Expert Systems with Applications*, *42*(3), 1314–1324. doi:10.1016/j.eswa.2014.09.024

Caron, F., Vanthienen, J., & Baesens, B. (2013). Comprehensive rule-based compliance checking and risk management with process mining. *Decision Support Systems*, *54*(3), 1357–1369. doi:10.1016/j.dss.2012.12.012

Carrier, L. M., Cheever, N. A., Rosen, L. D., Benitez, S., & Chang, J. (2009). Multitasking across generations: Multitasking choices and difficulty ratings in three generations of Americans. *Computers in Human Behavior*, *25*(2), 483–489. doi:10.1016/j.chb.2008.10.012

Carr, N. (2011). *The Shallows: What the Internet is doing to our brains*. New York: W.W. Norton Company.

Casaló, L. V., Flavián, C., & Guinalíu, M. (2010). Antecedents and consequences of consumer participation in on-line communities: The case of the travel sector. *International Journal of Electronic Commerce*, *15*(2), 137–167. doi:10.2753/JEC1086-4415150205

Casaló, L. V., Flavián, C., & Guinalíu, M. (2010). Determinants of the intention to participate in firm-hosted online travel communities and effects on consumer behavioral intentions. *Tourism Management*, *31*(6), 898–911. doi:10.1016/j.tourman.2010.04.007

Cascio, W. F., & Shurygailo, S. (2002). E-leadership and virtual teams. *Organizational Dynamics*, *31*(2), 362–376.

Cassell, J. (2002). "We have these rules inside": The effect of exercising voice in a children's online forum. In S. L. Calvert, A. B. Jordan, & R. R. Cocking (Eds.), *Children in the digital age: Influences of electronic media on development*. Westport, CT: Praeger.

Cassidy, J. (2006). Me media: How hanging out on the Internet became big business. *The New Yorker*, *82*(13), 50.

Cestre, G. (1996). Diffusion et Innovativité: Définition, Modélisation et Mesure. *Recherche et Applications en Marketing*, *11*(1), 69–88. doi:10.1177/076737019601100105

Chae, B., & Poole, M. S. (2005). Mandates and technology acceptance: A tale of two enterprise technologies. *The Journal of Strategic Information Systems*, *14*(2), 147–166. [REMOVED HYPERLINK FIELD] doi:10.1016/j.jsis.2005.04.001

Chan, C. K. K., & van Aalst, J. (2008). Collaborative inquiry and knowledge building in networked multimedia environments. In J. Voogt, & G. Knezek (Eds.), International handbook of information technology in primary and secondary education, 299–316. Dordrecht, The Netherlands: Springer. doi:10.1007/978-0-387-73315-9_18

Chapman, S. B., & Kirkland, S. (2013). *Make your brain smarter*. New York: Free Press.

Chen, B., & Marcus, J. (2012). Students' self-presentation on Facebook: An examination of personality and self-construal factors. *Computers in Human Behavior*, *28*(6), 2091–2099. doi:10.1016/j.chb.2012.06.013

Chen, F., Ruiz, N., Choi, E., Epps, J., Khawaja, M. A., Taib, R., & Wang, Y. et al. (2013). Multimodal behavior and interaction as indicators of cognitive load. *ACM Transactions on Interactive Intelligent Systems*, *2*(4), 1–36. doi:10.1145/2395123.2395127

Chen, H., Chiang, R. H. L., & Storey, V. C. (2012). Business Intelligence and Analytics: From Big Data to Big Impact. *Management Information Systems Quarterly*, *36*(4), 1165–1188. Retrieved from http://dblp.uni-trier.de/db/journals/misq/misq36.html#ChenCS12

Cheung, C. K., Chiu, P., & Lee, M. O. (2011). Online social networks: Why do students use Facebook? *Computers in Human Behavior*, *27*(4), 1337–1343. doi:10.1016/j.chb.2010.07.028

Chiu, C. M., Hsu, M. H., & Wang, E. T. G. (2006). Understanding knowledge sharing in virtual communities: An integration of social capital and social cognitive theories. *Decision Support Systems*, *42*(3), 1872–1888. doi:10.1016/j.dss.2006.04.001

Chowdhury, H. K., Parvin, N., Weitenberner, C., & Becker, M. (2006). Consumer Attitude toward Mobile Advertising in an Emerging Market: An Empirical Study. *International Journal of Mobile Marketing*, *1*(2).

Chulyadyo, R., & Leray, P. (2014). A Personalized Recommender System from Probabilistic Relational Model and Users' Preferences. *Procedia Computer Science*, *35*, 1063–1072. doi:10.1016/j.procs.2014.08.193

Chung, J. Y., & Buhalis, D. (2008). Information needs in online social networks. *Information Technology & Tourism*, *10*(4), 267–281. doi:10.3727/109830508788403123

Churchill, D. (2009). Educational applications of web 2.0: Using blogs to support teaching and learning. *British Journal of Educational Technology*, *40*(1), 179–183. doi:10.1111/j.1467-8535.2008.00865.x

Chu, S. K. W., Chan, C. K. K., & Tiwari, A. F. Y. (2012). Using blogs to support learning during internship. *Computers & Education*, *58*(3), 989–1000. doi:10.1016/j.compedu.2011.08.027

Chwolka, A., & Raith, M. G. (2012). The value of business planning before start-up — A decision-theoretical perspective. *Journal of Business Venturing*, *27*(3), 385–399. doi:10.1016/j.jbusvent.2011.01.002

Cialdini, R. B., & Goldstein, N. J. (2004). Social Influence: Compliance and Conformity. *Annual Review of Psychology*, *55*(1), 591–621. doi:10.1146/annurev.psych.55.090902.142015 PMID:14744228

Ciampa, K. K. (2014). Learning in a mobile age: An investigation of student motivation. *Journal of Computer Assisted Learning*, *30*(1), 82–96. doi:10.1111/jcal.12036

Cirucci, A. M. (2013). First person paparazzi: Why social media should be studied more like video games. *Telematics and Informatics*, *30*(1), 47–59. doi:10.1016/j.tele.2012.03.006

Cocea, M., & Weibelzahl, S. (2009). Log file analysis for disengagement detection in e-Learning environments. *User Modeling and User-Adapted Interaction*, *19*(4), 341–385. doi:10.1007/s11257-009-9065-5

Cockburn, A., Gutwin, C., & Greenberg, S. (2007, April). A predictive model of menu performance. In *Proceedings of the SIGCHI conference on Human factors in computing systems*. ACM. doi:10.1145/1240624.1240723

Coello, C. A. C., Lamont, G. B., & Van Veldhuizen, D. A. (2007). Evolutionary Algorithms for Solving Multi-Objective Problems (2nd ed.). Genetic and Evolutionary Computation. New York, NY: Springer.

Coffield, F., Moseley, D., Hall, E., & Ecclestone, K. (2004). *Learning styles and pedagogy in post-16 learning. A systematic and critical review*. London, England: Learning & Skills Research Centre.

Coleman, M. K., & Parker, D. S. (1996). Aesthetics-based graph layout for human consumption. *Software, Practice & Experience*, *26*(12), 1415–1438. doi:10.1002/(SICI)1097-024X(199612)26:12<1415::AID-SPE69>3.0.CO;2-P

Collet, C., Averty, P., & Dittmar, A. (2009). Autonomic nervous system and subjective ratings of strain in air-traffic control. *Applied Ergonomics*, *40*(1), 23–32. doi:10.1016/j.apergo.2008.01.019 PMID:18353277

Collins, J., Ketter, W., & Gini, M. (2010). Flexible decision support in dynamic inter-organisational networks. *European Journal of Information Systems*, *19*(4), 436–448. doi:10.1057/ejis.2010.24

Colombo, B., & Antonietti, A. (2013). Naïve theories about multimedia learning: A study on primary school textbooks. *Frontiers in Psychology*, *4*(450), 1–19. doi:10.3389/fpsyg.2013.00450 PMID:23382719

Colombo, B., Antonietti, A., Caravita, S., & Sala, R. (2013). Blog content and structure, cognitive style and metacognition. *International Journal of Technology and Human Interaction*, *9*(3), 1–17. doi:10.4018/jthi.2013070101

Colombo, B., & Sala, R. (2011). Blogosphere: cooperative and strategic aspects of self-regulated learning in formal and informal settings. In A. Bartolomé, P. Bergamin, D. Persico, K. Steffens, & J. Underwood (Eds.), *Self-regulated learning in technology enhanced learning environments: Problems and promises, 154-161*. Aachen: Shaker Verlag.

Context-Based Research Group. (2002). *The mobiles: Social evolution in a wireless society*. Retrieved from Context Based Research Group website: http://www.contextresearch.com/context/pdf/theMOBILES.pdf

Cook, D. A., Gelula, M. H., Dupras, D. M., & Schwartz, A. (2007). Instructional methods and cognitive and learning styles in web-based learning: Report of two randomised trials. *Medical Education*, *41*(9), 897–905. doi:10.1111/j.1365-2923.2007.02822.x PMID:17727530

Correa, T., Hinsley, A., & Gil De Zuniga, H. (2010). Who interacts on the Web? The intersection of users' personality and social media use (English). *Computers in Human Behavior*, *26*(2), 247–253. doi:10.1016/j.chb.2009.09.003

Cortese, T. (2011). Why niche social networks encourage more meaningful interaction: Opinion. Retrieved May 8, 2011, from http://mashable.com/2011/09/06/social-network-niche-interaction/

Costa, H., Benevenuto, F., & Merschmann, C. H. L. (2013). Detecting tip spam in location-based social networks. *Proceedings of the ACM Symposium on Applied Computing, Coimbra, Portugal*. Retrieved from http://homepages.dcc.ufmg.br/~fabricio/download/sac2013.pdf

Costa-Montenegro, E., Barragáns-Martínez, A. B., & Rey-López, M. (2012). Which App? A recommender system of applications in markets: Implementation of the service for monitoring users' interaction. *Expert Systems with Applications*, *39*(10), 9367–9375. doi:10.1016/j.eswa.2012.02.131

Cothrel, J. (2000). Measuring the success of an online community. *Strategy and Leadership*, *28*(2), 17–21. doi:10.1108/10878570010341609

Cotte, J., Chowdhury, T. G., Ratneshwar, S., & Ricci, L. M. (2006). Pleasure or utility? Time planning style and web usage behaviors. *Journal of Interactive Marketing*, *20*(1), 45–57. doi:10.1002/dir.20055

Council of Europe Convention on Cybercrime. (2001). *Treaties*. Retrieved from http://conventions.coe.int/Treaty/EN/Treaties/html/185.htm

Craig, E., & Wright, B. K. (2012). Computer-mediated relational development and maintenance on Facebook. *Communication Research Reports*, *29*(2), 119–129. doi:10.1080/08824096.2012.667777

Creativity is a key part of the educated mind. (2006). *Edutopia*. Retrieved May 17, 2011 from http://www.edutopia.org/take-chance-let-them-dance

Creswell, J. (1998). *Qualitative inquiry and research design: Choosing among five traditions*. Thousand Oaks, CA: Sage.

Crossler, R. E., Johnston, A. C., Lowry, P. B., Hu, Q., Warkentin, M., & Baskerville, R. (2013). Future directions for behavioral information security research. *Computers & Security*, (32), 90–101.

Cuevas, H. M., Fiore, S. M., Bowers, C. A., & Salas, E. (2004). Fostering constructive cognitive and metacognitive activity in computer-based complex task training environments. *Computers in Human Behavior*, *20*(2), 225–241. doi:10.1016/j.chb.2003.10.016

Cuevas, H. M., Fiore, S. M., & Oser, R. L. (2002). Scaffolding cognitive and metacognitive processes in low verbal ability learners: Use of diagrams in computer-based training environments. *Instructional Science*, *30*(6), 433–464. doi:10.1023/A:1020516301541

Cummings, J., Lee, J., & Kraut, R. (2006). Communication technology and friendship during the transition from high school to college. In R. E. Kraut, M. Brynin, & S. Kiesler (Eds.), *Computers, phones, and the Internet: Domesticating information technology* (pp. 265–278). New York, NY: Oxford University Press.

Cunha, M. P. E., Clegg, S. R., & Kamoche, K. (2006). Surprises in management and organization: Concept, sources and a typology. *British Journal of Management*, *17*(4), 317–329. doi:10.1111/j.1467-8551.2005.00470.x

Czerwinski, M., Smith, G., Regan, T., Meyers, B., Robertson, G., & Starkweather, G. (2003). Toward characterizing the productivity benefits of very large displays. *Proc. Interact*, *2003*, 9–16.

d'Hauteville, F. (1994). *Un Modèle d'Acceptation du Nouveau Produit par le Consommateur: Cas du Vin allégé en Alcool*. Unpublished Doctoral Dissertation, Université Montpellier II, France.

Dahal, K., Almejalli, K., & Hossain, M. A. (2013). Decision support for coordinated road traffic control actions. *Decision Support Systems*, *54*(2), 962–975. doi:10.1016/j.dss.2012.10.022

Davies, S. E., Bury, K. F., & Darnell, M. J. (1985). An experimental comparison of a windowed vs a non-windowed operating system environment. In *Proceedings of the Human Factors and Ergonomics Society Annual Meeting* (Vol. 29, No. 3, pp. 250–254). Santa Monica, CA: Human Factors and Ergonomics Society. doi:10.1177/154193128502900310

Davis, D. L., & Vitell, S. J. (1991). The ethical problems, conflicts, and beliefs of small business information personnel. *Journal of Computer Information Systems*, *22*(4), 53–57.

Davis, F. D. (1989). Perceived Usefulness, Perceived Ease of Use, and User Acceptance of Information Technology. *Management Information Systems Quarterly*, *13*(3), 319–340. doi:10.2307/249008

Davis, F. D., Bagozzi, R. P., & Warsaw, P. R. (1992). Extrinsic and intrinsic motivation to use computers in the work place. *Journal of Applied Social Psychology*, *22*(14), 1111–1132. doi:10.1111/j.1559-1816.1992.tb00945.x

Davis, F. D., Bagozzi, R. P., & Warshaw, P. R. (1989). User Acceptance of Computer Technology: A Comparison of Two Theoretical Models. *Management Science*, *35*(8), 982–1003. doi:10.1287/mnsc.35.8.982

de Tjerk, E. G., Henryk, F. R. A., & Neerincx, M. A. (2010). Adaptive automation based on an object-oriented task model: Implementation and evaluation in a realistic C2 environment. *Journal of Cognitive Engineering and Decision Making*, *4*(2), 152–182. doi:10.1518/155534310X522851

De Vries, L., Gensler, S., & Leeflang, P. (2012). Popularity of Brand Posts on Brand Fan Pages: An Investigation of the Effects of Social Media Marketing. *Journal of Interactive Marketing*, *26*(2), 83–91. doi:10.1016/j.intmar.2012.01.003

Deaudelin, C., & Richer, A. (1999). *A learning conversation approach integrating email: Its experiment to support the student learning process at college level*. Paper presented at the 8th EARLI Conference, Goteborg, Sweden.

Deci, E. L. (1975). *Intrinsic motivation*. New York: Plenum Publishing. doi:10.1007/978-1-4613-4446-9

del Olmo, F. H., & Gaudioso, E. (2008). Evaluation of recommender systems: A new approach. *Expert Systems with Applications*, *35*(3), 790–804. doi:10.1016/j.eswa.2007.07.047

Delaney, K. (2005, March 29). Identity theft made easier. *Wall Street Journal,* p. B1.

Deng, L., & Yuen, A. H. K. (2011). Towards a framework for educational affordances of blogs. *Computers & Education*, *56*(2), 441–451. doi:10.1016/j.compedu.2010.09.005

DeSanctis, G., & Gallupe, B. (1987). A foundation for the study of group decision support systems., *33*(12), 1589–1609.

Dholakia, U. M., Bagozzi, R. P., & Pearo, L. K. (2004). A social influence model of consumer participation in network- and small-group based virtual communities. *International Journal of Research in Marketing*, *21*(3), 241–263. doi:10.1016/j.ijresmar.2003.12.004

Dholakia, U. M., Blazevic, V., Wiertz, C., & Algesheimer, R. (2009). Communal service delivery: How customers benefit from participation in firm-hosted virtual P3 communities. *Journal of Service Research*, *12*(2), 208–226. doi:10.1177/1094670509338618

Dickinger, A., Arami, M., & Meyer, D. (2008). The role of perceived enjoyment and social norm in the adoption of technology with network externalities. *European Journal of Information Systems*, *17*(1), 4–11. doi:10.1057/palgrave.ejis.3000726

Dilla, W., Janvrin, D. J., & Raschke, R. (2010). Interactive Data Visualization: New Directions for Accounting Information Systems Research. *Journal of Information Systems*, *24*(2), 1–37. doi:10.2308/jis.2010.24.2.1

Dillon, C., & Greene, B. (2003). Learner differences in distance learning: Finding differences that matter. In M. G. Moore & W. G. Anderson (Eds.), Handbook of distance education, 235–244.

Dodds, W. B., Monroe, K. B., & Grewal, D. (1991). Effects of Price, Brand, and Store Information on Buyers' Product Evaluations. *JMR, Journal of Marketing Research*, *28*(3), 307–319. doi:10.2307/3172866

Dominick, J. (1999). Who do you think you are? Personal home pages and self-presentation on the World Wide Web. *Journalism & Mass Communication Quarterly*, *76*(4), 646–658. doi:10.1177/107769909907600403

Doolittle, P. E., & Mariano, G. J. (2008). Working memory capacity and mobile multimedia learning environments: Individual differences in learning while mobile. *Journal of Educational Multimedia and Hypermedia*, *17*(4), 511–530.

Dubey, G., & Moricot, C. (2006). *La polyvalence du Rafale ou l'objet total : la relation entre une technologie nouvelle et ses utilisateurs* (Vol. 81). C2SD.

Dunne, A., Lawlor, M.-A., & Rowley, J. (2010). Young people's use of online social networking sites – a uses and gratifications perspective. *Journal of Research in Interactive Marketing*, *4*(1), 46–58. doi:10.1108/17505931011033551

Dux, P. E., Tombu, M. N., Harrison, S., Rogers, B. P., Tong, F., & Marois, R. (2009). Training improves multitasking performance by increasing the speed of information processing in human prefrontal cortex. *Neuron*, *63*(1), 127–138. doi:10.1016/j.neuron.2009.06.005 PMID:19607798

Dye, J. (2007). Meet generation C: Creatively connecting through content. *EContent*, *30*(4), 38.

Eades, P., Wei, L., Misue, K., & Sugiyama, K. (1991). *Preserving the mental map of a diagram* (Technical Report No. IIAS-RR-91-16E). Retrieved from http://www.cs.tsukuba.ac.jp/~misue/publications/techreport/iias-rr-91-16e.pdf

Eastin, M. (2002). *From Internet use to the unique Internet user: Assessing information, entertainment and social cognitive models of Internet use*. Paper presented to the National Communication Association Conference, New Orleans, LA.

Edenius, M., & Rämö, H. (2011). An office on the go: Professional workers, smart phones and the return of place. *International Journal of Technology and Human Interaction*, *7*(1), 37–55. doi:10.4018/jthi.2011010103

Efklides, A. (2011). Interactions of metacognition with motivation and affect in self-regulated learning: The MASRL model. *Educational Psychologist*, *46*(1), 6–25. doi:10.1080/00461520.2011.538645

Eirinaki, M., & Vazirgiannis, M. (2007). Web site personalization based on link analysis and navigational patterns. *ACM Transactions on Internet Technology*, *7*(4), 21, es. doi:10.1145/1278366.1278370

Eisenhardt, K. M., & Graebner, M. E. (2007). Theory Building From Cases: Opportunities and Challenges. *Academy of Management Journal*, *50*(1), 25–32. doi:10.5465/AMJ.2007.24160888

Ellenberg, J. (2008). This psychologist might outsmart the math brains competing for the netflix prize. *Wired Magazine*, 114-122.

Elliott, A. M. E., & Hearst, M. A. (2002). A comparison of the affordances of a digital desk and tablet for architectural image tasks. *International Journal of Human-Computer Studies*, *56*(2), 173–197. doi:10.1006/ijhc.2001.0520

Ellis, R. (2010, August 16). Sextortion: Are your teens being "sexploited" online? *Examiner*. Retrieved from http://www.examiner.com/cyber-safety-in-national/sextortion-are-your-teens-being-sexploited-online

Ellison, N. B., Steinfield, C., & Lampe, C. (2007). The benefits of Facebook "friends": Social capital and college students' use of online social network sites. *Journal of Computer-Mediated Communication*, *12*(4), 1143–1168. http://jcmc.indiana.edu/ doi:10.1111/j.1083-6101.2007.00367.x

Ellison, N. B., Steinfield, C., & Lampe, C. (2011). Connection strategies: Social capital implications of Facebook-enabled communication practices. *New Media & Society*, *13*(6), 873–892. doi:10.1177/1461444810385389

Ellison, N., Lampe, C., & Steinfield, C. (2010). With a little help from my friends: social network sites and social capital. In Z. Papacharissi (Ed.), *The Networked Self: Identity, Community and Culture on Social Network Sites* (pp. 124–145). New York: Routledge.

Endsley, M. R. (1995). Toward a theory of situation awareness. *Human Factors*, *37*(1), 32–64. doi:10.1518/001872095779049543

Enteen, J. (2006). Spatial conceptions of URLs: Tamil Eelam networks on the World Wide Web. *New Media & Society*, *8*(2), 229–249. doi:10.1177/1461444806061944

Eom, S. B. (2004). *The Changing Structure of Decision Support Systems Research: An Empirical Investigation through Author Cocitation Mapping*. The 2004 IFIP International Conference on Decision Support Systems, Prato, Tuscany.

Eom, S. B., Lee, S., & Kim, E. B. (1998). A survey of decision support system applications (1988-1994)., *49*(2), 109–120.

Eric, B. (2009). Decisions 2.0: the Power of Collective Intelligence. *MIT Sloan Management Review*, *50*(2), 45. Retrieved from http://proquest.umi.com/pqdlink?did=1625861001&Fmt=7&clientId=9268&RQT=309&VName=PQD

Evers, J. (2006, May 2006). Teens arrested in alleged MySpace extortion scam. *Zdnet*. Retrieved from http://www.zdnet.com/news/teens-arrested-in-alleged-myspace-extortion-scam/148219

Fabbri, M., Antonietti, A., Giorgetti, M., Tonetti, L., & Natale, V. (2007). Circadian typology and style of thinking differences. *Learning and Individual Differences*, *17*(2), 175–180. doi:10.1016/j.lindif.2007.05.002

Facebook (2014). Facebook newsroom. http://newsroom.fb.com/company-info/

Facebook reports fourth quarter and full year 2012 results. (2014, May 30). *Facebook*. Retrieved from http://investor.fb.com/releasedetail.cfm?ReleaseID=736911

Facebook statistics by country. (n.d.). *Socialbakers*. Retrieved March 14, 2012, from http://www.socialbakers.com/Facebook-statistics/

Fafinski, S., Dutton, W. H., & Margetts, H. Z. (2010). Mapping and measuring cybercrime. *OII Working Paper, 18.*

Fang, L., & Ha, L. (2011). *Who are the heavy users of Social Network Sites among College Students? A Study of Social Network Sites and College Students*. Paper presented at the annual meeting of the Association for Education in Journalism and Mass Communication, St. Louis, MO.

Fan, W., Zhang, L., & Watkins, D. (2010). Incremental validity of thinking styles in predicting academic achievements: An experimental study in hypermedia learning environments. *Educational Psychology*, *30*(5), 605–623. doi:10.1080/01443410.2010.496899

Faraj, S., & Xiao, Y. (2006). Coordination in fast-response organizations. *Management Science*, *52*(8), 1155–1169. doi:10.1287/mnsc.1060.0526

FBI. (2011). *Manhattan U.S. attorney charges seven individuals for engineering sophisticated internet fraud scheme that infected millions of computers worldwide and manipulated internet advertising business*. [Press release]. Retrieved from http://www.fbi.gov/newyork/press-releases/2011/manhattan-u.s.-attorney-charges-seven-individuals-for-engineering-sophisticated-internet-fraud-scheme-that-infected-millions-of-computers-worldwide-and-manipulated-internet-advertising-business

Feaster, J. C. (2009). The repertoire niches of interpersonal media: Competition and coexistence at the level of the individual. *New Media & Society*, *11*(6), 965–984. doi:10.1177/1461444809336549

Feenburg, A., & Bakardjieva, M. (2004). Virtual community: No killer implication. *New Media & Society*, *6*(1), 37–43. doi:10.1177/1461444804039904

Feick, L. F., & Price, L. L. (1987). The market maven: A diffuser of market information. *Journal of Marketing*, *51*(1), 83–97. doi:10.2307/1251146

Feick, L. F., Price, L. L., & Higie, R. A. (1986). People who use people: The other side of opinion leadership. *Advances in Consumer Research. Association for Consumer Research (U. S.)*, *13*(1), 301–305.

Feng, Y., & Xie, W. (2014). Teens' concern for privacy when using social networking sites: An analysis of socialization agents and relationships with privacy-protecting behaviors. *Computers in Human Behavior*, *33*, 153–162. doi:10.1016/j.chb.2014.01.009

Fewster, S. (2010). Computer virus created by Anthony Scott Harrison sparked worldwide alert. *News*. Retrieved from http://www.news.com.au/technology/computer-virus-created-by-anthony-scott-harrison-sparked-worldwide-alert/story-e6frfrnr-1225897175334#ixzz1EiBSPNal

Findlater, L., & McGrenere, J. (2004, April). A comparison of static, adaptive, and adaptable menus. In *Proceedings of the SIGCHI conference on Human factors in computing systems* (pp. 89-96). ACM. doi:10.1145/985692.985704

Findlater, L., & McGrenere, J. (2008, April). Impact of screen size on performance, awareness, and user satisfaction with adaptive graphical user interfaces. In *Proceedings of the SIGCHI Conference on Human Factors in Computing Systems*. Florence (pp. 1247-1256). ACM. doi:10.1145/1357054.1357249

Findlater, L., Moffatt, K., McGrenere, J., & Dawson, J. (2009, April). Ephemeral adaptation: The use of gradual onset to improve menu selection performance. In *Proceedings of the SIGCHI Conference on Human Factors in Computing Systems* (pp. 1655-1664). ACM. doi:10.1145/1518701.1518956

Fiorina, L., Antonietti, A., Colombo, B., & Bartolomeo, A. (2007). Thinking style, browsing primes and hypermedia navigation. *Computers & Education*, *49*(3), 916–941. doi:10.1016/j.compedu.2005.12.005

Fisch, S. M. (2000). A capacity model of children's comprehension of educational content on television. *Media Psychology*, *2*(1), 63–91. doi:10.1207/S1532785XMEP0201_4

Fisher, F. M. (1979). Diagnosing monopoly. *The Quarterly Review of Economics and Business*, *19*(2), 7–33.

Fitchett, S., & Cockburn, A. (2012, May). Accessrank: predicting what users will do next. In *Proceedings of the SIGCHI Conference on Human Factors in Computing Systems* (pp. 2239-2242). ACM. doi:10.1145/2207676.2208380

Flavell, J. H. (1979). Metacognition and cognitive monitoring: A new area of cognitive-developmental inquiry. *The American Psychologist*, *34*(10), 906–911. doi:10.1037/0003-066X.34.10.906

Fleishman, E. A., Costanza, D. P., & Marshall-Mies, J. (1999). Abilities. In N. G. Peterson, M. D. Mumford, W. C. Borman, P. R. Jeanneret, & E. A. Fleishman (Eds.), *An occupational information system for the 21st century: The development of O*NET* (pp. 175–195). Washington, DC: American Psychological Association. doi:10.1037/10313-010

Florencio, D., & Herley, C. (2011) Sex, lies and cybercrime surveys.*Proceedings of the Tenth Workshop on the Economics of Information Security (WEIS)*, GeorgeMason University, USA.

Flynn, L. R., Goldsmith, R. E., & Eastman, J. K. (1996). Opinion leaders and opinion seekers: Two new measurement scales. *Journal of the Academy of Marketing Science*, *24*(2), 137–147. doi:10.1177/0092070396242004

Foehr, U. G. (2006). *Media multitasking among American Youth: Prevalence, predictors and pairings*. The Henry J. Kaiser Family Foundation.

Foerde, K., Knowlton, B. J., & Poldrack, R. A. (2006). Modulation of competing memory systems by distraction. *Proceedings of the National Academy of Sciences of the United States of America*. Retrieved on May 17, 2011 from http://www.pnas.org/content/103/31/11778.full

Fournier, L. R., Wilson, G. F., & Swain, C. R. (1999). Electrophysiological, behavioral, and subjective indexes of workload when performing multiple tasks: Manipulations of task difficulty and training. *International Journal of Psychophysiology*, *31*(2), 129–145. doi:10.1016/S0167-8760(98)00049-X PMID:9987059

Foxall, G., & Haskins, C. G. (1986). Cognitive Style and Consumer Innovativeness: An Empirical Test of Kirton's Adaptation-Innovation Theory in the Context of Food Purchasing. *European Journal of Marketing*, *20*(3/4), 63–80. doi:10.1108/EUM0000000004755

Freire, M., & Rodríguez, P. (2006). Preserving the mental map in interactive graph interfaces. In CelentanoA. (Ed.), *Proceedings of the Working Conference on Advanced Visual Interfaces* (pp. 270–273). New York, NY: ACM. doi:10.1145/1133265.1133319

Freitas, S., & Griffiths, M. (2008). The convergence of gaming practices with other media forms: What potential for learning? A review of the literature. *Learning, Media and Technology*, *33*(1), 11–20. doi:10.1080/17439880701868796

Fulgoni, G. (2011). *You've got to See it! What social media marketers can learn from broadway success*. Retrieved July 8, 2011, from http://blog.comscore.com/2011/09/what_social_marketers_can_learn_from_broadway.html

Funke, D. J., Neal, J. G., & Paul, R. D. (1993). An approach to intelligent automated window management. *International Journal of Man-Machine Studies*, *38*(6), 949–983. doi:10.1006/imms.1993.1044

Furnham, A., Christopher, A. N., Garwood, J., & Martin, G. N. (2007). Approaches to learning and the acquisition of general knowledge. *Personality and Individual Differences*, *43*(6), 1563–1571. doi:10.1016/j.paid.2007.04.013

Gabriel, K. R., & Neumann, J. (1962). A Markov chain model for daily rainfall occurrence at Tel Aviv. *Quarterly Journal of the Royal Meteorological Society*, *88*(375), 90–95. doi:10.1002/qj.49708837511

Gajos, K., Christianson, D., Hoffmann, R., Shaked, T., Henning, K., Long, J. J., & Weld, D. S. (2005). Fast and robust interface generation for ubiquitous applications. In UbiComp 2005: Ubiquitous Computing (pp. 37-55). Springer Berlin Heidelberg. doi:10.1007/11551201_3

Gajos, K. Z., Czerwinski, M., Tan, D. S., & Weld, D. S. (2006, May). Exploring the design space for adaptive graphical user interfaces. In *Proceedings of the working conference on Advanced visual interfaces* (pp. 201-208). ACM. doi:10.1145/1133265.1133306

Gallupe, R. B., & DeSanctis, G. (1988). Computer-based support for group problem-finding: An experimental investigation, *12*(2), 277–296.

Galy, E., Cariou, M., & Melan, C. (2012). What is the relationship between mental workload factors and cognitive load types? *International Journal of Psychophysiology*, *83*(3), 269–275. doi:10.1016/j.ijpsycho.2011.09.023 PMID:22008523

Gansner, E. R., Hu, Y., & North, S. (2013). Visualizing streaming text data with dynamic graphs and maps. In W. Didimo & M. Patrignani (Eds.), Lecture Notes in Computer Science: Vol. 7704. *Graph Drawing* (pp. 439–450). Berlin, Germany: Springer Berlin Heidelberg. doi:10.1007/978-3-642-36763-2_39

Gao, J., Zhang, M., Jiang, F., & Wang, X. (2005). Analysis of opinion leader in on-line communities. In R. Khosla, R. J. Howlett, & L. C. Jain (Eds.), *Lecture notes in artificial intelligence. Knowledge-based intelligent information and engineering systems, Pt. 1: Proceedings* (pp. 1153–1159). Berlin: Springer.

Gardner, J. S. (2008). *Simultaneous Media Usage: Effects on Attention.* Unpublished doctoral dissertation. Virginia Tech.

Gardner, H. (1993). *Multiple intelligences: The theory in practice.* New York: Basic Books.

Garza, G. (2002). The internet, narrative, and subjectivity. *Journal of Constructivist Psychology, 15*(3), 185–203. doi:10.1080/10720530290100433

Gatignon, H., & Robertson, T. S. (1985). A propositional Inventory for New Diffusion Research. *The Journal of Consumer Research, 11*(4), 849–867. doi:10.1086/209021

Gavalas, D., & Kenteris, M. (2011). A web-based pervasive recommendation system for mobile tourist guides. *Personal and Ubiquitous Computing, 15*(7), 759–770. doi:10.1007/s00779-011-0389-x

Gelderman, M. (2002). Task difficulty, task variability and satisfaction with management support systems. *Information & Management, 39*(7), 593–604. doi:10.1016/S0378-7206(01)00124-0

Ghazarian, A., & Noorhosseini, S. M. (2010). Automatic detection of users' skill levels using high-frequency user interface events. *User Modeling and User-Adapted Interaction, 20*(2), 109–146. doi:10.1007/s11257-010-9073-5

Gil de Zúñiga, H., Jung, N., & Valenzuela, S. (2012). Social Media Use for News and Individuals' Social Capital, Civic Engagement and Political Participation. *Journal of Computer-Mediated Communication, 17*(3), 319–336. doi:10.1111/j.1083-6101.2012.01574.x

Glogowski, K. (2009). Tracing the emergence of a blogging/writing community: Critical transformations in a grade eight classroom. *Dissertation Abstracts International* Section A, 69.

Goldman, R. H., Cohen, A. P., & Sheahan, F. (2008). Using seminar blogs to enhance student participation and learning in public health school classes. *American Journal of Public Health, 98*(9), 1658–1663. doi:10.2105/AJPH.2008.133694 PMID:18633075

Goldsmith, E. B., & Ronald, E. G. (1980). Dogmatism and confidence as related factors in evaluation of new products. *Psychological Reports, 47*(3), 1068–1070. doi:10.2466/pr0.1980.47.3f.1068

Goldsmith, R. E. (2006). Online consumer behavior. In S. Dasgupta (Ed.), *Encyclopedia of Virtual Communities and Technologies* (pp. 338–342). Chicago, IL: Rand McNally. doi:10.4018/978-1-59140-563-4.ch063

Goldsmith, R. E., d'Hauteville, F., & Flynn, L. R. (1998). Theory and Measurement of Consumer Innovativeness. *European Journal of Marketing, 32*(3/4), 340–353. doi:10.1108/03090569810204634

Goldsmith, R. E., & Hofacker, C. H. (1991). Measuring Consumer Innovativeness. *Journal of the Academy of Marketing Science, 19*(3), 209–221. doi:10.1007/BF02726497

Gonzales, A. L., & Hancock, J. T. (2011). Mirror, Mirror on my Facebook Wall: Effects of Exposure to Facebook on Self-Esteem. *Cyberpsychology, Behavior, and Social Networking, 14*(1/2), 79–83. doi:10.1089/cyber.2009.0411 PMID:21329447

Goodhue, D. L. (1995). Understanding user evaluations of information systems. *Management Science, 41*(12), 1827–1844. doi:10.1287/mnsc.41.12.1827

Goodhue, D. L., Klein, B. D., & March, S. T. (2000). User evaluations of IS as surrogates for objective performance. *Information & Management, 38*(2), 87–101. doi:10.1016/S0378-7206(00)00057-4

Goonetilleke, R. S., & Luximon, Y. (2010). The relationship between monochronicity, polychronicity and individual characteristics. *Behaviour & Information Technology, 29*(2), 187–198. doi:10.1080/01449290903222697

Gopher, D., Armony, L., & Greenspan, Y. (2000). Switching tasks and attention policies. *Journal of Experimental Psychology. General*, 229–308. PMID:11006903

Gorry, G. A., & Scott Morton, M. (1971). A Framework for Management Information Systems. *Sloan Management Review, 12*(1), 55–70.

Granberg, C. (2010). Social software for reflective dialogue: Questions about reflection and dialogue in student teachers' blogs. *Technology, Pedagogy and Education, 19*(3), 345–360. doi:10.1080/1475939X.2010.513766

Granovetter, M. S. (1973). The strength of weak ties. *American Journal of Sociology, 78*(6), 1360–1380. doi:10.1086/225469

Grasmuck, S., Martin, J., & Zhao, S. (2009). Ethno-racial identity displays on Facebook. *Journal of Computer-Mediated Communication, 15*(1), 158–188. http://jcmc.indiana.edu/ doi:10.1111/j.1083-6101.2009.01498.x

Greene, J. A., Choudhry, N. K., Kilabuk, E., & Shrank, W. H. (2011). Online Social Networking by Patients with Diabetes: A Qualitative Evaluation of Communication with Facebook. *Journal of General Internal Medicine, 26*(3), 287–292. doi:10.1007/s11606-010-1526-3 PMID:20945113

Greenhalgh, T., Robert, G., Macfarlane, F., Bate, P., & Kyriakidou, O. (2004). Diffusion of innovations in service organizations: Systematic review and recommendations. *The Milbank Quarterly, 82*(4), 581–629. doi:10.1111/j.0887-378X.2004.00325.x PMID:15595944

Greer, F. C., & Ferguson, A. D. (2011). Using Twitter for promotion and branding: A content analysis of local television Twitter sites. *Journal of Broadcasting & Electronic Media, 55*(2), 198–214. doi:10.1080/08838151.2011.570824

Greitemeyer, T., Mügge, D. O., & Bollermann, I. (2014). Having responsive Facebook friends affects the satisfaction of psychological needs more than having many Facebook friends. *Basic and Applied Social Psychology, 36*(3), 252–258. doi:10.1080/01973533.2014.900619

Grieve, R., Indian, M., Witteveen, K., Anne Tolan, G., & Marrington, J. (2013). Face-to-face or Facebook: Can social connectedness be derived online? *Computers in Human Behavior, 29*(3), 604–609. doi:10.1016/j.chb.2012.11.017

Griffith, D.M., Gunter, K., Watkins D.C. (2012). Measuring Masculinity in Research on Men of Color: Findings and Future Directions. *American Journal of Public Health, 102*(S2), S187-S194.

Griffith, H. (2005, June 7). eBay fraud boy con 'really easy.' *BBC News*. Retrieved from http://news.bbc.co.uk/1/hi/wales/4615117.stm

Gross, R., Acquisti, A., & Heinz, H. J., III. (2005). Information revelation and privacy in online social networks. *Proceedings of the 2005 ACM workshop on Privacy in the Electronic Society*, 71-80. New York. doi:10.1145/1102199.1102214

Guadagno, R. E., Okdie, B. M., & Eno, C. A. (2008). Who blogs? Personality predictors of blogging. *Computers in Human Behavior, 24*(5), 1993–2004. doi:10.1016/j.chb.2007.09.001

Guo, X., & Lim, J. (2007). Negotiation support systems and team negotiations: The coalition formation perspective. *Information and Software Technology, 49*(11-12), 1121–1127. doi:10.1016/j.infsof.2006.11.006

Gupta, S., & Mela, C. F. (2008, November). What is a free customer worth? *Harvard Business Review, 86*, 102–109, 138. Retrieved from http://hbr.org/magazine PMID:19009724

Hagel, J., & Armstrong, A. (1997). *Net gain: Expanding markets through virtual communities*. Boston, MA: Harvard Business School Press.

Hair, J. F., Anderson, R. E., Tatham, R. L., & Black, W. C. (1998). *Multivariate Data Analysis with Readings*. Englewood Cliffs, NJ: Prentice Hall.

Ha, L., & Hu, X. (2013). Social media involvement among college students and general population: Implications to media management. In M. Friedrichsen & W. Mühl-Benninghaus (Eds.), *Handbook of Social Media Management: Value Chain and Business Models in Changing Media Markets*. New York: Springer-Verlag. doi:10.1007/978-3-642-28897-5_44

Hall, A. (2009). *College students' motives for using social networking sites and their relationships to personality traits*. Conference paper – International Communication Association, 1 – 38.

Hall, E. T. (1959). *The Silent Language*. Garden City, NY: Doubleday.

Hall, H., & Davison, B. (2007). Social software as support in hybrid learning environments: The value of the blog as a tool for reflective learning and peer support. *Library & Information Science Research, 29*(2), 163–187. doi:10.1016/j.lisr.2007.04.007

Hanna, R., Rohm, A., & Crittenden, L. V. (2011). We're all connected: The power of social media ecosystem. *Business Horizons, 54*(3), 265–273. doi:10.1016/j.bushor.2011.01.007

Hanson, T. L., Drumheller, K., Mallard, J., McKee, C., & Schlegel, P. (2011). Cell phones, text messaging, and Facebook: Competing time demands of today's college students. *College Teaching*, *59*(1), 23–30. doi:10.1080/87567555.2010.489078

Hanson, W. (2000). *Principles of internet marketing*. Cincinnati, OH: South-Western College Publishing.

Hargittai, E. (2008). The digital reproduction of inequality. In D. Grusky (Ed.), *Social stratification*. Boulder, CO: Westview Press.

Hartley, A. A., Jonides, J., & Sylvester, C. (2011). Dual-task processing in younger and older adults: Similarities and differences revealed by fMRI. *Brain and Cognition*, *75*(3), 281–291. doi:10.1016/j.bandc.2011.01.004 PMID:21320741

Hartman, J. B., Shim, S., Barber, B., & O'Brien, M. (2006). Adolescents' utilitarian and hedonic Web consumption behavior: Hierarchical influence of personal values and innovativeness. *Psychology and Marketing*, *23*(10), 813–839. doi:10.1002/mar.20135

Hart, S. G., & Staveland, L. E. (1988). Development of NASA-TLX (Task Load Index): Results of empirical and theoretical research. In P. A. Hancock & N. Meshkati (Eds.), *Human Mental Workload* (pp. 139–183). Amsterdam: Elsevier Science. doi:10.1016/S0166-4115(08)62386-9

Hembrooke, H., & Gay, G. (2003). The Laptop and the lecture: The effects of multitasking in learning environments. *Journal of Computing in Higher Education*, *15*(1), 46–64. doi:10.1007/BF02940852

Henning, E. (2003). "I click, therefore I am (not)": Is cognition 'distributed' or is it 'contained' in borderless e-learning programmes? *International Journal of Training and Development*, *7*(4), 303–317. doi:10.1046/j.1360-3736.2003.00188.x

Henri, F. (1992). Computer conferencing and content analysis. In A. R. Kaye (Ed.), *Collaborative learning through computer conferencing: The Najaden papers* (pp. 115–136). New York: Springer. doi:10.1007/978-3-642-77684-7_8

Herlocker, J. L., Konstan, J. A., Terveen, L. G., & Riedl, J. T. (2004). Evaluating collaborative filtering recommender systems. *ACM Transactions on Information Systems*, *22*(1), 5–53. doi:10.1145/963770.963772

Herrmann, A., Huber, F., & Braunstein, C. (2000). Market-driven product and service design: Bridging the gap between customer needs, quality management, and customer satisfaction. *International Journal of Production Economics*, *66*(1), 77–96. doi:10.1016/S0925-5273(99)00114-0

Hills, P., & Argyle, M. (2002). The Oxford happiness questionnaire: A compact scale for the measurement of psychological well-being. *Personality and Individual Differences*, *33*(7), 1073–1082. doi:10.1016/S0191-8869(01)00213-6

Hilsen, A., & Helvik, T. (2014). The construction of self in social medias, such as Facebook. *AI & Society*, *29*(1), 3–10. doi:10.1007/s00146-012-0426-y

Hirshman, E. (1980). Innovativeness, Novelty Seeking and Consumer Creativity. *The Journal of Consumer Research*, *7*(3), 283–295. doi:10.1086/208816

Ho, C.-W. (2014). Consumer behavior on Facebook: Does consumer participation bring positive consumer evaluation of the brand? *EuroMed Journal of Business*, *9*(3), 252–267. doi:10.1108/EMJB-12-2013-0057

Hodgkinson, G. P., Bown, N. J., Maule, A. J., Glaister, K. W., & Pearman, A. D. (1999). Breaking the frame: An analysis of strategic cognition and decision making under uncertainty. *Strategic Management Journal*, *20*(10), 977–985. doi:10.1002/(SICI)1097-0266(199910)20:10<977::AID-SMJ58>3.0.CO;2-X

Hogg, M. A., & Abrams, D. (1988). *Social identifications: A social psychology of intergroup relations and group processes*. London, UK: Routledge.

Holbrook, M. B., & Havlena, W. J. (1986). The varieties of consumption experience: Comparing two typologies of emotion in consumer behavior. *The Journal of Consumer Research*, 394–404.

Holbrook, M. B., & Hirschman, E. C. (1982). The experiential aspects of consumption: Consumer fantasies, feelings, and fun. *The Journal of Consumer Research*, *9*(2), 132–140. doi:10.1086/208906

Holt, T. (2010). Examining the role of technology in the formation of deviant subcultures. *Social Science Computer Review*, *28*(4), 466–481. doi:10.1177/0894439309351344

Honeycutt, C., & Herring, S. C. (2009). Beyond microblogging: Conversation and collaboration via Twitter. *Proceedings of the 42nd Hawaii International Conference on System Sciences*, 1–10.

Hong, S., & Tam, K. Y. (2006). Understanding the Adoption of Multipurpose Information Appliances: The Case of Mobile Data Services. *Information Systems Research*, *17*(2), 162–179. doi:10.1287/isre.1060.0088

Horton, D., & Wohl, R. (1956). Mass communication and para-social interaction: Observations on intimacy at a distance. *Psychiatry*, *19*(3), 215–229. PMID:13359569

Hu, X., Ha, L., Mo, S. M., & Xu, Y. (2013). Who are the fans of Facebook fan pages? An electronic word-of-mouth communication perspective. A paper presented to the International Conference on Internet Studies, September 7-8, Hong Kong China.

Hudson, S., & Thal, K. (2013). The Impact of social media on the consumer decision process: Implications for tourism marketing. *Journal of Travel & Tourism Marketing*, *30*(1-2), 156–160. doi:10.1080/10548408.2013.751276

Huffaker, D. (2005). The educated blogger: Using weblogs to promote literacy in the classroom. *AACE Journal*, *13*(2), 91–98.

Hu, L. T., & Bentler, P. M. (1999). Cutoff criteria for fit indexes in covariance structure analysis: Conventional criteria versus new alternatives. *Structural Equation Modeling*, *6*(1), 21–55. doi:10.1080/10705519909540118

Hull, G. A., & Stornaiuolo, A. (2010). Literate arts in a global world: Reframing social networking as cosmopolitan practice. *Journal of Adolescent & Adult Literacy*, *54*(2), 85–97. doi:10.1598/JAAL.54.2.1

Hunt, D., Atkin, D., & Krisnan, A. (2012). The influence of computer-mediated communication apprehension on motives for Facebook use. *Journal of Broadcasting & Electronic Media*, *56*(2), 187–202. doi:10.1080/08838151.2012.678717

Hurst, A., Hudson, S. E., & Mankoff, J. (2007). *Dynamic detection of novice vs. skilled use without a task model.* Paper presented at the Proceedings of the SIGCHI conference on Human factors in computing systems, San Jose, California, USA. doi:10.1145/1240624.1240669

Hurt, H. T., Joseph, K., & Cook, C. D. (1977). Scales for the Measurement of Innovativeness. *Human Communication Research*, *4*(1), 58–65. doi:10.1111/j.1468-2958.1977.tb00597.x

Hutchings, D. R., & Stasko, J. (2002a). *New operations for display space management and window management* (Technical Report No. GIT-GVU-02-18). Retrieved from College of Computing / GVU Center, Georgia Institute of Technology website: https://smartech.gatech.edu/handle/1853/3270

Hutchings, D. R., & Stasko, J. (2004). *Revisiting display space management: understanding current practice to inform next-generation design.* Paper presented at the Proceedings of Graphics Interface 2004, London, Ontario, Canada.

Hutchings, D. R., & Stasko, J. (2002b). QuickSpace: new operations for the desktop metaphor. In L. Terveen & D. Wixon (Eds.), *CHI '02 Extended Abstracts on Human Factors in Computing Systems* (pp. 802–803). New York, NY: ACM. doi:10.1145/506443.506605

Hutchings, D. R., & Stasko, J. T. (2004). Shrinking window operations for expanding display space. In CostabileM. F. (Ed.), *Proceedings of the Working Conference on Advanced Visual Interfaces* (pp. 350–353). New York, NY: ACM. doi:10.1145/989863.989922

Hu, Y., Kobourov, S. G., & Veeramoni, S. (2012). Embedding, clustering and coloring for dynamic maps. In *Proceedings of the Pacific Visualization Symposium (PacificVis)* (pp. 33–40). Piscataway, NJ: IEEE. doi:10.1109/PacificVis.2012.6183571

Hyllegard, K. H., Ogle, J., Yan, R., & Reitz, A. R. (2011). An exploratory study of college students' fanning behavior on Facebook. *College Student Journal*, *45*(3), 601–616.

Ichino, J., Isoda, K., Hanai, A., & Ueda, T. (2013, April). Effects of the display angle in museums on user's cognition, behavior, and subjective responses. In *Proceedings of the SIGCHI Conference on Human Factors in Computing Systems* (pp. 2979-2988). ACM. doi:10.1145/2470654.2481413

IJsselsteijn, W., Poels, K., & de Kort, Y. A. W. (2008). The Game Experience Questionnaire: Development of a self-report measure to assess player experiences of digital games. *Eindhoven: TU Eindhoven. FUGA Deliverable D3.3. Technical Report.*

Im, J. H., & van Epps, P. D. (1991). Software piracy and software security in business schools: An ethical perspective. *Database*, *22*(3), 15–21.

Imperva (2011). Monitoring hacker forums. *Research Note, 5*, 1-14.

Indian, M., & Grieve, R. (2014). When Facebook is easier than face-to-face: Social support derived from Facebook in socially anxious individuals. *Personality and Individual Differences*, *59*, 102–106. doi:10.1016/j.paid.2013.11.016

INPES (French National Health Prevention and Education Institute). (2007), "Le tabac tue un fumeur sur deux. L'industrie du tabac compte sur vous pour les remplacer", *Espace Presse*, retrieved April 4, 2015 from www.inpes.sante.fr/70000/cp/07/cp070820.asp

Ishizaka, K., Marshall, S. P., & Conte, J. M. (2001). Individual differences in attentional strategies in multitasking situations. *Human Performance*, *14*(4), 339–358. doi:10.1207/S15327043HUP1404_4

Işık, Ö., Jones, M. C., & Sidorova, A. (2013). Business intelligence success: The roles of BI capabilities and decision environments. *Information & Management*, *50*(1), 13–23. doi:10.1016/j.im.2012.12.001

Jackson, L. A., Von Eye, A., Witt, E. A., Zhao, Y., & Fitzgerald, H. E. (2011). A longitudinal study of the effects of Internet use and videogame playing on academic performance and the roles of gender, race and income in these relationships. Computers in Human Behavior, 27(1), 228-239.

Jackson, M. (2008). *Distracted: The Erosion of Attention and the Coming Dark Age*. Amherst, NY: Prometheus Books.

Jacoby, J. (1971). Personality and innovation proneness. *JMR, Journal of Marketing Research*, *8*(2), 244–247. doi:10.2307/3149771

Jameson, A., Kiefer, J., Müller, C., Großmann-Hutter, B., Wittig, F., & Rummer, R. (2009). Assessment of a user's time pressure and cognitive load on the basis of features of speech. *Resource-Adaptive Cognitive Processes*, 171-204.

Jang, H., Olfman, L., Ko, I., Koh, J., & Kim, K. (2008). The influence of on-line brand community characteristics on community commitment and brand loyalty. *International Journal of Electronic Commerce*, *12*(3), 57–80. doi:10.2753/JEC1086-4415120304

Jarupathirun, S., & Zahedi, F. M. (2007). Exploring the influence of perceptual factors in the success of web-based spatial DSS. *Decision Support Systems*, *43*(3), 933–951. doi:10.1016/j.dss.2005.05.024

Jeanjean, L. (2012). 5 tips to increase your facebook fan base. *Journal of Financial Planning*, *1*(1), 20–21.

Jiang, Y., & de Bruijn, O. (2014). Facebook helps: A case study of cross-cultural social networking and social capital. *Information Communication and Society*, *17*(6), 732–749. doi:10.1080/1369118X.2013.830636

Jimison, H., Pavel, M., McKanna, J., & Pavel, J. (2004). Unobtrusive monitoring of computer interactions to detect cognitive status in elders. *IEEE Transactions on Information Technology in Biomedicine*, *8*(3), 248–252. doi:10.1109/TITB.2004.835539 PMID:15484429

Johnson-Laird, A. (1985). They look good in demos, but windows are a real pain. *Software News*, *42*, 36–37.

Johnson, S. (2006). *Everything Bad is Good for You: How Today's Popular Culture Is Actually Making Us Smarter.* Riverhead Books.

Johnston, D. W., Anastasiades, P., & Wood, C. (1990). The relationship between cardiovascular responses in the laboratory and in the field. *Psychophysiology*, *27*(1), 34–44. doi:10.1111/j.1469-8986.1990.tb02175.x PMID:2339186

Johnston, K., Tanner, M., Lalla, N., & Kawalski, D. (2013). Social capital: The benefit of Facebook 'friends'. *Behaviour & Information Technology*, *32*(1), 24–36. doi:10.1080/0144929X.2010.550063

Joinson, A. N. (2008). 'Looking at', 'Looking up' or 'Keeping up with' people? Motives and uses of Facebook. In Proceedings of the 26th annual SIGCHI conference on human factors in computing systems (pp. 1027–1036). New York: ACM.

Junco, R. (2012). In-class multitasking and academic performance. *Computers in Human Behavior*, *28*(6), 2236–2243. doi:10.1016/j.chb.2012.06.031

Junco, R. (2012a). Too much face and not enough books: The relationship between multiple indices of Facebook use and academic performance. *Computers in Human Behavior*, *28*(1), 187–198. doi:10.1016/j.chb.2011.08.026

Junco, R., Merson, D., & Salter, W. D. (2010). The effect of gender, ethnicity, and income on college students' use of communication technologies. *Cyberpsychology, Behavior, and Social Networking*, *13*(6), 619–627. doi:10.1089/cyber.2009.0357 PMID:21142986

Junglas, I., Abraham, C., & Ives, B. (2009). Mobile technology at the frontlines of patient care: Understanding fit and human drives in utilization decisions and performance. *Decision Support Systems*, *46*(3), 634–647. doi:10.1016/j.dss.2008.11.012

Just, M. A., Carpenter, P. A., Keller, T. A., Emery, L., Zajac, H., & Thulborn, K. (2001). Interdependence of non-overlapping cortical systems in dual cognitive tasks. *NeuroImage*, *14*(2), 417–426. doi:10.1006/nimg.2001.0826 PMID:11467915

Just, M. A., Keller, T. A., & Cynkar, J. A. (2008). A decrease in brain activation associated with driving when listening to someone speak. *Brain Research*, *1205*, 70–80. doi:10.1016/j.brainres.2007.12.075 PMID:18353285

Kahneman, D., & Klein, G. (2009). Conditions for intuitive expertise: A failure to disagree. *The American Psychologist*, *64*(6), 515–526. doi:10.1037/a0016755 PMID:19739881

Kallman, E. A., & Grillo, J. P. (1996). *Ethical decision making and information technology*. New York: McGraw Hill.

Kalyuga, S., & Plass, J. (2009)Evaluating and Managing Cognitive Load in Games. Handbook of research on effective electronic gaming in education, 2, 719-737.

Kamisaka, D., Muramatsu, S., Yokoyama, H., & Iwamoto, T. (2009, July). Operation prediction for context-aware user interfaces of mobile phones. In *Proceedings of the 2009 Ninth Annual International Symposium on Applications and the Internet* (pp.16-22). IEEE. doi:10.1109/SAINT.2009.12

Kandogan, E., & Shneiderman, B. (1997). Elastic windows: evaluation of multi-window operations. In PembertonS. (Ed.), *Proceedings of the ACM SIGCHI Conference on Human Factors in Computing Systems* (pp. 250–257). New York, NY: ACM. doi:10.1145/258549.258720

Kang, K., Tang, L., & Fiore, A. M. (2014). Enhancing consumer–brand relationships on restaurant Facebook fan pages: Maximizing consumer benefits and increasing active participation. *International Journal of Hospitality Management*, *36*, 145–155. doi:10.1016/j.ijhm.2013.08.015

Kang, S., & Jung, J. (2014). Mobile communication for human needs: A comparison of smartphone use between the US and Korea. *Computers in Human Behavior*, *35*, 376–387. doi:10.1016/j.chb.2014.03.024

Kanich, C., Kreibich, C., Levchenko, K., Enright, B., Voelker, G., Paxson, V., & Savage, S. (2009). Spamalytics: An empirical analysis of spam marketing conversion. *Proceedings of the 15th ACM Conference on Computer and Communications Security*. Alexandria, Virginia: ACM doi:10.1145/1562164.1562190

Kaplan, A. M., & Haenlein, M. (2010). Users of the world, unite! The challenges and opportunities of Social Media. *Business Horizons*, *53*(1), 59–68. doi:10.1016/j.bushor.2009.09.003

Karaiskos, D., Tzavellas, E., Balta, G., & Paparrigopoulos, T. (2010). Social network addiction: A new clinical disorder? *European Psychiatry*, *25*, 855. doi:10.1016/S0924-9338(10)70846-4

Karikoski, J., & Soikkeli, T. (2013). Contextual usage patterns in smartphone communication services. *Personal and Ubiquitous Computing*, *17*(3), 491–502. doi:10.1007/s00779-011-0503-0

Katz, E. (1959). Mass communication research and the study of popular culture: An editorial note on a possible future for this journal. *Studies in Public Communication*, *2*(1), 1–6.

Katz, E., Blumler, J. G., & Gurevitch, M. (1973). Uses and Gratifications research. *Public Opinion Quarterly*, *37*(4), 509–523. doi:10.1086/268109

Katz, E., Gurevitch, M., & Haas, H. (1973). On the use of the mass media for important things. *American Sociological Review*, *38*(2), 164–181. doi:10.2307/2094393

Katz, J. E., & Sugiyanma, S. (2005). Mobile phones as fashion statements: The co-creation of mobile communication's public meaning. In R. Ling & P. Pederson (Eds.), *Mobile communications: Renegotiation of the social sphere* (pp. 63–81). London: Springer. doi:10.1007/1-84628-248-9_5

Katz, M. L., & Shapiro, C. (1985). Network externalities, competition and compatibility. *The American Economic Review*, *75*, 424–440. Retrieved from http://www.jstor.org

Kelder, J. C. (2005). Using someone else's data: Problems, pragmatics and provisions. *Forum Qualitative Sozial Forschung*, *6*(1), 39.

Kelly, D. (2008). Adaptive versus learner control in a multiple intelligence learning environment. *Journal of Educational Multimedia and Hypermedia*, *17*(3), 307–336.

Kerawalla, L., Minocha, S., Kirkup, G., & Conole, G. (2009). An empirically grounded framework to guide blogging in higher education. *Journal of Computer Assisted Learning*, *25*(1), 31–42. doi:10.1111/j.1365-2729.2008.00286.x

Khan, U., Dhar, R., & Wertenbroch, K. (2005). A behavioral decision theory perspective on hedonic and utilitarian choice. *Inside consumption: Frontiers of research on consumer motives, goals, and desires*, 144-165.

Khawaja, M. Chen, Fang, Owen, Christine, & Hickey, Gregory. (2009). Cognitive Load Measurement from User's Linguistic Speech Features for Adaptive Interaction Design. In Human-Computer Interaction–INTERACT 2009 (pp. 485-489). Springer Berlin Heidelberg.

Kihyun, K., Gyenung-Min, K., & Eun Sook, K. (2009). Measuring the Compatibility Factors in Mobile Entertainment Service Adoption. *Journal of Computer Information Systems*, (Fall): 141–148.

Kim, J., & Lee, J. (2011). The Facebook paths to happiness: effects of the number of Facebook friends and self-presentation on subjective well-being. *Cyberpsychology, Behavior And Social Networking, 14*(6), 359-364.

Kim, H. J., Kim, M. S., & Nam, Y. (2010). An analysis of self-construals, motivations, Facebook use, and user satisfaction. *Journal of Human-Computer Interaction*, *26*(11-12), 11–12. doi:10.1080/10447318.2010.516726

Kim, J., & Lee, R. J. E. (2011). The Facebook paths to happiness: Effects on the number of Facebook friends and self-presentation on subjective well-being. *Cyberpsychology, Behavior, and Social Networking*, *14*(6), 359–364. doi:10.1089/cyber.2010.0374 PMID:21117983

Kim, T., & Biocca, F. (1997). Telepresence via television: Two dimensions of telepresence may have different connections to memory and persuasion. *Journal of Computer-Mediated Communication*, *3*(2), 1–22.

Kim, W., Jeong, O.-R., & Lee, S.-W. (2010). On social Web sites. *Information Systems*, *35*(2), 215–236. doi:10.1016/j.is.2009.08.003

Kim, Y. M., & Shim, K. Y. (2002). The influence of Internet shopping mall characteristics and user traits on purchase intent. *Irish Marketing Review*, *15*, 25–34.

Kim, Y. S., & Tran, V. L. (2013). Assessing the ripple effects of online opinion leaders with trust and distrust metrics. *Expert Systems with Applications*, *40*(9), 3500–3511. doi:10.1016/j.eswa.2012.12.058

King, W. R., & He, J. (2005). Understanding the role and methods of meta-analysis in IS research. *Communications of the Association for Information Systems*, *16*, 1.

Kishita, N., & Shimada, H. (2011). Effects of acceptance-based coping on task performance and subjective stress. *Journal of Behavior Therapy and Experimental Psychiatry*, *42*(1), 6–12. doi:10.1016/j.jbtep.2010.08.005 PMID:21074000

Klein, G. (1998). *Sources of Power How People Make Decisions*. Cambridge: MIT Press.

Koenigstorfer, J., & Groeppel-Klein, A. (2012). Consumer Acceptance of the Mobile Internet. *Marketing Letters*, *23*(4), 917–928. doi:10.1007/s11002-012-9206-1

Kohavi, R. (1995). A study of cross-validation and bootstrap for accuracy estimation and model selection. In Ijcai (Vol. 14, No. 2, pp. 1137-1145).

Koh, J., Kim, Y. G., & Kim, Y. G. (2003). Sense of virtual community: A conceptual framework and empirical validation. *International Journal of Electronic Commerce*, *8*(2), 75–94.

Konig, C. J., Buhner, M., & Murling, G. (2005).Working memory, fluid intelligence, and attention are predictors of multitasking performance, but polychronicity and extraversion are not. *Human Performance*, *18*(3), 243–266.

Koren, Y., Bell, R., & Volinsky, C. (2009). Matrix Factorization Techniques for Recommender Systems. *Computer*, *42*(8), 30–37. doi:10.1109/MC.2009.263

Kotler, P., & Armstrong, G. (2006). *Principles of Marketing* (11th ed.). Pearson Prentice-hall.

Kourouthanassis, P. E., Giaglis, G. M., & Vrechopoulos, A. P. (2007). Enhancing user experience through pervasive information systems: The case of pervasive retailing. *International Journal of Information Management*, *27*(5), 319–335. doi:10.1016/j.ijinfomgt.2007.04.005

Kozinets, R. V. (2002). The field behind the screen: Using netnography for marketing research in online communities. *JMR, Journal of Marketing Research*, *39*(1), 61–72. doi:10.1509/jmkr.39.1.61.18935

Kramarski, B., & Mevarech, Z. R. (2003). Enhancing mathematical reasoning in the classroom: Effects of cooperative learning and metacognitive training. *American Educational Research Journal*, *40*(1), 281–310. doi:10.3102/00028312040001281

Kramarski, B., & Ritkof, R. (2002). The effects of metacognition and email conversation on learning graphing. *Journal of Computer Assisted Learning*, *18*(1), 33–43. doi:10.1046/j.0266-4909.2001.00205.x

Kramarski, B., & Zeichner, O. (2001). Using technology to enhance mathematical reasoning: Effects of feedback and self regulation learning. *Educational Media International*, *38*(2-3), 77–82. doi:10.1080/09523980110041458

Kramer, A. F. (1991). Physiological metrics of mental workload: A review of recent progress. *Multiple-task performance*, 279-328.

Kraut, R., Patterson, M., Lundmark, V., Kiesler, S., Mukhopadhyay, T., & Scherlis, W. (1998). Internet paradox: A social technology that reduces social involvement and psychological well-being? *The American Psychologist*, *53*(9), 1017–1031. doi:10.1037/0003-066X.53.9.1017 PMID:9841579

Krebs, B. (2006, March 16). Hacking made easy. *Washington Post*. Retrieved from http://www.washingtonpost.com/wp-dyn/content/article/2006/03/16/AR2006031600916.html

Krebs, B. (2008, July 1). Amazon: Hey spammers, get off my cloud! *Washington Post*. Retrieved fromhttp://voices.washingtonpost.com/securityfix/2008/07/amazon_hey_spammers_get_off_my.html

Kshetri, N. (2006). The simple economics of cybercrimes. *IEEE Security and Privacy*, *4*(1), 33–39. doi:10.1109/MSP.2006.27

Kshetri, N. (2009). Positive externality, increasing returns, and the rise in cybercrimes. *Communications of the ACM*, *52*(12), 141. doi:10.1145/1610252.1610288

Kurihara, S., Moriyama, K., & Numao, M. (2013, November). Context-aware application prediction and recommendation in mobile devices. In *Proceedings of the 2013 IEEE/WIC/ACM International Joint Conferences on Web Intelligence (WI) and Intelligent Agent Technologies (IAT)-Volume 01*(pp. 494-500). IEEE Computer Society. doi:10.1109/WI-IAT.2013.69

Kuss, D. J., & Griffiths, M. D. (2011). Online social networking and addiction--a review of the psychological literature. *International Journal of Environmental Research and Public Health*, *8*(9), 3528–3552. doi:10.3390/ijerph8093528 PMID:22016701

Lampe, C., Ellison, N., & Steinfield, C. (2006). A Facebook in the crowd: Social searching vs. social browsing. In *Proceedings of the 2006 20th anniversary conference on computer supported cooperative work* (pp. 167–170). New York: ACM. doi:10.1145/1180875.1180901

Lang, P. J. (1980). Behavioral treatment and bio-behavioral assessment: Computer applications. *Technology in mental health care delivery systems*, 119 -137.

Lang, A. (2001).The limited capacity model of mediated message processing. *Journal of Communication*, , 46–70.

Langley, A. (1999). Strategies for theorizing from process data. *Academy of Management Review*, *24*(4), 691–710.

LaRue, M. E., Mitchell, M. A., Terhorst, L., & Karimi, A. H. (2010). Assessing mobile phone communication utility preferences in social support network. *Telematics and Informatics*, *27*(4), 363–369. doi:10.1016/j.tele.2010.03.002

Lazarus, R. S., & Folkman, S. (1984). *Stress, appraisal, and coping*. New York: Springer.

Le Louarn, P. (1997). La Tendance à Innover des Consommateurs: Analyse Conceptuelle et Proposition d'une Échelle de Mesure. *Recherche et Applications en Marketing*, *12*(1), 3–19. doi:10.1177/076737019701200101

Le Nagard-Assayag, E., & Manceau, D. (2011). *Le Marketing de l'Innovation, de la Création au Lancement de Nouveaux Produits* (2nd ed.). Paris: Dunod.

Lebraty, J. F., & Godé-Sanchez, C. (2010). Information System and Aerial Warfare: Enhancing Decision Making in Extreme Circumstances. In International Council of the Aeronautical Sciences 2010 (pp. 1–10). Nice, France.

Lebraty, J.-F., & Lobre, K. (2013). *Crowdsourcing: One Step Beyond*. Wiley-ISTE. doi:10.1002/9781118760765

Lee, D., Choi, J., Kim, J. H., Noh, S. H., Min, S. L., Cho, Y., & Kim, C. S. (1999, May). On the existence of a spectrum of policies that subsumes the least recently used (LRU) and least frequently used (LFU) policies. *Performance Evaluation Review*, *27*(1), 134–143. doi:10.1145/301464.301487

Lee, E., & MacGregor, J. (1985). Minimizing user search time in menu retrieval systems. *Human Factors: The Journal of the Human Factors and Ergonomics Society*, *27*(2), 157–162.

Lee, F. J., & Taatgen, N. A. (2002). Multi-tasking as Skill Acquisition.*Proceedings of the twenty-fourth annual conference of the cognitive science society*. pp. 572-577. Mahwah, NJ: Erlbaum.

Lee, G., Lee, J., & Kwon, S. (2011). Use of social-networking sites and subjective well-being: A study in South Korea. *Cyberpsychology, Behavior, and Social Networking*, *14*(3), 151–155. doi:10.1089/cyber.2009.0382 PMID:20649450

Lee, H. (2007). An empirical investigation of travel members' participation behaviors in the online travel community. *Journal of Tourism and Leisure Research*, *19*(2), 355–371.

Lee, J., Lin, L., & Robertson, T. (2011). The impact of media multitasking on learning. *Learning, Media and Technology*, *36*(3), 1–11.

Lee, K. M., & Peng, W. (2006). What Do We Know About Social and Psychological Effects of Computer Games? A Comprehensive Review of Current Literature. In P. Vorderer & J. Bryant (Eds.), *Playing video games. Motives, responses, and consequences* (pp. 325–346). Mahwah, NJ: Lawrence Erlbaum Associates.

Lee, L. (2009). Promoting intercultural exchanges with blogs and podcasting: A study of Spanish-American tele-collaboration. *Computer Assisted Language Learning*, *22*(5), 425–443. doi:10.1080/09588220903345184

Lee, S. K., Cho, Y. H., & Kim, S. H. (2010). Collaborative filtering with ordinal scale-based implicit ratings for mobile music recommendations. *Information Sciences*, *180*(11), 2142–2155. doi:10.1016/j.ins.2010.02.004

Lehmann, T., Hähnlein, I., & Ifenthaler, D. (2014). Cognitive, metacognitive and motivational perspectives on preflection in self-regulated online learning. *Computers in Human Behavior*, *32*, 313–323. doi:10.1016/j.chb.2013.07.051

Lenhart, A. (2013). *Young adults, mobile phones and social media: Technology and the transition to adulthood*. Retrieved June 10, 2013, from http://www.pewinternet.org/2013/05/07/young-adults-mobile-phones-and-social-media-technology-and-the-transition-to-adulthood/

Lessiter, J., Freeman, J., Keogh, E., & Davidoff, J. (2001). A Cross-Media Presence Questionnaire: The ITC-Sense of Presence Inventory. *Presence (Cambridge, Mass.)*, *10*(3), 282–297. doi:10.1162/105474601300343612

Le-tian, X., & Yuan-shen, Z. (2010). Study on Statistical Experiment of Markov Chain Prediction Methods. *Chinese Journal of Engineering Mathematics*, *27*(2), 313–320.

Leung, L. (2007). Unwillingness-to-communicate and college students' motives in SMS mobile messaging. *Telematics and Informatics*, *24*(2), 115–129. doi:10.1016/j.tele.2006.01.002

Leung, L., & Wei, R. (2000). More than just talk on the move: Uses and gratifications of the cellular phone. *Journalism & Mass Communication Quarterly*, *77*(2), 308–320. doi:10.1177/107769900007700206

Leung, R., Schuckert, M., & Yeung, E. (2013). Attracting user social media engagement: A Study of Three Budget Airlines Facebook Pages. In L. Cantoni & Z. Xiang (Eds.), *Information and Communication Technologies in Tourism 2013* (pp. 195–207). Wien: Springer.A. doi:10.1007/978-3-642-36309-2_17

Lewis, C. C., & George, J. F. (2008). Cross-cultural deception in social networking sites and face-to-face communication. *Computers in Human Behavior*, *24*(6), 2945–2964. doi:10.1016/j.chb.2008.05.002

Lewis, J., & West, A. (2009). 'Friending': London-based undergraduates' experience of Facebook. *New Media & Society*, *11*(7), 1209–1229. doi:10.1177/1461444809342058

Li, C., & Bernoff, J. (2008). *Groundswell: winning in a world transformed by social technologies*. Boston, MA: Harvard Business School Press.

Li, J., & Chignell, M. (2010). Birds of a feather: How personality influences blog writing and reading. *International Journal of Human-Computer Studies*, *68*(9), 589–602. doi:10.1016/j.ijhcs.2010.04.001

Lin, L. (2009). Breadth-biased versus focused cognitive control in media multitasking behaviors. *Proceedings of the National Academy of Sciences (PNAS) USA 106*: 15521-15522. doi:10.1073/pnas.0908642106

Lin, C. P., & Bhattacherjee, A. (2008). Elucidating individual intention to use interactive information technologies: The role of network externalities. *International Journal of Electronic Commerce*, *13*(1), 85–108. doi:10.2753/JEC1086-4415130103

Lin, C.-C., Lee, Y.-Y., & Yen, H.-C. (2011). Mental map preserving graph drawing using simulated annealing. *Information Sciences*, *181*(19), 4253–4272. doi:10.1016/j.ins.2011.06.005

Ling, R. (2000). "We will be reached": The use of mobile telephony among Norwegian youth. *Information Technology & People*, *13*(2), 102–120. doi:10.1108/09593840010339844

Ling, R. (2004). *The mobile connection: The cell phone's impact on society*. San Francisco: Morgan Kaufmann.

Ling, R. (2008). *New tech, new ties: How mobile communication is reshaping social cohesion*. Cambridge, MA: MIT Press.

Lin, K. Y., & Lu, H. P. (2011). Why people use social networking sites: An empirical study integrating network externalities and motivation theory. *Computers in Human Behavior*, *27*(3), 1152–1161. doi:10.1016/j.chb.2010.12.009

Lin, K.-Y., & Lu, H.-P. (2011). Intention to continue using Facebook fan pages from the perspective of social capital theory. *Cyberpsychology, Behavior, and Social Networking*, *14*(10), 565–570. doi:10.1089/cyber.2010.0472 PMID:21381968

Lin, L., & Bigenho, C. (2011). Note-taking and memory in different media environments. *Computers in the Schools*, *28*(3), 200–216. doi:10.1080/07380569.2011.594989

Lin, L., Lee, J., & Robertson, T. (2011). Reading while watching video: The effect of video content on reading comprehension and media multitasking ability. *Journal of Educational Computing Research*, *45*(2), 183–201. doi:10.2190/EC.45.2.d

Lin, L., Robertson, A., & Lee, J. (2012). The impact of outcome intentions on reading and multitasking performances. *Journal of Educational Technology Development and Exchange*, *5*(1), 77–94.

Lin, L., Robertson, T., & Lee, J. (2009). Reading performances between novices and experts in different media multitasking environments. *Computers in the Schools*, *26*(3), 169–186. doi:10.1080/07380560903095162

Lin, T., Hu, W. H., Imamiya, A., & Omata, M. (2006). Large display size enhances user experience in 3D games. *Smart Graphics. Proceedings*, *4073*, 257–262.

Lin, T., Imamiya, A., & Mao, X. (2008). Using multiple data sources to get closer insights into user cost and task performance. *Interacting with Computers*, *20*(3), 364–374. doi:10.1016/j.intcom.2007.12.002

Lipshitz, R., Klein, G., & Orasanu, J. (2001). Focus Article. *Taking Stock of Naturalistic Decision Making*, *352*, 331–352.

Liu, C., & Arnett, K. P. (2000). Exploring the factors associated with web site success in the context of electronic commerce. *Information & Management*, *38*(1), 23–33. doi:10.1016/S0378-7206(00)00049-5

Liu, C., Chapuis, O., Beaudouin-Lafon, M., Lecolinet, E., & Mackay, W. E. (2014). Effects of display size and navigation type on a classification task. In *Proceedings of the 32nd annual ACM conference on Human factors in computing systems* (pp. 4147-4156). ACM. doi:10.1145/2556288.2557020

Liu, L., Maddux, C., & Johnson, L. (2004). Computer attitude and achievement: Is time an intermediate variable? *Journal of Technology and Teacher Education*, *12*(4), 593–607.

Liu, X., Hu, Y., North, S., & Shen, H.-W. (2013). CompactMap: A mental map preserving visual interface for streaming text data. In *Proceedings of the IEEE International Conference on Big Data* (pp. 48–55). Piscataway, NJ: IEEE. doi:10.1109/BigData.2013.6691714

Li, X., & Chen, W. (2014). Facebook or Renren? A comparative study of social networking site use and social capital among Chinese international students in the United States. *Computers in Human Behavior*, *35*, 116–123. doi:10.1016/j.chb.2014.02.012

Li, Y., Ma, S., Zhang, Y., Huang, R., & Kinshuk, . (2013). An improved mix framework for opinion leader identification in online learning communities. *Knowledge-Based Systems*, *43*, 43–51. doi:10.1016/j.knosys.2013.01.005

Lobet-Marris, C. (2003). Mobile phone tribes: Youth and social identity. In L. Fortunati, J. Katz, & R. Riccini (Eds.), *Mediating the human body: Technology, communication and fashion* (pp. 87–92). Mawah, NJ: Lawrence Erlbaum.

Lombard, M., Reich, R. D., Grabe, M. E., Bracken, C. C., & Ditton, T. B. (2000). Presence and Television. *Human Communication Research*, *26*(1), 75–98.

Lometti, G. E., Reeves, B., & Bybee, C. R. (1977). Investigating the assumptions of uses and gratifications research. *Communication Review*, *4*(3), 321–338.

Long, J. (2005). *Google hacking for penetration testers*. Rockland, MA: Syngress Publishing.

Lüders, P., Ernst, R., & Stille, S. (1995). An approach to automatic display layout using combinatorial optimization algorithms. *Software Focus*, *25*(11), 1183–1202.

Luo, X., & Liao, Q. (2009). Ransomware: A new cyber hijacking threat to enterprises. In J. Gupta & S. Sharma (Eds.), Handbook of Research on Information Security and Assurance (1–6), Hershey, PA: IGI Global. doi:10.4018/978-1-59904-855-0.ch001

Maenpaa, K., Kanto, A., Kuusela, H., & Paul, P. (2006). More hedonic versus less hedonic consumption behaviour in advanced internet bank services. *Journal of Financial Services Marketing*, *11*(1), 4–16. doi:10.1057/palgrave.fsm.4760004

Mandryk, R. L., & Atkins, M. S. (2007). A fuzzy physiological approach for continuously modeling emotion during interaction with play technologies. *International Journal of Human-Computer Studies*, *65*(4), 329–347. doi:10.1016/j.ijhcs.2006.11.011

Mandryk, R. L., Atkins, M. S., & Inkpen, K. M. (2006). A continuous and objective evaluation of emotional experience with interactive play environments.*Proceedings of the SIGCHI conference on Human factors in computing systems*, 1027-1036. doi:10.1145/1124772.1124926

Mansson, D. H., & Myers, S. A. (2011). An initial examination of college students' expressions of affection through Facebook. *The Southern Communication Journal*, *76*(2), 155–168. doi:10.1080/10417940903317710

Manthiou, A., Tang, L., & Bosselman, R. (2014). Reason and reaction: The dual route of the decision-making process on Facebook fan pages. *Electronic Markets*, 1–39.

Mäntylä, T. (2013). Gender differences in multitasking reflect spatial ability. *Psychological Science*, *24*(4), 514–520. doi:10.1177/0956797612459660 PMID:23462757

Mäntymäki, M., & Islam, A. (2014). Social virtual world continuance among teens: Uncovering the moderating role of perceived aggregate network exposure. *Behaviour & Information Technology*, *33*(5), 536–547. doi:10.1080/0144929X.2013.872190

Martindale, C. (1999). Biological bases of creativity. In R. J. Sternberg (Ed.), Handbook of creativity (137–152). Cambridge, MA: Cambridge University Press.

Martínez, H. P., & Yannakakis, Georgios N. (2010). *Genetic search feature selection for affective modeling: a case study on reported preferences*. Paper presented at the Proceedings of the 3rd international workshop on Affective interaction in natural environments, Firenze, Italy. doi:10.1145/1877826.1877832

Martinez, H. P., Jhala, A., & Yannakakis, G. N. (2009, 10-12 Sept. 2009). *Analyzing the impact of camera viewpoint on player psychophysiology*. Paper presented at the Affective Computing and Intelligent Interaction and Workshops, 2009. ACII 2009. 3rd International Conference on.

Marwick, A., & Boyd, D. (2011). I Tweet Honestly, I tweet passionately: Twitter users, context collapse, and the imagined audience. *New Media & Society*, *13*(1), 114–133. doi:10.1177/1461444810365313

Mary, C., George, R., Brian, M., Greg, S., Daniel, R., & Desney, T. (2006). *Large display research overview*. Paper presented at the CHI '06 extended abstracts on Human factors in computing systems, Montreal, quebec, Canada.

Mashable. [Web log post]. (n.d.). *Smartphone*. Retrieved September 14, 2012, from http://mashable.com/follow/topics/smartphone/

Mashima, D., Kobourov, S., & Hu, Y. (2012). Visualizing Dynamic Data with Maps. *IEEE Transactions on Visualization and Computer Graphics*, *18*(9), 1424–1437. doi:10.1109/TVCG.2011.288 PMID:22184261

Massetti, B. (1996). An empirical Examiniation of the Value of Creativity Support Systems on Idea Generation. *Management Information Systems Quarterly*, *20*(1), 83–97. doi:10.2307/249543

Masson, J. (2010). Effets de la Modification d'un Attribut Constitutif d'un Produit Alimentaire sur son Adoption par les Consommateurs: Le cas du Vin à teneur réduite en Alcool. Unpublished Doctoral Dissertation, Montpellier Supagro, Centre International d'Etudes Supérieures en Sciences Agronomiques, France.

Mathieson, K. (1991). Predicting User Intentions: Comparing the Technology Acceptance Model with the Theory of Planned Behavior. *Information Systems Research*, *2*(3), 173–191. doi:10.1287/isre.2.3.173

Matsuba, M. (2006). Searching for self and relationships online. *Cyberpsychology & Behavior: The Impact Of The Internet, Multimedia And Virtual Reality On Behavior. Society*, *9*(3), 275–284.

Mauri, M., Cipresso, P., Balgera, A., Villamira, M., & Riva, G. (2011). Why is Facebook so successful? Psychophysiological measures describe a core flow state while using Facebook. *Cyberpsychology, Behavior, and Social Networking*, *14*(12), 723–731. doi:10.1089/cyber.2010.0377 PMID:21879884

Mayer, R. E. (1993). Illustrations that instruct. In R. Glaser (Ed.), Advances in instructional psychology, 4, 253–284. Hillsdale, NJ: Lawrence Erlbaum Associates.

Mayer, R. E. (2001). *Multimedia learning*. Cambridge, MA: Cambridge University Press. doi:10.1017/CBO9781139164603

Mayer, R. E., Fennell, S., Farmer, L., & Campbell, J. (2004). A personalization effect in multimedia learning: Students learn better when words are in conversational style rather than formal style. *Journal of Educational Psychology*, *96*(2), 389–395. doi:10.1037/0022-0663.96.2.389

Mazaeva, N., Ntuen, C., & Lebby, G. (2001, 25-28 July 2001). *Self-Organizing Map (SOM) model for mental workload classification*. Paper presented at the IFSA World Congress and 20th NAFIPS International Conference, 2001. Joint 9th.

McAfee. (2010). *Threats report: Third quarter 2010*. Retrieved from http://www.mcafee.com/us/resources/reports/rp-quarterly-threat-q3-2010.pdf

McArdle, M. (2014). Thank You For E-Smoking, Bloomberg Businessweek, 00077135, October 2, Edition 4366.

McGarry, C. (2013). Control who can find you on Facebook. *PC World*, *31*(11), 178–179.

McKena, K. Y. A., & Bargh, J. (2000). Plan 9 from cyberspace: The implications of the Internet for personality and social psychology. *Personality and Social Psychology Review*, *4*(1), 57–75. doi:10.1207/S15327957PSPR0401_6

McLuhan, M. (2002). Understanding Media: The Extensions of Man (10th, ed). Cambridge, MA: MIT Press.

McNally, K., O'Mahony, M. P., Coyle, M., Briggs, P., & Smyth, B. (2011). A case study of collaboration and reputation in social web search. *ACM Transactions on Intelligent Systems and Technology*, *3*(1), 4. doi:10.1145/2036264.2036268

McQuail, D., Blumler, J. G., & Brown, J. R. (1972). The television audience: A revised perspective. In D. McQuail (Ed.), *Sociology of mass communications* (pp. 135–164). Harmondsworth, UK: Penguin.

Meer, H., Arvanitis, N., & Slaviero, M. (2009). *Clobbering the cloud. Black Hat USA 2009*. Las Vegas, NV: Techweb.

Meerkerk, G.-J., Van Den Eijnden, R. J. J. M., Vermulst, A. A., & Garretsen, H. F. L. (2009). The compulsive internet use scale (CIUS): Some psychometric properties. *Cyberpsychology & Behavior*, *12*(1), 1–6. doi:10.1089/cpb.2008.0181 PMID:19072079

Mehdizadeh, S. (2010). Self-presentation 2.0: Narcissism and self-esteem on Facebook. *Cyberpsychology, Behavior, and Social Networking*, *13*(4), 357–364. doi:10.1089/cyber.2009.0257 PMID:20712493

Merrill, M. (2011a). HHS mHealth initiatives target smoking cessation, *Healthcare IT News*, retrieved April 4, 2015 from www.healthcareitnews.com/news/hhs-mhealth-initiatives-target-smoking-cessation

Merrill, M. (2011b). Study finds potential in social media tools for smoking cessation, *Healthcare IT News*, retrieved April 4, 2015 from www.healthcareitnews.com/news/study-find-potential-social-media-health-tools-smoking-cessation

Mesch, G. S., & Talmud, I. (2010). *Wired Youth: The social world of adolescence in the information age*. London, New York: Routledge.

Meyer, D. E., & Kieras, D. E. (1997). A Computational Theory of Executive Cognitive Processes and Multiple-task Performance: Part 1. Basic Mechanisms. *Psychological Review*, *104*, 229–233. PMID:9009880

Michikyan, M., Subrahmanyam, K., & Dennis, J. (2014). Can you tell who I am? Neuroticism, extraversion, and online self-presentation among young adults. *Computers in Human Behavior*, *33*, 179–183. doi:10.1016/j.chb.2014.01.010

Midgley, D. F., & Dowling, G. R. (1978). Innovativeness: The Concept and Its Measurement. *The Journal of Consumer Research*, *4*(4), 229–242. doi:10.1086/208701

Midgley, D. F., & Dowling, G. R. (1993). A Longitudinal Study of Product Form Innovation: The Interaction between Predispositions and Social Messages. *The Journal of Consumer Research*, *19*(4), 611–625. doi:10.1086/209326

Mikalef, P., Giannakos, M., & Pateli, A. (2013). Shopping and word-of-mouth intentions on social media. *Journal of Theoretical and Applied Electronic Commerce Research*, *8*(1), 17–34. doi:10.4067/S0718-18762013000100003

Minocha, S., & Roberts, D. (2008). Social, usability, and pedagogical factors influencing students' learning experiences with wikis and blogs. *Pragmatics & Cognition*, *16*(2), 272–306.

Mintzberg, H. (1994). The Fall and Rise of Strategic Planning. *Harvard Business Review*, *72*(1), 107–114. doi:10.1016/0024-6301(94)90173-2

Mintzberg, H., Raisinghani, D., & Théorêt, A. (1976). The structure of un-structured decision processes. *Administrative Science Quarterly*, *21*(2), 246–275. doi:10.2307/2392045

Mislove, A., Marcon, M., Gummadi, K. P., Druschel, P., & Bhattacharjee, B. (2007). Measurement and analysis of online social networks. In *Proceedings of the 7th ACM SIGCOMM conference on Internet measurement* (pp. 29-42). ACM. doi:10.1145/1298306.1298311

Misue, K., Eades, P., Lai, W., & Sugiyama, K. (1995). Layout adjustment and the mental map. *Journal of Visual Languages and Computing*, *6*(2), 183–210. doi:10.1006/jvlc.1995.1010

Mitchell, J., & Shneiderman, B. (1989). Dynamic versus static menus: An exploratory comparison. *ACM SIGCHI Bulletin*, *20*(4), 33–37. doi:10.1145/67243.67247

Miyake, S. (2001). Multivariate workload evaluation combining physiological and subjective measures. *International Journal of Psychophysiology*, *40*(3), 233–238. doi:10.1016/S0167-8760(00)00191-4 PMID:11228350

Mogford, R. X., Guttman, J. A., Morrow, S. L., & Kopardekar, P. (1995). *The Complexity Construct in Air Traffic Control: A Review and Synthesis of the Literature*. Atlantic City: FAA Technical Center.

Moller, A. C., Ryan, R. M., & Deci, E. L. (2006). Self-Determination Theory and Public Policy: Improving the Quality of Consumer Decisions without Using Coercion. *Journal of Public Policy & Marketing*, *25*(1), 104–116. doi:10.1509/jppm.25.1.104

Monsell, S. (2003). Task switching. *Trends in Cognitive Sciences*, *7*(3), 134–140. doi:10.1016/S1364-6613(03)00028-7 PMID:12639695

Morgan, R. M., & Hunt, S. D. (1994). The commitment–trust theory of relationship marketing. *Journal of Marketing*, *58*(3), 20–38. doi:10.2307/1252308

Mount, M., & Garcia Martinez, M. (2014). Social media: A tool for open innovation. *California Management Review*, *56*(4), 124–143. doi:10.1525/cmr.2014.56.4.124

Muniz, A. M. Jr, & O'guinn, T. C. (2001). Brand community. *The Journal of Consumer Research*, *27*(4), 412–432. doi:10.1086/319618

Muñiz, A. Jr, & Schau, H. J. (2005). Religiosity in the abandoned Apple Newton brand community. *The Journal of Consumer Research*, *31*(5), 737–747. doi:10.1086/426607

Murray, L., & Hourigan, T. (2008). Blogs for specific purposes: Expressivist or socio-cognitivist approach? *Journal of Eurocall*, *20*, 82–97.

Myers, B. A. (1988). A taxonomy of window manager user interfaces. *IEEE Computer Graphics and Applications*, *8*(5), 65–84. doi:10.1109/38.7762

Myers, J. H., & Robertson, T. S. (1972). Dimensions of opinion leadership. *JMR, Journal of Marketing Research*, *9*(1), 41–46. doi:10.2307/3149604

Nachreiner, F. (1995). Standards for ergonomics principles relating to the design of work systems and to mental workload. *Applied Ergonomics*, *26*(4), 259–263. doi:10.1016/0003-6870(95)00029-C PMID:15677027

Nam, C., & Zellner, R. D. (2011). The relative effects of positive interdependence and group processing on student achievement and attitude in online cooperative learning. *Computers & Education*, *56*(3), 680–688. doi:10.1016/j.compedu.2010.10.010

National Public Radio (NPR). (October 2, 2008). Think You're Multitasking? Think Again. *Morning Edition.* Retrieved on May 17, 2011 from http://www.npr.org/templates/story/story.php?storyId=95256794

Negash, S., Ryan, T., & Igbaria, M. (2003). Quality and effectiveness in Web-based customer support systems. *Information & Management*, *40*(4), 757–768. doi:10.1016/S0378-7206(02)00101-5

Newell, A., & Shaw, J. C. (1972). The process of creative thinking. In A. Newell & H. A. Simon (Eds.), *Human problem solving* (pp. 144–174). New Jersey, USA: Practice Hall, Englewood Cliff.

Newell, A., & Simon, H. A. (1976). Computer science as empirical inquiry: Symbols and search. *Communications of the ACM*, *19*(3), 113–126. doi:10.1145/360018.360022

Nielsen Company (The). (2012). *State of the media: Social media report*. Retrieved from The Nielsen Company website: http://blog.nielsen.com/nielsenwire/social/2012/

Norman, K. L. (1991). *The psychology of menu selection: Designing cognitive control at the human/computer interface*. Intellect Books.

Núñez-Valdéz, E. R., Lovelle, J. M. C., Martínez, O. S., García-Díaz, V., de Pablos, P. O., & Marín, C. E. M. (2012). Implicit feedback techniques on recommender systems applied to electronic books. *Computers in Human Behavior*, *28*(4), 1186–1193. doi:10.1016/j.chb.2012.02.001

Nunnally, J. C. (1978). *Psychometric theory*. New York, NY: McGraw- Hill.

O'Connor, A. J. (2013). The Power of Popularity An Empirical Study of the Relationship Between Social Media Fan Counts and Brand Company Stock Prices. *Social Science Computer Review*, *31*(2), 229–235. doi:10.1177/0894439312448037

O'Keefe, P. (2008). *Managing online forums*. New York, NY: AMACOM.

Oblinger, D. G., & Oblinger, J. L. (Eds.). (2005). *Educating the Net Generation*. Boulder, CO: EDUCAUSE. Retrieved on May 17, 2011 from http://www.educause.edu/educatingthenetgen

Ocepek, U., Bosnić, Z., Šerbec, N. I., & Rugelj, J. (2013). Exploring the relation between learning style models and preferred multimedia types. *Computers & Education*, *69*, 343–355. doi:10.1016/j.compedu.2013.07.029

Olander, M. V. (2008). Painting the voice: Weblogs and writing instruction in the high school classroom. ProQuest Information & Learning). *Dissertation Abstracts International. A, The Humanities and Social Sciences*, *68*, 4265–4265.

Olesen, J. K., Yannakakis, G. N., & Hallam, J. (2008, 15-18 Dec. 2008). *Real-time challenge balance in an RTS game using rtNEAT.* Paper presented at the Computational Intelligence and Games, 2008. CIG '08. IEEE Symposium On.

Olmsead, K., Mitchell, A., & Rosenstiel, T. (2011). Navigating news online: Here people go, how they get there and what lures them away. *Pew Research Journalism Project*. Retrieved from http://www.journalism.org/2011/05/09/navigating-news-online/

Olsen, B. I., Laeng, B., Kristiansen, K. A., & Hartvigsen, G. (2013). Size does matter: Women mentally rotate large objects faster than men. *Scandinavian Journal of Psychology*, *54*(3), 196–204. doi:10.1111/sjop.12043 PMID:23448540

Ong, W. J. (2002). *An Ong Reader: Challenges for Further Inquiry* (T. J. Farrell & P. A. Soukup, Eds.). Cresskill, NJ: Hampton Press Inc.

Ophir, E., Nass, C. I., & Wagner, A. D. (2009). Cognitive control in media multitaskers. *Proceedings of the National Academy of Sciences of the United States of America*, *106*(37), 15583–15587. doi:10.1073/pnas.0903620106 PMID:19706386

Ostlund, L. E. (1974). Perceived innovation attributes as predictors of innovativeness. *The Journal of Consumer Research*, *1*(2), 23–29. doi:10.1086/208587

Osyczka, A. (1985). Multicriteria optimization for engineering design. *Design Optimization*, 193-227.

Paas, F. G. (1992). Training strategies for attaining transfer of problem-solving skill in statistics: A cognitive-load approach. *Journal of Educational Psychology*, *84*(4), 429–434. doi:10.1037/0022-0663.84.4.429

Paas, F. G. W. C., & Van Merriënboer, J. J. G. (1994). Instructional control of cognitive load in the training of complex cognitive tasks. *Educational Psychology Review*, *6*(4), 351–371. doi:10.1007/BF02213420

Paas, F., Tuovinen, J. E., Tabbers, H., & Van Gerven, P. W. M. (2003). Cognitive load measurement as a means to advance cognitive load theory. *Educational Psychologist*, *38*(1), 63–71. doi:10.1207/S15326985EP3801_8

Pagani, M. (2007). A Vicarious Innovativeness Scale for 3G Mobile Services: Integrating the Domain Specific Innovativeness Scale with Psychological and Rational Indicators. *Technology Analysis and Strategic Management*, *19*(6), 709–728. doi:10.1080/09537320701711207

Palfrey, J., & Gasser, U. (2008). *Born digital: Understanding the first generation of digital natives*. New York: Basic Books.

Panda Security. (2009). *Studying or hacking? Today's adolescents could be the hackers of the future*. Retrieved from http://www.pitchengine.com/preview-release.php?id=11537

Papacharissi, Z. (2010). Uses and gratifications. In *D. Stacks & SalwenAn Integrated Approach to Communication Theory and Research. D. Stacks & Salwen* (2nd ed.). New York: Routledge.

Papacharissi, Z., & Rubin, A. (2000). Predictors of Internet usage. *Journal of Broadcasting & Electronic Media*, *44*(2), 175–196. doi:10.1207/s15506878jobem4402_2

Parasuraman, R., Mouloua, M., & Molloy, R. (1996). Effects of adaptive task allocation on monitoring of automated systems. *Human Factors: The Journal of the Human Factors and Ergonomics Society*, *38*(4), 665–679. doi:10.1518/001872096778827279 PMID:11536753

Paris, C. M., Lee, W., & Seery, P. (2010). The role of social media in promoting special events: Acceptance of Facebook Events. *Information and Communication Technologies in Tourism*, *14*, 531–541.

Parker, J. G., & Asher, S. R. (1993). Friendship and friendship quality in middle childhood: Links with peer group acceptance and feelings of loneliness and social dissatisfaction. *Developmental Psychology*, *29*(4), 611–621. doi:10.1037/0012-1649.29.4.611

Parkes, A. (2013). The effect of task–individual–technology fit on user attitude and performance: An experimental investigation. *Decision Support Systems*, *54*(2), 997–1009. doi:10.1016/j.dss.2012.10.025

Parkinson, S. R., Sisson, N., & Snowberry, K. (1985). Organization of broad computer menu displays. *International Journal of Man-Machine Studies*, *23*(6), 689–697. doi:10.1016/S0020-7373(85)80063-8

Park, N., Kee, K. F., & Valenzuela, S. (2009). Being immersed in social networking environment: Facebook groups, uses and gratifications, and social outcomes. *Cyberpsychology & Behavior*, *12*(6), 729–733. doi:10.1089/cpb.2009.0003 PMID:19619037

Park, N., & Lee, S. (2014). College students' motivations for Facebook use and psychological outcomes. *Journal of Broadcasting & Electronic Media*, *58*(4), 601–620. doi:10.1080/08838151.2014.966355

Parks, M. R., & Floyd, K. (1996). Making friends in cyberspace. *Journal of Communication*, *46*(1), 80–97. doi:10.1111/j.1460-2466.1996.tb01462.x

Partala, T., & Surakka, V. (2003). Pupil size variation as an indication of affective processing. *International Journal of Human-Computer Studies*, *59*(1-2), 185–198. doi:10.1016/S1071-5819(03)00017-X

Patton, M. (2001). *Qualitative Research and Evaluation Methods* (3rd ed.). London: Sage.

Pedersen, C., Togelius, J., & Yannakakis, G. N. (2009). *Modeling player experience in super mario bros.* In Computational Intelligence and Games, 2009. CIG 2009. IEEE Symposium on (pp. 132-139). IEEE. doi:10.1109/CIG.2009.5286482

Pempek, T. A., Yermolayeva, Y. A., & Calvert, S. L. (2009). College students' social networking experiences on Facebook. *Journal of Applied Developmental Psychology*, *30*(3), 227–238. doi:10.1016/j.appdev.2008.12.010

Pew Internet Research Project. (2012). *A comparison of results from surveys by the Pew Research Center and Google Consumer Surveys*. Retrieved June 10, 2014, from http://www.people-press.org/2012/11/07/a-comparison-of-results-from-surveys-by-the-pew-research-center-and-google-consumer-surveys/

Pew Internet Research Project. (2012). *Privacy management on social media sites*. Retrieved June 10, 2014 from http://www.pewinternet.org/2012/02/24/main-findings-12/

Pew Internet Research Project. (2012). *Teens and mobile app privacy*. Retrieved June 10, 2014 from http://www.pewinternet.org/2013/08/22/teens-and-mobile-apps-privacy/

Pew Research Center. (2010). *Social media & mobile Internet use among teens and young adults*. Retrieved March 2, 2010, from http://www.pewinternet.org/Reports/2010/Social- Media-and-Young-Adults.aspx

Pew Research Center. (2012). *The Demographics of Social Media Users*. Retrieved from http://www.pewinternet.org/Reports/2013/Social-media-users.aspx

Pew Research Center. (2013). Pew Research points out different (social) strokes for different folks. *Advertising Age*, *84*(7), 5.

Pew Research Internet Project. (2013). *Frequency of social media use*. Retrieved June 10, 2014 from http://www.pewinternet.org/2013/12/30/frequency-of-social-media-use/

Phukan, S. (2002). *IT ethics in the internet age: New dimensions. Proc. Informing Science & IT Education Conf. (InSITE)*. Santa Rosa, CA: Informing Science Press.

Picard, R. W., Vyzas, E., & Healey, J. (2001). Toward machine emotional intelligence: Analysis of affectivephysiological state. *IEEE Transactions on Pattern Analysis and Machine Intelligence*, *23*(10), 1175–1191. doi:10.1109/34.954607

Pilgrim, J. von. (2007). Mental map and model driven development. *Electronic Communications of the EASST*, *7*.

Plass, J. L., Moreno, R., & Brünken, R. (2010). *Cognitive load theory*. Cambridge, New York: Cambridge University Press. doi:10.1017/CBO9780511844744

Poe, M. (2014). 'The issues are complicated' with e-cigarettes, *The Times West Virginian*, March 9.

Poldrack, R. A., & Foerde, K. (2007). Category learning and the memory systems debate. *Neuroscience and Biobehavioral Reviews*, *32*(2), 197–205. doi:10.1016/j.neubiorev.2007.07.007 PMID:17869339

Powell, J. (2009). *33 million people in the room: how to create, influence, and run a successful business with social networking*. Upper Saddle River, NJ: FT Press.

Preece, J. (2000). *Online communities: Designing usability, supporting sociability*. New York, NY: John Wiley.

Prensky, M. (2001).Digital natives, digital immigrants. On the horizon. MCB University Press, Vol. 9 No. 5.

Prensky, M. (2001). *Digital Game-based Learning*. The McGraw-Hill Companies.

Protalinski, E. (2014). Facebook passes 1.23 billion monthly active users, 945 million mobile users, and 757 million daily users. Retrieved June 26, 2014, from http://thenextweb.com/facebook/2014/01/29/facebook-passes-1-23-billion- monthly-active-users-945-million-mobile-users-757-million-daily-users/

Puntambekar, S., & Kolodner, J. L. (2005). Toward implementing distributed scaffolding: Helping students learn science from design. *Journal of Research in Science Teaching*, *42*(2), 185–217. doi:10.1002/tea.20048

Purcell, K., Rainie, L., Mitchell, A., Rosenstiel, T., & Olmstead, K. (2010). Understanding the participatory news consumer. *Pew Internet and American Life Project*. Retrieved from http://www.pewinternet.org/Reports/2010/Online-News.aspx

Purchase, H., Hoggan, E., & Görg, C. (2007). How Important Is the "Mental Map"? – An Empirical Investigation of a Dynamic Graph Layout Algorithm. In Graph drawing (pp. 184-195). Springer Berlin /Heidelberg.

Purchase, H. C. (1998). Performance of layout algorithms: Comprehension, not computation. *Journal of Visual Languages and Computing*, *9*(6), 647–657. doi:10.1006/jvlc.1998.0093

Purchase, H. C., & Samra, A. (2008). Extremes are better: investigating mental map preservation in dynamic graphs. In G. Stapleton, J. Howse, & J. Lee (Eds.), Lecture Notes in Computer Science: Vol. 5223. *Diagrammatic representation and inference* (pp. 60–73). Berlin, Germany: Springer Berlin Heidelberg. doi:10.1007/978-3-540-87730-1_9

Putnam, R. (2000). *Bowling alone: Collapse and revival of American community*. New York: Simon & Schuster. doi:10.1145/358916.361990

PwC (PricewaterhouseCoopers) report June 2013, Socio-economic impact of m-health, An assessment report for the European Union, Retrieved November 2014, from http://www.pwc.fr/socio-economic-impact-of-mhealth-an-assessment-report-for-the-european-union.html

Qin, Han-shu, Wei, Yan, & Zeng, Shao-hua. (2012). Parameters optimization for SVM classification based on adaptive NGA. *International Journal of Advancements in Computing Technology, 4*(18), 490-497. doi: 18.5810.4156/ijact.vol4.issue

Qualman, E. (2009). *Socialnomics: how social media transforms the way we live and do business*. Hoboken, NJ: John Wiley & Sons, Inc.

Quan-Haase, A., & Young, A. L. (2010). Uses and gratifications of social media: A comparison of Facebook and Instant Messaging. *Technology & Society*, *30*(5), 350–361.

Qu, H., & Lee, H. (2011). Travelers' social identification and membership behaviors in online travel community. *Tourism Management*, *32*(6), 1262–1270. doi:10.1016/j.tourman.2010.12.002

Quinlan, J. R. (1993). *C4.5: programs for machine learning*. San Mateo, Calif.: Morgan Kaufmann Publishers.

Raacke, J., & Bonds-Raacke, J. (2008). MySpace and Facebook: Applying the uses and gratifications theory to exploring friend-networking sites. *Cyberpsychology & Behavior*, *11*(2), 169–174. doi:10.1089/cpb.2007.0056 PMID:18422409

Rafaeli, S. (1998). Interactivity. In R. P. Hawkins, J. M. Wiemann, & S. Pingree (Eds.), *Advancing Communication Science: Merging Mass and Interpersonal Processes* (pp. 110–134). Beverly Hills, California: Sage.

Ragan, E. D., Endert, A., Bowman, D. A., & Quek, F. (2012). *How spatial layout, interactivity, and persistent visibility affect learning with large displays*. In *Proceedings of the International Working Conference on Advanced Visual Interfaces* (pp. 91-98). ACM. doi:10.1145/2254556.2254576

Randall, N. (2014). Let your fingers do the talking: Students' social distance as revealed through cell phone contact lists. *Sociological Spectrum*, *34*(2), 136–142. doi:10.1080/02732173.2014.878608

Randel, J. M., Pugh, H. L., & Reed, S. K. (1996). Differences in expert and novice situation awareness in naturalistic decision making. *International Journal of Human-Computer Studies*, *45*(5), 579–597. doi:10.1006/ijhc.1996.0068

Ravaja, N. (2004). Contributions of Psychophysiology to Media Research: Review and Recommendations. *Media Psychology*, *6*(2), 193–235. doi:10.1207/s1532785xmep0602_4

Ravaja, N., Turpeinen, M., Saari, T., Puttonen, S., & Keltikangas-Jarvinen, L. (2008). The psychophysiology of James Bond: Phasic emotional responses to violent video game events. *Emotion (Washington, D.C.)*, *8*(1), 114–120. doi:10.1037/1528-3542.8.1.114 PMID:18266521

Ray, M. B. (2007). *Needs, Motives, and Behaviors in computer-mediated communication: An inductive exploration of social networking websites*. Conference paper—International Communication Association, 2007 Annual Conventioin.

Raymond, E. S., & Steele, G. S. (1991). *The new hacker's dictionary*. Cambridge, MA: MIT Press.

Recio-García, J. A., Quijano, L., & Díaz-Agudo, B. (2013). Including social factors in an argumentative model for Group Decision Support Systems. *Decision Support Systems*, *56*, 48–55. doi:10.1016/j.dss.2013.05.007

Reeves, B., Lang, A., Kim, E. Y., & Tatar, D. (1999). The effects of screen size and message content on attention and arousal. *Media Psychology*, *1*(1), 49–67. doi:10.1207/s1532785xmep0101_4

Reid, G. B., & Nygren, T. E. (1988). The Subjective Workload Assessment Technique: A Scaling Procedure for Measuring Mental Workload. In A. H. Peter & M. Najmedin (Eds.), Advances in Psychology (Vol. Volume 52, pp. 185-218): North-Holland. doi:10.1016/S0166-4115(08)62387-0

Reid, R. A., Thompson, J. K., & Logsdon, J. L. (1992). Knowledge and attitudes of management students toward software piracy. *Journal of Computer Information Systems*, *33*(1), 46–51.

Rheingold, H. (1993). *The virtual community: Homesteading on the electronic frontier. NewYork*. Addison Wesley.

Richet, J. L. (2012). Market Failure Mechanisms in Cybersecurity. *pre-ICIS workshop on Information Security and Privacy (SIGSEC)*. Paper 26. http://aisel.aisnet.org/wisp2012/26

Richet, J. L. (2013). Laundering Money Online: a review of cybercriminals methods. *arXiv preprint arXiv:1310.2368*.

Richet, J. L. (2013). From Young Hackers to Crackers. *International Journal of Technology and Human Interaction*, *9*(3), 53–62.

Rideout, V. J., Foehr, U. G., & Roberts, D. F. (2010). *Generation M^2 Media in the lives of 8- to 18-year-olds*. Menlo Park, CA: Kaiser Family Foundation.

Ristanpart, T., Tromer, E., Sacham, H., & Savage, S. (2009) Hey, you, get off of my cloud: Exploring information leakage in third-party compute clouds.*Proceedings of the 16th ACM Conference on Computer and Communications Security*. Chicago, IL: ACM. doi:10.1145/1653662.1653687

Roberts, D. F., & Foehr, U. G. (2008). Trends in Media Use. In J. Brooks-Gunn & E. Donahue (Ed.). Generation M: Media in the lives of 8-18 year olds. Menlo Park, CA: Kaiser Family Foundation.

Robertson, J. (2011). The educational affordances of blogs for self-directed learning. *Computers & Education*, *57*(2), 1628–1644. doi:10.1016/j.compedu.2011.03.003

Robertson, T. S., Zielinski, J., & Ward, S. (1984). *Consumer Behavior*. Glenview, IL: Scott, Foresman.

Robey, D. (1983). Cognitive style and DSS design: a comment on Huber's paper., *29*(5), 580–582.

Robinson, K. (2006). Take a chance . . . Let them dance: Validating artistic expression.

Rodgers, A., Corbett, T., Bramley, D., Riddell, T., Wills, M., Lin, R.-B., & Jones, M. (2005). Do u Smoke after txt? Results of a Randomised Trial of Smoking Cessation Using Mobile Phone Text Messaging. *Tobacco Control*, *14*(4), 255–261. doi:10.1136/tc.2005.011577 PMID:16046689

Roehrich, G. (1994). Innovativités Hédoniste et Sociale: Proposition d'une Échelle de Mesure. *Recherche et Applications en Marketing*, *9*(2), 19–42. doi:10.1177/076737019400900202

Rogers, E. M. (Ed.). (1962). *Diffusion of Innovations*. New York: The Free Press.

Rogers, M., Smoak, N., & Liu, J. (2006). Self-reported deviant computer behavior. *Deviant Behavior*, *27*(3), 245–268. doi:10.1080/01639620600605333

Rokeach, M. (1973). *The Nature of Human Values*. New York: The Free Press.

Rosen, L. D., Carrier, L. M., & Cheever, N. A. (2013). Facebook and texting made me do it: Media-induced task-switching while studying. *Computers in Human Behavior*, *29*(3), 948–958. doi:10.1016/j.chb.2012.12.001

Ross, C., Orr, S. E., Sisic, M., Arseneault, J. M., Simmering, M. G., & Orr, R. R. (2009). Personality and motivations associated with Facebook use. *Computers in Human Behavior*, *25*(2), 578–586. doi:10.1016/j.chb.2008.12.024

Roth, T. (2011). Breaking encryptions. *Black hat conference 2011*. Retrieved from https://media.blackhat.com/bh-dc-1/Roth/BlackHat_DC_2011_Roth_Breaking%20 encryptions-Slides.pdf

Rowe, D. W., Sibert, J., & Irwin, D. (1998). Heart Rate Variability: Indicator of User State as an Aid to Human-Computer Interaction.*Proceedings of the SIGCHI conference on Human factors in computing systems*, 480-487. doi:10.1145/274644.274709

Rowe, F. (2005). Are decision support systems getting people to conform? the impact of work organisation and segmentation on user behaviour in a French bank. *Journal of Information Technology*, *20*(2), 103–116. doi:10.1057/palgrave.jit.2000042

Roy, S. K., & Shekhar, V. (2010). Alternative models of trustworthiness of service providers. *Journal of Global Marketing*, *23*(5), 371–386. doi:10.1080/08911762.2010.521111

Rubin, A. (1994). Media uses and effects: A uses-and-gratifications perspective. In J. Bryant & D. Zillmann (Eds.), Media effects: Advances in theory and research (pp. 417-436). Hillsdale, N J: Lawrence Erlbaum Associates, Inc.

Rubin, A. (2002). The uses-and-gratifications perspective of media effects. In J. Bryant & D. Zillmann (Eds.), *Media effects: Advances in theory and research* (2nd ed., pp. 525–548). Hillsdale, NJ: Lawrence Erlbaum Associates, Inc.

Rubin, A. M. (1994). Media uses and effects: A uses-and-gratifications perspective. In J. Bryant & D. Zillmann (Eds.), *Media effects: Advances in theory and research* (pp. 417–436). Hillsdale, NJ: Lawrence Erlbaum Associates, Inc.

Rubinstein, J. S., Meyer, D. E., & Evans, J. E. (2001). Executive control of cognitive processes in task switching. *Journal of Experimental Psychology. Human Perception and Performance*, *27*(4), 763–797. doi:10.1037/0096-1523.27.4.763 PMID:11518143

Ruiz, N., Feng, Q. Q., Taib, R., Handke, T., & Chen, F. (2010). *Cognitive skills learning: pen input patterns in computer-based athlete training*. Paper presented at the International Conference on Multimodal Interfaces and the Workshop on Machine Learning for Multimodal Interaction, Beijing, China. doi:10.1145/1891903.1891955

Ruiz, N., Taib, R., & Chen, F. (2011). *Freeform pen-input as evidence of cognitive load and expertise*. Paper presented at the Proceedings of the 13th international conference on multimodal interfaces, Alicante, Spain. doi:10.1145/2070481.2070511

Ruiz, N., Taib, R., Shi, Y., Choi, E., & Chen, F. (2007). *Using pen input features as indices of cognitive load.* Paper presented at the Proceedings of the 9th international conference on Multimodal interfaces, Nagoya, Aichi, Japan. doi:10.1145/1322192.1322246

Ryan, J., & Jefferson, T. I. (2003). The use, misuse, and abuse of statistics in information security research.*Proc. 23rd ASEM National Conference*. St. Louis, MO: ASEM.

Sallaberry, A., Muelder, C., & Ma, K.-L. (2013). Clustering, Visualizing, and Navigating for Large Dynamic Graphs. In W. Didimo & M. Patrignani (Eds.), Lecture Notes in Computer Science: Vol. 7704. *Graph Drawing* (pp. 487–498). Berlin, Germany: Springer Berlin Heidelberg. doi:10.1007/978-3-642-36763-2_43

Salvucci, D., & Taatgen, N. (2008). Threaded cognition: An integrated theory of concurrent multitasking. *Psychological Review, 115*(1), 101–130. doi:10.1037/0033-295X.115.1.101 PMID:18211187

Salvucci, D., & Taatgen, N. (2011). *The Multitasking Mind. Oxford Series on Cognitive Models and Architectures.* Oxford: Oxford University Press.

Samarasinghe, S. (2007). *Neural networks for applied sciences and engineering: from fundamentals to complex pattern recognition.* Boca Raton, FL: Auerbach.

Sanchez-Franco, M. J., & Rondan-Cataluña, F. J. (2010). Virtual travel communities and customer loyalty: Customer purchase involvement and web site design. *Electronic Commerce Research and Applications, 9*(2), 171–182. doi:10.1016/j.elerap.2009.05.004

Sandberg-Diment, E. (1984). Value of windowing is questioned. *The New York Times*, p. 35.

Sarfo, F., & Elen, J. (2011). Investigating the impact of positive resource interdependence and individual accountability on students' academic performance in cooperative learning. *Electronic Journal of Research in Educational Psychology, 9*(1), 73–93.

Schilling, M. A. (2002). Technology success and failure in winner-take-all markets: The impact of learning orientation, timing and network externalities. *Academy of Management Journal, 45*(2), 387–398. doi:10.2307/3069353

Schmidt, S., Cantallops, A. S., & Santos, C. P. (2008). The characteristics of hotel websites and their implications for website effectiveness. *International Journal of Hospitality Management, 27*(3), 504–516. doi:10.1016/j.ijhm.2007.08.002

Schwarzer, R., Bäßler, J., Kwiatek, P., Schröder, K., & Zhang, J. (1997). The assessment of optimistic self-beliefs: Comparison of the German, Spanish, and Chinese versions of the General Self-efficacy Scale. *Applied Psychology, 46*(1), 69–88. doi:10.1111/j.1464-0597.1997.tb01096.x

Schweickert, R., & Boggs, G. J. (1984). Models of central capacity and concurrency. *Journal of Mathematical Psychology, 28*(3), 223–281. doi:10.1016/0022-2496(84)90001-4

Schwenk, C. R. (1984). Cognitive simplification processes in strategic decision-making. *Strategic Management Journal, 5*(2), 111–128. doi:10.1002/smj.4250050203

Scoble, R., & Israel, S. (2013). *Age of context: Mobile, sensors, data and the future of privacy.* Lexington, KY: Patrick Brewster Press.

Sears, A., & Shneiderman, B. (1994). Split menus: Effectively using selection frequency to organize menus. *ACM Transactions on Computer-Human Interaction, 1*(1), 27–51. doi:10.1145/174630.174632

Shaft, T. M., & Vessey, I. (2006). The Role of Cognitive Fit In The Relationship Between Software Comprehension And Modification. *Management Information Systems Quarterly, 30*(1), 29–55.

Shah, S., & Cole, D. (2011). Spyware/adware. *Black hat conference 2005.* Retrieved from http://www.blackhat.com/html/bh-japan-05/bh-jp-05-en-speakers.html

Shao, G. (2009). Understanding the appeal of user-generated media: A uses and gratification perspective. *Internet Research, 19*(1), 7–25. doi:10.1108/10662240910927795

Sheldon, K. M., Abad, N., & Hinsch, C. (2011). A two-process view of Facebook use and relatedness need-satisfaction: Disconnection drives use, and connection rewards it. *Journal of Personality and Social Psychology, 100*(4), 766–775. doi:10.1037/a0022407 PMID:21280967

Sheldon, P. (2008). Student favorite: Facebook and motives for its use. *Southwestern Mass Communication Journal, 23*(2), 39–53.

Shih, C. C. (2009). *The Facebook era: Tapping online social networks to build better products, reach new audiences, and sell more stuff.* Boston, MA: Pearson Education.

Shih, H. M., & Goonetilleke, R. S. (1998). Effectiveness of menu orientation in Chinese. *Human Factors: The Journal of the Human Factors and Ergonomics Society, 40*(4), 569–576. doi:10.1518/001872098779649373

Shiv, B., & Fedorikhin, A. (1999). Heart and mind in conflict: The interplay of affect and cognition in consumer decision making. *The Journal of Consumer Research, 26*(3), 278–292. doi:10.1086/209563

Shneiderman, B. (1992). Tree visualization with tree-maps: 2-d space-filling approach. *ACM Transactions on Graphics*, *11*(1), 92–99. doi:10.1145/102377.115768

Shoham, A., & Ruvio, A. (2008). Opinion leaders and followers: A replication and extension. *Psychology and Marketing*, *25*(3), 280–297. doi:10.1002/mar.20209

Sicilia, M., & Palazón, M. (2008). Brand communities on the internet: A case study of Coca-Cola's Spanish virtual community. Corporate Communications. *International Journal (Toronto, Ont.)*, *13*(3), 255–270.

Simon, H. A. (1960). *The New Science of Management Decision*. Harper Brothers. doi:10.1037/13978-000

Simon, H. A., Dantzig, A., Hogarth, G. B., Plott, C. R., Raiffa, H., Schelling, T. C., & Winter, S. et al. (1987). Decision Making and Problem Solving. *Interfaces*, *17*(5), 11–32. doi:10.1287/inte.17.5.11

Slater, S., & Narver, J. (2000). Intelligence generation and superior customer value. *Journal of the Academy of Marketing Science*, *28*(1), 120–127. doi:10.1177/0092070300281011

Smith, A. (2013). *Pew research Internet project*. Retrieved June 5, 2014, from http://www.pewinternet.org/2013/06/05/smartphone-ownership-2013/

Smith, B. E. (2011). *Sams teach yourself Facebook for business in 10 minutes*. San Antonio, TX: Pearson Education press.

Smith, S., Winchester, D., Bunker, D., & Jamieson, R. (2010). Circuits of Power: A Study of Mandated Compliance to an Information Systems Security De Jure Standard in a Government Organization. *Management Information Systems Quarterly*, *34*(3), 463–486.

Snyder, F. (2001). Sites of criminality and sites of governance. *Social & Legal Studies*, *10*, 251–256.

Socialmediaexaminer. Social media marketing industry report 2013, Retrieved December 1st 2013, from http://www.socialmediaexaminer.com/SocialMediaMarketingIndustryReport2013.pdf

Song, K., Nam, S., Lim, H. & Kim, J. (2013). Analysis of youngsters' media multitasking behaviors and effect on learning. *International Journal of Multimedia and Ubiquitous Engineering, 8* (4), 191-198.

Song, I., LaRose, R., Eastin, M. S., & Lin, C. A. (2004). Internet gratifications and internet addiction: On the uses and abuses of new media. *Cyberpsychology & Behavior*, *7*(4), 384–394. doi:10.1089/cpb.2004.7.384 PMID:15331025

Spiegleberg, H. (1982). *The Phenomenological Moment: A Historical Introduction* (3rd ed.). The Hague: Martinus Nijhoff. doi:10.1007/978-94-009-7491-3

Spies Shapiro, L., & Margolin, G. (2014). Growing up wired: Social networking sites and adolescent psychosocial development. *Clinical Child and Family Psychology Review*, *17*(1), 1–18. doi:10.1007/s10567-013-0135-1 PMID:23645343

Spink, A., & Park, M. (2005). Information and non-information multitasking interplay. *The Journal of Documentation*, *61*(4), 548–554. doi:10.1108/00220410510607516

Springer, S. P., & Deutsch, G. (1997). *Left brain, right brain: Perspectives from cognitive neuroscience*. New York: Freeman.

Stafford, T., & Gonier, D. (2004). What Americans like about being online. *Communications of the ACM*, *47*(11), 107–112. doi:10.1145/1029496.1029502

Standing, C., Sims, I., & Love, P. (2009). IT non-conformity in institutional environments: E-marketplace adoption in the government sector. *Information & Management*, *46*(2), 138–149. doi:10.1016/j.im.2008.10.001

Stankov, U., Lazić, L., & Dragićević, V. (2010). The extent of use of basic Facebook user-generated content by the national tourism organizations in Europe. *European Journal of Tourism Research*, *3*(2), 105–113.

Steenkamp, J.-B. E. M., & Gielens, K. (2003). Consumer and Market Drivers of the Trial Probability of New Consumer Packaged Goods. *The Journal of Consumer Research*, *29*(December), 368–384. doi:10.1086/378615

Steinfield, C., Ellison, N. B., & Lampe, C. (2008). Social capital, self-esteem, and use of online social network sites: A longitudinal analysis. *Journal of Applied Developmental Psychology*, *29*(6), 434–445. doi:10.1016/j.appdev.2008.07.002

Sterling, B. (1993). *The hacker crackdown: Law and disorder on the electronic frontier*. New York: Bantam.

Sternberg, R. J. (1997). *Thinking styles*. New York: Cambridge University Press. doi:10.1017/CBO9780511584152

Stern, T., & Kumar, N. (2014). Improving privacy settings control in online social networks with a wheel interface. *Journal of the Association For Information Science & Technology*, *65*(3), 524–538. doi:10.1002/asi.22994

Storey, M.-A. D., Fracchia, F. D., & Müller, H. A. (1999). Customizing a Fisheye View Algorithm to Preserve the Mental Map. *Journal of Visual Languages and Computing*, *10*(3), 245–267. doi:10.1006/jvlc.1999.0124

Strader, T. J., Ramaswani, S. N., & Houle, P. A. (2007). Perceived network externalities and communication technology acceptance. *European Journal of Information Systems*, *16*(1), 54–65. doi:10.1057/palgrave.ejis.3000657

Strauss, A., & Corbin, J. (2008). *Basics of Qualitative Research Techniques and Procedures for Developing Grounded Theory* (3rd ed.). London: Sage Publication.

Subashini, S., & Kavitha, V. (2011). A survey on security issues in service delivery models of cloud computing. *Journal of Network and Computer Applications*, *34*(1), 1–11. doi:10.1016/j.jnca.2010.07.006

Subin, I., Mason, C. H., & Houston, M. B. (2007). Does Innate Consumer Innovativeness relate to New Product/Service Adoption Behavior? The Intervening Role of Social Learning via Vicarious Innovativeness? *Journal of the Academy of Marketing Science*, *35*(1), 63–75. doi:10.1007/s11747-006-0007-z

Summers, J. O. (1970). The identity of women's clothing fashion opinion leaders. *JMR, Journal of Marketing Research*, *7*(2), 178–185. doi:10.2307/3150106

Sun, T., & Wu, G. (2012). Traits, Predictors, and Consequences of Facebook Self- Presentation. *Social Science Computer Review*, *30*(4), 419–433. doi:10.1177/0894439311425978

Sutko, D. M., & de Souza e Silva, A. (2011). Location-aware mobile media and urban sociability. *New Media & Society*, *13*(5), 807–823. doi:10.1177/1461444810385202

Sweeney, M., Maguire, M., & Shackel, B. (1993). Evaluating user-computer interaction: A framework. *International Journal of Man-Machine Studies*, *38*(4), 689–711. doi:10.1006/imms.1993.1032

Sweller, J., Ayres, P., & Kalyuga, S. (2011). Measuring Cognitive Load.[). New York, NY: Springer New York.]. *Cognitive Load Theory*, *1*, 71–85.

Syed, S. F., & Nurullah, A. S. (2011). Use of Mobile Phones and the Social Lives of Urban Adolescents: A Review of Literature. *Trends in Information Management*, *7*(1), 1–18.

Tak, S., Scarr, J., Gutwin, C., & Cockburn, A. (2011). Supporting window switching with spatially consistent thumbnail zones: design and evaluation. In P. Campos, N. Graham, J. Jorge, N. Nunes, P. Palanque, & M. Winckler (Eds.), Lecture Notes in Computer Science: Vol. 6946. *Human-Computer Interaction – INTERACT 2011* (pp. 331–347). Berlin, Germany: Springer Berlin Heidelberg.

Tamborini, R., & Skalski, P. (2006). The role of presence in the experience of electronic games. In P. Vorderer & J. Bryant (Eds.), *Playing video games: Motives, responses, and consequences* (pp. 225–240). Lawrence Erlbaum Associates.

Tan, D. S. (2004). *Exploiting the Cognitive and Social Benefits of Physically Large Displays (thesis).* Carnegie Mellon University.

Tan, D. S., Czerwinski, M., & Robertson, G. (2003). *Women go with the (optical) flow*. In *Proceedings of the SIGCHI conference on Human factors in computing systems*, 209-215.

Tan, D. S., Gergle, D., Scupelli, P., & Pausch, R. (2003). With similar visual angles, larger displays improve spatial performance. In *Proceedings of the SIGCHI conference on Human factors in computing systems* (pp. 217-224). ACM. doi:10.1145/642611.642650

Tan, D. S., Gergle, D., Scupelli, P., & Pausch, R. (2004). Physically large displays improve path integration in 3D virtual navigation tasks.*InProceedings of the CHI 2004 Conference on Human Factors in Computing Systems*, 439-446. doi:10.1145/985692.985748

Tan, D. S., Meyers, B., & Czerwinski, M. (2004). WinCuts: manipulating arbitrary window regions for more effective use of screen space. In E. Dykstra-Erickson & M. Tscheligi (Eds.), *CHI '04 Extended Abstracts on Human Factors in Computing Systems* (pp. 1525–1528). New York, NY: ACM. doi:10.1145/985921.986106

Tang, T. (2013). *The role of structure in social media use: Locating synthesis in measuring social media structures*. Paper presented in 2013 Eastern Communication Association, Theory and Methodology session.

Taylor, C. (2011). *Twitter has 100 million active users*. Retrieved July 28th, 2013 from http://mashable.com/2011/09/08/twitter-has-100-million-active-users/

Taylor, J. W. (1977). A striking characteristic of innovators. *JMR, Journal of Marketing Research*, *14*(1), 104–107. doi:10.2307/3151066

Tazghini, S., & Siedlecki, K. L. (2013). A mixed method approach to examining Facebook use and its relationship to self-esteem. *Computers in Human Behavior*, *29*(3), 827–832. doi:10.1016/j.chb.2012.11.010

Te Wang, K., Huang, Y., Jeng, Y., & Wang, T. (2008). A blog-based dynamic learning map. *Computers & Education*, *51*(1), 262–278. doi:10.1016/j.compedu.2007.06.005

Techopedia, (2014a). *Facebook Page*. Retrieved November 15, 2014, from http://www.techopedia.com/definition/15441/facebook-page

Techopedia, (2014b). *Facebook Fan*. Retrieved November 15, 2014, from http://www.techopedia.com/definition/15436/facebook-fan

The Matrix. (1999). Retrieved at http://www.imdb.com/title/tt0133093/?ref_=nv_sr_1

Thompson, C., & Wembridge, M. (2012). Big Tobacco Push for Cigarette Alternatives. *Financial Times, August* 12.

Thompson, R. L., Higgins, C. A., & Howell, J. M. (1991). Personal Computing: Towards a Conceptual Model of Utilization. *Management Information Systems Quarterly*, *15*(1), 125–143. doi:10.2307/249443

Thompson, S. A., & Sinha, R. K. (2008). Brand communities and new product adoption: The influence and limits of oppositional loyalty. *Journal of Marketing*, *72*(6), 65–80. doi:10.1509/jmkg.72.6.65

Tidwell, L. C., & Walther, J. B. (2002). Computer-mediated communication effects on disclosure, impressions, and interpersonal evaluations: Getting to know one another a bit at a time. *Computers in Human Behavior*, *28*, 317–348.

Tifferet, S., & Vilnai-Yavetz, I. (2014). Gender differences in Facebook self-presentation: An international randomized study. *Computers in Human Behavior*, *35*, 388–399. doi:10.1016/j.chb.2014.03.016

Tiggemann, M., & Slater, A. (2014). NetTweens: The internet and body image concerns in preteenage girls. *The Journal of Early Adolescence*, *34*(5), 606–620. doi:10.1177/0272431613501083

Todd, P., & Benbasat, I. (2000). The Impact of Information Technology on Decision Making: A cognitive Perspective. In R. Zmud (Ed.), *Framing the Domains of IT Management* (pp. 1–14). Pinaflex.

Toma, C. L., & Hancock, J. T. (2013). Self-Affirmation Underlies Facebook Use. *Personality and Social Psychology Bulletin*, *39*(3), 321–331. doi:10.1177/0146167212474694 PMID:23359086

Tong, S. T., Van der Heide, B., Langwell, L., & Walther, J. B. (2008). Too much of a good thing? The relationship between number of friends and interpersonal impressions on Facebook. *Journal of Computer-Mediated Communication*, *13*(3), 531–549. doi:10.1111/j.1083-6101.2008.00409.x

Top Dining Brands. (2014). *Fan page list*. Retrieved June 26, 2014, from http://fanpagelist.com/category/brands/dining/

Torrance, E. P. (1988). *Style of Learning and Thinking: Administrator's manual*. Bensenville, IL: Scholastic Testing Service.

Trend Micro. (2009). Brits breeding the next-generation of computer hackers. *Global Security Mag*. Retrieved from http://www.globalsecuritymag.com/Trend-Micro-Brits-Breeding-the,20090403,8329

Trend Micro. (2011). *Big botnet busts*. Retrieved from http://blog.trendmicro.com/big-botnet-busts/

Trivedi, N., Lai, W., & Zhang, Z. (2001). Optimizing windows layout by applying a genetic algorithm. In *Proceedings of the IEEE Congress on Evolutionary Computation* (Vol. 1, pp. 431–435). Piscataway, NJ: IEEE. doi:10.1109/CEC.2001.934423

Trusov, M., Bucklin, R., & Pauwels, K. (2009). Effects of Word-of-Mouth Versus Traditional Marketing: Findings from an Internet Social Networking Site. *Journal of Marketing, 73*(5), 90–102. doi:10.1509/jmkg.73.5.90

Tsandilas, T. (2005, April). An empirical assessment of adaptation techniques. In CHI'05 Extended Abstracts on Human Factors in Computing Systems (pp. 2009-2012). ACM. doi:10.1145/1056808.1057079

Tsandilas, T. (2007, April). Bubbling menus: a selective mechanism for accessing hierarchical drop-down menus. In *Proceedings of the SIGCHI conference on Human factors in computing systems* (pp. 1195-1204). ACM. doi:10.1145/1240624.1240806

Tsiklis, T. (2010). *Avira-Ransomware threatens with official complaint of piracy*. Retrieved from http://www.virus.gr/portal/en/content/avira-ransomware-threatens-official-complaint-piracy

Tufin Technologies. (2010). *Survey of hacking habits in New York*. Retrieved from http://www.tufin.com/news_events_press_releases.php?index=2010-04-14

Turban, E. (1995). *Decision Support and Expert Systems : Management Support Systems*. New Jersey, USA: Prentice Hall PTR.

Turkle, S. (2011). *Alone together: Why we expect more from technology and less from each other*. Perseus Printing, New York: Basic Books.

Turner, J., & Croucher, S. (2014). An examination of the relationships among United States college students' media use habits, need for cognition, and grade point average. *Learning, Media and Technology, 39*(2), 199–214. doi:10.1080/17439884.2013.777349

Tuten, T., & Solomon, M. (2014). Social Media Marketing. 1st edition. Pearson Education limited, Edinburgh.

Tversky, A., & Kahneman, D. (1986). *Rational Choice and the Framing of Decisions, 59*(4), 251–278.

Ulusu, Y. (2010). Determinant factors of time spent on Facebook: Brand community engagement and usage types. *Journal of Yasar University, 18*(5), 2949–2957.

Underwood, J. D. M., Kerlin, L., & Farrington-Flint, L. (2011). The lies we tell and what they say about us: Using behavioural characteristics to explain Facebook activity. *Computers in Human Behavior, 27*(5), 1621–1626. doi:10.1016/j.chb.2011.01.012

Urista, M. A., Dong, Q., & Day, K. N. (2009). Explaining why young adults use MySpace and Facebook through uses and gratifications theory. *Human Communication, 12*(2), 215–229.

Urista, M. A., Qingwen, D., & Day, K. D. (2009). Explaining Why Young Adults Use MySpaceand Facebook Through Uses and Gratifications Theory. *Human Communication, 12*(2), 215–229.

Utz, S., Tanis, M., & Vermeulen, I. (2012). It is all about being popular: The effects of needs for popularity on social network site use. *Cyberpsychology, Behavior, and Social Networking, 15*(1), 37–42. doi:10.1089/cyber.2010.0651 PMID:21988765

Valenzuela, S., Park, N., & Kee, K. F. (2009). Is there social capital in a social network site?: Facebook use and college students' life satisfaction, Trust, and Participation. *Journal of Computer-Mediated Communication, 14*(4), 875–901. doi:10.1111/j.1083-6101.2009.01474.x

Valette-Florence, P. (1989). Conceptualisation et Mesure de l'Implication. *Recherche et Applications en Marketing, 4*(1), 57–78. doi:10.1177/076737018900400104

Valkenburg, P. M., & Peter, J. (2007). Internet communication and its relationship to wellbeing: Identifying some underlying mechanisms. *Media Psychology, 9*(1), 43–58. doi:10.1080/15213260709336802

Van Gennep, A. (1960). *The rites of passage*. Chicago, IL: The University of Chicago Press.

Van Maanen, J., Sorensen, J., & Mitchell, T. R. (2007). Introduction to special topic forum: The interplay between theory and method. *Academy of Management Review, 32*(4), 1145–1154. doi:10.5465/AMR.2007.26586080

van Manen, M. (1990). *Researching lived experience: Human science for an action sensitive pedagogy*. Albany: State University of New York Press.

Vanderberg, R., & Lee Swanson, H. (2007). Which components of working memory are important in the writing process? *Reading and Writing*, *20*(7), 721–752. doi:10.1007/s11145-006-9046-6

Vankatesh, V., & Davis, F. (2000). A Theoretical Extension of the Technology Acceptance Model: Four Longitudinal Field Studies. *Management Science*, *46*(2), 186–204. doi:10.1287/mnsc.46.2.186.11926

Vargas, S., & Castells, P. (2011, October). Rank and relevance in novelty and diversity metrics for recommender systems. In *Proceedings of the fifth ACM conference on Recommender systems* (pp. 109-116). ACM. doi:10.1145/2043932.2043955

Varshney, U. (2014). A model for improving quality of decisions in mobile health. *Decision Support Systems*, *62*, 66–77. doi:10.1016/j.dss.2014.03.005

Venkatesh, V., & Brown, S. (2001). A Longitudinal Investigation of Personal Computers in Homes: Adoption Determinants and Emerging Challenges. *Management Information Systems Quarterly*, *25*(1), 71–102. doi:10.2307/3250959

Venkatesh, V., & Morris, M. (2000). Why don't Men ever Stop to Ask for Directions? Gender, Social Influence and their Role in Technology Acceptance and Usage Behavior. *Management Information Systems Quarterly*, *24*(1), 115–139. doi:10.2307/3250981

Venkatraman, M. P. (1989). Opinion leaders, adopters, and communication adopters: A role analysis. *Psychology and Marketing*, *6*(1), 51–68. doi:10.1002/mar.4220060104

Venkatraman, M. P., & Price, L. (1990). Differentiating between Cognitive and Sensory Innovativeness: Concepts, Measurement, and Implications. *Journal of Business Research*, *20*(4), 293–315. doi:10.1016/0148-2963(90)90008-2

Verkasalo, H., Lopez-Nicolas, C., Molina-Castillo, F. J., & Bouwman, H. (2010). Analysis of users and non-users of smartphone applications. *Telematics and Informatics*, *27*(3), 242–255. doi:10.1016/j.tele.2009.11.001

Vessey, I. (1991). Cognitive fit: A theory based analysis of the graphs versus tables literature. *Decision Sciences*, *22*(2), 219–241. doi:10.1111/j.1540-5915.1991.tb00344.x

Vitak, J., & Ellison, N. B. (2013). 'There's a network out there you might as well tap': Exploring the benefits of and barriers to exchanging informational and support-based resources on Facebook. *New Media & Society*, *15*(2), 243–259. doi:10.1177/1461444812451566

Vizer, L. M., Zhou, L., & Sears, A. (2009). Automated stress detection using keystroke and linguistic features: An exploratory study. *International Journal of Human-Computer Studies*, *67*(10), 870–886. doi:10.1016/j.ijhcs.2009.07.005

Vogt, C., & Fesenmaier, D. R. (1998). Expanding the functional tourism information search model: Incorporating aesthetic, hedonic, innovation and sign dimensions. *Annals of Tourism Research*, *25*(5), 551–579. doi:10.1016/S0160-7383(98)00010-3

Voss, K. E., Spangenberg, E. R., & Grohmann, B. (2003). Measuring the hedonic and utilitarian dimensions of consumer attitude. *JMR, Journal of Marketing Research*, *40*(3), 310–320. doi:10.1509/jmkr.40.3.310.19238

Wagner, C. (2003). Putting another (B)Log on the wire: Publishing learning logs as weblogs. *Journal of Information Systems Education*, *14*(2), 131–132.

Waldner, M., Steinberger, M., Grasset, R., & Schmalstieg, D. (2011). Importance-driven compositing window management. In TanD.FitzpatrickG.GutwinC.BegoleB.KelloggW. A. (Eds.), *Proceedings of the SIGCHI Conference on Human Factors in Computing Systems* (pp. 959–968). New York, NY: ACM.

Wallace, E., Buil, I., de Chernatony, L., & Hogan, M. (2014). Who "Likes" You … and Why? A Typology of Facebook Fans. *Journal of Advertising Research*, *54*(1), 92–109. doi:10.2501/JAR-54-1-092-109

Wallace, L., Hui, L., & Cefaratti, M. A. (2011). Information Security and Sarbanes-Oxley Compliance: An Exploratory Study. *Journal of Information Systems*, *25*(1), 185–211. doi:10.2308/jis.2011.25.1.185

Wall, D. (2001). Cybercrimes and the internet. In D. Wall (Ed.), *Crime and the internet*. London: Routledge. doi:10.4324/9780203164501_chapter_1

Wall, D. S. (2007). *Cybercrime: The transformation of technology in the networked age*. Cambridge: Polity Press.

Wang, C. C., Hsu, Y., & Fang, W. (2004). Acceptance of technology with network externalities: An empirical study of internet instant messaging services. *Journal of Information Technology Theory and Application, 6*, 15–28. Retrieved from http://aisel.aisnet.org/jitta/

Wang, C. C., Lo, S. K., & Fang, W. (2008). Extending the technology acceptance model to mobile telecommunication innovation: The existence of network externalities. *Journal of Consumer Behaviour, 7*(2), 101–110. doi:10.1002/cb.240

Wang, N. N., Kosinski, M. M., Stillwell, D. D., & Rust, J. J. (2014). Can well-being be measured using Facebook status updates? Validation of Facebook's gross national happiness index. *Social Indicators Research, 115*(1), 483–491. doi:10.1007/s11205-012-9996-9

Wang, T. (2009). Rethinking teaching with information and communication technologies (ICTs) in architectural education. *Teaching and Teacher Education, 25*(8), 1132–1140. doi:10.1016/j.tate.2009.04.007

Wang, Y. C., & Fesenmaier, D. (2004a). Towards understanding members' general participation in and active contribution to an online travel community. *Tourism Management, 25*(6), 709–722. doi:10.1016/j.tourman.2003.09.011

Wang, Y., & Fesenmaier, D. R. (2004). Modeling participation in an online travel community. *Journal of Travel Research, 42*(3), 261–270. doi:10.1177/0047287503258824

Wang, Y., Yu, Q., & Fesenmaier, D. R. (2002). Defining the virtual tourist community: Implications for tourism marketing. *Tourism Management, 23*(4), 407–417. doi:10.1016/S0261-5177(01)00093-0

Watson, H. J., Rainer, R. K. Jr, & Koh, C. E. (1991). Executive Information Systems: A Framework for Development and a Survey of Current Practices. *Management Information Systems Quarterly, 15*(1), 13–30. doi:10.2307/249431

Weber, L. (2009). *Marketing to the social Web: How digital customer communities build your business*. Hoboken, NJ: John Wiley & Sons. doi:10.1002/9781118258125

Webster, J., & Watson, R. T. (2002). Analyzing the past to prepare for the future: Writing a literature review. *Management Information Systems Quarterly, 26*(2), iii–xiii.

Weick, K. E., & Sutcliffe, K. M. (2007). *Managing the unexpected : resilient performance in an age of uncertainty* (2nd ed.). San Francisco: Jossey-Bass.

Weimann, G., Tustin, D. H., Vuuren, D. V., & Joubert, J. P. R. (2007). Looking for opinion leaders: Traditional vs. modern measures in traditional societies. *International Journal of Public Opinion Research, 19*(2), 173–190. doi:10.1093/ijpor/edm005

Wei, P., & Lu, H. (2014). Why do people play mobile social games? An examination of network externalities and of uses and gratifications. *Internet Research, 24*(3), 313–331. doi:10.1108/IntR-04-2013-0082

Wei, R. (2008). Motivations for using the mobile phone for mass communications and entertainment. *Telematics and Informatics, 25*(1), 36–46. doi:10.1016/j.tele.2006.03.001

Wei, R., & Lo, V. H. (2006). Staying connected while on the move: Cell phone use and social connectedness. *New Media & Society, 8*(1), 53–72. doi:10.1177/1461444806059870

Weiser, J. (2001). The functions of Internet use and their social and psychological consequence. *Cyberpsychology & Behavior, 4*(6), 723–743. doi:10.1089/109493101753376678 PMID:11800180

Weissman, D. H., Gopalakrishnan, A., Hazlett, C. J., & Woldorff, M. G. (2005). Dorsal anterior cingulate cortex resolves conflict from distracting stimuli by boosting attention toward relevant events. *Cerebral Cortex, 15*(2), 229–237. doi:10.1093/cercor/bhh125 PMID:15238434

Wembridge, M., & Thompson, C. (2012), Big Tobacco bets on e-cigarette future, Financial Times, August 12. Retrieved November, 2014, from http://www.ft.com/cms/s/0/cb76997e-e0a5-11e1-b465-00144feab49a.html#ixzz23RJXFAJZ

Wertenbroch, K., & Dhar, R. (2000). Consumer choice between hedonic and utilitarian goods. *JMR, Journal of Marketing Research*, 60–71.

West, C. R., Kahn, J. H., & Nauta, M. M. (2007). Learning styles as predictors of self-efficacy and interest in research: Implications for graduate research training. *Training and Education in Professional Psychology, 1*(3), 174–183. doi:10.1037/1931-3918.1.3.174

Whittaker, R., Maddison, R., McRobbie, H., Bullen, C., Denny, S., Dorey, E., & Rodgers, A. et al. (2008). A Multimedia Mobile Phone-Based Youth Smoking Cessation Intervention: Findings from Content Development and Piloting Studies. *Journal of Medical Internet Research*, *10*(5), e49. doi:10.2196/jmir.1007 PMID:19033148

Wickens, C. D. (1984). Processing resources in attention. In R. Parasuraman, J. Beatty, & R. Davies (Eds.), *Varieties of Attention* (pp. 63–101). New York: Wiley.

Wierwille, Walter W, & Eggemeier, F Thomas. (1993). Recommendations for mental workload measurement in a test and evaluation environment. *Human Factors: The Journal of the Human Factors and Ergonomics Society*, *35*(2), 263–281.

Willison, R., & Warkentin, M. (2013). Beyond deterrence: An expanded view of employee computer abuse. *Management Information Systems Quarterly*, *37*(1).

Wilson, G. F., & Russell, C. A. (2003). Real-time assessment of mental workload using psychophysiological measures and artificial neural networks. *Human Factors: The Journal of the Human Factors and Ergonomics Society*, *45*(4), 635–644. doi:10.1518/hfes.45.4.635.27088 PMID:15055460

Wilson, G. M., & Sasse, A. (2004). From doing to being: Getting closer to the user experience. *Interacting with Computers*, *16*(4), 697–705. doi:10.1016/j.intcom.2004.06.001

Wise, K., Alhabash, S., & Park, H. (2010). Emotional responses during social information seeking on Facebook. *Cyberpsychology, Behavior, and Social Networking*, *13*(5), 555–562. doi:10.1089/cyber.2009.0365 PMID:20950180

Witmer, B. G., & Singer, M. J. (1998). Measuring presence in virtual environments- A Presence questionnaire. *Presence (Cambridge, Mass.)*, *7*(3), 225–240. doi:10.1162/105474698565686

Wohn, D., Ellison, N., Khan, M., Fewins-Bliss, R., & Gray, R. (2013). The role of social media in shaping first-generation high school students' college aspirations: A social capital lens. *Computers & Education*, *63*, 424–436. doi:10.1016/j.compedu.2013.01.004

Woisetschläger, D. M., Hartleb, V., & Blut, M. (2008). How to make brand communities work: Antecedents and consequences of consumer participation. *Journal of Relationship Marketing*, *7*(3), 237–256. doi:10.1080/15332660802409605

Wolf, M. (2008). *Proust and the squid: the story and science of the reading brain*. New York: Harper Collins Publishers.

Wright, K. (2012). Similarity, network convergence, and availability of emotional support as predictors of strong-tie/weak-tie support network preference on Facebook. *The Southern Communication Journal*, *77*(5), 389–402. doi:10.1080/1041794X.2012.681003

Wright, K., Rosenberg, J., Egbert, N., Ploeger, N., Bernard, D., & King, S. (2013). Communication competence, social support, and depression among college students: A model of Facebook and face-to-face support network influence. *Journal of Health Communication*, *18*(1), 41–57. doi:10.1080/10810730.2012.688250 PMID:23030518

Xie, Y., Ke, F., & Sharma, P. (2008). The effect of peer feedback for blogging on college students' reflective learning processes. *The Internet and Higher Education*, *11*(1), 18–25. doi:10.1016/j.iheduc.2007.11.001

Xie, Y., Ke, F., & Sharma, P. (2010). The effects of peer-interaction styles in team blogs on student's cognitive thinking and blog participation. *Journal of Educational Computing Research*, *42*(4), 459–479. doi:10.2190/EC.42.4.f

Xie, Y., & Sharma, P. (2011). Exploring evidence of reflective thinking in student artifacts of blogging-mapping tool: A design-based research approach. *Instructional Science*, *39*(5), 695–719. doi:10.1007/s11251-010-9149-y

Xu, C., Ryan, S., Prybutok, V., & Wen, C. (2012). It is not for fun: An examination of social network site usage. *Information & Management*, *49*(5), 210–217. doi:10.1016/j.im.2012.05.001

Xu, K. S., Kliger, M., & Hero, A. O. III. (2013). A regularized graph layout framework for dynamic network visualization. *Data Mining and Knowledge Discovery*, *27*(1), 84–116. doi:10.1007/s10618-012-0286-6

Yang, Y., Wang, X., & Li, L. (2013). Use mobile devices to wirelessly operate computers. *International Journal of Technology and Human Interaction*, *9*(1), 64–67. doi:10.4018/jthi.2013010105

Yar, M. (2005). The novelty of cybercrime: An assessment in light of routine activity theory. *European Journal of Criminology*, *2*(4), 407–427. doi:10.1177/147737080556056

Yin, B., Chen, F., Ruiz, N., & Ambikairajah, E. (2008). *Speech-based cognitive load monitoring system.* Paper presented at the International Conference on Acoustics, Speech and Signal Processing.

Yin, B., Ruiz, N., & Chen, F., & Khawaja, M. Asif. (2007). *Automatic cognitive load detection from speech features.* Paper presented at the Proceedings of the 19th Australasian conference on Computer-Human Interaction: Entertaining User Interfaces, Adelaide, Australia. doi:10.1145/1324892.1324946

Yin, B., Ruiz, N., Chen, F., & Ambikairajah, E. (2008). *Investigating speech features and automatic measurement of cognitive load.* Paper presented at the IEEE 10th Workshop on Multimedia Signal Processing. doi:10.1109/MMSP.2008.4665218

Young, A. L., & Quan-Haase, A. (2009). Information revelation and internet privacy concerns on social network sites: A case study of Facebook. In *Proceedings of the Fourth international conference on communities and technologies* (pp. 265–274). doi:10.1145/1556460.1556499

Young, A., & Quan-Haase, A. (2013). Privacy protection strategies on Facebook. *Information Communication and Society*, *16*(4), 479–500. doi:10.1080/1369118X.2013.777757

Young, K. (2011). Social ties, social networks and the Facebook experience. *International Journal of Emerging Technologies & Society*, *9*(1), 20–34.

Yu, K., Epps, J., & Chen, F. (2011). *Cognitive load evaluation of handwriting using stroke-level features.* Paper presented at the Proceedings of the 16th international conference on Intelligent user interfaces, Palo Alto, CA, USA. doi:10.1145/1943403.1943481

Yu, A. Y., Tian, S. W., Vogel, D., & Chi-Wai Kwok, R. (2010). Can learning be virtually boosted? An investigation of online social networking impacts. *Computers & Education*, *55*(4), 1494–1503. doi:10.1016/j.compedu.2010.06.015

Yuan, Y., Archer, N., Connelly, C. E., & Zheng, W. (2010). Identifying the ideal fit between mobile work and mobile work support. *Information & Management*, *47*(3), 125–137. doi:10.1016/j.im.2009.12.004

Zack, M. H. (2007). The role of decision support systems in an indeterminate world. *Decision Support Systems*, *43*(4), 1664–1674. doi:10.1016/j.dss.2006.09.003

Zhang, J., & Norman, D. A. (1994). Representations in distributed cognitive tasks. *Cognitive Science*, *18*(1), 87–122. doi:10.1207/s15516709cog1801_3

Zhang, L. (2006). Relationships between hemispheric dominance cognitive styles and big five personality traits. *Chinese Journal of Clinical Psychology*, *14*, 354–356.

Zhang, Y., Tang, L., & Leung, L. (2011). Gratifications, collective self-esteem, online emotional openness, and traitlike communication apprehension as predictors of Facebook uses. *Cyberpsychology, Behavior, and Social Networking*, *14*(12), 733–739. doi:10.1089/cyber.2010.0042 PMID:21651417

Zhou, T., & Lu, Y. (2011). Examining mobile instant messaging user loyalty from the perspectives of network externalities and flow experience. *Computers in Human Behavior*, *27*(2), 883–889. doi:10.1016/j.chb.2010.11.013

Zigurs, I., & Khazanchi, D. (2008). From Profiles to Patterns: A New View of Task-Technology Fit. *Information Systems Management*, *25*(1), 8–13. doi:10.1080/10580530701777107

Zimmerman. (2007). Metatasking vs. Multitasking. *Library Journal.* Retrieved on May 17, 2011 from http://www.libraryjournal.com/article/CA6431788.html

Zumback, J. (2006). Cognitive overhead in hypertext learning reexamined: Overcoming the myths. *Journal of Educational Multimedia and Hypermedia*, *15*(4), 411–432.

Zywica, J., & Danowski, J. (2008). The Faces of Facebookers: Investigating Social Enhancement and Social Compensation Hypotheses; Predicting Facebook™ and Offline Popularity from Sociability and Self-Esteem, and Mapping the Meanings of Popularity with Semantic Networks. *Journal of Computer-Mediated Communication*, *14*(1), 1–34. doi:10.1111/j.1083-6101.2008.01429.x

About the Contributors

Anabela Mesquita is a professor at the School of Accounting and Administration of Porto (ISCAP)/ Polytechnic School of Porto (IPP), Portugal. She is the Director of the Research Centre for Communication and Education and member of Algoritmi Research Centre (Minho University). She is also the Editor-in-Chief of the International Journal of Technology and Human Interaction, which is an academic journal indexed by EI-Compendex. In addition, Dr. Mesquita has published numerous papers in various international journals and conferences. She also serves as one of Organizing Committee Chairs of The Third International Conference on E-Technologies and Business on the Web (EBW2015) that will be held at the ESG Management School, Paris, France in 2015. Moreover, Dr. Mesquita is an Associate Editor of the Information Resources Management Journal and referee for the Journal of Cases of Information Technology. She has also been evaluator and reviewer for European commission projects.

Chia-Wen Tsai is an associate professor in the Department of Information Management, Ming Chuan University. Dr. Tsai is one of the Editors-in-Chief of International Journal of Online Pedagogy and Course Design, and International Journal of Technology and Human Interaction. He is also the Associate Editor of Cyberpsychology, Behavior, and Social Networking, and Taiwan Corresponding Editor of British Journal of Educational Technology. He is interested in online teaching methods and knowledge management.

* * *

Mohammad Hatim Abuljadail is a third year PhD student at the School of Media and Communication at Bowling Green State University in Ohio. He is also currently a member of the Emerging Media Research Cluster at Bowling Green State University. Several projects of his have been presented at the Broadcast and Education Association's annual conference and the annual conference of the Association of Education in Journalism and Mass Communication. His primary research interests are media management, international advertising, online communities and the marketing applications of new media.

Alesssandro Antonietti is full professor of Cognitive Psychology at the Catholic University of the Sacred Heart in Milano (Italy). He is the head of the Service of Learning and Educational Psychology, the coordinator of the Cognitive Psychology Lab and of the master in Cognitive Assessment and Training. He is member of the board of various international and national journals and academic editor of PlosONe, as well as reviewer of many scientific journals. He carried out experimental studies about creativity, problem-solving, decision-making, mental imagery, and analogy. He investigated the role played by media in cognition. He is interested in the applications of cognitive issues in the field of instruction and rehabilitation.

Silvia Cacho-Elizondo is currently an Associate Professor of Marketing and Associate Academic Director at IPADE Business School in México. Prior to that, she was Professor-Researcher at the European Business School (EBS Paris) and at the Institut Supérieur de Commerce (ISC Paris) and lecturer at HEC Paris. She has been lecturer and guest lecturer in several business schools in France, Spain and Mexico. Dr. Cacho-Elizondo started her professional career as a Research and Commercial Engineer at the Instituto de Investigaciones Eléctricas (I.I.E.) and as a Marketing Professor at the ITESM Campus Morelos. She has also gained professional experience working for Procter & Gamble and collaborating with e-Medicis, a French start-up specialized in mobile health services. Dr. Cacho-Elizondo earned her PhD in Management with a major in Marketing at HEC Paris (France), her thesis analysed: "The impact of online services on the consumer-brand relationship". She holds a M.Sc. in Marketing & Strategy (Université Paris-IX Dauphine, France), an MBA (IPADE Business School, México), a M.Sc. in Management of Technology (Sussex University, United Kingdom) and a bachelor degree in Electronic Systems Engineering (ITESM Campus Monterrey, Mexico). Dr. Silvia Cacho-Elizondo´s research interests include areas like: Consumer Behaviour, Brand Relationships, CRM, Innovation Adoption, Management of Technology, Online & Mobile Services and Corporate Social Responsibility. Her research has appeared in journals like: Journal of Retailing and Consumer Services, Journal of Health Marketing Quarterly. International Journal of Technology and Human Interaction, ISTMO and International Business Research Journal, among others. She is a member of the AFM (Association Française de Marketing), the EMAC (European Marketing Association) and the ACR (Association for Consumer Research).

Simona C. S. Caravita is an Associate Professor at the Department of Psychology, Catholic University of the Sacred Heart (Brescia, Italy), and she is member of the Centro di Ricerca delle Dinamiche Evolutive ed Educative (C.R.I.d.e.e.) at the Catholic University of the Sacred Heart. She received her doctorate in Psychology of communication and linguistic processes from the Catholic University of the Sacred Heart. Her major research interests include moral cognition, moral development, aggression, school bullying, social status within the peer group, peer group dynamics, peer relationships.

Rui Chen received the B.Sc. and M.Sc. degrees in Computer Science and Technology from Sichuan University, Chengdu, China, in 2006 and 2010, respectively. He is currently working towards the Ph.D. degree in Computer Applied Technology at the same school. His current research interests include automatic evaluation of interface aesthetics and usability, interaction design, adaptive user interface and intelligence in games.

Yu Chen received the MS degree in computer science from Sichuan University, China. Currently, He is working at Sichuan University for nationalities,China. He has broad research interests in the areas of computational intelligence, human-computer interaction, and software engineering.

Lanming Chen M.A. in Media & Communication, Bowling Green State University.

Lanlung Luke Chiang's hospitality work experiences are varied and diverse. He has served at different management levels in the businesses of both the lodging and the electronic engineering industries for over 5.5 years in Taiwan and United States. Besides gaining administrative experience, he also had opportunities to present many strategic management and marketing training programs for the staff of these businesses. Moreover, he received PhD degree from Iowa State University. His areas of research have focused on the areas of marketing and entrepreneurship in the service industry.

Barbara Colombo, assistant professor in General Psychology at Università Cattolica del Sacro Cuore, in Milan, Italy. Her theoretical and research background is in cognitive psychology. During the last few years she has been studying cognitive processes like attention, mental imagery, decision making and problem solving. She also has been exploring the role of individual differences (such as cognitive styles and personality traits) and emotions in the elaboration of multimedia stimuli. During the last years she has been focusing on cognitive neuroscience, trying to enrich the traditional research linked to cognitive science with data derived from new neuro-technologies such as eye-tracking, biofeedback and transcranical stimulation.

Ling Fang is a fifth-year doctoral student from the School of Media and Communication at Bowling Green State University (BGSU). She received her Master's Degree in Communication Studies from BGSU and Bachelor's Degree in Public Management from Xiamen University in China. Her primary field of study is new media effects and is currently working on her dissertation research on American and Chinese college students' social media involvement in relation to stress coping.

Cécile Godé is currently full professor at Lyon 2 University, COACTIS research Department, and associated researcher at the Research Center of the French Air Force (CReA). Her privileged research fields are information systems management and technological change. Her current contributions focus both on decision/coordination processes in extreme environments and cultural aspects of technological change. She published research papers in different academic journals such as International Journal of Technology and Human Interaction, European Management Journal, Project Management Journal.

Louisa Ha is a professor in the Department of Telecommunications at Bowling Green State University, Ohio, U.S.A. She is also the editor of Journalism and Mass Communication Quarterly, the flagship journal of the Association for Education in Journalism and Mass Communication. She is the founder and chair of the Emerging Media Research Cluster in the School of Media and Communication there. She created the International Advertising Resource Center Web Site since 1996. Her research interests are media management, media technologies, new media business models and audience research.

Jean-Fabrice Lebraty is currently full Professor at Lyon3 University. He is member of a management science laboratory named Magellan EA3713. He teaches information systems in many French universities and other countries. He is one of the French specialists of naturalistic decision making approach and crowdsourcing.

Lin Lin, Ed.D., is an Associate Professor of Learning Technologies at the University of North Texas. Her research lies in the intersections of technology, cognition, and education. She has conducted research on (1) the design of effective online and blended learning, (2) the impact of media multitasking on cognition and learning, and (3) the power, motivation, ethics, and usability issues related to game-based learning. Lin has published at dozens of academic journals, including the Proceedings of the National Academy of Sciences and Computers and Education. She has been cited and interviewed by magazines and newspapers such as Le Temps, Ingenioren, the Scientific American, and Fort Worth Star-Telegram. Lin received her doctoral degree from Teachers College Columbia University. One can find more info about Lin at http://courseweb.unt.edu/llin/.

Tao Lin received the MS degree in computer science from Sichuan University, China, in 2003, and the PhD degree in information science from University of Yamanashi, Japan, in 2007. He also worked at Waseda University, Japan, as a visiting lecturer. Currently, He is an associate professor at Sichuan University, China.

William M. Lipsmeyer is an accomplished Sr. Systems Engineer in the U.S. aviation industry. He received his MS in Systems Engineering from the Thomas J. Watson School of Engineering and Applied Science, State University of New York, Binghamton, NY, and his BS in Management from the U.S. Naval Academy, Annapolis, MD. He is a former US Navy commander and was Commanding Officer of TACRON ELEVEN, an Air Traffic Control squadron located in San Diego, CA. Bill's expertise lies in the field of aviation requirements engineering, specifically, requirements management for fixed wing and rotary wing aircraft. He has a wealth of integrative design experience on the F-35, MH-60R/S, and the V-22 aircraft. He has a history of contributions to the training and education of both aviation and design units, as a US Navy replacement crew flight instructor, an engineering technical training analyst, workforce Business System Modernization consultant, and certified Six Sigma Green Belt and Continuous Improvement instructor. He has performed as a contributing senior Systems Engineer at Lockheed Martin, Bell Helicopter - Textron, and Sikorsky Aircraft Corporation. He can be reached at: lipsme007@gmail.com.

Aikaterini Manthiou, Assistant Professor (NEOMA Business School, Department of Marketing, Rouen, France). Her research interest is service management and social media marketing.

Jean-Loup Richet is Information Systems Service Manager at Orange and Research Fellow at ESSEC Business School - Institute for Strategic Innovation & Services. He graduated from the French National Institute of Telecommunications, Telecom Business School, and holds a master of research from IAE/HEC Paris. Expert in IS Security, Jean-Loup Richet has published numerous articles in scientific and trade journals and is a frequent speaker at national and international conferences in Information Systems. He contributes regularly to Infosecurity Magazine, ComputerWorld, and Herdict (a project of the Berkman Center for Internet & Society at Harvard University). He is also a lecturer in Information Systems at Sorbonne Graduate Business School (MBA) and guest lecturer at HEC Paris.

Niousha Shahidi, PhD in Sciences in Applied Mathematics area (Université Paris Dauphine, France), is a lecturer-researcher at EDC Paris Business School (Ecole des Dirigeants et Créateurs d'entreprise) and a member of OCRE (The Observatory and Centre for Research in Entrepreneurship). Her thesis dealt with optimization problems in insurance models (Ceremade). She worked at the Laboratory of Econometrics of Ecole Polytechique Paris. She has published in international journals and conferences on research themes which include theory of the decision, Information asymmetry models in insurance, Entrepreneurial intention and data analysis.

Luisa Simonelli graduated in Psychology at the Catholic University of the Sacred Heart (Italy). She currently collaborates with the Department of Psychology of the Catholic University of the Sacred Heart, working on research projects on moral reasoning.

Ningjiu Tang received the PhD degree in computer science from Sichuan University, China, in 2009. He is a professor at Sichuan University, China. He has broad research interests in the areas of computational intelligence, human-computer interaction, and software engineering. He is a member of the IEEE and ACM.

Liang Rebecca Tang got her Ph. D. degree of tourism management from Purdue University. Her primary research interest is tourism marketing and destination planning. She has produced over 80 journal papers and conference papers. Many of her research papers have been accepted by top tourism and hospitality journals such as Tourism Management, Journal of Travel Research, International Journal of Hospitality Management, Journal of Hospitality & Tourism Research, etc. Three of her conference papers have been selected as a "best paper" or a "most recommended paper" in the national and international tourism and hospitality conferences. Moreover, one of her papers published in the Tourism Management was nominated for W. Bradford Wiley Memorial Best Paper of 2012 Award at International Council Hotel, Restaurant, Institutional Education (CHRIE).

Vesselina Tossan graduated HEC Paris in 1982. After 18 years of various executive positions mostly in marketing in international groups like Procter & Gamble, Unilever, Mattel and Sara Lee, she got a research masters degree in Marketing and Strategy in the University of Paris 9 Dauphine . She got a PhD in Management Science with Centre de Gestion Scientifique of Ecole des Mines de Paris. Her action research delt with management of innovation in multidivisional service groups. Her publications are in various French and International Journals . After lecturing in various business schools in France she became Associate Professor in University of Haute Alsace in Mulhouse. She is now Associate Professor in Marketing and Management in Cnam and adjunct professor at EDC Paris Business School.

Fang Wang, Ph. D. Candidate in School of Media and Communication in at Bowling Green State University, Bowling Green, Ohio. Lecturer in College of Foreign Languages at Guangxi University, Nanning, Guangxi, China, 530004.

Zhiming Wu is a postgraduate in Sichuan University, graduated from East China Normal University with a degree of bachelor in Software Engineering, and is studying Human-Computer Interaction to complete a doctor degree of Computer Science in Sichuan University.

Shaomei Wu received the MS degree in computer science from Sichuan University, China in 2013. Currently, she is working at Journal of Sichuan University (Natural Science Edition). She has broad research interests in the areas of human-computer interaction, and software engineering.

Tian-Tian Xie (1989-) received his bachelor's degree in Software Engineering from Sichuan University, Chengdu, China, in 2011. He also received his master degree in Computer Technology from Sichuan University in 2014. His main research interests include Multi-objective Optimization, HCI, user experience modeling of video game, computational intelligence in games and game design.

Chen Yang is a Ph.D. candidate in the School of Media & Communication, Bowling Green State University. He has received his MS degree in Applied Statistics from College of Business, Bowling Green State University and also hold a MA degree in Linguistics from Soochow University, Suzhou, China. His research interests include social media, Internet marketing, public opinion, public relations, information technology, and intercultural communication.

Liu Yang is a Ph.D candidate in the Department of Media and Communication at Bowling Green State University, Ohio, U.S.A. Her research interests are in advertising, e-commerce, social media and online marketing.

Franklin Nii Amankwah Yartey, Ph.D. (Bowling Green State University), is an Assistant Professor of Communication at the University of Dubuque, Dubuque, Iowa. Yartey held graduate teaching and research assistantships at Indiana State University and Bowling Green State University. His research focuses on Digital Media and Globalization/Social Media, with a secondary focus on Intercultural Communication. Other research interests include online microfinance and its impact on third world women, health communication, media ethics, and globalization. Yartey has won several academic awards, including the National Communication Association's Doctoral Honors Seminar and a Dissertation Fellowship at Bowling Green State University. Yartey also received the Iowa Communication Association (ICA) Outstanding New Teacher Award for 2014.

Gi Woong Yun, Associate Professor in School of Media & Communication, BGSU.

Yuxi Zhu (1990-) received the B.Sc. degrees in Software Engineering from Sichuan University, Chengdu, China, in 2012. She is currently working towards the M.Sc. degree in Computer Applied Technology at the same school. Her current research interests include automatic evaluation of software learnablility and game interface aesthetics.

Index

A

Absolute Distribution 164
Adaptive Menu 162-163, 187
Aircrafts 50, 55, 212
Air Traffic Control 130, 145, 210, 212, 215

B

Behavioral Features 211, 228
Blog 21-39, 44, 97, 260
Botnet 233, 240
brain capacity 1-2, 13
Brand Community 209, 280-281
Breaking Addictions 91

C

Cognitive Fit 47-48, 54, 61, 68
Cognitive Load 2-3, 12, 148, 210-214, 216-217, 219-222, 228
Cognitive load evaluation 211-212, 221
Cognitive Overload 211, 222, 228
Cognitive Style 21-23, 25-30, 44
Collaboration Work 130, 145
collaborative learning 22, 30, 38
college students 95-98, 100, 102, 104-106, 196, 203-204, 241-242, 244, 246-247, 251-253, 259-262, 264-265, 267, 279, 285-286
Crackers 230, 232-233, 235-236, 240
credit card 232, 235
cybercrime 229-236

D

DDoS Attacks 240
Decision Support Systems 45, 47-48, 50, 68
deviant behavior 230, 234, 236
diffusion 73-75, 86, 229-231, 233-234, 236, 245
digital natives 81, 233-234
Dual Task 2, 20
Dumpster Diving 235, 240

E

empty screen 146, 148, 151, 155
Expertise 1-3, 26, 47, 49, 213, 231, 233, 236, 257

F

Facebook Fan 189-196, 201-202, 209, 279-288, 293
Facebook Friend 257, 267
Facebook Page 98, 196, 202, 209, 280, 286, 288
fan pages 189-204, 279-288
Fit Models 46-48, 63, 68
Focus Area 153, 155, 160
friend types 266-267, 269
Functional Need 190, 193, 209

G

Genetic Algorithm 148, 151, 160, 162, 164-166, 187

H

hacker 230-236
Healthcare 47, 49, 72, 87
Health & Human Services (HHS) 73, 91
Hedonic Motivations 283-285, 287-288, 293
Hedonic Need 190, 194, 209
High-frequency interaction events 210
Highlighting Menu 164, 166, 175, 187

I

Identity 96-98, 100, 105, 122, 191, 194-196, 199-202, 230, 235-236, 240, 243-244
Immersion into Virtual Reality 145
Information Broadcasting 241, 244-246, 251, 253, 257
Information Seeking 241, 244-246, 251-253, 257, 261, 280, 283-284
Information source 193, 195-196, 200, 202

L

Large display 130-135, 137-141
layout optimization 147, 150-152, 155, 157
Link 16 45-46, 50-55, 57-60, 63

M

machine learning 212, 216, 219-222
Managerial Perspective 68
marital status 248-251
Markov chain 162, 164-165, 168-170, 173, 175-176, 179-180, 182-183
mass media 46, 230, 233, 243, 260, 262, 283
Media Multitasking 20
Media Psychology 44
Mental Map 146-153, 155-156, 160
Menu Item 163-164, 166-167, 169, 175-178, 180, 182-184, 187
Menu Item Prediction Model 187
Metacognition 23-26, 34-35, 37-38, 44
metacognitive awareness 22, 25, 28, 30, 33-35, 38
Military Applications 68
Mobile Apps 94, 103
Mobile Coaching Services 73-76, 85, 87
Mobile Health or mHealth 92
motivation theory 96, 98
Multimedia Learning 23, 44
Multi-Objective Optimization 146, 148, 154, 160
Multitasking 1-14, 20, 145

N

Naturalistic Decision Making 49, 61, 69
Navigation Time 162-164, 182-184, 187
network externalities theory 96
Network Externality 99, 122

O

Online Brand Communities 204, 281, 293
Online Community 190, 193-194, 203, 209, 281
online databases 231, 235
Opinion Leader 253, 257

P

Perceived Enjoyment 72, 75, 77, 82, 85-87, 92
Perceived Irritation 72, 75, 81, 92
Perceived Monetary Value 72, 75, 83-84, 86, 92
Performance Measures 211, 213, 228
Peripheral Awareness 132, 145
physical screen size 141
Physiological Measures 130-131, 133, 135-136, 213, 221, 228
Polychronicity 2, 4, 20
prediction model 163-164, 169, 173, 175-176, 181-182, 187
Probability Summation 164-165
Psychological Need 190, 194-195, 201, 209

R

Ransomware 235, 240
Recognition Memory 132, 145
recommendation system 182
Relational Maintenance 241, 244-245, 247, 250-253, 257

S

screen space 146-148, 151-152, 155-157, 160
Screen Space Recycling 156, 160
Self-Broadcasting 95-97, 99-100, 102, 104-106, 122, 259, 261-263, 282
Self-Esteem 96-98, 105, 122, 242, 260, 264, 266
self-presentation 96-97, 100, 261, 263
self-promotion 99, 242
Smoking Cessation 73, 87, 93-94
Social Cognitive Theory 259-260, 264-266, 270, 276
Social identity 100, 194-196, 199-202
Social Influence 72, 75-76, 82, 87, 92
Social Media 11, 94-106, 124, 189-193, 201, 209, 241-242, 244-245, 251-253, 257-268, 270, 280, 282, 284-285, 287-288
Social Needs 194, 202, 209
Social Networking Service 258
Social Networking Sites 95-96, 106, 190-191, 252, 259-260, 267, 276, 280
Social Network Site 293
Social Network Size 103, 241, 258
Social Plugin 123
Surveillance 130, 145, 243-244

T

Task Switch 20
Task Technology Fit 69
Technology Acceptance Model (TAM) 75, 79, 92
Twitter 106, 190, 241-243, 245-251, 253, 258, 260, 267-268, 276, 279
Twitter Follower 258

U

usage patterns 241-242, 252-253
user behavior 216, 218, 220
User Experience (UX) 145
user interface 228
Uses and Gratifications 242-243, 253, 258, 260-261, 270, 279-280, 283, 293
utilitarian 193, 201, 279-280, 282-288, 293
Utilitarian Motivations 279-280, 282-285, 287-288, 293

V

Vicarious Innovativeness 72, 74, 78-79, 85, 87, 92
Visual Analytics 130, 145, 147, 157
Visual Angle 130-134, 137-141, 145
visual search 132, 148, 162-164, 166, 175, 177-179, 183, 187
Visual Search Time 162-164, 166, 175, 177-178, 183, 187

W

Web 2.0 35, 38, 100, 106, 123, 190, 209
Window Importance 146, 161
Window Layout 146-148, 151-152, 154, 157, 161

www.ingramcontent.com/pod-product-compliance
Lightning Source LLC
LaVergne TN
LVHW060408300926
849162LV00006B/1375